Advances in Quantum Systems in Chemistry, Physics, and Biology

Progress in Theoretical Chemistry and Physics

VOLUME 32

More information about this series at http://www.springer.com/series/6464

Liliana Mammino · Davide Ceresoli ·
Jean Maruani · Erkki Brändas
Editors

Advances in Quantum Systems in Chemistry, Physics, and Biology

Selected Proceedings of QSCP-XXIII (Kruger Park, South Africa, September 2018)

Springer

Editors
Liliana Mammino
School of Mathematical and
Natural Sciences
University of Venda
Thohoyandou, South Africa

Jean Maruani
Laboratoire de Chimie Physique
CNRS & UPMC
Paris, France

Davide Ceresoli
Istituto di Scienze e Tecnologie Moleculari
Consiglio Nazionale delle Ricerche
Milan, Italy

Erkki Brändas
Department of Chemistry
Uppsala University
Uppsala, Sweden

ISSN 1567-7354 ISSN 2215-0129 (electronic)
Progress in Theoretical Chemistry and Physics
ISBN 978-3-030-34940-0 ISBN 978-3-030-34941-7 (eBook)
https://doi.org/10.1007/978-3-030-34941-7

This Springer imprint is published by the registered company Springer Nature Switzerland AG
The registered company address is: Gewerbestrasse 11, 6330 Cham, Switzerland

PTCP Aim and Scope

Progress in Theoretical Chemistry and Physics

A series reporting advances in theoretical molecular and material sciences, including theoretical, mathematical and computational chemistry, physical chemistry and chemical physics and biophysics.

Aim and Scope

Science progresses by a symbiotic interaction between theory and experiment: Theory is used to interpret experimental results and may suggest new experiments; the experiment helps to test theoretical predictions and may lead to improved theories. Theoretical chemistry (including physical chemistry and chemical physics) provides the conceptual and technical background and apparatus for the rationalization of phenomena in the chemical sciences. It is, therefore, a wide-ranging subject, reflecting the diversity of molecular and related species and processes arising in chemical systems. The book series *Progress in Theoretical Chemistry and Physics* aims to report advances in methods and applications in this extended domain. It will comprise monographs as well as collections of papers on particular themes, which may arise from proceedings of symposia or invited papers on specific topics as well as from initiatives from authors or translations.

The basic theories of physics—classical mechanics and electromagnetism, relativity theory, quantum mechanics, statistical mechanics, quantum electrodynamics—support the theoretical apparatus which is used in molecular sciences. Quantum mechanics plays a particular role in theoretical chemistry, providing the basis for the valence theories, which allow to interpret the structure of molecules, and for the spectroscopic models, employed in the determination of structural information from spectral patterns. Indeed, quantum chemistry often appears synonymous with theoretical chemistry; it will, therefore, constitute a major part of this book series. However, the scope of the series will also include other areas of theoretical chemistry, such as mathematical chemistry (which involves the use of algebra and topology in the analysis of molecular structures and reactions); molecular mechanics, molecular dynamics and chemical thermodynamics, which play an important role in rationalizing the geometric and electronic structures of molecular assemblies and polymers, clusters and crystals; surface, interface, solvent and solid-state effects; excited-state dynamics, reactive collisions and chemical reactions.

Recent decades have seen the emergence of a novel approach to scientific research, based on the exploitation of fast electronic digital computers. Computation provides a method of investigation which transcends the traditional division between theory and experiment. Computer-assisted simulation and design may afford a solution to complex problems which would otherwise be intractable to theoretical analysis and may also provide a viable alternative to difficult or costly laboratory experiments. Though stemming from theoretical chemistry, computational chemistry is a field of research in its own right, which can help to test theoretical predictions and may also suggest improved theories.

The field of theoretical molecular sciences ranges from fundamental physical questions relevant to the molecular concept, through the statics and dynamics of isolated molecules, aggregates and materials, molecular properties and interactions, to the role of molecules in the biological sciences. Therefore, it involves the physical basis for geometric and electronic structure, states of aggregation, physical and chemical transformations, thermodynamic and kinetic properties, as well as unusual properties such as extreme flexibility or strong relativistic or quantum field effects, extreme conditions such as intense radiation fields or interaction with the continuum, and the specificity of biochemical reactions.

Theoretical chemistry has an applied branch (a part of molecular engineering), which involves the investigation of structure–property relationships aiming at the design, synthesis, and application of molecules and materials endowed with specific functions, now in demand in such areas as molecular electronics, drug design or genetic engineering. Relevant properties include conductivity (normal, semi- and super-), magnetism (ferro- and ferri-), optoelectronic effects (involving nonlinear response), photochromism and photoreactivity, radiation and thermal resistance, molecular recognition and information processing, biological and pharmaceutical activities, as well as properties favoring self-assembling mechanisms and combination properties needed in multifunctional systems.

Progress in Theoretical Chemistry and Physics is made at different rates in these various research fields. The aim of this book series is to provide timely and in-depth coverage of selected topics and broad-ranging yet detailed analysis of contemporary theories and their applications. The series will be of primary interest to those whose research is directly concerned with the development and application of theoretical approaches in the chemical sciences. It will provide up-to-date reports on theoretical methods for the chemist, thermodynamician or spectroscopist, the atomic, molecular or cluster physicist, and the biochemist or molecular biologist who wish to employ techniques developed in theoretical, mathematical and computational chemistry in their research programs. It is also intended to provide the graduate student with a readily accessible documentation on various branches of theoretical chemistry, physical chemistry, and chemical physics.

Preface

This volume collects 16 selected papers from the scientific contributions presented at the Twenty-Third International Workshop on Quantum Systems in Chemistry, Physics, and Biology (QSCP-XXIII), organized by Pr Liliana Mammino from the University of Venda at Mopani Camp, Kruger National Park, South Africa, on September 23–29, 2018. Close to 70 scientists from 15 countries attended this meeting. The participants discussed the state of the art, new trends, and future evolution of methods in molecular quantum mechanics and their applications to a broad variety of problems in chemistry, physics, and biology.

The high-level attendance attained in this conference was particularly gratifying. It is the renowned interdisciplinary nature and friendly feeling of QSCP meetings that make them so successful discussion forums.

The workshop took place at Mopani Camp, one of the coziest of the nine camps in Kruger National Park and the closest to the Phalaborwa airport gate. From the large balcony of the restaurant one could watch, after the talks, a crocodile or a hippopotamus freely swimming, or a giraffe or elephant herd coming to drink in nearby waters. The Kruger National Park, which became South Africa's first national park in 1926, is one of the largest game reserves in Africa, extending over 360 km from North to South and 65 km from West to East. It is located in the west of Mozambique and south of Zimbabwe and is part of the Kruger to Canyons International Man and Biosphere UNESCO Reserve.

Details of the Kruger Park meeting, including the scientific and social programs, can be found on the web site: https://sites.google.com/view/qscp-2018. Altogether, there were 12 morning and afternoon sessions, where 28 plenary lectures and invited talks were given, and three evening poster sessions, with 18 posters being displayed. We are grateful to all the participants for making the QSCP-XXIII workshop a stimulating experience and a great success.

QSCP-XXIII followed the traditions established at previous workshops:

QSCP-I, organized by Roy McWeeny in 1996 at San Miniato (Pisa, Italy);
QSCP-II, by Stephen Wilson in 1997 at Oxford (England);
QSCP-III, by Alfonso Hernandez-Laguna in 1998 at Granada (Spain);

QSCP-IV, by Jean Maruani in 1999 at Marly-le-Roi (Paris, France);
QSCP-V, by Erkki Brändas in 2000 at Uppsala (Sweden);
QSCP-VI, by Alia Tadjer and Yavor Delchev in 2001 at Sofia (Bulgaria);
QSCP-VII, by Ivan Hubac in 2002 near Bratislava (Slovakia);
QSCP-VIII, by Aristides Mavridis in 2003 at Spetses (Athens, Greece);
QSCP-IX, by J.-P. Julien in 2004 at Les Houches (Grenoble, France);
QSCP-X, by Souad Lahmar in 2005 at Carthage (Tunisia);
QSCP-XI, by Oleg Vasyutinskii in 2006 at Pushkin (St. Petersburg, Russia);
QSCP-XII, by Stephen Wilson in 2007 near Windsor (London, England);
QSCP-XIII, by Piotr Piecuch in 2008 at East Lansing (Michigan, USA);
QSCP-XIV, by G. Delgado-Barrio in 2009 at El Escorial (Madrid, Spain);
QSCP-XV, by Philip Hoggan in 2010 at Cambridge (England);
QSCP-XVI, by Kiyoshi Nishikawa in 2011 at Kanazawa (Japan);
QSCP-XVII, by Matti Hotokka in 2012 at Turku (Finland);
QSCP-XVIII, by M. A. C. Nascimento in 2013 at Paraty (Rio, Brazil);
QSCP-XIX, by Cherri Hsu in 2014 at Taipei (Taiwan);
QSCP-XX, by Alia Tadjer and Rossen Pavlov in 2015 at Varna (Bulgaria);
QSCP-XXI, by Yan Wang in 2016 at Vancouver (BC, Canada);
QSCP-XXII, by S. Jenkins and S. Kirk in 2017 at Changsha (China).

The lectures presented at QSCP-XXIII were grouped into six areas, in the field of *Quantum Systems in Chemistry, Physics, and Biology*, ranging from Concepts and Methods in Quantum Chemistry through Relativistic Effects in Quantum Chemistry, Atoms and Molecules in Strong Electric and Magnetic Fields, Reactive Collisions and Chemical Reactions, Molecular Structure, Dynamics and Spectroscopy, and Molecular and Nano materials, to Computational Chemistry, Physics, and Biology.

The width and depth of the topics discussed at QSCP-XXIII are reflected in the contents of this volume of proceedings in *Progress in Theoretical Chemistry and Physics*, which includes four parts:

I. Exotic Atomic Systems (4 papers);
II. Clusters and Molecules Interactions (4 papers);
III. Biochemistry and Biophysics (4 papers);
IV. Fundamental Theory (4 papers).

In addition to the scientific program, the workshop had its usual share of cultural events. There was an evening wildlife exploration in Kruger Park and a night dance and music show by the Mopani Camp Folklore Group.

The award ceremony of the CMOA Prize and Medal took place during the banquet. The CMOA Prize for junior scientists was shared between the three selected nominees: Jian Liu (Peking, China), Daniel Roca-Sanjuan (Valencia, Spain), and Mwombeki Kabanda (Mmabatho, South Africa). The prestigious CMOA Medal for senior scientists was awarded to Prof. Alia Tadjer (Sofia, Bulgaria).

The venue of the next QSCP workshop was set to be at Odessa, Ukraine, on August 2019.

We are most grateful to the Local Patronizing and Organizing Committees for their help and dedication, which made the stay and work of participants so pleasant and fruitful. We would like to express our gratitude to the personnel of Mopani Camp for their efficient logistics. Last but not least, we thank the International Scientific and Honorary Committees for their invaluable expertise and advice.

We hope the readers will find as much interest in consulting these proceedings as the participants in attending the meeting.

Thohoyandou, South Africa — Liliana Mammino
Milan, Italy — Davide Ceresoli
Paris, France — Jean Maruani
Uppsala, Sweden — Erkki Brändas

Obituary: Gerardo Delgado-Barrio (1946–2018)

Picture provided by *GEFAM*

Gerardo Delgado-Barrio (GDB) was a Spanish theoretical chemical physicist. He was born on 9 April 1946 in Santiago de Compostela to a family descendant of the Spanish Grandees, which had a long tradition in education, medicine, and law.

GDB received his Bachelor's degree in the physical sciences at *La Universidad Complutense de Madrid* (UCM) in June 1968. He started working in the biophysics group of Carlos Dávila at the Nuclear Energy Board, where he met the medical physicist Marina Téllez de Cepada, who later became his wife. Shortly after, he joined the research group of Yves Gabriel Smeyers-Guillemin, a pioneer in quantum chemistry in Spain, at *Rocasolano Institute*, which was part of the *Consejo Superior de Investigaciones Científicas* (CSIC). He obtained his Ph.D. degree in April 1973, and in June 1973 he also received the title of *el Técnico de Sistemas* from *La Universidad Politécnica de Madrid* (UPM).

Then Gerardo went for a postdoctoral stay to *Laboratoire de Photophysique Moléculaire* at Orsay (France), where he collaborated with Roger Prat on H-F instability. He also interacted with Roland Lefebvre's group, especially Osman

Atabek and Juan Alberto Beswick, on problems of molecular dynamics collisions, photo-fragmentation, etc. Following one of his visits to Orsay, in teamwork with Alberto Beswick and Joshua Jortner, one of the first numerical simulations of the vibrational predissociation of the van der Waals molecule HeI_2 was published, which paved the way to numerous papers on molecular clusters of increasing size and complexity, involving up to 15 Ph.D. theses.

In 1975, Gerardo returned to Madrid and, after one more year of postdoctorate at *Rocasolano Institute*, he was hired as Associate Professor at the Department of Quantum Chemistry in *Universidad Autónoma de Madrid* (UAM), from 1976 to 1979. From 1979 to 1987, he held various positions at *Instituto de Estructura de la Materia* (IEM), except for a one-year leave in 1982–1983 as Associate Professor at UCM. In 1986, he contributed to founding what was to become the *Instituto de Matemáticas y Física Fundamental* (IMAFF) within CSIC. There he became *Profesor de Investigación* (Senior Researcher), and eventually Director. After retirement, from 2016 onwards, he held the title of *Doctor ad Honorem* at CSIC.

GDB was world-renowned for his research in theoretical atomic and molecular physics, especially on the structure and dynamics of atomic clusters, a topic that he introduced in Spain. He was invited to various foreign institutes, in France as we know but also in Italy, Israel, the USA, the USSR, and especially the Latin American countries, where he was acknowledged as a leader in theoretical chemical physics. He published over 250 papers in scientific journals, gave over 100 talks at scientific meetings, and acted as a coeditor of about 20 volumes. He was a member of the Board of publications such as FCTL, IJQC, PTCP, EPL, and in 2000–2002 he was the Chairman of the Board of the *European Physical Journal*.

He has been the President or a member of the Board of about 20 Spanish, European or International scientific bodies, within which he helped organize some 20 scientific meetings. Among these meetings were a score of QSCP workshops, especially QSCP-XIV at El Escorial in 2009, where the CMOA Medal was awarded, for the first time, to Alberto Beswick, Gerardo's old friend.

GDB held a number of responsibilities in the world of research and education in physics. He was the first chairman of the Atomic and Molecular Physics Group within the Spanish Royal Physical Society, and then became the President of this Society from 1997 through 2005. In 2001–2004, he was in the Executive Committee of CSIC, and in 2006, he became the Director of IMAFF. He was also the President of the Iberian-American Federation of Physical Societies and a representative of Spain at various international bodies, including IUPAP, COST, and the European Physical Society in which he was in charge of international relations.

GDB received a number of honorary distinctions, including the Golden Medal of the (Spanish) Royal Physical Society in 1983 and a Doctorate Honoris Causa from the ISCTN at La Habana (Cuba) in 1998. He could meet King Juan Carlos, to whom he endeavored to promote a consistent and dynamic policy of research in pure and applied physics. During GDB's presidency of the Royal Physical Society, this latter experienced an amazing growth in number of members, financial resources, and impact on civil society.

GDB is acknowledged by his colleagues and coworkers as a person of passion, empathy, and honesty, whose dynamism, open-mindedness, and readiness to share his expertise was a great incentive to those who worked at his side. He was the advisor of 15 Ph.D. students and helped many more young researchers find their way in science. I remember the nice time we had with our wives at ISTCP and QSCP meetings, especially in Japan and in Taiwan. The last time we met was at QSCP-XX at Varna (Bulgaria) in 2015, where he already showed signs of progressive weakness. But he managed to publish a paper in the proceedings, with Pablo Villareal, on '*Rubidium dimer interacting with Helium atoms.*'

In addition to being an outstanding physicist, Gerardo had a very broad range of interests—ranging from History, Philosophy, Politics and Religion to Music and Theater. Discussions with him were pleasant and fruitful. For many years, he suffered from the same illness as Yves Smeyers, whom he highly respected and had replaced in the Editorial Board of this book series. Yves and Gerardo shared other similarities: both had numerous daughters, loved fine arts and enjoyed painting.

Gerardo Delgado-Barrio was delicate and tactful, noble and loyal. His decease came as a great loss not only for his family, but also for all chemical physicists, particularly in Europe and Latin America.

Gerardo Delgado-Barrio passed away in Madrid on July 26, 2018, at the age of 72. He is survived by his wife Marina, four daughters (Marina, Ana, Laura, and Mar), and six grandchildren.

Gerardo Delgado-Barrio was a real *caballero*, both in life and in the community of theoretical chemical physicists. He will be remembered and missed.

Requiesce in Pace ....

July 2019 Jean Maruani

With the cooperation of Pablo Villareal,
Alia Tadjer, Roland Lefebvre, and Erkki Brändas

Contents

Exotic Atomic Systems

Advanced Relativistic Energy Approach in Spectroscopy of Autoionization States of Multielectron Atomic Systems

Alexander V. Glushkov

Abstract We present the generalized energy approach to relativistic calculation of the autoionization decay (resonances) energies and probabilities (widths) in the neutral multielectron atomic systems and multicharged ions. The approach is based on the Gell-Mann and Low S-matrix formalism and the relativistic many-body perturbation theory (PT) with using the optimized one-quasiparticle representation and an accurate account of the relativistic and correlation. In relativistic case the Gell-Mann and Low formula expresses an energy shift ΔE through the electrodynamical scattering matrix including the interaction with as the laser field as the photon vacuum field. The last case is corresponding to definition of the radiative and autoionization decays probabilities for atomic systems. As illustration, the results of relativistic calculation of the autoionization states energies and widths are presented for a number of atoms (helium, barium) and discussed from viewpoint of the correct accounting for the relativistic and exchange-correlation effects.

Keywords Autoionization states · Multielectron atomic systems · Advanced relativistic energy approach · Resonances energies and widths

1 Introduction

Accurate radiative and autoionization decay widths and probabilities for atomic systems are needed for example in astrophysics and quantum physics, atomic and molecular spectroscopy, laboratory and thermonuclear plasma diagnostics, fusion research, laser physics, quantum electronics etc. [1–113]. Traditionally, advanced modeling is important and of current interest in the theory of atomic spectra and associated spectral lines. Spectral lines are usually characterized by their wavelength and width. Typically, autoionization and radiative decay probabilities and widths are known less accurately than wavelengths. Moreover, for many spectral lines of heavy atoms and especially multicharged ions the radiative and autoionization decay probabilities are not reliably known at all.

A. V. Glushkov (✉)
Odessa State Environmental University, L'vovskaya str., bld. 15, Odessa 65016, Ukraine

L. Mammino et al. (eds.), *Advances in Quantum Systems in Chemistry, Physics, and Biology*, Progress in Theoretical Chemistry and Physics 32,
https://doi.org/10.1007/978-3-030-34941-7_1

Generally speaking, an autoionization falls within the general class of phenomena known as the Auger effect [1]. In the Auger effect an atomic system "seemingly" spontaneously decays into a partition of its constituent parts [6, 7]. Traditionally the phenomenon of autoionization, or more particularly the autoionization state decay itself is treated as decay of a bound state. Nevertheless, many scientists have indicated that the process is rigorously a part of the scattering continuum. Due mostly to the known work of Feshbach [5], a rigorous formulation can be established whereby the main element of the theory can be made into a bound state problem with the scattering elements built around it. The major constituent of both these features is accomplished with projection operators. In our opinion [2, 3], an autoionization state should be considered as a resonance in a continuum and we will treat it for the most part in this Chapter. As it is well known the autoionization states can be formed by scattering processes and photoabsorption. These are the inverses of the autoionization and photon emission processes by which they can decay. In the scattering process, formation of the autoionization state corresponds to a resonance in the scattering cross section. Autoionization is the process which corresponds to the decay of the resonance. The decay of the resonance (autoionization) is then seen to be the last half of the resonant scattering process. Numerous studies of autoionization states in spectra of different atoms and ions have shown that the resonant or autoionization state, although it may be long-lived, is not completely stationary. According to Ref. [4, 5] that is the reason that the word "seemingly" was used to describe the Auger process. Since the early work of Beutler the interaction of a discrete (bound) atomic state with a continuum channel has received a great deal of attention. A detailed treatment of such autoionization processes by configuration interaction has been given by Fano [4].

In recent years a rapid progress has been made in studying and understanding autoionization processes in two-quasiparticle atomic systems. Experimentally, multistep laser excitation of the heavier earth-alkaline atoms resulted in a wealth of data on doubly-excited configurations, particularly for the barium atom (e.g. [21–33]). Multi-channell quantum defect theory (MQDT) in combination with eigenchannel R-matrix calculations is quite successful in describing autoionization spectra of these atoms even in cases involving large numbers of continuum channels [21–31]. In a semiempirical approach model potentials are used that are adjusted to the ionic spectrum for each value of the orbital angular momentum l. Much progress has been achieved in the theoretical and computational treatment of electron-impact ionization and autoionization processes due to using such methods as exterior complex scaling (ECS)], time-dependent close-coupling (TDCC), convergent close coupling (CCC) or R-matrix with pseudo-states (RMPS) [2–60]. Nicolaides et al. (e.g. Refs. in [29, 30]) have applied the time-dependent Schrödinger equation approach to treating the doubly excited autoionizing states in simple atomic systems. Nikitin, Ostrovsky et al. developed an analytical framework to compute autoionization rates in two-electron systems using single configuration wavefunctions (e.g. [73]). Its applicability is limited to doubly excited states where the outer electron has a large orbital angular momentum $l > 3$. Poirier [35] included core polarization effects into the Nikitin-Ostrovsky model [31] and performed computing radial matrix elements. This model has been also tested in the 6pnf manifold of barium for J = 1–5 total angular momentum Rydberg series by Abutaleb et al. (1991); experimentally the autoinization rates of the 6pnf J states were found to be up to a factor of two larger than calculated values. Accounting for $6p_{3/2}nf$-$6p_{1/2}\varepsilon l$ fine structure autoinization resulted in only minor changes of 10–20% in the rates of autoionization. Poirier compared his computation of autoionization rates of $6p_{3/2}ng$ J = 5 levels with experimental linewidths measured by Jaffe et al. (e.g. [22–24, 35]). For the K = 9/2 level good agreement was found but for K = 11/2 level the calculated autoionization rates was found to be three times lower than the experimental value. Van Leuwen et al. [22] investigated 5dng states in barium considering a possibility to use these states as intermediates for the excitation of higher lying doubly-excited states with large orbital angular momentum (Jones 1991). Luc Koenig et al. [23, 24] has performed accurate measurements of the autoionization widths of 5d5g levels of barium coupled to theoretical computation, based on the eigenchannel R-matrix method and multichannel quantum defect theory. Besides, this investigation allowed to understand the quadrupolar autoionization mechanism. It was found that the direct polarization of the inner electron by the outer one was important correction to the Coulomb repulsion that causes autoionization. Besides, it is worth to note that these authors proved that the dielectronic polarization interaction significantly influences the autoionization widths. According to Ref. [23], the occurrence of these polarization effects in the 5d5g double excited states with nonoverlapping valence electron results in a significant narrowing of their widths and explains the failure of the single-configuration model for high-l Rydberg levels. Ivanova, Ivanov et al. [113–120] have performed the detailed computing energies and widths of the autoionization resonances, Rydberg levels for ytterbium and thallium. The positions and widths of the autoionization states belonging to the 7s6p, 6p5d, $6p^2$, and $5d^2$ configurations were calculated using the method of relativistic perturbation theory (PT) with the model potential zeroth approximation [113–135]. The advanced version of this approach has been presented and used in Refs. [113, 118–122] to get more exact data for radiative and autoionization characteristics in the lanthanides atoms. In Refs. [131–135] it has been discovered a principally new spectroscopic effect of a giant broadening autoionization resonances of the lanthanides atoms (thulium, gadolinium) in sufficiently weak external electric (laser) field. In general, a problem of studying highly excited Rydberg atomic states and autoionization resonances in an external electromagnetic field attracts a great interest and importance. In fact, a problem of treating parameters of autoionization resonances in a field is obviously close to the problem of describing resonances in general and the Stark resonances in particular. So, here it is worth to remind about a great number of different approaches such as the Pade and then Borel summation of the divergent Rayleigh-Schrödinger perturbation theory (PT) series (Franceschini et al. 1985; Popov et al. 1990), expansions of the wave function over finite basis (Benassi ans Grecchi 1980; Maquet et al. 1983; Kolosov 1987; Telnov 1989; Anokhin-Ivanov 1994), operator perturbation method by Glushkov-Ivanov, complex-scaling method by Reinhardt, Chu, Rao, Liu and Li, Rao-Li et al. [64–110], the Weyl's theory by Hehenberger-McIntosh-Brändas and the Weyl's theory and the complex-rotation method by Brändas, Rittby, Elander and Froelich (1977)

Weyl's theory by Hehenberger-McIntosh-Brändas and the Weyl's theory [67–70]. In any case one could wait for discovery of very interesting, non-trivial features of the autoionization states in an external electric (laser) field especially for heavy atoms and ions.

The purpose of this chapter is to present a generalized relativistic energy approach to calculation of the autoionization decay (resonances) energies and probabilities (widths) in the neutral multielectron atomic systems and multicharged ions and illustrative data on the autoionization states parameters for some interesting complex atomic systems. The basis of an energy approach to one-electron ions has been considered by Labzovsky et al. [112]. Originally the energy approach to radiative and autoionization processes in multielectron atoms and ions has been developed by Ivanova-Ivanov et al. [113–130] (the PC code "Superatom-ISAN"). More accurate, advanced version of the relativistic energy approach has been further developed in Refs. [118–122, 136–141].

The energy approach is based on the Gell-Mann and Low S-matrix formalism combined with the relativistic perturbation theory (PT). In relativistic case the Gell-Mann and Low formula expressed an energy shift ΔE through the electrodynamical scattering matrix including interaction with as the photon vacuum field as a laser field. The first case is corresponding to determination of the radiative and autoionization decays probabilities for atomic systems.

Earlier we have applied the corresponding generalized versions of the energy approach to many problems of atomic, nuclear and even molecular spectroscopy, including computing energy spectra and radiative decay (oscillator strengths) probabilities, energy and spectral characteristics of the cooperative electron-gamma-nuclear "shake-up" processes in atomic and molecular systems, multiple topics of electron-muon-beta-gamma-nuclear spectroscopy, spectroscopy of atoms in a laser field etc. [96, 135–179].

2 Relativistic Energy Approach to Calculation of Autoionization Decay Processes in Multielectron Atoms

2.1 An Energy Approach. General Remarks

In non-relativistic theory of multi-electron atoms it is known an effective field approach for computing the electron energy shift ΔE of the degenerate states, which are usually present in the dense spectra of the complex relativistic atomic multi-electron systems (Tolmachev-Ivanov-Ivanova 1969–1974). The key algorithm of this approach includes construction of the secular matrix M [114–118] with using the known Gell-Mann and Low adiabatic formula and its further diagonalization.

The analogous approach using the Gell-Mann and Low formula with an electrodynamic scattering matrix has been developed in a theory of the relativistic atom

[118–120], however, the M matrix elements in the relativistic representation are complex; the corresponding imaginary parts determine values of radiation widths.

The total energy shift of the state is usually presented in the form:

$$\begin{aligned} \delta E &= \mathrm{Re}\,\delta E + i\,\mathrm{Im}\delta E \\ \mathrm{Im}\,\delta E &= -\Gamma/2 \end{aligned} \tag{1}$$

where Γ is interpreted as the level width including the radiation and autoionization widths simultaneously. In this approach, the whole calculation of the energies and decay probabilities of a non-degenerate excited state is reduced to the calculation and diagonalization of the complex matrix M. In the papers of different authors, the $\mathrm{Re}\delta E$ calculation procedure has been generalized for the case of nearly degenerate states, whose levels form a more or less compact group. One of these variants has been previously [114, 115] introduced: for a system with a dense energy spectrum, a group of nearly degenerate states is extracted and their matrix M is calculated and diagonalized. If the states are well separated in energy, the matrix M reduces to one term, equal to δE. The non-relativistic secular matrix elements are expanded in a PT series for the interelectron interaction. The complex secular matrix M is represented in the form [114, 115]:

$$M = M^{(0)} + M^{(1)} + M^{(2)} + M^{(3)}. \tag{2}$$

Here $M^{(0)}$ is the energy contribution of the vacuum diagrams of all order of PT, and $M^{(1)}$, $M^{(2)}$,$M^{(3)}$ those of the one-, two- and three- quasiparticle diagrams respectively. $M^{(0)}$ is a real matrix, proportional to the unit matrix. It determines only the general level shift. It is usually assumed $M^{(0)} = 0$. The diagonal matrix $M^{(1)}$ can be presented as a sum of the independent one-quasiparticle contributions. For simple systems (such as alkali atoms and ions) the one-quasiparticle energies can be taken from the experiment. Substituting these quantities into (5) one could have summarized all the contributions of the one-quasiparticle diagrams of all orders of the formally exact relativistic PT. However, the necessary experimental quantities are not often available.

The first two order corrections to $\mathrm{Re}M^{(2)}$ have been analyzed previously [113–116] using the Feynman diagrams technique. The contributions of the first-order diagrams have been completely calculated. In the second order, there are two kinds of diagrams: polarization and ladder ones. The polarization diagrams take into account the quasiparticle interaction through the polarizable core, and the ladder diagrams account for the immediate quasiparticle interaction. An effective forms for the two-particle polarizable operator have been proposed in Refs. [115, 180–182]. The technique of determination of the matrix elements of these polarization potentials has been presented in [114–135].

As usual, a multielectron atom is described by the Dirac relativistic Hamiltonian (the atomic units are used):

$$H = \sum_{i} h(r_i) + \sum_{i>j} V(r_i r_j). \tag{3}$$

Here, $h(r)$ is one-particle Dirac Hamiltonian for electron in a field of the finite size nucleus and V is potential of the inter-electron interaction. In order to take into account the retarding effect and magnetic interaction in the lowest order on parameter α^2 (α is the fine structure constant) one could write [2]:

$$V(r_i r_j) = exp(i\omega_{ij} r_{ij}) \cdot \frac{(1 - \alpha_i \alpha_j)}{r_{ij}}, \tag{4}$$

where ω_{ij} is the transition frequency; α_i, α_j are the Dirac matrices.

Further we will use the relativistic many-body PT with the Dirac-Kohn-Sham (DKS) zeroth Hamiltonian [4, 120, 141, 144, 148–150, 152, 155]. The mean-field potential in the zeroth –order Hamiltonian is as follows:

$$V_{MF} = V^{DKS}(r) = [V^D_{Coul}(r) + V_X(r) + V_C(r|b)] \tag{5}$$

Here $V^D_{Coul}(r)$ is the standard Coulomb potential, $V_X(r)$ is the Kohn-Sham (KS) exchange:

$$V_X[\rho(r), r] = V_X^{KS}(r) \cdot \{\frac{3}{2} \ln \frac{[\beta + (\beta^2 + 1)^{1/2}]}{\beta(\beta^2 + 1)^{1/2}} - \frac{1}{2}\}, \tag{6}$$

where $\beta = [3\pi^2 \rho(r)]^{1/3}/c$.

The corresponding correlation functional $V_C(r|b)$ is taken in the following form [4]:

$$V_C[\rho(r), r|b] = -0.0333 \cdot b \cdot \ln[1 + 18.3768 \cdot \rho(r)^{1/3}], \tag{7}$$

where b is the optimization parameter (for details, see below and Refs. [118–121] too). Naturally, the potential (6) is subtracted from the interelectron potential in (3) in the perturbation operator.

The most complicated problem of the relativistic PT computing the radiative and collisional characteristics of the multielectron atomic systems is in an accurate, precise accounting for the exchange-correlation effects (including polarization and screening effects, a continuum pressure etc.) as the effects of the PT second and higher orders. Using the standard Feynman diagrammatic technique one should consider two kinds of diagrams (the polarization and ladder ones), which describe the polarization and screening exchange-correlation effects. The polarization diagrams take into account the quasiparticle (external electrons or vacancies) interaction through the polarizable core, and the ladder diagrams account for the immediate quasiparticle interaction. The detailed description of the polarization diagrams and the corresponding analytical expressions for matrix elements of the polarization quasiparticles interaction (through the polarizable core) potential are presented in Refs. [96, 114–135, 141–179]. An effective approach to accounting for the polarization diagrams contributions is in adding the effective two- quasiparticle polarizable operator into the PT first order matrix elements. In Ref. [114] the corresponding non-relativistic polarization functional has been derived. More correct relativistic expression has been presented in the Refs. [120, 121, 180–182].

2.2 *Radiative Decay in an Energy Approach and the Optimized One-Electron Representation*

It is well known that the topic of the searching and construction of the optimal one-electron (one-quasiparticle) representation is one of the oldest in the theory of multielectron atoms and generally speaking, in quantum chemistry. A great number of different approaches have been developed (c.g. [39–63]). One could mention the known Davidson's "natural orbitals" representation. The density functional method represents one of the simplified recipes too. It is well-known that the density functional theory is very effective approach in a modern quantum chemistry, however, generally speaking, it doesn't provide a regular and precise refinement procedure in the case of the complicated atom with few quasiparticles (electrons or vacancies above a core of the closed electronic shells). Other approaches are in details described in Refs. [4, 120].

An effective ab initio approach to construction of the optimized PT basis has been developed in Refs. [118–120] and reduced to consistent treating gauge-dependent multielectron contributions $\mathrm{Im}\delta E_{\mathrm{ninv}}$ of the lowest relativistic PT corrections to the atomic level radiation width and their further functional minimization. The only one-quasiparticle Feynman diagram a (Fig. 1), contributing the $\mathrm{Im}\Delta E$ (the radiation decay width) is in the lowest (second) order of the QED PT for the δE there. The diagrams, whose contribution into the ImΔE accounts for the core polarization effects, appear in the next (fourth) order of the QED PT (second order of standard atomic PT). Their contribution is dependent upon the photon propagators or electromagnetic potentials gauge (we call it as the gauge non-invariant contribution). Further usually one should consider the multielectron atom with one quasiparticle in the first excited state, connected with the ground state by the radiation transition.

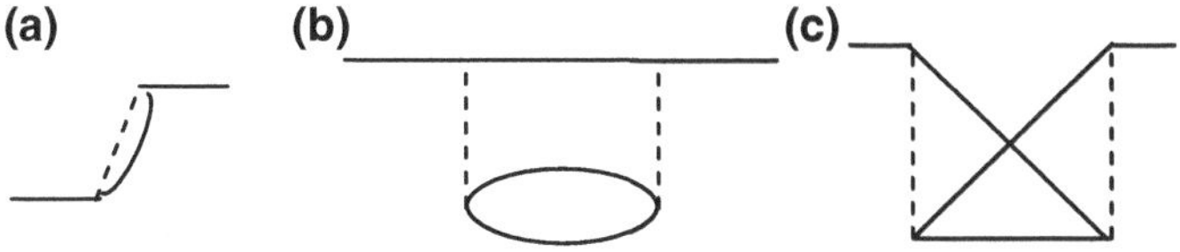

Fig. 1 **a** second other PT diagram contributing the imaginary energy part related to the radiation transitions; **b** and **c** fourth order polarization diagrams

In the PT zeroth approximation one can use the one-electron potential (5). As usually, in terms of the second quantization representation the perturbation looks as follows:

$$-V_{MF}(r)\psi^{+}(r)\psi(r) - j_{\mu}(x)A^{\mu}(x). \tag{8}$$

One should treat the lowest order multielectron effects, in particular, the gauge dependent radiative contribution for the certain class of the photon propagator or electromagnetic potentials gauge. This contribution should be considered as the typical representative of the interelectron correlation effects. So, it is natural that its minimization is an effective and reasonable criteria of the searching for the optimal one-electron basis of the effective relativistic many-body PT.

According to Ref. [120], the diagram a (Fig. 1) imaginary part contribution can be presented as a sum of the partial contributions of α-n transitions from the initial state α to the final state n:

$$\mathrm{Im}\,\delta E_{\alpha}(\mathrm{a}) = \sum_{S} \mathrm{Im}\,\delta E(\alpha - n; \mathrm{a}). \tag{9}$$

In an energy approach the radiative decay probability is directly connected with imaginary part of electron energy of the system, which is defined in the lowest (second) order of the PT as follows [114, 120]:

$$\mathrm{Im}\delta E = -\frac{1}{4\pi} \sum_{\substack{\alpha>n>f \\ [\alpha<n\le f]}} V_{\alpha n\alpha n}^{|\omega_{\alpha n}|}, \tag{10}$$

where $\sum_{\alpha>n>f}$ – for electron and $\sum_{\alpha<n\le f}$ – for vacancy. The potential V is as follows:

$$V_{ijkl}^{|\omega|} = \iint dr_1 dr_2 \Psi_i^*(r_1)\Psi_j^*(r_2)\frac{\sin|\omega|r_{12}}{r_{12}}(1-\alpha_1\alpha_2)\Psi_k^*(r_2)\Psi_l^*(r_1). \tag{11}$$

It is worth to note that here one should use the corresponding expansion for $\sin|\omega|r_{12}/r_{12}$ on spherical harmonics for computing the matrix elements (11). This expansion is corresponding to the standard multipole expansion for probability of the radiative transition. According to Refs. [114–116], the following expression is obtained after substitution of the expansion (11) to matrix element of interaction:

$$V_{1234}^{\omega} = [(j_1)(j_2)(j_3)(j_4)[^{1/2} \sum_{\lambda\mu} (-1)^{\mu} \begin{pmatrix} j_1 j_3 & \lambda \\ m_1 - m_3 & \mu \end{pmatrix} \times \mathbf{Im}\, Q_{\lambda}(1234)$$

$$Q_{\lambda} = Q_{\lambda}^{\mathrm{Qul}} + Q_{\lambda}^{\mathrm{Br}} \tag{12}$$

where j_i are the entire single electron momentums, m_i—their projections; Q_{λ}^{Cul} and Q_{λ}^{Br} are connected with the Coulomb and Breit magnetic parts of the operator (4).

Two fourth order polarization diagrams b, c (Fig. 1) should be considered further. The corresponding contributions are gauge-dependent and the typical representatives of the interelectron correlation effects. It is self-understood that the results of the exact calculation of any physical quantity must be gauge independent.

The detailed consideration and calculation of the direct polarization diagram b (Fig. 1) contribution results in the following expression [118]:

$$
\begin{aligned}
Im\delta E_{ninv}(\alpha - s|b) = -C\frac{e^2}{4\pi}\int\int\int\int dr_1dr_2dr_3dr_4 \\
\times \sum_{n>f,m\leq f}\left(\frac{1}{\omega_{mn}+\omega_{\alpha s}}+\frac{1}{\omega_{mn}-\omega_{\alpha s}}\right) \\
\times \Psi_{\alpha}^{+}(r_1)\Psi_{m}^{+}(r_2)\Psi_{s}^{+}(r_4)\Psi_{n}^{+}(r_3) \\
\cdot [(1-\alpha_1\alpha_2)/r_{12}]\cdot\{[\alpha_3\alpha_4-(\alpha_3n_{34})(\alpha_4n_{34})]/r_{34} \\
\times \sin[\omega_{\alpha n}(r_{12}+r_{34})] \\
+[1+(\alpha_3n_{34})(\alpha_4n_{34})]\omega_{\alpha n}\cos[\omega_{\alpha n}(r_{12}+r_{34})]\} \\
\times \Psi_m(r_3)\Psi_{\alpha}(r_4)\Psi_n(r_2)\Psi_s(r_1) \qquad (13)
\end{aligned}
$$

where C is the gauge constant; $m \leq f$ indicates the finite number of states in the core and the states of the negative continuum (accounting for the electron vacuum polarization). The remaining expression includes summation over the bound and upper continuum atomic states.

The next key step is the minimization of the functional Im δE_{ninv} (b + c), which leads to the integro-differential equation for the ρ_c (the Dirac-Fock or DKS -like equations for the electron density) that are numerically solved. In result we obtain the optimal one-quasiparticle representation, which is further used in calculation of the radiative and autoionization decay characteristics. The detailed description of the whole procedure can be found in Refs. [118–122]. More consistent version is presented in Ref. [123]. It is important to note further that usually a problem of accounting for continuum states can be effectively solved within the Sturm expansions method. It is well known [114] that the space of functions of the atomic states can "pull" on the space Sturm orbitals, which is both discrete and countable. This idea underlies the relatively efficient and formally accurate approach, in order to eliminate a problem of accounting the continuous spectrum.

Naturally, the set of orbitals is introduced assault with a specially prescribed asymptotics that is crucial for the convergence of the spectral decomposition, including spectral decomposition of the Green's function. Among examples of the effective using the Sturm expansions in various problems in atomic and molecular physics one should indicate the works by Dalgarno et al., Buchachenko et al., Ivanova-Ivanov et al., Gruzdev et al., Glushkov et al. etc. (e.g., [1–4, 11, 12, 114]). In a practical implementation the Sturm expansions method is as follows. In the first phase one should solve the system of relativistic DKS equation with respect to the Dirac radial functions and diagonal Lagrange parameters ε^{nlj}. In the second stage it is numerically solved the system of equations, which is equivalent to the Dirac-like equation:

$$(-i\alpha c\nabla + V^D_{Coul}(r) + V_X(r) + \delta_i V_C(r|b) - \varepsilon_i)\phi_i = 0. \tag{14}$$

Two parameters ε_i, δ_i correspond to each orbital of the real or Sturm state. The parameter $\delta_i = 1$ for orbitals of the real states. It is also important to emphasize that all the orbitals of the Sturm supplement Eq. (14) have an exponential asymptotic behavior for $r \to \infty$. It coincides with the asymptotic behavior of the last real state orbital in the corresponding basis of the real states orbitals.

2.3 Autoionization Resonances in Spectrum of Relativistic Multi-electron Atom: Theory

The important characteristics of any autoionization state (resonance) are an energy and width. The autoionization decay in the one-particle approximation can be represented as follows: $\alpha_1\alpha_2 \to \alpha_3 k$, where α_i (i ==l, 2, 3) describes a set of quantum numbers of the bound states, k—state of the free electron. Obviously, an autoionization state decay is possible only into the continuous spectrum state of the same parity and the total angular momentum J. Then, the level width Γ associated with autoionization decay is determined by its relation with the states of the continuous spectrum:

$$\Gamma = 2\pi|\langle i|V|f\rangle|^2 \infty |V_M(\alpha_1\alpha_2, \alpha_3 k)|^2, \tag{15}$$

where $\langle i|$ is an initial, $|f\rangle$ is the final state, V—is the electron-electron interaction operator. Such an approach is usually used in almost all of modern theories. Of course, the basis for this approach is the well-known, remarkable work by Fano.

Within a relativistic energy approach [114–123], the autoionization and radiation widths can be determined using an adiabatic Gell-Mann and Low formulae for an energy shift δE (1). An expansion of an energy shift into the PT series allows to construct a regular method for computing the autoionizing states width. It is worth to remind that the autoionization width firstly appears in the QED PT fourth (second order of an atomic PT) order. The corresponding correction for definite state, say, $n_1^0 j_1^0 n_2^0 j_2^0[J]$, can be represented as follows:

$$\delta E^{(4)} = \underset{k_1 k_2}{S} \sum_{\substack{\beta_1\beta_2\\ \beta_1'\beta_2'}} C^J(\beta_1\beta_2) \frac{V_{\beta_1\beta_2;k_1k_2} V_{k_2k_1;\beta_1'\beta_2'}}{E\left(n_1^0 j_1^0 n_2^0 j_2^0\right) - E(k_1k_2) + i0} C^J\left(\beta_1'\beta_2'\right) \tag{16}$$

where V is matrix element of interelectron interaction, $\underset{k_1k_2}{S}$ means a double sum and an integral over the entire spectrum of one-electron Dirac functions; the coefficients $C^J(\beta_1\beta_2)$ provide a right angle symmetry. It is important to note that we will take into account all harmonics of the interelectron interaction potential.

An autoionizing state width in an energy approach is defined as follows:

$$\Gamma(n_1^0 j_1^0, n_2^0 j_2^0; J) = \frac{2\pi\varepsilon}{K_0} \sum_{\beta_1\beta_2} \sum_{\beta_1'\beta_2'} C^J(\beta_1\beta_2) C^J(\beta_1'\beta_2') \sum_{\beta\beta_K} V_{\beta_1\beta_2;\beta\beta_K} V_{\beta_K\beta;\beta_1'\beta_2'} \tag{17a}$$

$$V_{\beta_1\beta_2;\beta_4\beta_3} = \int\int dr_1 dr_2 \Psi_{\beta_1}(r_1)\Psi_{\beta_2}(r_2) \frac{\cos|\omega| r_{12}(1-\alpha_1\alpha_2)}{r_{12}} \Psi_{\beta_4}(r_2)\Psi_{\beta_3}(r_1) \tag{17b}$$

$$C^J(\beta_1\beta_2) = C^J(n_1 j_1 n_1^0 j_1^0; n_2 j_2 n_2^0 j_2^0) A(j_1 m_1; j_2 m_2; JM) \tag{17c}$$

$$A(j_1m_1, j_2m_2JM) = (-1)^{j_1-j_2+M} \begin{pmatrix} j_1 & j_2 & J \\ m_1 & m_2 & -M \end{pmatrix} \sqrt{2J+1} \tag{17d}$$

$$\begin{aligned} C^J(n_1 j_1 n_1^0 j_1^0; n_2 j_2 n_2^0 j_2^0) = {} & N(n_1^0 j_1^0, n_2^0 j_2^0)[\delta(n_1^0 j_1^0 n_1 j_1)\delta(n_2^0 j_2^0 n_2 j_2) \\ & + (-1)^{j_1+j_2+J+1} \delta\left(n_1^0 j_1^0 n_2 j_2\right)\delta\left(n_2^0 j_2^0 n_1 j_1\right)] \end{aligned} \tag{17e}$$

$$N\left(n_1^0 j_1^0; n_2^0 j_2^0\right) = \begin{cases} \frac{1}{\sqrt{2}} & n_1^0 j_1^0 = n_2^0 j_2^0 \\ 1 & n_1^0 j_1^0 \neq n_2^0 j_2^0 \end{cases} \tag{17f}$$

where $\beta = nljm$, Ψ_β—the one-electron Dirac functions, $K = \frac{1}{\alpha z}\sqrt{1-\xi}$, ξ is an energy of an injected electron.

The secondary quantification representation and the standard momenta coupling procedure (Tolmachev-Ivanov-Ivanova, 1969–1974) present the decay width in the lowest non-vanishing PT order as a sum of products of radial integrals and angular coefficients:

$$S_a^J(n_1 j_1 l_1\ n_2 j_2 l_2;\ n_4 j_4 l_4\ n_3 j_3 l_3) = (-1)^{j_4 - j_2 + J + 1} \begin{Bmatrix} j_3 & j_4 & J \\ j_2 & j_1 & a \end{Bmatrix}$$
$$\times \sqrt{(2j_1+1)(2j_2+1)(2j_3+1)(2j_4+1)}$$
$$\times \begin{pmatrix} j_1 & j_3 & a \\ 1/2 & -1/2 & 0 \end{pmatrix} \begin{pmatrix} j_2 & j_4 & a \\ 1/2 & -1/2 & 0 \end{pmatrix}$$
$$\times \left[\{a l_1 l_3\}\{a l_2 l_4\} F_a(n_1 j_1 l_1\ n_2 j_2 l_2;\ n_4 j_4 l_4\ n_3 j_3 l_3) \right.$$
$$+ \sum_l (-1)^{l+a}$$
$$\left. \times\ (2a+1) M_a^l(n_1 j_1 l_1\ n_2 j_2 l_2;\ n_4 j_4 l_4\ n_3 j_3 l_3) \right]. \quad (18)$$

Here $jl \neq j'l'$ and $\{a_1\ l_1\ l_3\}$ denotes the triangle condition for moments a_1, l_1, l_3. The detailed expressions for the Coulomb and Breit parts can be found in Refs. [4, 114–120, 128–130]. The quantity F is connected with the Coulomb part of the interelectron interaction:

$$F_a(n_1 j_1 l_1\ n_2 l_2 j_2;\ n_4 l_4 j_4\ n_3 j_3 l_3) = R_a(n_1 j_1 l_1\ n_2 j_2 l_2;\ n_4 j_4 l_4\ n_3 j_3 l_3)$$
$$+ R_a\left(n_1 j_1 l_1\ n_2\ \tilde{j}_2\, l_2;\ n_4\ \tilde{j}_4\, l_4\ n_3 j_3 l_3\right)$$
$$+ R_a\left(n_1\ \tilde{j}_1\, l_1\ n_2 j_2 l_2;\ n_4 j_4 l_4\ n_3\ \tilde{l}_3\ j_3\right)$$
$$+ R_a\left(n_1\ \tilde{j}_1\, l_1\ n_2\ \tilde{j}_2\, l_2;\ n_4\ \tilde{j}_4\, l_4\ n_3\ \tilde{j}_3\, l_3\right). \quad (19)$$

The quantity M is connected with the Breit part of interelectron interaction:

$$M_a^l(n_1 j_1 l_1\ n_2 j_2 l_2;\ n_4 j_4 l_4\ n_3 j_3 l_3) = R_l\left(n_1 j_1 l_1\ n_2 j_2 l_2;\ n_4\ \tilde{j}_4\, l_4\ n_3\ \tilde{j}_3\, l_3\right)$$
$$\times X_a^l(j_1 l_1;\ j_3 l_3')X_a^l(j_2 l_2;\ j_4 l_4')$$
$$+ R_l\left(n_1\ \tilde{j}_1\, l_1\ n_2\ \tilde{j}_2\, l_2;\ n_4 j_4 l_4\ n_3 j_3 l_3\right)$$
$$\times X_a^l(j_1 l_1';\ j_3 l_3)X_a^l(j_2 l_2';\ j_4 l_4)$$
$$+ R_l\left(n_1\ \tilde{j}_1\, l_1\ n_2 j_2 l_2;\ n_4\ \tilde{j}_4\, l_4\ n_3 j_3 l_3\right)$$
$$\times X_a^l(j_1 l_1';\ j_3 l_3)X_a^l(j_2 l_2;\ j_4 l_4')$$
$$+ R_l\left(n_1 j_1 l_1\ n_2\ \tilde{j}_2\, l_2;\ n_4 j_4 l_4\ n_3\ \tilde{j}_3\, l_3\right)$$
$$\times X_a^l(j_1 l_1;\ j_3 l_3')X_a^l(j_2 l_2';\ j_4 l_4). \quad (20)$$

The formulas for the angular coefficients are given in the Refs. [2, 6, 26]. As an example, we consider briefly the integral:

$$R_\lambda\left(n_1 j_1 l_1\ n_2 j_2 l_2;\ n_4 j_4 l_4\ n_3 j_3 l_3\right) = \int\limits_0^\infty dr_1 r_1^2 \int\limits_0^\infty dr_2 r_2^2 U_\lambda(r_1, r_2; \omega)$$

$$R_{n_1 j_1 l_1}(r_1) R_{n_2 j_2 l_2}(r_2) R_{n_4 j_4 l_4}(r_2) R_{n_3 j_3 l_3}(r_1) \tag{21a}$$

$$U_\lambda\left(r_1 r_2,\ \omega\right) = \frac{r_<^\lambda}{r_>^{\lambda+1}} Z_\lambda^{(1)}(\omega r_<) Z_\lambda^{(2)}(\omega r_>), \tag{21b}$$

and functions $Z^{(1)}$ and $Z^{(2)}$ can be expressed thorough the standard Bessel functions by the standard way [115].

According to the very effective Ivanova-Ivanov differential equations method [114–116], the calculation procedure for the radial integrals can be reduced to numerical solution of the ordinary differential equations system. The functions $Z^{(1)}$ and $Z^{(2)}$ are calculated from the system of standard differential equations:

$$\begin{aligned} &\ddot{Z}_\lambda^{(1)} + \frac{2(\lambda+1)}{r}\dot{Z}_\lambda^{(1)} + (\alpha z)^2 E_{12}^2 Z_\lambda^{(1)} = 0 \\ &\ddot{Z}_\lambda^{(2)} - \frac{2\lambda}{r}\dot{Z}_\lambda^{(2)} + (\alpha z)^2 E_{12}^2 Z_\lambda^{(2)} = 0 \end{aligned} \tag{22a}$$

with boundary conditions like:

$$Z_\lambda^{(1)}(r_0) = 1 - (\alpha z)^2 E_{12}^2 r_0^2 / \left(2(2\lambda+3)\right) \tag{22b}$$

It is worth noting that the functions $Z_\lambda^{(1)}$, $Z_\lambda^{(2)} \to 0$ when $r_0 \to 0$. This is convenient for numerical integration of equations. The difference of the functions $Z_\lambda^{(1)}$, $Z_\lambda^{(2)}$ from unity for wholly is associated with relativistic effects.

The calculation of the integral $R_l\left(n_1 j_1 l_1\ n_2 j_2 l_2;\ n_4 j_4 l_4\ n_3 j_3 l_3\right)$ is reduced to solution of the following equations system:

$$\begin{cases} y_1' = -\lambda y_1 \big/ r + R_{n_2 j_2 l_2} R_{n_4 j_4 l_4} Z_\lambda^{(1)}(\omega r) \\ y_2' = -\lambda y_2 \big/ r + R_{n_1 j_1 l_1} R_{n_3 j_3 l_3} Z_\lambda^{(1)}(\omega r) \\ y_3' = \frac{Z^{(2)}(\omega r)}{r}\left[y_1 R_{n_1 j_1 l_1} R_{n_3 j_3 l_3} + y_2 R_{n_2 j_2 l_2} R_{n_4 j_4 l_4}\right] \end{cases} \tag{23}$$

with the corresponding border conditions. It is easily to show that:

$$R_l\left(n_1 j_1 l_1\ n_2 j_2 l_2;\ n_4 j_4 l_4\ n_3 j_3 l_3\right) = \lim_{r\to\infty} y_3. \tag{24}$$

When calculating the widths of autoionization resonances, the important question is the correct calculation of the normalization factor. It is worth to remind about the correct calculation of the normalization factor N for the continuum functions. Naturally, the correct functions have the following asymptotic (up to the first term in the expansion):

$$\left.\begin{matrix} F \\ G \end{matrix}\right\} \to (\pi\xi)^{-\frac{1}{2}} \begin{cases} \left(\xi + (\alpha z)^{-2}\right)^{1/2} \sin(kr + \sigma) \\ \left(\xi - (\alpha z)^{-2}\right)^{1/2} \cos(kr + \sigma) \end{cases} \tag{25}$$

It is easy to show that the normalization factor is as follows:

$$N = \lim_{r \to \infty} N(r)$$

$$N^2(r) = \pi\xi\left[F^2\left[\xi + (\alpha z)^{-2}\right] + G^2\left[\xi - (\alpha z)^{-2}\right]\right]$$

Note that, as a rule, a function $N^2(r)$ tends to its stationary value, experiencing slowly damped oscillations at $r \to \infty$.

Other details can be found in Refs. [4, 11, 12, 96, 114–130, 141–155]. All computing is performed with using the modified PC code "Superatom-ISAN" (the modified version 93).

3 Spectroscopy of Autoionization States of Complex Atomic Systems: Illustrative Theoretical Results

3.1 Autoionization Resonances in Spectrum of Helium

In Fig. 2 it is presented the typical fragment of spectrum of photoionization (absorption), according to [17] (e.g. [17–19]). The spectral range contains the AS, which lie on average 35–40 eV above the first ionization potential (24.58 eV).

One of the first members in the AS series is associated with the permitted transition to the double-excited level 2s2p $^1P^0_1$. In general, two series of resonances 2snp, 2pns are identified, both of which have the first term of 2s2p and converge to 189.6A.

In Table 1 we present the experimental data on the energy and width of the AS $^1P_0$, which lies below the ionization threshold n = 2, as well as theoretical results - one of the most accurate theory of the Fano type (Bhatia-Temkin: Th1) and our theory, based on relativistic energy approach (REA) combined with the relativistic many-body PT with the DKS approximation (Th2). The comparison of the presented data demonstrates a physically reasonable agreement between theory and experiment. More details can be found in Ref. [4, 6, 7].

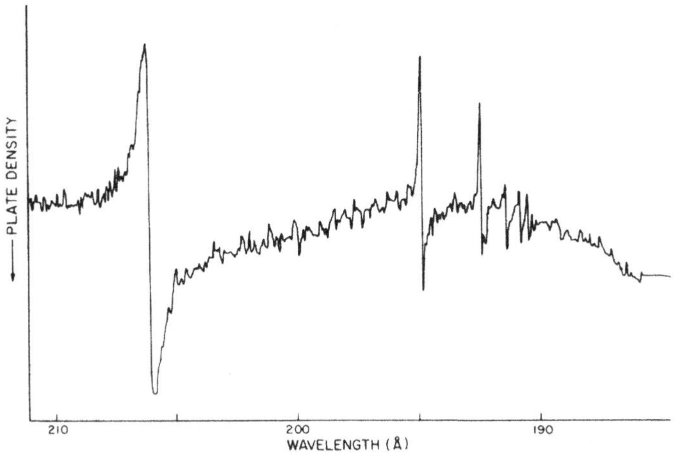

Fig. 2 The typical fragment of the He absorption spectrum (from Ref. [17]; see text)

Table 1 The energy and width of the AS He $^1P_0$ (see text)

	Th.1	Th.2 (our)	Exp. (NIST)
E	60.1444	60.1392	60.133±0.015 60.151±0.0103
Γ	0.0369	0.0374	0.038±0.004 0.038±0.002

Another important test of any theory is the calculation of the AS 3s3p $^1P_0$ parameters. In Table 2 we present the experimental and theoretical data for the AS 3s3p $^1P_0$. (e.g. [4–7, 17–19, 64]).

The results of our approach are compared with the corresponding results of other theories, including the method of complex rotation by Ho, the algebraic approach by Wakid-Callaway, the diagonalization method by Senashenko-Wague, the Nicolaides-Komninos approach (relativistic Hartree-Fock method (RHF)), the R-matrix method by Hayes- Scott, the method of the adiabatic potential curves by Koyoma-Takafuji-Matsuzawa and Sadeghpour, the L^2 technique with using the Sturm expansions by Broad-Gershacher and Moccia-Spizzo, the Feshbach method by Wu-Xi and the experimental data: NIST (NBS 2SO-MeV electron synchrotron storage ring SURF-II), Wisconsin Tantalus storage ring, Stanford Synchrotron Radiation Laboratory (SSRL), Berlin electron storage ring (BESSY), Daresbury Synchrotron Radiation Source (DSRS).

An analysis of the presented data shows that from the one hand, there is a physically reasonable agreement between our theory and experiment. From the second side, it should be noted that the most of the cited methods are developed specifically for the study a few-electron atomic systems and their generalization on the case of essentially multielectron systems requires a serious modification.

Table 2 The energy and width of the AS He 3s3p $^1P_0$ (see text)

Method	Authors	E_r (Ry)	Γ/2 (Ry)
REA	Glushkov et al.	−0.668802	0.006814
Complex-rotation	Ho	−0.671252	0.007024
Algebraic close coupling	Wakid-Callaway	−0.670	0.00695
Diagonalization method	Senashenko-Wague	−0.6685	0.00548
RHF	Nicolaides-Komninos	−0.671388	–
R-matrix calculation	Hayes-Scott	−0.6707	0.00660
Adiabatic potential curves	Koyoma et al.	−0.6758	–
Adiabatic potential curves	Sadeghpour	−0.67558	–
L^2 tech. + Sturmian func.	Gershacher-Broad	−0.67114	0.00704
Feshbach method	Wu-Xi	−0.669 27	0.00420
K-matrix L^2 basis-set	Moccia-Spizzo	−0.670 766	0.00676
Method	E_r (eV)	Γ/2 (eV)	
REA	69.9055	0.1854	
Complex-rotation	69.8722	0.1911	
Multicontigurational HF	69.8703	–	
R-matrix calculation	69.8797	0.1796	
Exp.- NBS-I (1973)	69.919 ± 0.007	0.132 ± 0.014	
Exp.- Wisconsin (1982)	69.917 ± 0.012	0.178 ± 0.012	
Exp.- SSRL (1987)	69.917 ± 0.012	0.178 ± 0.012	
Exp.- BESSY (1988)	69.914 ± 0.015	0.200 ± 0.020	
Exp.- DSRS (2009)	69.880 ± 0.022	0.180 ± 0.015	

Note for the transfer of units of Ry-eV, the value of the energy of the He ground state E = −5.80744875 Ry is used and Rydrberg reduced constant 1Ry = 13.603876 eV

3.2 Some Autoionization States in Spectrum of Barium

The barium atom is of considerable interest both from the point of view of atomic spectroscopy and theoretical atomic physics, since it is essentially relativistic and multielectron system, and it has an extremely complex and interesting spectrum of both discrete and AS [4, 22, 23]. On the other hand, information about its spectroscopic parameters is important for further research using laser spectroscopy methods, and finally, an understanding of the physics and mechanisms of interatomic interactions.

The study of the spectrum of this atom is of tremendous interest from the point of view of the relevance of studying the shift and broadening of the transition lines between low- and highly-excited speakers in the spectrum of barium in the atmosphere of inert buffer gases (e.g. [4]). The 5d5g states are among the most interesting and important barium AS, these resonances correspond to transitions $5d6p^3F_4$-$5d_{5/2}5g$ [K = 9/2,7/2,5/2], $5d^{2\,3}F_3$-$5d_{5/2}5g$ – $5d^{2\,3}F_3$-$5d_{5/2}5g$ [22–24].

In Table 3 we present the measured and calculated AS width (cm^{-1}) for 5d5g AS Ba with J < 6: (3) Experiment; (4) Theory: HF method with partial accounting for the correlations (Van Leeuwen et al. [53]); (5) R-matrix method without accounting for the polarization effect of the shells (Luc-Koenig et al.); (6) R-method with accounting the polarization of shells and without accounting for dielectronic interaction (Luc-Koenig et al.); (7) R-method with full accounting for the polarization and dielectronic correlation effects (Luc-Koenig et al. [23, 24]); (8,9) our theory without accounting for the correlation (polarization) effects and with full accounting the cited corrections.

In Table 4 we present the measured and calculated AS energies (theoretical data by Van Leeuwen et al. [22] and this approach). The analysis of 5d5g barium levels, the channels $6s_{1/2}\varepsilon l_1 j_1$ are open, $55d_{j2}n_1l_1j_1$ are closed, and $p_{j2}np_{j1}$ are closed. The double excited states of 6p2 and 6p7p lie on the energy scale far from 5dng states, respectively below the $6s_{1/2}$ threshold and above the $5d_{5/2}$ threshold. Therefore, it is quite enough to take into account the general connection of 5dng-6pnp channels. All 6pnf J = 1–5 states are located at least 8000 cm^{-1} above the $5d_{5/2}$ threshold and, thus, the $6p_{j2}nf_{j1}$ channels are closed. The R-matrix method of Luc-Koenig et al. [24], taking into account polarization effects, is based on the use of the empirical single-particle polarization potential of the Dalgarno form. In our theory, the essentially multi-particle polarization operators of [182] act as an analogue of the Dalgarno polarization potential. These potentials are derived by summing the sequence of Feynman diagrams corresponding to the second and higher orders of Rayleigh-Schrödinger PT in the density functional approximation (Thomas-Fermi) and, accordingly, are initially more correct.

In any case to get a physically reasonable agreement between theoretical and experimental data one should accurately take the complex exchange-correlation (polarization, dielectronic interaction etc.) effects into account under using the optimized one-quasiparticle representation (with adding the Sturm expansions) in the relativistic many-body PT theory zeroth approximation.

Very interesting in Ba spectroscopy is the study of the Rydberg AS that correspond to transitions in the state 4fnf, in particular, $4f_{5/2}$n'f J = 6, 5, 4, excited from the initial $5d_{3/2}$15f J = 5 state [28, 64]. The method of laser spectroscopy method is applied to study the corresponding $4f_{5/2}$n'f J = 6, 5, 4, states, excited from the initial $5d_{3/2}$15f J = state, and $4f_{5/2,7/2}$n'f J = 6, 5, 4 one speakers (the initial state is $5d_{3/220\ 20}$f J = 4).

As illustration, in Table 5 we present experimental energy values (cm^{-1}) of 4fnf, n = 15 autoionization resonances, n = 15 (averaged over the fine structure) measured experimentally, as well as calculated on the basis of relativistic energy approach (REA; combined with the PT with the DKS approximation) [4, 64] and the multichannel quantum defect method (MCQD) with empirical fitting (de Graaff et al.) [28].

One should note that the use of the MCQD is possible provided that its parameters are reliably determined (quantum defects either on the basis of empirical information or the simplest screening models).

An analysis of the data presented in Table 5 shows that both the methods (MCQD with an empirical fitting and our theory with a correct accounting for the exchange-polarization effects), provide a physically reasonable description of the energies of

Table 3 Measured and calculated AS width (cm^{-1}) for 5d5g AS Ba with $J < 6$: (3) Experiment; (4) Theory: HF method with partial accounting for the correlations (Van Leeuwen et al.; (5) R-matrix method without accounting for the polarization effect of the shells (Luc-Koenig et al.); (6) R-method with accounting the polarization of shells and without accounting for dielectronic interaction (Luc-Koenig et al.); (7) R-method with full accounting for the polarization and dielectronic correlation effects (Luc-Koenig et al.); (8, 9) our theory without accounting for the correlation (polarization) effects and with full accounting the cited corrections (see text)

1:j	K	J	2: εl	(3)	(4)	(5)	(6)	(7)	(8)	(9)
$5d_{3/2}$	5/2	2	εd	0.062(4)	0.0074	0.077	0.072	0.018	0.071	0.058
$5d_{3/2}$	5/2	3	εd	0.056(5)	0.0074	0.048	0.046	0.025	0.044	0.055
$5d_{3/2}$	9/2	4	εg	0.18(2)	1.00	1.11	0.11	0.23	0.10	0.20
$5d_{3/2}$	9/2	5	εg	0.18(2)	1.00	1.19	0.11	0.23	0.09	0.19
$5d_{3/2}$	11/2	5	εi	–	2.90	3.19	1.44	1.74	1.39	1.52
$5d_{3/2}$	11/2	6	εi	–	2.90	3.20	1.44	1.73	1.38	1.51
$5d_{5/2}$	3/2	1	εd	0.046(3)	0.012	0.038	0.066	0.041	0.065	0.044
$5d_{5/2}$	7/2	3	εg	0.29(3)	1.43	1.61	0.13	0.29	0.15	0.28
$5d_{5/2}$	9/2	4	εg	0.43(4)	2.29	2.57	0.22	0.48	0.23	0.45
$5d_{5/2}$	9/2	5	εg	0.43(4)	2.29	2.71	0.22	0.48	0.24	0.46
$5d_{5/2}$	11/2	5	εi	0.28(3)	0.45	0.56	0.25	0.30	0.21	0.28
$5d_{5/2}$	11/2	6	εi	0.32(3)	0.45	0.55	0.25	0.30	0.22	0.29
$5d_{5/2}$	13/2	6	εi	–	3.37	3.95	1.83	2.15	1.85	2.17

Table 4 Measured and calculated AS energies (cm^{-1}): theoretical data by Van Leeuwen et al. and this approach (see text)

j_{sd}	K	J	Exp. energy	[22]	This approach
3/2	5/2	2, 3	42509.9	42511.2	42509.8
3/2	7/2	3, 4	42480.1	42480.3	42480.2
3/2	9/2	4, 5	42470.8	42469.5	42470.6
3/2	11/2	5, 6	–	42500.4	42499.7
5/2	3/2	1, 2	43323.4	43322.5	43323.4
5/2	5/2	2, 3	43302.8	43302.5	43302.6
5/2	7/2	3, 4	43282.1	43283.5	43282.2
5/2	9/2	4, 5	43271.02	43271.8	43270.9
5/2	11/2	5, 6	43275.6	43274.8	43275.5
5/2	13/2	6, 7	–	43306.4	43307.0

Table 5 Experimental energy values (cm^{-1}) for the 4fnf (n = 15) AS (averaged over the fine structure) measured experimentally, as well as calculated on the basis of our theory and the multichannel quantum defect method (MCQD) with empirical fitting (de Graaff et al.) [28] (see text)

AC	J	Exp.	MCQD	REA
$4f_{5/2}15f_{7/2}$	6	89758.4 ± 0.5	89759.1	89758.8
$4f_{7/2}15f_{5/2}$	6	89993.6 ± 0.5	89992.4	89993.6
$4f_{7/2}15f_{7/2}$	6	89926.6 ± 5.0	89937.1	89926.8
$4f_{5/2}15f_{7/2}$	5	89726.3 ± 1.0	89718.7	89726.9
$4f_{5/2}15f_{5/2}$	5	89749.2 ± 0.5	89748.6	89749.7
$4f_{7/2}15f_{7/2}$	5	89951.0 ± 0.5	89952.9	89951.6
$4f_{7/2}15f_{5/2}$	5	–	89943.6	89942.3
$4f_{5/2}15f_{5/2}$	4	89705.6 ± 0.5	89706.8	89705.4
$4f_{5/2}15f_{7/2}$	4	–	89720.0	89718.5
$4f_{7/2}15f_{5/2}$	4	89937.8 ± 2.0	89937.2	89937.6
$4f_{7/2}15f_{7/2}$	4	89951.0 ± 2.0	89951.8	89951.5
$4f_{5/2}15f_{5/2}$	3	–	89725.9	89728.5
$4f_{5/2}15f_{7/2}$	3	89741.5 ± 2.0	89738.0	89740.9
$4f_{7/2}15f_{7/2}$	3	89969.3 ± 2.0	89972.0	89969.8
$4f_{7/2}15f_{5/2}$	3	–	89955.9	89953.4
$4f_{5/2}15f_{7/2}$	2	89766.5 ± 5.0	89774.7	89767.8

the studied states. Let us also note that according to our data the widths of the resonances indicated in the table vary extremely irregularly and are in the range of $1.64 \times 10^{-6} - 1.92 \times 10^{-5}$ Ry.

4 Conclusions

To conclude, we presented the generalized energy approach to relativistic calculation of the autoionization decay (resonances) energies and probabilities (widths) in the neutral multielectron atomic systems. The approach is based on the Gell-Mann and Low S-matrix formalism and the relativistic many-body PT with using the optimized one-quasiparticle representation. In relativistic case the Gell-Mann and Low formula expresses an energy shift ΔE through the electrodynamical scattering matrix including the interaction with as the laser field as the photon vacuum field. The last case is corresponding to definition of the radiative and autoionization decays probabilities for atomic systems. The optimized one-electron representation in the PT zeroth approximation is constructed by means of the correct treating the gauge dependent multielectron contribution of the lowest PT corrections to the radiation widths of atomic levels. It is important to note that an approach is universal and, generally speaking, can be applied to quantum systems of other nature (see, for example, [57–66] and Refs. therein).

As illustration, the results of relativistic calculation of the autoionization states energies and widths are presented for a number of atoms (helium, barium) and discussed from point of view of the correct accounting for the relativistic and exchange-correlation effects. The calculations encourage us to believe that using energy approach combined with the relativistic many-body PT with the optimal one-electron basis is quite consistent and effective tool from the point of view of the theory correctness and results exactness. This fact is confirmed by other calculations of the oscillator strengths, radiative widths, hyperfine structure constants for atoms and multicharged ions (see Refs. [4, 6, 27–35, 82, 88, 163–183]). The obtained data can be used in different applications, namely, astrophysical analysis, laboratory, thermonuclear plasmas diagnostics, fusion research, laser physics, quantum electronics etc.

Acknowledgements The author would like to thank Professor Jean Maruani and Professor Liliana Mammino (the organizer of the QSCP-XXIII Workshop) for their invitation to submit contributions in these Proceedings.

References

1. Grant IP (2007) Relativistic quantum theory of atoms and molecules, theory and computation. Springer Series on Atomic, Optical, and Plasma Physics, vol 40. Springer, Berlin, pp 587–626

2. Glushkov AV (2008) Relativistic quantum theory. Quantum mechanics of atomic systems. Astroprint, Odessa
3. Khetselius OY (2008) Hyperfine structure of atomic spectra. Astroprint, Odessa
4. Fano U (1960) Effects of configuration interaction on intensities and phase shifts. Phys Rev 124:1866–1878
5. Feshbach H (1967) The unified theory of nuclear reactions: III. Overlapping resonances. Ann Phys (N Y) 43:410–420
6. Temkin A, Bhatia AK (1985) Theory and calculation of resonances and autoionization of two-electron atoms and ions. In: Temkin A (ed) Autoionization. Physics of atoms and molecules. Springer, Boston, pp 1–34
7. Temkin A, Bhatia AK (1996) Autoionization. In: Drake GW (ed) Atomic, molecular and optical handbook. AIP Press, New York, pp 391–399
8. Chung KT, Davis BF (1985) Hole-projection method for calculating Feshbach resonances and inner-shell vacancies. In: Temkin A (ed) Autoionization. Physics of atoms and molecules. Springer, Boston, pp 73–102
9. George A, Doschek GA (1985) Diagnostics of solar and astrophysical plasmas dependent on autoionization phenomena. In: Temkin A (ed) Autoionization. Physics of atoms and molecules. Springer, Boston, pp 171–256
10. Oks E (2010) Relation between theories, experiments, and simulations of spectral line shapes. Int J Spectrosc 10:852581
11. Aglitsky EV, Safronova UI (1985) Spectroscopy of autoionization states of atomic systems. Energoatomizdat, Moscow
12. Glushkov AV (2006) Relativistic and correlation effects in spectra of atomic systems. Astroprint, Odessa
13. Glushkov AV, Khetselius OY, Svinarenko AA, Buyadzhi VV (2015) Spectroscopy of autoionization states of heavy atoms and multiply charged ions. TEC, Odessa
14. Glushkov AV, Malinovskaya SV, Loboda AV, Shpinareva IM, Gurnitskaya EP, Korchevsky DA (2005) Diagnostics of the collisionally pumped plasma and search of the optimal plasma parameters of x-ray lasing: calculation of electron-collision strengths and rate coefficients for Ne-like plasma. J Phys: Conf Ser 11:188–198
15. Glushkov AV, Malinovskaya SV, Chernyakova YuG, Svinarenko AA (2004) Cooperative laser-electron-nuclear processes: QED calculation of electron satellites spectra for multi-charged ion in laser field. Int J Quant Chem 99(5):889–893
16. Johnson W, Sapistein J, Blundell S (1988) Finite basis sets for the Dirac equation constructed from B splines. Phys Rev A 37:307–315
17. Madden R, Codling K (1963) New autoionization atomic energy levels in He, Ne, Ar. Phys Rev Lett 10:516–520
18. Ho YK (1991) Autoionizing $^1P^0$ states of He between the N = 2 and 3 threshold of He^+. Phys Rev A 44:4154–4161
19. Sakho I, Konté K, Ndao AS, Biaye M, Wagué A (2010) Calculations of $(nl)^2$ and (3lnl') autoionizing states in two-electron systems. Phys Scr 82:035301
20. Aggarwal KM, Keenan FP (2012) Energy levels, radiative rates and electron impact excitation rates for transitions in He-like Kr XXXV. Phys Scr 86:035302
21. Greene CH, Aymar M (1991) Spin-orbit effects in the heavy alkaline-earth atoms. Phys Rev A 44(3):1773–1790
22. Van Leuwen R, Ubachs W, Hogervorst W (1994) Autoionization of low-lying 5dng states in Barium. J Phys B: Atom Mol Opt Phys 27:3891–3904
23. Luc-Koenig E, Aymar M, Van Leeuwen R, Ubachs W, Hogervorst W (1995) Polarization effects in autoionization processes: the 5d5g states in Barium. Phys Rev A 52:208–215
24. Luc-Koenig E, Lyras A, Lecomte J-M et al (1997) Eigenchannel R-matrix study of two-photon processes including above-threshold ionization in Mg. J Phys B: At Mol Opt Phys 30:5213–5232
25. Wesdorp C, Noordam LD, Robicheaux F (1999) Dynamics of forced autoionization. Phys Rev A 60:R3377–R3380

26. Klose JZ, Fuhr JR, Wiese WL (2002) Critically evaluated atomic transition probabilities for Ba I and Ba II. J Phys Chem Ref Data 31:217–230
27. Bokor J, Freeman R, Cooke W (1982) Autoionization—pumped laser. Phys Rev Lett 48:1242–1247
28. De Graaff RJ, Ubachs W, Hogervorst W (1992) 4fnf doubly excited autoioinizing states in Barium. Phys Rev 45(1):166–178
29. Nicolaides CA (1992) Hole-projection, saddle points and localization in the theory of autoionizing states. Phys Rev A 46:690–698
30. Nicolaides CA, Piangos NA (2001) State-specific approach and computation of resonance states. Identification and properties of the lowest ^2P_o and 2D triply excited states of He^-. Phys Rev A 64:052505
31. Nikitin SI, Ostrovsky VN (1980) The autoionization of high Rydberg atomic states with large orbital momentum. J Phys B: Atom Mol Opt Phys 13:1961–1984
32. Yi Jong-hoon, Lee J, Kong HJ (1995) Autoionizing states of the ytterbium atom by three-photon polarization spectroscopy. Phys Rev A 51:3053–3057
33. Yi J, Park H, Lee J (2001) Investigation of even parity autoionizing states of Ytterbium atom by two-photon ionization spectroscopy. J Korean Phys Soc 39:916–920
34. Bylicki M (1998) Methods involving complex coordinates applied to atoms. Adv Quant Chem 32:207–226
35. Poirier M (1997) Analysis of correlation effects in autoionizing doubly excited states of barium using Coulomb Green's function. Z Phys D 39:189–193
36. Chernenko AA, Beterov IM, Permyakova OI (2000) Modeling of amplification without inversion near transitions from autoionization levels of ytterbium atom. Laser Phys 10:133–138
37. Buyadzhi VV, Chernyakova YuG, Smirnov AV, Tkach TB (2016) Electron-collisional spectroscopy of atoms and ions in plasma: Be-like ions. Photoelectronics 25:97–101
38. Buyadzh VV, Chernyakova YuG, Antoshkina OA, Tkach TB (2017) Spectroscopy of multicharged ions in plasmas: oscillator strengths of Be-like ion Fe. Photoelectronics 26:94–102
39. Laughlin C, Victor GA (1989) Model-potential methods. Adv Atom Mol Phys 25:163
40. Cheng K, Kim Y, Desclaux J (1979) Electric dipole, quadrupole, and magnetic dipole transition probabilities of ions isoelectronic to the first-row atoms, Li through F. Atom Data Nucl Data Tabl 24:111
41. Indelicato P, Desclaux JP (1993) Projection operator in the multiconfiguration Dirac-Fock method. Phys Scr 46:110
42. Bieron J, Pyykkö P, Jonsson P (2005) Nuclear quadrupole moment of ^{201}Hg. Phys Rev A 7:012502
43. Lund AP, Ralph TC (2005) Coherent-state linear optical quantum computing gates using simplified diagonal superposition resource states. Phys Rev A 71:032502
44. Feller D, Davidson ER (1989) An approximation to frozen natural orbitals through the use of the Hartree-Fock exchange potential. J Chem Phys 74:3977–3979
45. Dietz K, Heß BA (1989) Single particle orbitals for configuration interaction derived from quantum electrodynamics. Phys Scr 39:682–688
46. Glushkov AV, Malinovskaya SV, Filatov VV (1989) S-Matrix formalism calculation of atomic transition probabilities with inclusion of polarization effects. Sov Phys J 32(12):1010–1014
47. Khetselius OY (2008) Relativistic calculating the spectral lines hyperfine structure parameters for heavy ions. AIP Conf Proc 1058:363–365
48. Glushkov AV, Lovett L, Khetselius OY, Gurnitskaya EP, Dubrovskaya YuV, Loboda AV (2009) Generalized multiconfiguration model of decay of multipole giant resonances applied to analysis of reaction (μ-n) on the nucleus ^{40}Ca. Int J Mod Phys A 24(2–3):611–615
49. Glushkov AV, Malinovskaya SV, Sukharev DE, Khetselius OY, Loboda AV, Lovett L (2009) Green's function method in quantum chemistry: new numerical algorithm for the Dirac equation with complex energy and Fermi-model nuclear potential. Int J Quant Chem 109:1717–1727

50. Khetselius OY (2009) Relativistic perturbation theory calculation of the hyperfine structure parameters for some heavy-element isotopes. Int J Quant Chem 109:3330–3335
51. Khetselius OY (2009) Relativistic calculation of the hyperfine structure parameters for heavy elements and laser detection of the heavy isotopes. Phys Scr 135:014023
52. Glushkov AV, Khetselius OY, Gurnitskaya EP, Loboda AV, Sukharev DE (2009) Relativistic quantum chemistry of heavy ions and hadronic atomic systems: spectra and energy shifts. Theory and applications of computational chemistry. AIP Conf Proc 1102:168–171
53. Khetselius OY (2012) Quantum geometry: new approach to quantization of the quasistationary states of Dirac equation for super heavy ion and calculating hyper fine structure parameters. Proc Intern Geom Center 5(3–4):39–45
54. Quinet P, Argante C, Fivet V et al (2007) Atomic data for radioactive elements Ra I, Ra II, Ac I and Ac II and application to their detection in HD 101065 and HR 465. Astrophys Astron 474:307–314
55. Biémont E, Fivet V, Quinet P (2004) Relativistic Hartree-Fock and Dirac-Fock atomic structure calculations in Fr-like ions Ra+, Ac2+, Th3+ and U5+. J Phys B: At Mol Opt Phys 37:4193–4204
56. Froese Fischer C, Tachiev G (2004) Breit-Pauli energy levels, lifetimes, and transition probabilities for the beryllium-like to neon-like sequences. Atom Data Nucl Data Tables 87:1
57. Sapirstein J, Cheng KT (2005) Calculation of radiative corrections to E1 matrix elements in the neutral alkali metals. Phys Rev A 71:022503
58. Glushkov AV, Svinarenko AA, Khetselius OY, Buyadzhi VV, Florko TA, Shakhman AN (2015) Relativistic quantum chemistry: An advanced approach to the construction of the Green function of the Dirac equation with complex energy and mean-field nuclear potential. In: Nascimento M, Maruani J, Brändas E, Delgado-Barrio G (eds) Frontiers in Quantum Methods and Applications in Chemistry and Physics. Series: Progress in Theoretical Chemistry and Physics, vol 29. Springer, Cham, pp197–217
59. Yerokhin V, Artemyev AN, Shabaev VM (2007) QED treatment of electron correlation in Li-like ions. Phys Rev A 75:062501
60. Khetselius OY, Florko TA, Svinarenko AA, Tkach TB (2013) Radiative and collisional spectroscopy of hyperfine lines of the Li-like heavy ions and Tl atom in an atmosphere of inert gas. Phys Scr T 153:014037
61. Kohn W, Sham LJ (1965) Self-consistent equations including exchange and correlation effects. Phys Rev A 140:1133
62. Hohenberg P, Kohn W (1964) Inhomogeneous electron gas. Phys Rev B 136:864
63. Buyadzhi VV, Zaichko PA, Antoshkina OA, Kulakli TA, Prepelitsa GP, Ternovsky VB, Mansarliysky VF (2017) Computing of radiation parameters for atoms and multicharged ions within relativistic energy approach: advanced Code. J Phys: Conf Ser 905:012003
64. Ternovsky EV, Buyadzhi VV, Tsudik AV, Svinarenko AA (2018) Relativistic calculation of Rydberg autoionization states parameters in spectrum of barium. Photoelectronics 27:34–43
65. Rao J, Liu W, Li B (1994) Theoretical complex Stark energies of hydrogen by a complex-scaling plus B-spline approach. Phys Rev A 50:1916–1919
66. Rao J, Li B (1995) Resonances of the hydrogen atom in strong parallel magnetic and electric fields. Phys Rev A 51:4526–4530
67. Meng H-Y, Zhang Y-X, Kang S et al (2008) Theoretical complex Stark energies of lithium by a complex scaling plus the B-spline approach. J Phys B: At Mol Opt Phys 41:155003
68. Brändas E, Froelich P (1977) Continuum orbitals, complex scaling problem, and the extended virial theorem. Phys Rev A 16(6):2207–2210
69. Rittby M, Elander N, Brändas E (1981) Weyl's theory and the complex-rotation method applied to phenomena associated with a continuous spectrum. Phys Rev A 24(3):1636–1639
70. Froelich P, Davidson ER, Brändas E (1983) Error estimates for complex eigenvalues of dilated Schrödinger operators. Phys Rev A 28(5):2641–2645
71. Lipkin N, Moiseyev N, Brändas E (1989) Resonances by the exterior-scaling method within the framework of the finite-basis-set approximation. Phys Rev A 40(2):549–553

72. Simon B (1979) The definition of molecular resonance curves by the method of exterior complex scaling. Phys Lett A 71(2–3):211–214
73. Resonances (1989) The unifying route towards the formulation of dynamical processes—foundations and applications. In: Brändas E, Elander N (eds) Nuclear, atomic and molecular physics. Series: Lecture Notes in Physics, vol 325. Springer, Berlin, pp 1–564
74. Ostrovsky VN, Elander N (2005) Scattering resonances and background associated with an asymmetric potential barrier via Siegert pseudostates. Phys Rev A 71:052707
75. Sigal IM (1988) Geometric theory of Stark resonances in multielectron systems. Commun Math Phys 119(2):287–314
76. Herbst IW, Simon B (1978) Stark effect revisited. Phys Rev Lett 41(2):67–69
77. Silverstone HJ, Adams BG, Cizek et al (1979) Stark effect in hydrogen: dispersion relation, asymptotic formulas, and calculation of the ionization rate via high-order perturbation theory. Phys Rev Lett 43(20):1498–1501
78. Cerjan C, Hedges R, Holt C et al (1978) Complex coordinates and the Stark effect. Int J Quant Chem 14(4):393–418
79. Luc-Koenig E, Bachelier A (1980) Systematic theoretical study of the Stark spectrum of atomic hydrogen. I. Density of continuum states. J Phys B: At Mol Phys 13:1743–1756
80. Buyadzhi VV, Zaichko P A, Gurskaya M Y, Kuznetsova AA, Ponomarenko EL, Ternovsky VB (2017) Relativistic theory of excitation and ionization of Rydberg atomic systems in a Black-body radiation field. J Phys: Conf Ser 810:012047
81. Maquet A, Chu SI, Reinhardt WP (1983) Stark ionization in dc and ac fields: an L2 complex-coordinate approach. Phys Rev A 27(6):2946–2970
82. Reinhardt WP (1982) Padé summations for the real and imaginary parts of atomic stark eigenvalues. Int J Quant Chem 21(1):133–146
83. Franceschini V, Grecchi V, Silverstone H (1985) Complex energies from real perturbation series for the LoSurdo-Stark effect in hydrogen by Borel-Padé approximants. J Phys Rev A 32(3):1338
84. Benassi L, Grecchi V (1980) Resonances in the Stark effect and strongly asymptotic approximants. J Phys B: At Mol Phys 13(5):911
85. Farrelly D, Reinhardt WP (1983) Uniform semiclassical and accurate quantum calculations of complex energy eigenvalues for the hydrogen atom in a uniform electric field. J Phys B: At Mol Phys 16(12):2103
86. Filho O, Fonseca A, Nazareno H et al (1990) Different approach to the Stark effect: application to the hydrogen ground state. Phys Rev A 42(7):4008–4014
87. Kondratovich VD, Ostrovsky VN (1984) Resonance and interference phenomena in the photoionisation of a hydrogen atom in a uniform electric field. II. Overlapping resonances and interference. J Phys B: At Mol Phys 17(10):2011
88. Telnov DA (1989) DC Stark effect in a hydrogen atom via Sturmian expansions. J Phys B: At Mol Opt Phys 22(14):L399–L403
89. Ho Y-K (1983) The method of complex coordinate rotation and its applications to atomic collision processes. Phys Rep 99(1):1–68
90. Ivanov IA, Ho Y-K (2004) Complex rotation method for the Dirac Hamiltonian. Phys Rev A 69:023407
91. González-Férez R, Schweizer W (2000) In: Hernández-Laguna A, Maruani J, McWeeny R, Wilson S (eds) Quantum systems in chemistry and physics. Series: Progress in Theoretical Chemistry and Physics, vol 23. Springer, Berlin, p 17
92. Sahoo S, Ho Y-K (2000) Stark effect on the low-lying excited states of the hydrogen and the lithium atoms. J Phys B: At Mol Opt Phys 33:5151–5164
93. Sahoo S, Ho Y-K (2000) The complex absorbing potential method (CAP) to study the Stark effect in hydrogen and lithium. J Phys B: At Mol Opt Phys 33:2195–2206
94. Zimmerman ML, Littman MG, Kash MM et al (1979) Stark structure of the Rydberg states of alkali-metal atoms. Phys Rev A 20:2251–2275
95. Harmin DA (1982) Theory of the Stark effect. Phys Rev A 26:2656–2681

96. Buyadzhi VV (2015) Laser multiphoton spectroscopy of atom embedded in Debye plasmas: multiphoton resonances and transitions. Photoelectronics 24:128–133
97. Kuznetsova AA, Glushkov AV, Ignatenko AV, Svinarenko AA, Ternovsky VB (2019) Spectroscopy of multielectron atomic systems in a DC electric field. Adv Quantum Chem 78:287–306 (Elsevier). https://doi.org/10.1016/bs.aiq.2018.06.005
98. Glushkov AV (2005) Atom in electromagnetic field. KNT, Kiev
99. Glushkov AV, Ivanov LN (1993) DC strong-field Stark effect: consistent quantum-mechanical approach. J Phys B: At Mol Opt Phys 26:L379–L386
100. Glushkov AV (2013) Operator perturbation theory for atomic systems in a strong DC electric field. In: Hotokka M, Brändas E, Maruani J, Delgado-Barrio G (eds) Advances in quantum methods and applications in chemistry, physics, and biology. Series: Progress in Theoretical Chemistry and Physics, vol 27. Springer, Cham, pp 161–177
101. Glushkov AV, Ambrosov SV, Ignatenko AV, Korchevsky DA (2004) DC strong field Stark effect for non-hydrogenic atoms: new consistent quantum mechanical approach. Int J Quant Chem 99(5):936–939
102. Glushkov AV, Malinovskaya SV, Loboda AV, Shpinareva IM, Prepelitsa GP (2006) Consistent quantum approach to new laser-electron-nuclear effects in diatomic molecules. J Phys: Conf Ser 35:420–424
103. Glushkov AV, Ambrosov SV, Ignatenko AV (2001) Non-hydrogenic atoms and Wannier-Mott excitons in a DC electric field: photoionization, Stark effect, resonances in ionization continuum and stochasticity. Photoelectronics 10:103–106
104. Ignatenko AV (2007) Probabilities of the radiative transitions between Stark sublevels in spectrum of atom in an DC electric field: new approach. Photoelectronics 16:71–74
105. Benvenuto F, Casati G, Shepelyansky DL (1994) Rydberg stabilization of atoms in strong fields: "magic" mountain in chaotic sea. J Phys B 94:481–486
106. Buchleitner A, Delande D (1997) Secular motion of three-dimensional Rydberg states in a microwave field. Phys Rev A 55:R1585
107. Gallagher TF, Noel M, Griffith MW (2000) Classical subharmonic resonances in microwave ionization of lithium Rydberg atoms. Phys Rev A 62:063401
108. Grutter M, Zehnder O, Softley T et al (2008) Spectroscopic study and multichannel quantum defect theory analysis of the Stark effect in Rydberg states of neon. J Phys B: At Mol Opt Phys 41:115001
109. Glushkov AV, Malinovskaya SV, Svinarenko AA, Vitavetskaya LA (2005) Sensing spectral hierarchy, quantum chaos, chaotic diffusion and dynamical stabilisation effects in a multiphoton atomic dynamics with intense laser field. Sens. Electr. and Microsyst. Tech. 2(2):29–36
110. Glushkov AV, Ternovsky VB, Buyadzhi VV, Prepelitsa GP (2014) Geometry of a relativistic quantum chaos: new approach to dynamics of quantum systems in electromagnetic field and uniformity and charm of a chaos. Proc Intern Geom Center 7(4):60–71
111. Khetselius OY (2011) Quantum structure of electroweak interaction in heavy finite Fermi-systems. Astroprint, Odessa
112. Braun MA, Dmitriev YuYu, Labzovsky LN (1969) Relativistic theory of the heavy atom. JETP 57:2189
113. Ivanov LN, Ivanova EP (1996) Method of sturm orbitals in calculation of physical characteristics of radiation from atoms and ions. JETP 83:258–266
114. Ivanov LN, Ivanova EP (1979) Atomic ion energies for Na-like ions by a model potential method Z = 25-80. Atom Data Nucl Data Tabl 24(2):95–109
115. Ivanova EP, Ivanov LN, Glushkov AV, Kramida A (1985) High order corrections in the relativistic perturbation theory with the model zeroth approximation, Mg-like and Ne-like ions. Phys Scr 32:513–522
116. Ivanova EP, Glushkov AV (1986) Theoretical investigation of spectra of multicharged ions of F-like and Ne-like isoelectronic sequences. J Quant Spectr Rad Transfer 36:127–145
117. Ivanov LN, Ivanova EP, Aglitsky EV (1988) Modern trends in the spectroscopy of multicharged ions. Phys Rep 164(6):315–375

118. Glushkov AV, Ivanov LN., Ivanova EP (1986) Radiation decay of atomic states. Generalized energy approach. In: Autoionization phenomena in atoms. Moscow State Univ., Moscow, p 58
119. Glushkov AV, Ivanov LN (1992) Radiation decay of atomic states: atomic residue polarization and gauge noninvariant contributions. Phys Lett A 170:33–36
120. Glushkov AV (2012) Advanced relativistic energy approach to radiative decay processes in multielectron atoms and multicharged ions. In: Nishikawa K, Maruani J, Brandas E, Delgado-Barrio G, Piecuch P (eds) Quantum systems in chemistry and physics: progress in methods and applications. Series: Progress in Theoretical Chemistry and Physics, vol 26. Springer, Dordrecht, pp 231–252
121. Glushkov AV (2019) Multiphoton spectroscopy of atoms and nuclei in a laser field: relativistic energy approach and radiation atomic lines moments method. Adv Quant Chem, vol 78:253–285 (Elsevier). https://doi.org/10.1016/bs.aiq.2018.06.004
122. Khetselius OY (2019) Optimized relativistic many-body perturbation theory calculation of wavelengths and oscillator strengths for Li-like multicharged ions. Adv Quant Chem 78:223–251 (Elsevier). https://doi.org/10.1016/bs.aiq.2018.06.001
123. Ivanov LN, Letokhov VS (1985) Spectroscopy of autoionization resonances in heavy elements. Com Mod Phys D: Atom Mol Phys 4:169–184
124. Ivanova EP, Grant IP (1998) Oscillator strength anomalies in the neon isoelectronic sequence with applications to x-ray laser modelling. J Phys B: At Mol Opt Phys 31(13):2871–2883
125. Ivanova EP, Zinoviev NA (1999) Calculation of the vacuum-UV radiation gains in transitions of Ne-like argon in capillary discharges. Quant Electr 29(6):484–492
126. Ivanova EP, Zinoviev NA (2001) The possibility of X-ray lasers based on the innershell transitions of Ne-like ions. Phys Lett A 274(5-6):239–246
127. Bekov GI, Vidolova-Angelova E, Ivanov LN, Letokhov VS, Mishin V (1981) Laser spectroscopy of narrow doubly excited autoionizing states of Ytterbium atoms. JETP 53(3):441–447
128. Vidolova-Angelova E, Ivanov LN, Ivanova EP et al (1986) Relativistic perturbation method for investigating the radiation decay of highly excited many electron atoms: Tm atom. J Phys B: At Mol Opt Phys 19:2053–2069
129. Vidolova-Angelova E, Ivanov LN (1991) Autoionizing Rydberg states of thulium. Re-orientation decay due to monopole interaction. J Phys B: At Mol Opt Phys 24:4147–4158
130. Ivanov LN, Ivanova EP, Knight L (1993) Energy approach to consistent QED theory for calculation of electron-collision strengths: Ne-like ions. Phys Rev A 48:4365–4378
131. Glushkov AV, Ivanov LN, Letokhov VS (1991) Nuclear quantum optics. Preprint of Institute for Spectroscopy of the USSR Academy of Sciences. ISAN, Moscow-Troitsk, AS-4
132. Glushkov AV, Ivanov LN, Letokhov VS (1992) Laser separation of heavy lanthanides and actinides isotopes: Autoionization resonances and decay in electric field. Preprint of Institute for Spectroscopy of the USSR Academy of Sciences. ISAN, Moscow-Troitsk, AS N5
133. Glushkov AV, Ivanov LN (1992) A broadening of the thulium atom autoionization resonances in a weak electric field. Preprint of Institute for Spectroscopy of the USSR Academy of Sciences. ISAN, Moscow-Troitsk, AS N2
134. Glushkov AV, Ivanov LN (1992) Shift and deformation of radiation atomic lines in the laser emission field. Multiphoton processes. Preprint of Institute for Spectroscopy of the USSR Academy of Sciences. ISAN, Moscow-Troitsk, AS N3
135. Glushkov AV, Loboda AV (2007) Calculation of the characteristics of radiative multiphoton absorption and emission lines when an atom interacts with pulsed laser radiation. J Appl Spectr 74:305–309 (Springer)
136. Glushkov AV (2008) QED theory of radiation emission and absorption lines for atoms and ions in a strong laser field. AIP Conf Proc 1058:134–136
137. Khetselius OY, Glushkov AV, Gurskaya M Yu, Kuznetsova AA, Dubrovskaya Yu V, Serga IN, Vitavetskaya LA (2017) Computational modelling parity nonconservation and electroweak interaction effects in heavy atomic systems within the nuclear-relativistic many-body perturbation theory. J Phys.: Conf Ser 905:012029

138. Glushkov AV (2012) Spectroscopy of cooperative muon-gamma-nuclear processes: energy and spectral parameters. J Phys: Conf Ser 397:012011
139. Glushkov AV (2014) Spectroscopy of atom and nucleus in a strong laser field: Stark effect and multiphoton resonances. J Phys: Conf Ser 548:012020
140. Khetselius OY (2012) Spectroscopy of cooperative electron-gamma-nuclear processes in heavy atoms: NEET effect. J Phys: Conf Ser 397:012012
141. Khetselius OY (2012) Relativistic energy approach to cooperative electron-γ-nuclear processes: NEET effect. In: Nishikawa K, Maruani J, Brändas E, Delgado-Barrio G, Piecuch P (eds) Quantum systems in chemistry and physics. Series: Progress in Theoretical Chemistry and Physics, vol 26. Springer, Dordrecht, pp 217–229
142. Glushkov AV, Kondratenko PA, Buyadgi VV, Kvasikova AS, Sakun TN, Shakhman AN (2014) Spectroscopy of cooperative laser electron-γ-nuclear processes in polyatomic molecules. J Phys: Conf Ser 548:012025
143. Buyadzhi VV, Glushkov AV, Lovett L (2014) Spectroscopy of atoms and nuclei in a strong laser field: AC Stark effect and multiphoton resonances. Photoelectronics 23:38–43
144. Malinovskaya SV, Glushkov AV, Khetselius OY (2008) New laser-electron nuclear effects in the nuclear γ transition spectra in atomic and molecular systems. In: Wilson S, Grout P, Maruani J, Delgado-Barrio G, Piecuch P (eds) Frontiers in quantum systems in chemistry and physics. Series: Progress in Theoretical Chemistry and Physics, vol 18. Springer, Dordrecht, pp 525–541
145. Glushkov AV, Khetselius OY, Malinovskaya SV (2008) Optics and spectroscopy of cooperative laser-electron nuclear processes in atomic and molecular systems—new trend in quantum optics. Eur Phys J ST 160:195–204
146. Glushkov AV, Malinovskaya SV, Ambrosov SV, Shpinareva IM, Troitskaya OV (1997) Resonances in quantum systems in strong external fields consistent quantum approach. J Techn Phys 38(2):215–218
147. Glushkov AV, Khetselius OY, Malinovskaya SV (2008) Spectroscopy of cooperative laser-electron nuclear effects in multiatomic molecules. Mol Phys 106:1257–1260
148. Glushkov AV, Khetselius OY, Svinarenko AA (2012) Relativistic theory of cooperative muon-γ-nuclear processes: negative muon capture and metastable nucleus discharge. In: Hoggan P, Brändas E, Maruani J, Delgado-Barrio G, Piecuch P (eds) Advances in the theory of quantum systems in chemistry and physics. Series: Progress in Theoretical Chemistry and Physics, vol 22. Springer, Dordrecht, pp 51–68
149. Glushkov AV, Khetselius OY, Lovett L (2009) Electron-β-nuclear spectroscopy of atoms and molecules and chemical bond effect on the β-decay parameters. In: Piecuch P, Maruani J, Delgado-Barrio G, Wilson S (eds) Advances in the theory of atomic and molecular systems dynamics, spectroscopy, clusters, and nanostructures. Series: Progress in Theor. Chem. and Phys., vol 20. Springer, Dordrecht, pp 125–152
150. Glushkov AV, Khetselius OY, Loboda AV, Svinarenko AA (2008) QED approach to atoms in a laser field: multi-photon resonances and above threshold ionization In: Wilson S, Grout P, Maruani J, Delgado-Barrio G, Piecuch P (eds) Frontiers in quantum systems in chemistry and physics. Series: Progress in Theoretical Chemistry and Physics, vol 18. Springer, Dordrecht, pp 543–560
151. Glushkov AV, Khetselius OY, Svinarenko AA, Prepelitsa GP, Shakhman AN (2010) Spectroscopy of cooperative laser-electron nuclear processes in diatomic and multiatomic molecules. AIP Conf Proc 1290(1):269–273
152. Khetselius OY, Glushkov AV, Dubrovskaya YV, Chernyakova YG, Ignatenko AV, Serga IN, Vitavetskaya LA (2018) Relativistic quantum chemistry and spectroscopy of exotic atomic systems with accounting for strong interaction effects. In: Wang YA, Thachuk M, Krems R, Maruani J (eds) Concepts, methods and applications of quantum systems in chemistry and physics. Series: Progress in Theoretical Chemistry and Physics, vol 31. Springer, Cham, pp 71–91
153. Khetselius OY (2010) Relativistic hyperfine structure spectral lines and atomic parity non-conservation effect in heavy atomic systems within QED theory. AIP Conf Proc 1290(1):29–33

154. Glushkov AV, Rusov VD, Ambrosov SV, Loboda AV (2003) Resonance states of compound super-heavy nucleus and EPPP in heavy nucleus collisions. In: Fazio G, Hanappe F (eds) New projects and new lines of research in nuclear physics. World Scientific, Singapore, pp 126–132
155. Glushkov AV, Khetselius OY, Gurnitskaya EP, Loboda AV, Florko TA, Sukharev DE, Lovett L (2008) Gauge-invariant QED perturbation theory approach to calculating nuclear electric quadrupole moments, hyperfine structure constants for heavy atoms and ions. In: Wilson S, Grout P, Maruani J, Delgado-Barrio G, Piecuch P (eds) Frontiers in quantum systems in chemistry and physics. Series: Progress in Theoretical Chemistry and Physics, vol 18. Springer, Dordrecht, pp 507–524
156. Malinovskaya SV, Glushkov AV (1992) Calculation of the spectra of potassium-like multicharged ions. Russ Phys J 35(11):999–1004
157. Glushkov AV, Butenko YuV, Serbov NG, Ambrosov SV, Orlova VE, Orlov SV, Balan AK, Dormostuchenko GM (1996) Calculation of the oscillator strengths in Fr-like multiply charged ions. J Appl Spectrosc 63(1):28–30
158. Glushkov AV, Kondratenko PA, YaI Lepikh, Fedchuk AP, Svinarenko AA, Lovett L (2009) Electrodynamical and quantum-chemical approaches to modelling the electrochemical ancatalytic processes on metals, metal alloys and semiconductors. Int J Quant Chem 109(14):3473–3481
159. Florko TA, Ambrosov SV, Svinarenko AA, Tkach TB (2012) Collisional shift of the heavy atoms hyperfine lines in an atmosphere of the inert gas. J Phys: Conf Ser 397:012037
160. Glushkov AV (1992) Negative ions of inert gases. JETP Lett 55(2):97–100
161. Glushkov AV, Khetselius OY, Svinarenko AA (2013) Theoretical spectroscopy of autoionization resonances in spectra of lanthanide atoms. Phys Scr T 153:014029
162. Glushkov AV, Ambrosov SV, Loboda AV, Gurnitskaya EP, Khetselius OY (2005) QED calculation of heavy multicharged ions with account for correlation, radiative and nuclear effects. In: Julien J-P, Maruani J, Mayou D, Wilson S, Delgado-Barion G (eds) Recent advances in the theory of chemical and physical systems. Series: Progress in Theoretical Chemistry and Physics, vol 15. Springer, Dordrecht, pp 285–299
163. Malinovskaya SV, Glushkov AV, DubrovskayaYV, Vitavetskaya LA (2006) Quantum calculation of cooperative muon-nuclear processes: discharge of metastable nuclei during negative muon capture. In: Julien J-P, Maruani J, Mayou D, Wilson S, Delgado-Barion G (eds) Recent advances in the theory of chemical and physical systems. Series: Progress in Theoretical Chemistry and Physics, vol 15. Springer, Dordrecht, pp 301–307
164. Malinovskaya SV, Glushkov AV, Khetselius OY, Svinarenko AA, Mischenko EV, Florko TA (2009) Optimized perturbation theory scheme for calculating the interatomic potentials and hyperfine lines shift for heavy atoms in the buffer inert gas. Int J Quant Chem 109:3325–3329
165. Glushkov AV, Khetselius OY, Lopatkin YM, Florko TA, Kovalenko OA, Mansarliysky VF (2014) Collisional shift of hyperfine line for rubidium in an atmosphere of the buffer inert gas. J Phys: Conf Ser 548:012026
166. Khetselius OY, Lopatkin YuM, Dubrovskaya YuV, Svinarenko AA (2010) Sensing hyperfine-structure, electroweak interaction and parity non-conservation effect in heavy atoms and nuclei: New nuclear-QED approach. Sensor Electr and Microsyst Techn 7(2):11–19
167. Glushkov AV, Dan'kov SV, Prepelitsa G, Polischuk VN, Efimov AV (1997) QED theory of nonlinear interaction of the complex atomic systems with laser field multi-photon resonances. J Techn Phys 38(2):219–222
168. Glushkov AV, Malinovskaya SV, Gurnitskaya EP, Khetselius OY, Dubrovskaya YuV (2006) Consistent quantum theory of recoil induced excitation and ionization in atoms during capture of neutron. J Phys: Conf Ser 35:425–430
169. Glushkov AV, Malinovskay SV, Prepelitsa GP, Ignatenko V (2005) Manifestation of the new laser-electron nuclear spectral effects in the thermalized plasma: QED theory of co-operative laser-electron-nuclear processes. J Phys: Conf Ser 11:199–206
170. Glushkov AV, Ambrosov SV, Loboda AV, Gurnitskaya EP, Prepelitsa GP (2005) Consistent QED approach to calculation of electron-collision excitation cross sections and strengths: Ne-like ions. Int J Quant Chem 104:562–569

171. Glushkov AV, Khetselius OY, Loboda AV, Ignatenko AV, Svinarenko AA, Korchevsky DA, Lovett L (2008) QED approach to modeling spectra of the multicharged ions in a plasma: oscillator and electron-ion collision strengths. AIP Conf Proc 1058:175–177
172. Glushkov AV (1988) True effective molecular valency hamiltonian in a logical semiempricial theory. J Struct Chem 29(4):495–501
173. Glushkov AV (1990) Correction for exchange and correlation effects in multielectron system theory. J Struct Chem 31(4):529–532
174. Glushkov AV, Efimov VA, Gopchenko ED, Dan'kov SV, Polishchyuk VN, Goloshchak OP (1998) Calculation of spectroscopic characteristics 4 of alkali-metal dimers on the basis of a model perturbation theory. Opt Spectr 84(5):670–678
175. Svinarenko AA, Glushkov AV, Khetselius OY, Ternovsky VB, Dubrovskaya YuV, Kuznetsova AA, Buyadzhi VV (2017) Theoretical Spectroscopy of Rare-Earth Elements: Spectra and Autoionization Resonance. In: Jose EA (ed) Rare Earth Element. InTech, Orjuela, pp 83–104. https://doi.org/10.5772/intechopen.69314
176. Buyadzhi VV, Glushkov AV, Mansarliysky VF, Ignatenko AV, Svinarenko AA (2015) Spectroscopy of atoms in a strong laser field: new method to sensing AC Stark effect, multiphoton resonances parameters and ionization cross-sections. Sens Electr Microsys Techn 12(4):27–36
177. Glushkov AV, Yu Gurskaya M, Ignatenko AV, Smirnov AV, Serga IN, Svinarenko AA, Ternovsky EV (2017) Computational code in atomic and nuclear quantum optics: advanced computing multiphoton resonance parameters for atoms in a strong laser field. J Phys: Conf Ser 905(1):012004
178. Dubrovskaya YV, Khetselius OY, Vitavetskaya LA, Ternovsky VB, Serga IN (2019) Quantum chemistry and spectroscopy of pionic atomic systems with accounting for relativistic, radiative, and strong interaction effects. Adv Quantum Chem 78:193–222 (Elsevier). https://doi.org/10.1016/bs.aiq.2018.06.003
179. Glushkov AV, Khetselius OY, Svinarenko AA, Buyadzhi VV, Ternovsky VB, Kuznetsova AA, Bashkarev PG (2017) Relativistic Perturbation Theory Formalism to Computing Spectra and Radiation Characteristics: Application to Heavy Elements. In: Dimo I. Uzunov (ed) Recent Studies in Perturbation Theory, InTech, pp 131–150. https://doi.org/10.5772/intechopen.69102
180. Glushkov AV, Khetselius OY, Svinarenko AA, Buyadzhi VV (2015) Methods of computational mathematics and mathematical physics. P.1. TES, Odessa
181. Glushkov AV (1992) Oscillator strengths of Cs and Rb-like ions. J Appl Spectrosc 56(1):5–9
182. Glushkov AV (1990) Relativistic polarization potential of a many-electron atom. Sov Phys J 33(1):1–4
183. Svinarenko AA, Khetselius OY, Buyadzhi VV, Florko TA, Zaichko PA, Ponomarenko EL (2014) Spectroscopy of Rydberg atoms in a black-body radiation field: relativistic theory of excitation and ionization. J Phys: Conf Ser 548:012048

Relativistic Quantum Chemistry and Spectroscopy of Kaonic Atomic Systems with Accounting for Radiative and Strong Interaction Effects

Olga Yu. Khetselius, Andrey A. Svinarenko, Valentin B. Ternovsky, Yuliya V. Dubrovskaya and Inga N. Serga

Abstract It is presented a consistent relativistic approach to calculation of spectra and spectroscopic characteristics of the kaonic multielectron atomic systems on the basis of the Klein–Gordon–Fock equation with simultaneous accounting for the electromagnetic and strong kaon-nuclear interactions. The theoretical study of different characteristics of kaonic atoms is of a great interest both for the development of modern relativistic quantum chemistry of the systems with electromagnetic and strong interactions, as well as for further understanding and development of advanced concepts of atomic and nuclear structures. In our approach the known modified method of optical potential is used to take a strong kaon-nuclear interaction into consideration. The consistent procedures are used to take the nuclear (the finite nuclear size effect) and QED (radiative) corrections into account. In particular, the advanced Uehling–Serber approach is applied to calculate the Lamb shift polarization part contribution. As illustration, we present some results of calculation of the energy and spectroscopic characteristics for kaonic atoms of the He, Li, K, ^{184}W, ^{207}Pb, ^{238}U etc. An analysis of the electromagnetic and strong interactions contributions to the transitions energies in X-ray spectra of the kaonic helium and of the "kaonic helium puzzle" is presented. The results of calculation of the transition energies in the heavy kaonic atoms are presented and compared with the experimental and alternative theoretical data, in particular, obtained with using the simplest cascade model by Fermi–Teller and consistent method by Indelicato et al. The largest contribution to the transition energy is determined by the main Coulomb term. The contribution of radiation QED corrections, in particular, corrections for a vacuum polarization (including higher orders), is essential for the precise determination of the corresponding transition energy. The calculated and measured strong interaction shifts and widths for the kaonic atom X-ray transitions are listed and analysed.

Keywords Quantum chemistry and spectroscopy · Kaonic atoms · Relativistic many-body perturbation theory · Klein–Gordon–Fock equation · Strong kaon-nuclear optical potential · Radiative corrections

---

O. Yu. Khetselius · A. A. Svinarenko (✉) · V. B. Ternovsky · Y. V. Dubrovskaya · I. N. Serga
Odessa State Environmental University, L'vovskaya Str., 15, Odessa 65016, Ukraine

L. Mammino et al. (eds.), *Advances in Quantum Systems in Chemistry, Physics, and Biology*, Progress in Theoretical Chemistry and Physics 32,
https://doi.org/10.1007/978-3-030-34941-7_2

1 Introduction

The theoretical and experimental study of the spectra, energy and spectroscopic characteristics of exotic quantum systems, in particular, hadronic (kaonic, pionic and other) atoms is of great interest both for the development of quantum physics and chemistry of finite Fermi systems with electromagnetic and strong interactions, as well as for further understanding and development of advanced concepts of atomic and nuclear structures (e.g., Refs. [1–90]). It is well known that the concept of exotic atom was first introduced in 1947 by Fermi, Teller, Wheeler to explain experiments on the absorption of negative muons in matter. In recent years, the study of kaonic atoms has become especially relevant in the light of the well-known progress of experimental studies (at meson factories in the laboratories of LAMPF (USA), PSI (Switzerland), TRIUMF (Canada), IFF (Russia), RIKEN (KEK, Japan), RAL (United Kingdom), DEAR at the DAPNE (Italy) and further substantial development of modern nuclear theory, quantum mechanics of atoms etc.

The study of spectra and spectroscopic, radiation and other properties of the kaonic atoms is a unique tool for studying fundamental interactions, including validation of the Standard Model, giving extremely important data on properties of a nucleus and hadrons themselves, the nature of their interaction with nucleons. In principle, it allows to determine the masses and magnetic moments of the particles (kaon, pion, antiproton etc.), which are the most accurate so far. The light kaonic atoms are candidates for the creation of new, efficient low-energy X-ray standards. Moreover, the experiments with different hadronic atoms allow to determine the spins and pairs of the hadrons. Since the products of the reactions of hadrons (kaons) with neutrons and protons are different, this allows to obtain the quantitative data on the distribution of protons and neutrons in nuclei from measurements of the widths of X-ray transitions and the relative probabilities of different reactions.

One of the fundamental questions in the modern hadron's physics is connected with the hadron masses being much higher than the mass of their quark content. The current mass of the up (u) and down (d) quarks is two orders of magnitude smaller than a typical hadron's mass of about 1 GeV. This extraordinary phenomenon is proposed to originate from spontaneous breaking of chiral symmetry of massless quarks in strong interaction physics [1–5]. One of the most sensitive tests for the chiral symmetry breaking scenario in the modern hadron's physics is provided by studying the exotic (in particular, kaonic) atomic systems. It should be reminded that the most comprehensive theory of the hadronic multielectron atoms must be based on the principles of a quantum chromodynamics (QCD) and quantum electrodynamics (QED).

From the modern viewpoint, QCD is the most fundamental gauge theory of strong interactions with the coloured quarks and gluons. Nowadays the energy and spectral properties of the hadronic atoms are determined with an unprecedented precision and it is possible to study the strong interaction at low energies measuring the energy and natural widths of the levels with a precision of few meV [1–27]. For a long time the similar experimental investigations have been carried out in the laboratories of Berkley, Virginia (USA), CERN (Switzerland) etc.

The mechanism of creation of the hadronic (kaonic) atoms is well known now (e.g., [1–5]). Really, such an atom is formed when a negative kaon (pion) enters a medium, looses its kinetic energy through excitation and ionization channels and eventually is captured, replacing the electron in an excited atomic orbit. The further de-excitation scenario includes the different cascade processes such as the Auger transitions, Coulomb de-excitation, scattering etc. When a kaon reaches a low-n state with the little angular momentum, strong interaction with the nucleus causes its absorption.

The strong interaction is the reason for a shift in the energies of the low-lying levels from the purely electromagnetic values and the finite lifetime of the state corresponds to an increase in the observed level width. At present time several highly precise measurements are carried out for the kaonic (pionic) hydrogen, helium and other elements, including heavy systems. The E570 experiment [18, 19] allowed to make the precise measurement of the X-ray energies in the kaonic helium atom. For this system during the last decades a very complicated situation takes place. Speech is about a large discrepancy between the theories and experiments on the kaonic helium 2p state.

At the beginning of the 1970s and 80s several experimental groups (WG71-Wiegand–Pehl (1971), BT79-Batty et al. (1989); e.g. [1–26]) declared relatively large repulsive shift (~−40 eV), while the physically reasonable optical models calculations give shift less on the order. This sharp disagreement between the experimental and theoretical results received a status of the "kaonic helium puzzle". In Fig. 1 the Experimental K–He spectrum (Okada et al. 2007; E570 exp. At KEK 12 GeV proton synchrotron RIKEN Nishina Center, JAPAN) is presented (from Refs. [18, 19]).

Much larger shift was predicted in the theory with assuming the existence of the deeply bound kaonic nuclear states. Several theoretical estimates of the last years (look e.g. [18, 19]) did not confirm the large shift in the kaonic helium. The new measurement of the kaonic helium X-ray was performed using the KEK-PS K5 kaon beam channel in 2005.

The widely applied theoretical approaches to study of the hadronic (kaonic and other) atomic systems are described in Refs. [1–56]. In Refs. [43–56] the effective ab initio schemes to the Klein-Gordon-Fock equation solution and further determination of the X–ray spectra for multi-electron kaonic atoms are presented with the different schemes for accounting of the nuclear, QED, interparticle correlation effects. The theoretical studying the strong interaction shifts and widths from X-ray spectroscopy of kaonic atoms (U, Pb etc.) was fulfilled.

The most difficult aspects of the theoretical studying are reduced to the correct description of the kaon (pion)-strong interaction as the electromagnetic part of the problem is reasonably taken into consideration [2, 41–45, 57–68]. Besides, quite new aspect is linked with the possible, obviously, very tiny electroweak and hyperfine interactions.

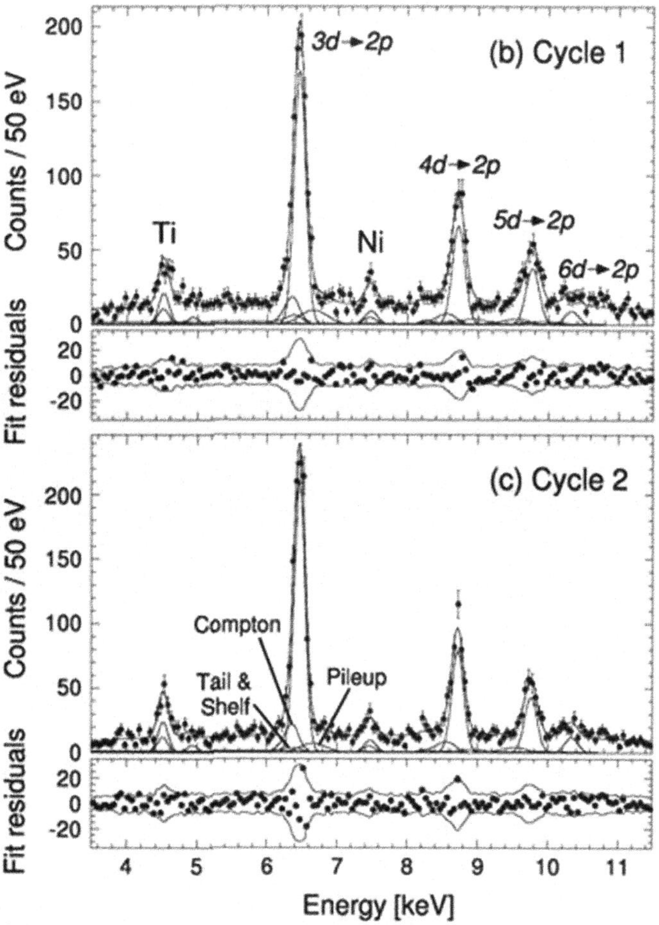

Fig. 1 Experimental K–He spectrum (Okada et al. 2007; E570 exp. At KEK 12 GeV proton synchrotron RIKEN Nishina Center, JAPAN) (from Refs. [18, 19])

In this chapter we present an effective relativistic approach to calculation of spectra and the spectroscopic properties of the kaonic multielectron atomic systems. The approach is based on the Klein–Gordon–Fock equation solution with simultaneous treatment of the electromagnetic and strong kaon-nuclear interactions. The modified method of optical potential is used to take a strong kaon-nuclear interaction into consideration. The consistent procedures are applied to take the nuclear (the finite nuclear size effect) and QED corrections into consideration. In particular, the advanced Uehling–Serber approach is applied to estimate a vacuum polarization contribution.

As illustration we present some results of calculation of the energy and spectral parameters for the kaonic atoms of He, Li, K, ^{184}W, ^{207}Pb, ^{238}U, with taking the radiation (vacuum polarization), nuclear (finite size of a nucleus) and the strong kaon-nuclear interaction corrections into account. An analysis of different physical contributions to the transition energies in the X-ray spectra of the kaonic atoms and kaonic helium puzzle is given. The available experimental results (DAΦNE EAR, Frascatti, Italy, 2004; E570-KEK, RIKEN, Nishina Centre, JAPAN etc.) and alternative theoretical data are listed too.

2 Relativistic Theory of Kaonic Atoms with Accounting for Nuclear, Radiative and Strong Interaction Effects

2.1 Electromagnetic Interactions and Quantum Electrodynamics Effects in Kaonic Atoms

The fundamentals of our theoretical approach to study of the exotic atomic systems are presented in detail in Refs. [51–55, 91–96]. So, here we are limited to presenting only the main elements of a general approach and some new elements, related to kaonic atoms. Let us remind that a kaon is the Boson with spin 0 and here we use the following values for its mass and radius:

$$m_{K^-} = 493.667 \pm 0.013\,\text{MeV},$$
$$\mathrm{r}_{\mathrm{K}^-} = 0.560 \pm 0.031\,\text{fm}.$$

As usually, the relativistic particle wave functions can be determined from solution of the Klein–Gordon–Fock equation (e.g. [92]):

$$\left\{\frac{1}{c^2}[E + eV_0(r)]^2 + \hbar^2\nabla^2 - m^2c^2\right\}\varphi(x) = 0. \tag{1}$$

Here c is a speed of light, h is the Planck constant, E is the total energy of the system.

The V_0 in Eq. (1) represents the sum of the electric potential of a nucleus, the strong interaction potential and some radiation potential. The electric potential of a nucleus is determined by the standard way (e.g. [91–93]). In order to describe the charge distribution in a nucleus here we use the known Fermi-model approximation [91], though one could use the known Gauss-model and relativistic mean-field model approximations too.

The corresponding equations for the Gaussian model of a charge distribution are presented in detail in Refs. [92, 93]. Some authors prefer to use the uniformly charged sphere model too.

The important feature of our approach to determination of the electric potential of a nucleus is connected with using the effective algorithm based on the differential equations method. This is the method originally proposed by Ivanova and Ivanov

[86], which has been used with success in many problems of classical atomic and molecular spectroscopy. The detailed description of the method is presented in Refs. [38, 39, 86, 87, 91–96].

Unlike leptons such as electrons or muons, a kaon is the composite particle with a nonzero radius of charge distribution. To describe the electrical interaction of a nucleus of finite size (with a radius of R_1) with a kaon (of finite size with a radius of R_2), one should consider the potential by Indelicato–Desclaux et al. [44] (e.g. Refs. [42, 43, 49, 50] too).

The next principally important aspect of a precise theory of the kaonic atomic systems is connected with a consistent and accurate treatment of the radiative or QED effects. Naturally, at present, the theory of QED effects has been worked out at a very good level (e.g., [41, 57–68]).

In this work we use the procedures, which are presented in detail in Refs. [37–39, 86–90]. These procedures allow to take effectively the main radiative effects into account. In particular, the contributions, provided by the standard Uehling–Serber term and the known Källen–Sabry and Wichmann–Kroll corrections, are taken into consideration. The modification of the generalized Uehling–Serber potential allows to account for the high-order radiation corrections in the first perturbation theory order.

It should be noted that another important QED effect (not important for kaonic atoms), namely, the self-energy part of the Lamb shift can be determined in our approach within the procedures [91–93]. This method generalizes the known formalism by Mohr and radiation model potential method by Flambaum–Ginges (look details in Refs. [41, 57, 58, 85–87]).

It is worth also mentioning other precision corrections that should be considered when describing the energy spectra of kaonic atoms. First, for deep-lying levels in the kaonic atomic system, the contribution will be made by the vacuum polarization due to the formation of virtual muon pairs, which is naturally taken into account in the modified Uehling–Serber approximation. This correction turns out to be numerically smaller than all the others and in any case it turns out to be significantly less than the energy shift, say 1 s level in the kaonic atom, due to the strong kaon-nucleon interaction.

Secondly, the correction for the reduced mass $(1 + m_{\bar{K}}/M_{KA})$, where $m_{\bar{K}}$ is the kaon mass, M_{KA} is the mass of the entire atomic system, is actually already present in the energy calculated from the solution of the Klein–Gordon–Fock equation. The relativistic recoil correction can be elementarily estimated as $B^2/2M_{KA}$, where B is the energy level.

Note that similar corrections of the following orders for fermions are known and analyzed in detail (e.g. [1–5]). For bosons, the situation with these effects is more complicated. In any case, accounting for the relativistic corrections sought is beyond the scope of the experiment. The other details of our approach can be found in Refs. [51–55, 90–140]. All calculations are performed with using the numeral code Superatom (version 98).

2.2 Strong Interactions in Kaonic Atomic System

The most difficult aspect of the problem is an adequate accounting for the strong kaon-nuclear interaction in the exotic system. Now it is well known that the most fundamental and consistent microscopic theory of the strong interactions is provided by the modern quantum chromodynamics. One should be reminded that here speech is about a gauge theory based on the representation of the confined coloured quarks and gluons. Naturally one could consider the regimes of relatively low and high energies (asymptotic freedom). In a case of the low energies so called coupling constant increases to the order 1 and, therefore, this perturbation methods fail to describe the interaction of strongly interacting hadrons (including kaons).

Naturally, to describe the strong kaon-nuclear interaction (even at relatively low energies) microscopically, a different approaches can be developed (look details in Refs. [1–56]).

In a case of the kaonic atomic systems the most popular approach to treatment of the strong interaction between the nucleus and orbiting kaon is the phenomenological optical potential of model, such as [45]:

$$V_N = -\frac{2\pi}{\mu}\left[1 + \frac{M_K}{M_N}\right]\left[A_{Kp}\rho_p(r) + A_{Kn}\rho_n(r)\right], \tag{2}$$

where M_K and M_N are the kaon and nucleon masses and μ is the kaon-nucleus reduced mass, $\rho_p(r)$, $\rho_n(r)$ are the proton and neutron densities in the nucleus and A_{Kp}, A_{Kn} are the complex effective *Kp* and *Kn* scattering lengths. The known Batty approximation [45] reduces Eq. (2) to the next expression:

$$V_N = -\frac{2\pi}{\mu}\left[1 + \frac{M_K}{M_N}\right][a\rho(r)], \tag{3}$$

where the effective averaged K-nucleon scattering length:

$$a = [(0.34 \pm 0.03) + \mathrm{i}(0.84 \pm 0.03)]\,(\mathrm{fm}).$$

The presented value of the length has been indeed chosen to describe the low and middle Z nuclei [11]. The disadvantage of the usually used approach is connected with approximate definition of the proton and neutron densities and using the effective averaged K-nucleon scattering length (e.g., Refs. [45, 51–53]).

An energy of the kaonic atom can be determined as follows:

$$E = E_{KGF} + E_{FS} + E_{QED} + E_N + E_{other}. \tag{4}$$

Here E_{KGF} is an energy of kaon in a nucleus with the point-like charge, E_{FS} is the contribution to energy, provided by the nucleus finite size effect, E_N is the energy shift, provided by the strong interaction V_N, E_{QED} is the correction due to the QED effects (vacuum-polarization effect), E_{other} —other corrections.

A direct estimate of a shift of the energy levels in a kaonic atom, due to the strong kaon-nucleus interaction, can be obtained from the relation:

$$E_N = E - (E_{KGF} + E_{FS} + E_{QED} + E_{other}), \tag{5}$$

where $E = E_{exp}$ is the experimental value of energy, and the sum in brackets in the right-hand side of (5) is actually the exact value of the "electromagnetic" contribution to energy, i.e. contribution due to all electromagnetic interactions.

The total Klein–Gordon–Fock equation taking into account the potential of a strong kaon-nuclear interaction V_N is written in the following form:

$$\left[\hbar^2\nabla^2 + c^{-2}(E - V_{FS})^2 - \mu^2c^2\right]\psi = 2\mu V_N\psi. \tag{6}$$

The elementary assessment of a strong kaon-nuclear interaction contributions is determined as follows:

$$E_N(n,l) \sim \int V_N(r)\psi_{nl}^2(r)dr. \tag{7}$$

It should be noted that even in [43], using the example of studying the effects of strong interaction in kaon atoms of heavy elements, it was shown that expressions (2, 3), calibrated on the nuclei of light elements, turn out to be, generally speaking, not sufficiently correct for use with respect to heavy atoms.

In addition, it should be recalled that because of the Coulomb barrier, the proton density decreases at the periphery of the nucleus ("nuclear stratosphere") faster than the neutron density (neutron halo), which has now been well studied experimentally. In heavy nuclei, hadron absorption occurs precisely at the periphery.

3 Some Results and Conclusions

Below we present some important our results of calculation of the energy and spectroscopic characteristics for kaonic atoms of the He, Li, K, ^{184}W, ^{207}Pb, ^{238}U etc. In Table 1 the experimental and theoretical X-ray energies (in keV) for the 2–1 transition of the kaonic helium are presented (taken from Refs. [18, 19, 41, 51–53]). The experimental K–He spectrum (Okada et al. 2008; E570 exp. At KEK 12 GeV proton synchrotron RIKEN Nishina Center, JAPAN; [18, 19]) is presented in Fig. 1.

The corresponding strong interaction shift is defined by Eq. (5). It is interesting to present an analysis of the different theoretical and experimental estimates. According to the WG71, BT79, BR83 experiments (e.g. [1, 2]) the strong interaction contribution to the transition energy in the K^{-4} He is −40 eV. The corresponding value from the

Table 1 Measured and calculated K^{-4} He X-ray energies (eV) of 3d–2p, 4d–2p and 5d–2p transitions

Transition	3d–2p	4d–2p	5d–2p
Experiment [18, 19]	6466.7 ± 2.5	8723.3 ± 4.6	9760.1 ± 7.7
Theory [41]	6463.50	8721.70	9766.80
Theory [51]	6463.00	8722.00	–
This work	6464.03	8721.10	9766.54

E570 experiment is 1.9 eV, the model potential theoretical value by Friedman et al. is 0.4 eV; our value is 1.57 eV [12, 18, 19, 51–53].

The "electromagnetic" value of the transition energy calculated by us and further comparison with the experimental value of the transition made it possible to obtain a theoretical estimate of the "strong" shift in kaonic helium, which is in good agreement with the KEK experimental shift.

In Table 2 the theoretical (E_c) and measured (E_m) X-ray energies for some kaonic atomic systems (in particular, the Li-, K-, W-, U-kaonic atoms) are listed (from Refs. [43–53]).

Note that in Table 2 there are presented the electromagnetic (EM) X-ray energies of K^- atoms for transitions between circular levels and the kaon mass was assumed to be 493.677 ± 0.013 MeV.

In Table 3 we list the energy contributions (in keV) to the transition energy $12o \rightarrow 11n$ in the kaonic lead spectrum: data from theories by Indelicato et al., Cheng et al. and Kunzelman et al. (cascade models) [41, 51, 53] and our theory; experiment—Cheng et al. [46]. As can be seen from the above data, the agreement of the theoretical data with the experimental data is generally acceptable. The largest contribution to the transition energy is determined by the main Coulomb term.

The contribution of radiation QED corrections, in particular, corrections for vacuum polarization (including higher orders), etc., is essential for the precise determination of the corresponding transition energy.

Table 2 Calculated (E_c) and measured (E_m) kaonic atoms X-ray energies (in keV): the theoretical data by Batty et al. (theory EM1,2 [45] with using the simplest cascade model by Fermi–Teller [40] and Leon–Seki [11]), the results of Indelicato et al. (theory EM3 [43]) and data of our theory (theory EM4)

Atom	Transition	E_c, EM4	E_c, EM1	E_c, EM2	E_c, EM3	E_m [41, 45]
Li	3–2	15.335	15.392	15.319	15.330	15.320 (24) 15.00 (30)
K	5–4	105.962	105.970	–	105.952	105.86 (28)
W	8–7	346.586	346.54	–	346.571	346.624 (25)
W	7–6	535.180	535.24	–	535.240	534.886 (92)
Pb	8–7	426.1748	426.15	426.201	426.180	426.221 (57)
U	8–7	538.520	538.72	538.013	537.44	538.315 (100)

Table 3 The energy contributions (in keV) to the transition energy $12o \rightarrow 11n$ in the kaonic lead spectrum: data from theories by Indelicato et al., Cheng et al. and Kunzelman et al. (cascade models) [41, 51, 53] and our theory; experiment—Chen et al. [46]

Contribution	Indelicato et al.	This work	Cheng et al.	Kunzelman et al.
Coulomb term	116.5666	116.5644	116.575	116.600
Polarization of vacuum	0.4134	0.4067	0.412	0.410
$\alpha(Z\alpha)$	0.4203	–	0.421	–
$\alpha(Z\alpha)^3 + \alpha^2(Z\alpha)$	−0.0069	–	−0.009	–
Other corrections	0.0004	−0.0126	−0.044	−0.050
Total:	116.9804	116.9585	116.943	116.960
Exp.	–	–	116.952 (10)	–

Table 4 Calculated (C) and measured (M) strong interaction shifts ΔE and widths G for K^- -atoms X-ray transitions: a—shift, estimated with Miller et al. measured energy [2]; b—shift estimated with Chen et al. measured energy [46]; c—theory by Batty et al. [45]; d—this work

Atom Trans.	ΔE_C (d)	G_C (d)	ΔE_C(c)	G_C (c)	ΔE_M	G_M
W, 8–7	0.038	0.072	−0.003	0.065	0.079[c] 0.052[d]	0.070 (15)
W, 7–6	-0.294	3.85	−0.967	4.187	−0.353[c] −0.250[d]	3.72 (35)
Pb, 8–7	0.035	0.281	−0.023	0.271	0.072[c] 0.047[d]	0.284 (14) 0.370 (150)[a]
U, 8–7	-0.205	2.620	−0.189	2.531	0.120[a]; 0.032[b] −0.40[c]; – 0.213[d]	2.67 (10) 1.50 (75)[a]

In Table 4 we present the calculated (C) and measured (M) strong interaction shifts ΔE and widths G (in keV) for the kaonic atoms X-ray transitions, taken from Ref. [2, 44–46, 51–53].

The width G is the strong width of the lower level which was obtained by subtracting the electromagnetic widths of the upper and lower level from the measured value (e.g. [2, 9]). The shift ΔE is defined as difference between the measured E_M and calculated E_{EM} (electromagnetic) values of transition energies. The calculated value is obtained by direct solving the Eq. (3) with the optical model kaon-nucleon potential. It is easily to understand that when there is close agreement between theoretical and experimental shifts, the corresponding energy levels are not significantly sensitive to strong nuclear interaction, i.e. the electromagnetic contribution is dominative. In the opposite situation the strong-interaction effect is very significant.

For highly lying kaon states in an atom, when a strong interaction is negligibly small, the precision QED theory of the spectra of kaonic atoms provides valuable information on the electromagnetic parameters of the system.

To conclude, let us underline that in general, theoretical spectroscopy of kaonic atoms gives an opportunity to obtain new information on the properties of nuclei, kaons (hadrons), the nature of their interaction with nucleons as a result of measurements of the energies of X-ray quanta emitted during transitions of hadrons between Rydberg states, and hence, as mentioned, allows to determine the masses and magnetic moments of hadrons. A physically reasonable agreement between experimental and theoretical data for the kaonic atomic systems can be achieved only with simultaneous accurate and correct consideration of relativistic, radiation, and nuclear effects. In order to provide the further improvement of the theoretical approach and to increase the calculational data accuracy, it is necessary to use more consistent approach to the strong kaon-nuclear interaction. In this sense, it seems extremely important more exact information about the electromagnetic interaction parameters for a kaonic atomic system as it allows to find more true values for the kaon-nuclear potential parameters. The subsequent more detailed consideration of more subtle effects such as, for example, the nuclear-polarization corrections etc. (for example, within the Woods-Saxon model or relativistic mean filed theory), a spatial distribution of the magnetic moment inside a nucleus (the Bohr–Weisskopf effect) etc., is extremely useful and interesting too.

Acknowledgements The authors are grateful to the Chair of QSCP-XXIII, Prof. Liliana Mammino, and the Cochair Prof. Jean Maruani, for their generous invitation to present this work in the Proceedings of the XXIII International workshop on Quantum Systems in Chemistry, Physics and Biology.

References

1. Ito TM, Hayano RS, Nakamura SN et al (1998) Observation of kaonic hydrogen atom x rays. Phys Rev C 58:2366–2382
2. Deloff A (2003) Fundamentals in hadronic atomic theory. World Scientific, New Jersey
3. Ishiwatari T (2007) On behalf of the siddharta collaboration, silicon drift detectors for the kaonic atom x-ray measurements in the siddharta experiment. Nucl Instr Methods Phys Res Sec A 581(1–2):326–329
4. Cargnelli M, Ishiwatari T, Kienle P et al (2007) Kaonic hydrogen X rays—experiments at DAFNE. In: Proceedings KAON international conference, Laboratori Nazionali di Frascati dell'INFN, Rome, Italy
5. Feshbach H, Villars F (1958) Elementary relativistic wave mechanics of spin 0 and spin 1/2 particles. Rev Mod Phys 30:24
6. Lensky V, Baru V, Haidenbauer J et al (2006) Towards a field theoretic understanding of NN $\rightarrow$ NNπ. Eur Phys J A 27:37–45
7. Deslattes RD, Kessler EG, Indelicato P et al (2003) X-ray transition energies: new approach to a comprehensive evaluation. Rev Mod Phys 75:35–99
8. Gall KP, Austin E, Miller JP et al (1998) Precision measurements of K^- and & #x03A3;-masses. Phys Rev Lett 60:186–190
9. Menshikov LI, Evseev MK (2001) Some problems of the physics of exotic atoms. Phys Uspekhi 44:135–180
10. Scherer S (2003) Introduction to chiral perturbation theory. In: Negele JW, Vogt EW (eds) Advances in nuclear physics, vol 27. Springer, Heidelberg, pp 277–538
11. Leon M, Seki R (1989) Atomic capture of negative mesons. Phys Rev Lett 32:132

12. Batty CJ, Eckhause M, Gall KP et al (1989) Strong interaction effects in high-ZK atoms. Phys Rev C 40:2154–2162
13. Chen MY, Asano Y, Cheng SC et al (1975) Dynamic mixing in p^- and k^- atoms of ^{238}U. Nucl Phys A 254:413–421
14. Olaniyi B, Shor A, Cheng S et al (1982) Electric quadrupole moments and strong interaction effects in piomic atoms of ^{165}Ho, ^{175}Lu, ^{176}Lu,^{179}Hf, ^{181}Ta. Nucl Phys A 403:572
15. Ericson M, Ericson T (1966) Optical properties of low energy pions in nuclei. Ann Phys 36:323
16. Batty CJ, Biagi SF, Friedman E, Hoath SD (1978) Shifts and widths of 2p levels in pionic atoms. Phys Rev Lett 40:931
17. Batty CJ, Friedman E, Gal A (1983) Saturation effects in pionic atoms and the π^- -nucleus optical potential. Nucl Phys A 402:411–428
18. Okada S, Beer G, Bhang H et al (2007) Precision measurement of the 3d → 2p x-ray energy in kaonic ^4He. Phys Lett B 653(5–6):387–391
19. Okada S, Beer G, Bhang H et al (2007) Precision spectroscopy of kaonic helium 3d → 2p X-rays. Nucl Phys A 790(1–4):663–666
20. Anagnostopoulos D, Biri S, Boisbourdain V et al (2003) Low-energy X-ray standards from pionic atoms. Nucl Inst Meth B 25:9
21. Wiegand CE, Penl RH (1971) Measurement of Kaonic X Rays from 4He. Phys Rev Lett 27(21):1410–1412
22. Nagels MM, de Swart J, Nielsen H et al (1976) Compilation of coupling constants and low-energy parameters. Nucl Phys B 109:1–90
23. Lauss B (2009) Fundamental measurements with muons—view from PSI. Nucl Phys A 827:401. PSI experiment R-98.01. http://pihydrogen.psi.ch; http://www.fz-juelich.de/ikp/exotic-atoms
24. CERN DIRAC Collaboration (2011) Search for long-lived states of $\pi^+\pi^-$ and π^-K atoms. CERN-SPSLC-2011-001 SPSLC-P-284-ADD, 22
25. Umemoto Y, Hirenzaki S, Kume K et al (2001) Deeply bound pionic atoms on β-unstable nuclei. Nucl Phys A 679:549–562
26. Nose-Togawa N, Hirenzaki S, Kume K (1999) Pion-nucleus potential parameters and quadrupole effect in deeply bound pionic atom. Nucl Phys A 646:467–478
27. Ninomiya K, Nakagaki R, Nakatsuka T et al (2004) Initial angular momentum distribution of atomic pions in the formation process of pionic atoms. Radiochem Acta 93:513
28. Hatsuda T, Kunihiro T (1994) Observed hadron properties for the study of QCD vacuum structure. Phys Rept 247:221
29. Ikeno N, Kimura R, Yamagata-Sekihara et al (2011) Precision spectroscopy of deeply bound pionic atoms and partial restoration of chiral symmetry in medium. arXiv:1107.5918v1 [nucl-th]
30. Kolomeitsev EE, Kaiser N, Weise W (2003) Chiral dynamics of deeply bound pionic atoms. Phys Rev Lett 90:092501
31. Lyubovitskij V, Rusetsky A (2000) π^-patom in ChPT: strong energy-level shift. A Phys Lett B 494:9–18
32. Schlesser S, Le Bigot E-Q, Indelicato P et al (2011) Quantum-electrodynamics corrections in pionic hydrogen. Phys Rev C 84:015211
33. Sigg D, Badertscher A, Bogdan M et al (1996) The strong interaction shift and width of the ground state of pionic hydrogen original research article. Nuc Phys A 609:269–436
34. Gotta D, Amaro F, Anagnostopoulos D et al (2008) Conclusions from recent pionic-atom experiments. In: Precision physics of simple atoms and molecules, Series in Lecture notes in physics, vol 745. Springer, Heidelberg, pp 165–186
35. Glushkov AV, Khetselius OY, Gurnitskaya EP, Loboda AV, Sukharev DE (2009) Relativistic quantum chemistry of heavy ions and hadronic atomic systems: spectra and energy shifts. AIP Conf Proc 1102:168–171
36. Khetselius OY, Lopatkin YuM, Dubrovskaya YV, Svinarenko AA (2010) Sensing hyperfine-structure, electroweak interaction and parity non-conservation effect in heavy atoms and nuclei: new nuclear-QED approach. Sens Electr Microsyst Technol 7(2):11–19

37. Glushkov AV, Ivanov LN (1986) Radiation decay of atomic states: atomic residue polarization and gauge noninvariant contributions. Phys Lett A 170:33–36
38. Glushkov AV, Ivanov LN, Ivanova EP (1986) Autoionization phenomena in atoms. Moscow University Press, Moscow, pp 58–160
39. Ivanova EP, Glushkov AV (1986) Theoretical investigation of spectra of multicharged ions of F-like and Ne-like isoelectronic sequences. J Quant Spectrosc Radiat Transfer 36:127–145
40. Fermi E, Teller E (1947) Atomic capture of negative mesons. Phys Rev 72:399–407
41. Mohr PJ (1983) Energy levels of H-like atoms predicted by quantum electrodynamics, $10 < Z<40$. Atomic Data Nucl Data Tables 24:453
42. Indelicato P (1996) Relativistic effects in few-electron heavy ions. Ab initio evaluation of levels energy and transitions probabilities. Phys Scripta 65:57
43. Indelicato P, Trassinelli M (2005) From heavy ions to exotic atoms. arXiv:physics, 0510126v1
44. Santos J, Parente F, Boucard S et al (1985) X-ray energies of circular transitions and electron screening in kaonic atoms. Phys Rev A 71:032501
45. Batty CJ, Eckhause M, Gall KP et al (1989) Strong interaction effects in high Z-K-atoms. Phys Rev C 40:2154–2160
46. Chen MY, Asano Y, Chen SC et al (1975) E2 dynamic mixing in p and K-atoms of 207Pb, 238U. Nucl Phys A 254:413–421
47. Serot B, Walecka J (1986) The relativistic nuclear many body problem. In: Advances in nuclear physics, vol 16. Plenum Press, New York
48. Zmescal Z (2004) On behalf of the DEAR Collob. New measurements of the kaonic hydrogen X-ray spectra at DEAR. Proc Int Conf Nucl Phys Geteborg, Sweden, W31
49. Indelicato P (2004) Exotic atoms. arXiv:physics, 0409058v1
50. Trassinelli M, Indelicato P (2007) Relativistic calculations of pionic and kaonic atoms hyperfine structure. arXiv:physics, 0611263v2
51. Sukharev DE, Dubrovskaya YV, Svinarenko AA (2010) Sensing strong interaction effects in x-ray spectroscopy of hadronic atoms. Photoelectronics 19:103–106
52. Florko TA, Khetselius OY, Dubrovskaya YV, Sukharev DE (2009) Bremsstrahlung and X-ray spectra for kaonic and pionic hydrogen and nitrogen. Photoelectronics 18:16–20
53. Khetselius OY, Turin AV, Sukhare DE, Florko TA (2009) Estimating of X-ray spectra for kaonic atoms as tool for sensing the nuclear structure. Sens Electr Microsyst Technol 1:30–35
54. Dubrovskaya YV, Serga IN, Shakhman AN, Sukharev DE (2013) An advanced approach to quantization of quasi-stationary states of Klein-Gordon-Fock equation: kaonic systems. Proc Int Geom Cent 6(4):28–32
55. Sukharev DE, Khetselius OY, Dubrovskaya YV (2009) Sensing strong interaction effects in spectroscopy of hadronic atoms. Sensor Electr Microsyst Technol 3:16–21
56. Glushkov AV, Khetselius OY, Dubrovskaya YV, Loboda AV (2006) Sensing the capture of negative muon by atoms: energy approach. Sensor Electr Microsyst Technol 3(4):31–35
57. Mohr PJ (1982) Self-energy of the n = 2 states in a strong Coulomb field. Phys Rev A 26:2338
58. Flambaum VV, Ginges JSM (2005) Radiative potential and calculations of QED radiative corrections to energy levels and electromagnetic amplitudes in many-electron atoms. Phys Rev A 72:052115
59. Safranova UI, Safranova MS, Johnson WR (2005) Excitation energies, hyperfine constants, E1, E2, and M1 transition rates, and lifetimes of $6s^2nl$ states in $T1_I$ and $T2_{II}$. Phys Rev A 71:052506
60. Yerokhin VA, Artemyev AN, Shabaev VM (2007) QED treatment of electron correlation in Li-like ions. Phys Rev A 75:062501
61. Tomaselli M, Schneider SM, Kankeleit E et al (1995) Ground-state magnetization of ^{209}Bi in a dynamic-correlation model. Phys Rev C 51:2989
62. Bieron J, Pyykkő P, Jonsson P (2005) Nuclear quadrupole moment of ^{201}HgP. Phys Rev A 71:012502
63. Johnson WR, Sapirstein J, Blundell SA (1993) Atomic structure calculations associated with PNC experiments in atomic cesium. Phys Scripta 46:184

64. Persson H, Lindgren I, Salomonson S et al (1996) Two-electron lamb-shift calculations on heliumlike ions. Phys Rev Lett 76:204–207
65. Nagasawa T, Naga A, Nakano M (2004) Hyperfine splitting of hydrogen like atoms based on relativistic mean field theory. Phys Rev C 69:034322
66. Shahbaz A, Muller C, Staudt A et al (2007) Nuclear signatures in high-order harmonic generation from laser-driven muonic atoms. Phys Rev Lett 98:263901
67. Glushkov AV (2008) Relativistic quantum theory. Quantum mechanics of atomic systems. Astroprint, Odessa
68. Khetselius OY (2011) Quantum structure of electroweak interaction in heavy finite fermi-systems. Astroprint, Odessa
69. Glushkov AV, Rusov VD, Ambrosov SV, Loboda AV (2003) Resonance states of compound super-heavy nucleus and EPPP in heavy nucleus collisions. In: Fazio G, Hanappe F (eds) New projects and new lines of research in nuclear physics. World Scientific, Singapore, pp 126–132
70. Glushkov AV (2005) Energy approach to resonance states of compound superheavy nucleus and EPPP in heavy nuclei collisions. In: Grzonka D, Czyzykiewicz R, Oelert W et al (eds) Low energy antiproton physics, vol 796. AIP Conf Proc, New York, pp 206–210
71. Glushkov AV (2012) Spectroscopy of cooperative muon-gamma-nuclear processes: energy and spectral parameters. J Phys Conf Ser 397:012011
72. Khetselius OY (2012) Relativistic energy approach to cooperative electron-γ-nuclear processes: NEET effect. In: Nishikawa K, Maruani J, Brändas E, Delgado-Barrio G, Piecuch P (eds) Quantum systems in chemistry and physics: progress in methods and applications. Series: progress in theoretical chemistry and physics, vol 26. Springer, Dordrecht, pp 217–229
73. Khetselius OY (2009) Relativistic perturbation theory calculation of the hyperfine structure parameters for some heavy-element isotopes. Int J Quant Chem 109:3330–3335
74. Khetselius OY (2009) Relativistic calculation of the hyperfine structure parameters for heavy elements and laser detection of the heavy isotopes. Phys Scripta T135:014023
75. Khetselius OY (2015) Optimized perturbation theory for calculating the hyperfine line shift and broadening of heavy atoms in a buffer gas. In: Nascimento M, Maruani J, Brändas E, Delgado-Barrio G (eds) Frontiers in quantum methods and applications in chemistry and physics. Series: progress in theoretical chemistry and physics, vol 29. Springer, Cham, pp 55–76
76. Khetselius OY (2008) Hyperfine structure of atomic spectra. Astroprint, Odessa
77. Khetselius OY (2009) Atomic parity non-conservation effect in heavy atoms and observing P and PT violation using NMR shift in a laser beam: to precise theory. J Phys Conf Ser 194:022009
78. Khetselius OY (2010) Relativistic hyperfine structure spectral lines and atomic parity nonconservation effect in heavy atomic systems within QED theory. AIP Conf Proc 1290:29–33
79. Glushkov AV (1992) Oscillator strengths of Cs and Rb-like ions. J Appl Spectrosc 56(1):5–9
80. Glushkov AV (1990) Relativistic polarization potential of a many-electron atom. Sov Phys J 33(1):1–4
81. Serga IN, Dubrovskaya YV, Kvasikova AS, Shakhman AN, Sukharev DE (2012) Spectroscopy of hadronic atoms: energy shifts. J Phys Conf Ser 397:012013
82. Khetselius OY, Glushkov AV, Gurskaya MY, Kuznetsova AA, Dubrovskaya YV, Serga IN, Vitavetskaya LA (2017) Computational modelling parity nonconservation and electroweak interaction effects in heavy atomic systems within the nuclear-relativistic many-body perturbation theory. J Phys Conf Ser 905:012029
83. Serga IN, Khetselius OY, Vitavetskaya LA, Bystryantseva AN (2017) Relativistic theory of spectra of pionic atomic system ^{208}Pb with account of strong pion-nuclear interaction effects. Photoelectronics 26:68–77
84. Bystryantseva AN, Yu Khetseliu O, Dubrovskaya YV, Vitavetskaya LA, Berestenko AG (2016) Relativistic theory of spectra of heavy pionic atomic systems with account of strong pion-nuclear interaction effects: ^{93}Nb, ^{173}Yb, ^{181}Ta, ^{197}Au. Photoelectronics 25:56–61

85. Khetselius OY, Zaichko PA, Smirnov AV, Buyadzhi VV, Ternovsky VB, Florko TA, Mansarliysky VF (2017) Relativistic many-body perturbation theory calculations of the hyperfine structure and oscillator strength parameters for some heavy element atoms and ions. In: Tadjer A, Pavlov R, Maruani J, Brändas E, Delgado-Barrio G (eds) Quantum systems in physics, chemistry, and biology. Series: progress in theoretical chemistry and physics, vol 30. Springer, Cham, pp 271–281
86. Ivanov LN, Ivanova EP (1996) Method of sturm orbitals in calculation of physical characteristics of radiation from atoms and ions. J Exp Theor Phys 83:258–266
87. Ivanova EP, Ivanov LN, Glushkov AV, Kramida AE (1985) High order corrections in the relativistic perturbation theory with the model zeroth approximation, Mg-like and Ne-like ions. Phys Scripta 32:513
88. Ivanov LN, Letokhov VS (1985) Doubly excited autoionization states of multielectron atoms. Com Mod Phys D Mol Phys 4:169–184
89. Khetselius OY (2011) Determination of the hyperfine and electroweak interactions parameters and parity non-conservation amplitudes in heavy atoms and nuclei within nuclear-qed theory. Photoelectronics 20:12–17
90. Svinarenko AA (2014) Study of spectra for lanthanides atoms with relativistic many—body perturbation theory: rydberg resonances. J Phys Conf Ser 548:012039
91. Glushkov AV, Khetselius OY, Gurnitskaya EP, Loboda AV, Florko TA, Sukharev DE, Lovett L (2008) Gauge-invariant qed perturbation theory approach to calculating nuclear electric quadrupole moments, hyperfine structure constants for heavy atoms and ions. In: Wilson S, Grout PJ, Maruani J, Delgado-Barrio G, Piecuch P (eds) Frontiers in quantum systems in chemistry and physics. Series: progress in theoretical chemistry and physics, vol 18. Springer, Dordrecht, pp 507–524
92. Khetselius OY, Glushkov AV, Dubrovskaya YV, Chernyakova YuG, Ignatenko AV, Serga IN, Vitavetskaya LA (2018) Relativistic quantum chemistry and spectroscopy of exotic atomic systems with accounting for strong interaction effects. In: Wang YA, Thachuk M, Krems R, Maruani J (eds) Concepts, methods and applications of quantum systems in chemistry and physics. Series: progress in theoretical chemistry and physics, vol 31. Springer, Cham, pp 71–91
93. Dubrovskaya YV, Khetselius OY, Vitavetskaya LA, Ternovsky VB, Serga IN (2019) Quantum chemistry and spectroscopy of pionic atomic systems with accounting for relativistic, radiative, and strong interaction effects. Advances in quantum chemistry, vol 78. Elsevier, Amsterdam, pp 193–222. https://doi.org/10.1016/bs.aiq.2018.06.003
94. Glushkov AV, Svinarenko AA, Ternovsky VB, Smirnov AV, Zaichko PA (2015) Spectroscopy of the complex autoionization resonances in spectrum of helium: test and new spectral data. Photoelectronics 24:94–102
95. Svinarenko AA, Glushkov AV, Khetselius OY, Ternovsky VB, Dubrovskaya YV, Kuznetsova AA, Buyadzhi VV (2017) Theoretical spectroscopy of rare-earth elements: spectra and autoionization resonances. In: Jose EAO (ed) Rare earth element, InTech, pp 83–104 (https://doi.org/10.5772/intechopen.69314)
96. Glushkov AV, Khetselius OY, Svinarenko AA, Buyadzhi VV, Ternovsky VB, Kuznetsova AA, Bashkarev PG (2017) Relativistic perturbation theory formalism to computing spectra and radiation characteristics: application to heavy elements. In: Dimo IU (ed) Recent studies in perturbation theory, InTech, pp 131–150 (https://doi.org/10.5772/intechopen.69102)
97. Glushkov AV, Malinovskaya SV, Filatov VV (1989) S-matrix formalism calculation of atomic transition probabilities with inclusion of polarization effects. Sov Phys J 32(12):1010–1014
98. Glushkov AV, Butenko YuV, Serbov NG, Ambrosov SV, Orlova VE, Orlov SV, Balan AK, Dormostuchenko GM (1996) Calculation and extrapolation of oscillator strengths in Rb-like, multiply charged ions. Russian Phys J 39(1):81–83
99. Mohr PJ (1993) Quantum electrodynamics calculations in few-electron systems. Phys Scripta 46:44
100. Glushkov AV, Loboda AV, Gurnitskaya EP, Svinarenko AA (2009) QED theory of radiation emission and absorption lines for atoms in a strong laser field. Phys Scripta 135:014022

101. Glushkov AV (2008) QED theory of radiation emission and absorption lines for atoms and ions in a strong laser field. AIP Conf Proc 1058:134–136
102. Glushkov AV, Loboda AV (2007) Calculation of the characteristics of radiative multiphoton absorption and emission lines when an atom interacts with pulsed laser radiation. J Appl Spectr 74:305–309
103. Glushkov AV (2014) Spectroscopy of atom and nucleus in a strong laser field: stark effect and multiphoton resonances. J Phys Conf Ser 548:012020
104. Glushkov AV (2012) Advanced relativistic energy approach to radiative decay processes in multielectron atoms and multicharged ions. In: Nishikawa K, Maruani J, Brändas E, Delgado-Barrio G, Piecuch P (eds) Quantum systems in chemistry and physics: progress in methods and applications. series: progress in theoretical chemistry and physics, vol 26. Springer, Dordrecht, pp 231–252
105. Glushkov AV (2013) Operator perturbation theory for atomic systems in a strong DC electric field. In: Hotokka M, Brändas E, Maruani J, Delgado-Barrio G (eds) Advances in quantum methods and applications in chemistry, physics, and biology. Series: progress in theoretical chemistry and physics, vol 27. Springer, Cham, pp 161–177
106. Glushkov AV, Malinovskaya SV, Loboda AV, Shpinareva IM, Gurnitskaya EP, Korchevsky DA (2005) Diagnostics of the collisionally pumped plasma and search of the optimal plasma parameters of x-ray lasing: calculation of electron-collision strengths and rate coefficients for Ne-like plasma. J Phys Conf Ser 11:188–198
107. Glushkov AV, Malinovskaya SV, Chernyakova YuG, Svinarenko AA (2004) Cooperative laser-electron-nuclear processes: QED calculation of electron satellites spectra for multi-charged ion in laser field. Int J Quant Chem 99(5):889–893
108. Glushkov AV, Ambrosov SV, Ignatenko AV, Korchevsky DA (2004) DC strong field stark effect for non-hydrogenic atoms: new consistent quantum mechanical approach. Int J Quant Chem 99(5):936–939
109. Glushkov AV, Malinovskaya SV, Loboda AV, Shpinareva IM, Prepelitsa GP (2006) Consistent quantum approach to new laser-electron-nuclear effects in diatomic molecules. J Phys Conf Ser 35:420–424
110. Glushkov AV, Kondratenko PA, Buyadgi VV, Kvasikova AS, Sakun TN, Shakhman AN (2014) Spectroscopy of cooperative laser electron-γ-nuclear processes in polyatomic molecules. J Phys Conf Ser 548:012025
111. Malinovskay SV, Glushkov AV (1992) Calculation of the spectra of potassium-like multi-charged ions. Russian Phys J 35(11):999–1004
112. Glushkov AV, Butenko YuV, Serbov NG, Ambrosov SV, Orlova VE, Orlov SV, Balan AK, Dormostuchenko GM (1996) Calculation of the oscillator strengths in Fr-like multiply charged ions. J Appl Spectrosc 63(1):28–30
113. Buyadzhi VV, Glushkov AV, Lovett L (2014) Spectroscopy of atoms and nuclei in a strong laser field: Ac Stark effect and multiphoton resonances. Photoelectronics 23:38–43
114. Glushkov AV, Kondratenko PA, Lepikh YI, Fedchuk AP, Svinarenko AA, Lovett L (2009) Electrodynamical and quantum-chemical approaches to modelling the electrochemical ancatalytic processes on metals, metal alloys and semiconductors. Int J Quant Chem 109(14):3473–3481
115. Florko TA, Ambrosov SV, Svinarenko AA, Tkach TB (2012) Collisional shift of the heavy atoms hyperfine lines in an atmosphere of the inert gas. J Phys Conf Ser 397:012037
116. Khetselius OYu (2012) Spectroscopy of cooperative electron-gamma-nuclear processes in heavy atoms: NEET effect. J Phys: Conf Ser 397:012012
117. Glushkov AV, Malinovskaya SV, Prepelitsa GP, Ignatenko VM (2005) Manifestation of the new laser-electron nuclear spectral effects in the thermalized plasma: QED theory of co-operative laser-electron-nuclear processes. J Phys Conf Ser 11:199–206
118. Glushkov AV (1988) True effective molecular valency hamiltonian in a logical semiempricial theory. J Struct Chem 29(4):495–501
119. Glushkov AV (1990) Correction for exchange and correlation effects in multielectron system theory. J Struct Chem 31(4):529–532

120. Glushkov AV, Efimov VA, Gopchenko ED, Dan'kov SV, Polishchyuk VN, Goloshchak OP (1998) Calculation of spectroscopic characteristics 4 of alkali-metal dimers on the basis of a model perturbation theory. Optics Spectr 84(5):670–678
121. Buyadzhi VV (2015) Laser multiphoton spectroscopy of atom embedded in debye plasmas: multiphoton resonances and transitions. Photoelectronics 24:128–133
122. Glushkov AV, Ambrosov SV, Loboda AV, Gurnitskaya EP, Prepelitsa GP (2005) Consistent QED approach to calculation of electron-collision excitation cross-sections and strengths: Ne-like ions. Int J Quant Chem 104(4):562–569
123. Svinarenko AA, Nikola LV, Prepelitsa GP, Tkach TB, Mischenk EV (2010) The Auger (autoionization) decay of excited states in spectra of multicharged ions: relativistic theory. AIP Conf Proc 1290:94–98
124. Glushkov AV, Svinarenko AA, Ignatenko AV (2011) Spectroscopy of autoionization resonances in spectra of the lanthanides atoms. Photoelectronics 20:90–94
125. Glushkov AV, Ambrosov SV, Ignatenko AV (2001) Non-hydrogenic atoms and Wannier-Mott excitons in a DC electric field: photoionization, stark effect, resonances in continuum and stochasticity. Photoelectronics 10:103–106
126. Khetselius OY (2019) Optimized relativistic many-body perturbation theory calculation of wavelengths and oscillator strengths for Li-like multicharged ions. Advances in quantum chemistry, vol 78, Elsevier, Amsterdam, pp 223–251. https://doi.org/10.1016/bs.aiq.2018.06.001
127. Glushkov AV (2019) Multiphoton spectroscopy of atoms and nuclei in a laser field: relativistic energy approach and radiation atomic lines moments method. Advances in quantum chemistry, vol 78. Elsevier, Amsterdam, pp 253–285. https://doi.org/10.1016/bs.aiq.2018.06.004
128. Kuznetsova AA, Glushkov AV, Ignatenko AV, Svinarenko AA, Ternovsky VB (2019) Spectroscopy of multielectron atomic system in a DC electric field. Advances in quantum chemistry, vol 78. Elsevier, Amsterdam, pp 287–306
129. Ignatenko AV, Buyadzhi AA, Buyadzhi VV, Kuznetsova AA, Mashkantsev AA, Ternovsky EV (2019) Nonlinear chaotic dynamics of quantum systems: molecules in an electromagnetic field. Adv Quantum Chem 78:149–170
130. Glushkov AV, Gurskaya MYu, Ignatenko AV, Smirnov AV, Serga IN, Svinarenko AA, Ternovsky EV (2017) Computational code in atomic and nuclear quantum optics: advanced computing multiphoton resonance parameters for atoms in a strong laser field. J Phys Conf Ser 905:012004
131. Malinovskaya SV, Dubrovskaya Yu, Zelentzova TN (2004) The atomic chemical environment effect on the β decay probabilities: relativistic calculation. Herald Kiev Nat Univ Ser Phys-Math 4:427–432
132. Glushkov AV, Malinovskaya SV, Dubrovskaya YV (2005) Sensing the atomic chemical composition effect on the beta decay probabilities. Sensor Electr Microsyst Technol 2(1):16–20
133. Buyadzhi VV, Zaichko PA, Antoshkina OA, Kulakli TA, Prepelitsa GP, Ternovsky VB, Mansarliysky VF (2017) Computing of radiation parameters for atoms and multicharged ions within relativistic energy approach: advanced code. J Phys Conf Ser 905:012003
134. Glushkov AV, Svinarenko AA, Khetselius OY, Buyadzhi VV, Florko TA, Shakhman AN (2015) Relativistic quantum chemistry: an advanced approach to the construction of the green function of the Dirac equation with complex energy and mean-field nuclear potential. In: Nascimento M, Maruani J, Brändas E, Delgado-Barrio G (eds) Frontiers in quantum methods and applications in chemistry and physics. Series: progress in theoretical chemistry and physics, vol 29. Springer, Cham, pp 197–217
135. Glushkov AV, Khetselius OY, Svinarenko AA, Buyadzhi VV (2015) Spectroscopy of autoionization states of heavy atoms and multiply charged ions. TEC, Odessa
136. Glushkov AV, Lovett L, Khetselius OY, Gurnitskaya EP, Dubrovskaya YuV, Loboda AV (2009) Generalized multiconfiguration model of decay of multipole giant resonances applied to analysis of reaction (μ–n) on the nucleus ^{40}Ca. Int J Modern Phys A 24(2–3):611–615

137. Gubanova ER, Glushkov AV, Khetselius OY, Bunyakova YuYa, Buyadzhi VV, Pavlenko EP (2017) New methods in analysis and project management of environmental activity: Electronic and radioactive waste. FOP, Kharkiv
138. Glushkov AV, Khetselius OY, Svinarenko AA, Buyadzhi VV (2015) Methods of computational mathematics and mathematical physics, P.1. TES, Odessa
139. Glushkov AV (2005) Atom in electromagnetic field. KNT, Kiev
140. Glushkov AV (2006) Relativistic and correlation effects in spectra of atomic systems. Astroprint, Odessa

Spectroscopy of Rydberg Atomic Systems in a Black-Body Radiation Field

Alexander V. Glushkov, Valentin B. Ternovsky, Anna A. Kuznetsova and Andrey V. Tsudik

Abstract It is presented a consistent relativistic approach to calculation of the energy, spectroscopic, radiation decay (excitation, ionization) characteristics of the Rydberg atomic systems in a Black-body radiation field. The approach is based on an advanced relativistic energy approach (in a single-electron approximation realization) and formalism of the relativistic many-body perturbation theory with the zeroth density functional approximation. The key features of the approach are connected with an accurate treatment of the complex exchange-correlation effects (interelectron polarization interaction through the Fermi sea, continuum pressure, the non-Coulomb grouping of levels in the heavy Rydberg spectra and others) and application of the optimized bases of relativistic wave functions, and correspondingly, fulfilling the principle of gauge invariance in calculation of the radiative decay characteristics. As illustration, we present the results of calculation of the spectroscopic characteristics (ionization rate, effective lifetime values etc.) for sodium Rydberg atoms in a Black-body radiation field for different states and temperatures. The obtained spectroscopic data are compared with available experimental and alternative theoretical results.

Keywords Rydberg atomic systems · Black-body radiation excitation and ionization · Relativistic quantum theory · Energy approach · Many-body exchange-correlation effects · Atomic core polarization effect

1 Introduction

At the present time, the study of Rydberg atoms (molecules) is definitely one of the most popular and very interesting directions of modern quantum physics and chemistry, atomic optics and spectroscopy. The huge relevance of the investigation of the energy and spectral properties of the Rydberg atoms (molecules) is, of course, due to the standard requirements for spectroscopic information of a number of applications and related physical disciplines, which include physics and chemistry of laboratory, astrophysical plasma, astrophysics and radioastronomy, atomic and molecular optics

A. V. Glushkov · V. B. Ternovsky (✉) · A. A. Kuznetsova · A. V. Tsudik
Odessa State Environmental University, L'vovskaya str., bld. 15, Odessa 65016, Ukraine

L. Mammino et al. (eds.), *Advances in Quantum Systems in Chemistry, Physics, and Biology*, Progress in Theoretical Chemistry and Physics 32,
https://doi.org/10.1007/978-3-030-34941-7_3

and spectroscopy, laser physics and quantum electronics and many others (e.g., [1–93]). By the way, it is known that a number of astrophysical processes in interstellar gas that interacts with fragments of supernova explosions, collisions of interstellar clouds, and supersonic gas flows in the formation of stars, lead to the formation of the Rydberg atoms with very large n (~1000).

The reasons for the significant interest in study of the Rydberg atoms RA are well known and related, first, to their extraordinary properties, namely, sufficiently long life ($\tau \sim n^3$ for $l \sim 1$, $\tau \sim n^5$ for $n \sim l$; l is the orbital quantum number) and geometric sizes (the typical Rydberg orbit size, $n^2 a_0/Z$, $a_0 = \hbar^2/me^2$), significant polarizabilities ($\sim n^2$), dipole moments of radiation transitions, high sensitivity to external electromagnetic fields (Stark shift ~ $\sim n^7 l^5$, Zeeman effect $\sim n^2$). Table 1 contains the qualitative values of the basic physical parameters of the atoms that are in the highly excited Rydberg states.

The idea of the existence of a new state of matter, namely, the Rydberg matter, is quite justified. As a result, intensive studies have been stimulated in recent years in the field of standard fundamental spectroscopy of Rydberg atoms related to the calculation of the spectral properties of these atoms, important for the general development of the relativistic (QED) theory of atomic spectra, and applied research in quantum optics, computer science, cryptography, quantum computing (c.g. Los Alamos Nat. Security, LLC, at http://qist.lanl.gov/), interferometry, astrophysical and laboratory plasma, etc., including the study of the Bose condensate in the Rydberg atoms vapours, cold atoms fountains, Carnot atomic radiation machines, etc.

The relevance of investigation of the spectroscopic properties of Rydberg atoms for solving many problems remains high, not only in the above cited fields, but also in relation to new applications, such as the development of fundamentally new experimental methods of laser spectroscopy, magnetic-optical traps, up-tracking of the properties of cold Rydberg atoms, in general, ultra-cold Rydberg plasma, which, in particular, occurs in ionization RA by laser or thermal (black-body radiation, BBR) radiation [1–12].

Table 1 Typical values of the basic physical parameters for the Rydberg atoms

Physical parameter	Analytical estimate	Typical value
The binding energy of Rydberg electron	$Z^2 R_\infty/n^2$, $R_\infty = 13.6058$ eV	1.36×10^{-3} eV
Typical size of Rydberg orbit	$n^2 a_0/Z$ $a_0 = \hbar^2/me^2 = 0.5291773$ Å	$\sim 0.53 \times 10^{-4}$ cm
Geometric cross-section		0.88×10^{-8} cm^2
Frequency of transitions between adjacent levels	$2Z^2 R_\infty/\hbar n^3$	4.13×10^7 s^{-1}
Strength of electric field acting on Rydberg electron	$Z^3 E_0/n^4$	51.4 V/cm
Strength of electric field, which corresponds to ionization threshold of Rydberg state	$Z^3 E_0/16n^4$, $E_0 = 5.142 \times 10^9$ V/cm	3.2 V/cm

From a physical point of view, there is a spontaneous transformation of the cloud of cold Rydberg atoms into supercooled plasma in magneto-optical traps due to avalanche ionization by thermal BBR radiation. Naturally, the study of different elementary processes involving Rydberg atoms in the BBR radiation field has been considered in a fairly large number of works.

The vast majority of existing papers on the description of Rydberg atoms in the thermal radiation field (c.g. [1–32]) are based on the Coulomb hydrogen-like approximation, different versions of the quantum defect method, classical and quasiclassical model approaches, the model and pseudo—potential methods. The authors of the papers [3–10] applied the Coulomb approximation, quantum defect formalism, different versions of the model and pseudo-potential method etc. (as a rule, the non-relativistic versions are used) to determine the spectral and radiative properties of different Rydberg atoms and ions.

It should be noted separately the cycles of theoretical and experimental works by Ryabtsev-Beterov et al. [1–3, 13–16], as well as theoretical works of Gorelovsky et al., Dyachkov-Pankratov and others (c.g. [7]), in which the advanced versions of a quasi-classical approach to the calculation of radiation amplitudes, oscillator strengths, and cross-sections for the Rydberg atoms in the BBR radiation field were actually developed. In the papers [1–3, 13–16] the authors present the calculational data on the ionization rates for Rydberg atoms of alkali elements (lithium, sodium, potassium, caesium) by a BBR radiation field. The calculations were carried out for the nS, nP, and nD states in the wide range of principal quantum numbers and temperatures. The above theoretical works and relevant models were substantially based on non-relativistic approximation.

At the same time one should note that for heavy Rydberg atoms (both in the free state and in an external electromagnetic field) it is fundamentally important to accurately account for both relativistic and exchange-correlation effects. The quality and consistency of accounting for these effects also determine the accuracy of description of the energy and spectroscopic parameters of the heavy Rydberg atoms, including these atoms in a thermal radiation field.

Among the fundamentally important exchange-correlation effects for essentially many-electron systems, one should single out such effects as polarization interaction and screening, continuum pressure, the non-Coulomb grouping of levels in the heavy Rydberg atoms spectra etc. It should be noted that these effects are not correctly considered, for example, within simplified Coulomb approximation or quantum defect models (c.g. [50, 51, 62–64]).

Another important point in studying different properties of heavy multi-electron atoms is related to the need to use optimized bases of relativistic wave functions, which is directly related to fulfilling the principle of gauge invariance. In fact this point is directly connected with an accurate consideration of the multielectron exchange-correlation effects. Here one should single out the gauge dependent contributions into the imaginary part of the electron energy shift (radiation width or transition probability or oscillator strength) for the certain class of the photon propagators (c.g. [56–61]).

In this chapter we present a consistent relativistic approach to computing the energy, spectroscopic, radiation decay (excitation, ionization) characteristics of the Rydberg atomic systems in a BBR radiation field. The approach is based on an advanced relativistic energy formalism (in a single-electron approximation) and the method of relativistic many-body perturbation theory (RMBPT) with the zeroth density functional approximation [56–62, 86–95].

As illustration, we list some results of computing spectroscopic characteristics (ionization rate, effective lifetime values etc.) of the Rydberg atoms (sodium) in a Black-body radiation field for different electron states and temperatures.

2 Theoretical Method

The fundamentals of our theoretical approach to study of the Rydberg atoms in a BBR radiation field are earlier presented in detail in Refs. [51–55, 90–95]. So, here we are limited to presenting only the main blocks of the approach and some principal new elements, related to the Rydberg multielectron atoms. The approach represents the combination of an advanced relativistic energy approach and formalism of the relativistic many-body perturbation theory with the zeroth density functional approximation (see details in Refs. [40, 51–95, 96–100]).

According to Ref. [58, 62], the RMBPT zeroth order Hamiltonian of the Rydberg atomic system is as follows:

$$H_0 = \sum_i \{\alpha c p_i - \beta mc^2 + [-Z/r_i + U_{MF}(r_i|b) + V_{XC}(r_i)]\}, \tag{1}$$

where c is the velocity of light, α_i, α_j—the Dirac matrices, ω_{ij}—the transition frequency, Z is a charge of atomic nucleus. The general potential in (1) includes self-consistent Coulomb-like mean-field potential $U_{MF}(r_i|b)$, ab ibitio one-particle exchange-correlation (relativistic generalized exchange Kohn-Sham potential plus generalized correlation Lundqvist-Gunnarsson potential) $V_{XC}(r_i|b)$ with the gauge calibrated parameter b (it is determined within special relativistic procedure on the basis of relativistic energy approach; c.g. [56–61]).

The perturbation operator is as follows:

$$H^{PT} = \sum_{i>j} \exp(i\omega_{ij} r_{ij}) \cdot \frac{(1 - \alpha_i \alpha_j)}{r_{ij}} - \sum_i [U_{MF}(r_i) + V_{XC}(r_i|b)]. \tag{2}$$

The multielectron interelectron exchange-correlation effects (the core polarization and screening effects, continuum pressure etc.) are taken into consideration as the RMBPT second and higher orders contributions. The details of calculation of the corresponding matrix elements of the polarization and screening interelectron interaction potentials are described in Refs. [40, 52–69].

The key idea to take the polarization interelectron interaction into consideration is to include an effective ab initio two-quasiparticle polarization interaction functional to operator (2). The correct non-relativistic expression for this functional is obtained in Ref. [52]. In this work we use more consistent relativistic expression, which is derived in Ref. [81] and applied in this work.

An account for the screening interelectron effects is fulfilled by modification of the RMBPT zeroth approximation mean-field potential (see details in Refs. [58, 62, 93]). Simultaneously we use the optimized one-quasiparticle representation in the RMBPT zeroth perturbation theory order. Under its construction we use an effective, ab initio relativistic optimization procedure [68] (see also [58–61]). This procedure is directly related to the condition of fulfilling the principle of gauge invariance and, in particular, reduced to minimization of the gauge-noninvariant contributions into the imaginary part of electron energy ImΔE for the certain class of the electromagnetic field potentials (see also [56–61]).

The procedure works within a relativistic energy formalism [49–53, 56–61], where the radiation decay (transition, excitation, ionization) probability Γ is determined through an imaginary part of the electron energy shift ImΔE, i.e. $\Gamma \sim$ ImΔE and the corresponding matrix elements of the imaginary part of the operator (2), which contain the sum of the Coulomb and Breit terms. The detailed procedures of computing the radial and angular integrals in these matrix elements are described in Refs. [45–59].

From physical viewpoint, the process of the BBR Rydberg atoms excitation and ionization is sufficiently easily understandable [2–5]. Moreover, for the interested interval of temperatures one could implement a single-electron approximation for calculating the ionization rates (cross sections) within relativistic energy approach. The important point is in appearance of a product with the Planck's distribution for the thermal photon number density. Corresponding decay (ionization) rate results in the integral over the Blackbody radiation frequencies (see details in Refs. [1–19]).

The total ionization rate of the Rydberg atomic system in the BBR radiation field is usually determined as the sum of direct BBR ionization rate of the initially excited state, the ionization (field ionization) rate of highly excited states, which are populated from the initial Rydberg state via absorption of the BBR photons, the rate of direct BBR-induced ionization of atoms from the neighbouring Rydberg states and the rate of field ionization of high-lying Rydberg states (with populating through so called two-step process via the BBR photons absorption).

The total width of the Rydberg state (naturally isolated from all external electromagnetic fields except BBR one) consists, apparently, of natural, spontaneous radiation width and BBR-induced (thermal) width:

$$\Gamma_{nl}^{tot} = \Gamma_{nl}^{sp} + \Gamma_{nl}^{BBR}(T). \tag{3a}$$

Accordingly, the effective lifetime of the Rydberg state is inversely proportional to the total decay rate as a result of spontaneous transitions and transitions induced by the BBR radiation:

$$\frac{1}{\tau_{eff}} = \Gamma_0 + \Gamma_{\mathrm{BBR}} = \frac{1}{\tau_0} + \frac{1}{\tau_{BBR}}. \tag{3b}$$

The detailed procedures of calculation of the radial and angular integrals (amplitudes) in the matrix elements are described in Refs. [45–62]. All computing is performed on the basis of the modified quantum PC code "Superatom-ISAN" (version 93).

3 Results and Conclusions

Below we present some results of calculation of the spectroscopic characteristics (ionization rates, effective lifetimes) of the sodium atom in a Black-body radiation field for different states and temperatures.

In Fig. 1, we list our data for temperature dependence of the sodium ionization rate (state 17D), the model calculational results by Kleppner et al. within the Coulomb approximation and corresponding experimental data [3].

The analysis of the data in Fig. 1 shows that at temperatures above 150 K the data of our theory and the Coulomb model calculational data by Kleppner et al. are very close to each other and are sufficiently consistent with the experimental measurements, but at temperatures below 150 K there is some deviation of the experimental data from the Coulomb model calculational results.

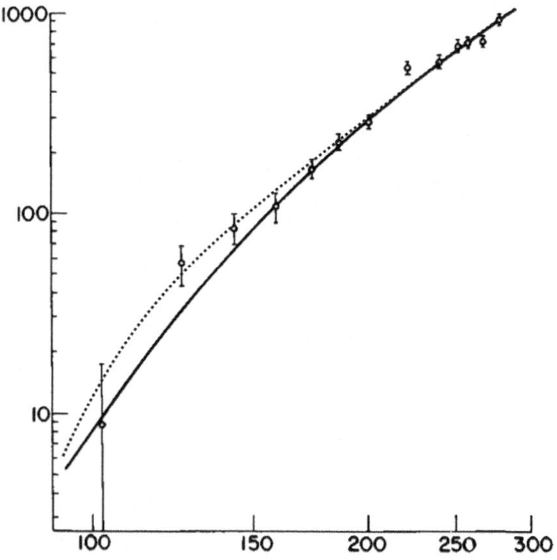

Fig. 1 Temperature dependence of the 17D sodium ionization rate: our data (interruptible line); the model calculation data by Kleppner et al. (continuous line); experimental data by Kleppner et al. (circles)

Table 2 Effective lifetime (μs) of the nP, nD Rydberg states in the sodium spectrum for the temperatures T = 300, 600 K: [2]—theory by Beterov et al. and this work

N	$T = 300\ K$ $P_{1/2}$ $P_{3/2}$ [1]	$T = 300\ K$ $P_{1/2}$ $P_{3/2}$ This work	$T = 600\ K$ $P_{1/2}$ $P_{3/2}$ This work	$T = 300\ K$ $D_{3/2}$ $D_{5/2}$ This work	$T = 600\ K$ $D_{3/2}$ $D_{5/2}$ This work
10	4.80 4.76	4.84 4.81	2.84 2.83	0.913 0.914	0.837 0.838
15	11.59 11.51	11.64 11.57	6.40 6.38	2.847 2.848	2.457 2.458
20	20.99 20.89	21.06 20.96	11.42 11.38	6.263 6.266	5.164 5.167
25	33.33 33.21	33.42 33.32	17.96 17.86	11.465 11.468	9.095 9.098
30	48.71 48.56	48.84 48.70	26.03 25.97	18.602 18.609	14.281 14.285

In Table 2, we present our theoretical data on the effective lifetime of the sodium nP, nD Rydberg states and some theoretical data by Beterov et al. [1, 2] for comparison for temperatures T = 300, 600 K.

Next, in Fig. 2, we present the experimental and theoretical data on the total rate of BBR-induced ionization for the Rydberg (a) nS states and (b) nD states of the sodium atom (T = 300 K): Experiment (circles and squares); Theory: improved quasiclassical model data by Beterov et al. (continuous line) [1]; our theory (interrupted line).

The analysis of the data presented in Fig. 2 shows that, firstly, both theories quite well describe the experimental data for the nS and nD states of the sodium atom for

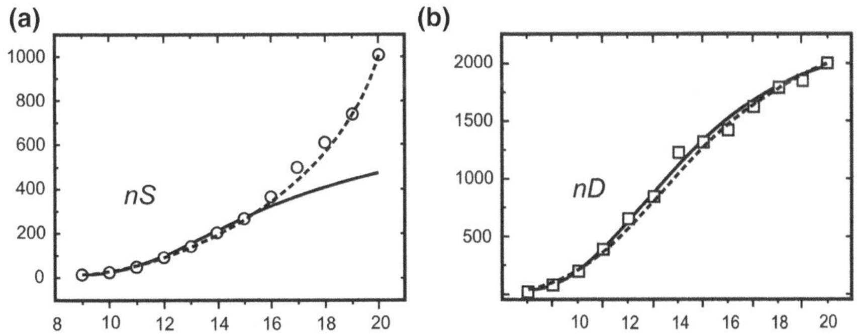

Fig. 2 **a** Total rate of BBR-induced ionization for the sodium Rydberg nS states (T = 300 K): Experiment (circles, squares); Theory: advanced quasiclassical model data by Beterov et al. model [1] (continuous line); our theory (interrupted line). **b** Total rate of BBR-induced ionization for the sodium Rydberg nD states (T — 300 K): Experiment (circles and squares); Theory advanced quasiclassical model data by Beterov et al. model [1] (continuous line); our theory (interrupted line)

n till 165, but there is still a significant deviation of the data of the quasiclassical calculation from the experiment (see Fig. 2a).

The most adequate explanation for this important fact was correctly provided by Beterov et al. in the Refs. [1, 2]. The corresponding deviation of the quasiclassical theory from the experiment is provided by the over-importance value of exchange-correlation corrections, in particular, for nS states. In fact, for nS states there is a certain anomaly, since the orbitals for the desired states deep enough penetrate into the atomic core, which causes a very strong interaction with it. An additional factor is the known circumstance, namely, the quantum defects for these states are quite large, while for the nD states of the sodium atom the picture is fundamentally different. In our theory, the corresponding effects are rather thoroughly and correctly taken into account, therefore, the theory yields results in physically reasonable agreement with available experiment.

To conclude, we presented a consistent relativistic quantum approach to determination of the elementary atomic processes parameters of the Rydberg atomic systems in a Black-body radiation field. The approach is based on an advanced relativistic energy approach and formalism of the relativistic many-body perturbation theory with the zeroth density functional approximation. The important features of our approach are connected with an accurate treatment of the complex exchange-correlation effects (interelectron polarization interaction through the Fermi sea, continuum pressure, the non-Coulomb grouping of levels in the heavy Rydberg spectra etc.) and application of the optimized bases of relativistic wave functions, and correspondingly. We listed some results of calculation of the spectroscopic characteristics (ionization rate, effective lifetime values etc.) for the sodium Rydberg atom in a Black-body radiation field for different states and temperatures and compared the obtained data with available experimental and alternative theoretical results.

Acknowledgements The authors are grateful to the Chair of QSCP-XXIII, Prof. Liliana Mammino, and the Cochair Prof. Jean Maruani, for their generous invitation to present this work in the Proceedings of the XXIII International workshop on Quantum Systems in Chemistry, Physics and Biology.

References

1. Beterov II, Tretyakov DV, Ryabtsev II et al (2009) Ionization of Rydberg atoms by blackbody radiation. New J Phys 11:013052
2. Beterov II, Ryabtsev II, Tretyakov DB, Entin VM (2009) Quasiclassical calculations of blackbody-radiation-induced depopulation rates and effective lifetimes of Rydberg *nS*, *nP*, and *nD* alkali-metal atoms with $n \sim 80$. Phys Rev A 79:052504
3. Spencer WP, Vaidyanathan AG, Kleppner D, Ducas TW (1982) Photoionization by blackbody radiation. Phys Rev A 26:1490–1493
4. Burkhardt CE, Corey RL, Garver WP, Leventhal JJ, Allegrini M, Moi L (1986) Ionization of Rydberg atoms. Phys Rev A 34:80–88
5. Lehman GW (1983) Rate of ionisation of H and Na Rydberg atoms by black-body radiation. J Phys B: At Mol Phys 16:2145

6. Farley JW, Wing WH (1981) Accurate calculation of dynamic stark shifts and depopulation rates of Rydberg energy levels induced by blackbody radiation. Hydrogen, helium, alkali-metal atoms. Phys Rev A 23:2397–2410
7. Dyachkov LG, Pankratov PM (1994) On the use of the semiclassical approximation for the calculation of oscillator strengths and photoionization cross sections. J Phys B 27(3):461
8. Hoogenraad JH, Noordam LD (1998) Rydberg atoms in far-infrared radiation fields. I. Dipole matrix elements of H, Li, Rb. Phys Rev A 57:4533–4545
9. Glukhov IL, Nikitina EA, Ovsiannikov VD (2016) Blackbody-radiation-induced shifts and the broadening of Rydberg states in the ions of group IIa elements. J Phys B: Atom Mol Opt Phys 49(3):035003
10. Glukhov IL, Ovsiannikov VD (2009) Blackbody-radiation-induced decay and excitation of Rydberg states in sodium. Acta Phys Pol 116(4):528–531
11. Nascimento V, Caliri L, de Oliveira A et al (2006) Measurement of the lifetimes of S and D states below n = 31 using cold Rydberg gas. Phys Rev A 74:054501
12. Piotrowicz MJ, MacCormick C, Kowalczyk A et al (2011) Measurement of the electric dipole moments for transitions to rubidium Rydberg states via autler-townes splitting. New J Phys 13:093012
13. Beterov II, Tretyakov DB, Ryabtsev II, Ekers A, Bezuglov NN (2007) Ionization of sodium and rubidium *nS*, *nP*, and *nD* Rydberg atoms by blackbody radiation. Phys Rev A 75:052720
14. Ryabtsev II, Tretyakov DB, Beterov II, Bezuglov NN, Miculis K, Ekers A (2005) Collisional and thermal ionization of sodium Rydberg atoms: I. Experiment for *n*S and *n*D atoms with $n = 8$–20. J Phys B: Atom Mol Opt Phys 38(2):379–386
15. Miculis K, Beterov II, Bezuglov NN, Ryabtsev II, Tretyakov DB, Klucharev AN (2005) Collisional and thermal ionization of sodium Rydberg atoms: II. Theory for *n*S, *n*P and *n*D states with $n = 5$–25. J Phys B: Atom Mol Opt Phys 38(11):1811–1822
16. Beterov II, Tretyakov DB, Ryabtsev II, Bezuglov NN, Miculis K, Ekers A (2005) Collisional and thermal ionization of sodium Rydberg atoms III. Experiment and theory for *n*S and *n*D states with $n = 8$–20 in crossed atomic beams. J Phys B: Atom Mol Opt Phys 38(24):4349–4363
17. Li W, Noel MW, Robinson MP et al (2004) Evolution dynamics of a dense frozen Rydberg gas to plasma. Phys Rev A 70:042713
18. Buyadzhi VV, Zaichko PA, Gurskaya MY, Kuznetsova AA, Ponomarenko EL, Ternovsky VB (2017) Relativistic theory of excitation and ionization of Rydberg atomic systems in a black-body radiation field. J Phys: Conf Ser 810:012047
19. Svinarenko AA, Khetselius OY, Buyadzhi VV, Florko TA, Zaichko PA, Ponomarenko EL (2014) Spectroscopy of Rydberg atoms in a black-body radiation field: relativistic theory of excitation and ionization. J Phys: Conf Ser 548:012048
20. Malinovskaya SV, Glushkov AV, Khetselius OY, Loboda AV, Lopatkin Y, Nikola L, Svinarenko AA, Perelygina T (2011) Generalized energy approach to calculating electron collision cross sections for multicharged ions in a plasma: debye shielding model. Int J Quant Chem 111:288–296
21. Khetselius OY, Florko TA, Svinarenko AA, Tkach TB (2013) Radiative and collisional spectroscopy of hyperfine lines of the Li-like heavy ions and Tl atom in an atmosphere of inert gases. Phys Scr T 153:014037
22. Wenhui L, Tanner PL, Gallagher TF (2005) Dipole-dipole excitation and ionization in an ultracold gas of Rydberg atoms. Phys Rev Lett 94:173001
23. Viteau M, Bason MG, Radogostowicz J, Malossi N, Ciampini D, Morsch O, Arimondo E (2011) Rydberg excitations in bose-einstein condensates in quasi-one-dimensional potentials and optical lattices. Phys Rev Lett 107:060402
24. Galvez EJ, MacGregor CW, Chaudhuri B, Gupta S (1997) Blackbody-radiation-induced resonances between Rydberg-Stark states of Na. Phys Rev A 55:3002–3006
25. Safronova UI, Johnson WR, Derevianko A (1999) Relativistic many-body calculations of energy levels, hyperfine constants, electric-dipole matrix elements, and static polarizabilities for alkali-metal atoms. Phys Rev A 60:4476

26. Safronova UI, Safronova MS (2009) Third-order relativistic many-body calculations of energies, transition rates, hyperfine constants, and blackbody radiation shift in $^{171}Yb^+$. Phys Rev A 79:022512
27. Glushkov AV, Khetselius OY, Loboda AV, Ignatenko AV, Svinarenko AA, Korchevsky DA, Lovett L (2008) QED approach to modeling spectra of the multicharged ions in a plasma: oscillator and electron-ion collision strengths. AIP Conf Proc 1058:175–177
28. Buyadzhi VV, Zaichko PA, Antoshkina OA, Kulakli TA, Prepelitsa GP, Ternovsky VB, Mansarliysky VF (2017) Computing of radiation parameters for atoms and multicharged ions within relativistic energy approach: advanced code. J Phys: Conf Ser 905:012003
29. Buyadzhi VV (2015) Laser multiphoton spectroscopy of atom embedded in debye plasmas: multiphoton resonances and transitions. Photoelectronics 24:128–133
30. Buyadzhi VV, Chernyakova YG, Smirnov AV, Tkach TB (2016) Electron-collisional spectroscopy of atoms and ions in plasma: be-like ions. Photoelectronics 25:97–101
31. Buyadzh VV, Chernyakova YG, Antoshkina OA, Tkach TB (2017) Spectroscopy of multicharged ions in plasmas: Oscillator strengths of Be-like ion Fe. Photoelectronics 26:94–102
32. Florko TA, Ambrosov SV, Svinarenko AA, Tkach TB (2012) Collisional shift of the heavy atoms hyperfine lines in an atmosphere of the inert gas. J Phys: Conf Ser 397(1):012037
33. Buyadzhi VV, Glushkov AV, Mansarliysky VF, Ignatenko AV, Svinarenko AA (2015) Spectroscopy of atoms in a strong laser field: new method to sensing AC stark effect, multiphoton resonances parameters and ionization cross-sections. Sens Electr Microsys 12(4):27–36
34. Glushkov AV, Gurskaya MY, Ignatenko AV, Smirnov AV, Serga IN, Svinarenko AA, Ternovsky EV (2017) Computational code in atomic and nuclear quantum optics: advanced computing multiphoton resonance parameters for atoms in a strong laser field. J Phys: Conf Ser 905(1):012004
35. Dubrovskaya YV, Khetselius OY, Vitavetskaya LA, Ternovsky VB, Serga IN (2019) Quantum chemistry and spectroscopy of pionic atomic systems with accounting for relativistic, radiative, and strong interaction effects. Adv Quantum Chem 78:193–222 (Elsevier). https://doi.org/10.1016/bs.aiq.2018.06.003
36. Kuznetsova AA, Glushkov AV, Ignatenko AV, Svinarenko AA, Ternovsky VB (2019) Spectroscopy of multielectron atomic systems in a DC electric field. Adv Quant Chem 78:287–306 (Elsevier)
37. Glushkov AV (2005) Atom in electromagnetic field. KNT, Kiev
38. Gross EG, Kohn W (2005) Exchange-correlation functionals in density functional theory. Plenum, New York
39. Glushkov AV, Lovett L, Khetselius OY, Gurnitskaya EP, Dubrovskaya YV, Loboda AV (2009) Generalized multiconfiguration model of decay of multipole giant resonances applied to analysis of reaction (μ-n) on the nucleus ^{40}Ca. Int J Modern Phys A 24(2–3):611–615
40. Glushkov AV, Malinovskaya SV, Sukharev DE, Khetselius OY, Loboda AV, Lovett L (2009) Green's function method in quantum chemistry: new numerical algorithm for the dirac equation with complex energy and Fermi-model nuclear potential. Int J Quant Chem 109:1717–1727
41. Khetselius OY (2009) Relativistic perturbation theory calculation of the hyperfine structure parameters for some heavy-element isotopes. Int J Quant Chem 109:3330–3335
42. Khetselius OY (2009) Relativistic calculation of the hyperfine structure parameters for heavy elements and laser detection of the heavy isotopes. Phys Scr T 135:014023
43. Serga IN, Dubrovskaya YV, Kvasikova AS, Shakhman AN, Sukharev DE (2012) Spectroscopy of hadronic atoms: energy shifts. J Phys: Conf Ser 397:012013
44. Svinarenko AA (2014) Study of spectra for lanthanides atoms with relativistic many-body perturbation theory: Rydberg resonances. J Phys: Conf Ser 548:012039
45. Indelicato P, Desclaux JP (1993) Projection operator in the multiconfiguration Dirac-Fock method. Phys Scr T46:110
46. Sapirstein J (1998) Theoretical methods for the relativistic atomic many-body problem. Rev Mod Phys 70:55

47. Glushkov AV, Ambrosov SV, Ignatenko AV, Korchevsky DA (2004) DC strong field stark effect for non-hydrogenic atoms: consistent quantum mechanical approach. Int J Quant Chem 99:936–939
48. Glushkov AV, Ternovsky VB, Buyadzhi VV, Prepelitsa GP (2014) Geometry of a relativistic quantum chaos: new approach to dynamics of quantum systems in electromagnetic field and uniformity and charm of a chaos. Proc Intern Geom Cent 7(4):60–71
49. Ivanov LN, Ivanova EP (1979) Atomic ion energies for Na-like ions by a model potential method Z = 25–80. Atom Data Nucl Data Tabl 24:95–109
50. Vidolova-Angelova E, Ivanov LN (1991) Autoionizing Rydberg states of thulium. Re-orientation decay due to monopole interaction. J Phys B: At Mol Opt Phys 24:4147–4158
51. Ivanov LN, Ivanova EP (1996) Method of Sturm orbitals in calculation of physical characteristics of radiation from atoms and ions. JETP 83:258–266
52. Ivanova EP, Ivanov LN, Glushkov AV, Kramida AE (1985) High order corrections in the relativistic perturbation theory with the model zeroth approximation, Mg-Like and Ne-Like Ions. Phys Scr 32:513–522
53. Ivanova EP, Glushkov AV (1986) Theoretical investigation of spectra of multicharged ions of F-like and Ne-like isoelectronic sequences. J Quant Spectr Rad Transf 36:127–145
54. Ivanov LN, Ivanov EP, Knight L (1993) Energy approach to consistent QED theory for calculation of electron-collision strengths: ne-like ions. Phys Rev A 48:4365–4374
55. Glushkov AV, Ivanov LN, Ivanova EP (1986) Autoionization phenomena in atoms. Moscow University Press, Moscow
56. Glushkov AV, Ivano LN (1992) Radiation decay of atomic states: atomic residue polarization and gauge noninvariant contributions. Phys Lett A 170:33–36
57. Glushkov AV (2008) Relativistic quantum theory. Quantum mechanics of atomic systems. Astroprint, Odessa
58. Glushkov AV (2012) Advanced relativistic energy approach to radiative decay processes in multielectron atoms and multicharged ions. In: Nishikawa K, Maruani J, Brandas E, Delgado-Barrio G, Piecuch P (eds) Advances in the theory of quantum systems in chemistry and physics. Springer, Berlin, pp 231–252 (Ser. Progress in Theor. Chemistry and Physics)
59. Khetselius OY (2019) Optimized relativistic many-body perturbation theory calculation of wavelengths and oscillator strengths for li-like multicharged ions. Adv Quant Chem 78:223–251 (Elsevier). https://doi.org/10.1016/bs.aiq.2018.06.001
60. Glushkov AV, Multiphoton spectroscopy of atoms and nuclei in a laser field: relativistic energy approach and radiation atomic lines moments method Adv Quant Chem 78:253–285 (Elsevier). https://doi.org/10.1016/bs.aiq.2018.06.004
61. Glushkov AV, Khetselius OY, Svinarenko AA, Buyadzhi VV (2015) Spectroscopy of autoionization states of heavy atoms and multiply charged ions. TEC, Odessa
62. Ternovsky VB, Glushkov AV, Khetselius OY, Gurskaya MY, Kuznetsova AA (2018) Spectroscopy of radiative decay processes in heavy Rydberg alkali atomic systems. In: Wang Yan A, Thachuk Mark, Krems Roman, Maruani Jean (eds) Concepts, methods and applications of quantum systems in chemistry and physics, vol 31. Springer, Cham, pp 229–241 (Series: Progress in Theoretical Chemistry and Physics)
63. Glushkov AV, Khetselius OY, Svinarenko AA (2013) Theoretical spectroscopy of autoionization resonances in spectra of lanthanide atoms. Phys Scr T153:014029
64. Malinovskaya SV, Glushkov AV, Khetselius OY, Svinarenko AA, Mischenko EV, Florko TA (2009) Optimized perturbation theory scheme for calculating the interatomic potentials and hyperfine lines shift for heavy atoms in the buffer inert gas. Int J Quant Chem 109:3325–3329
65. Glushkov AV, Khetselius OY, Lopatkin YM, Florko TA, Kovalenko OA, Mansarliysky VF (2014) Collisional shift of hyperfine line for rubidium in an atmosphere of the buffer inert gas. J Phys: Conf Ser 548:012026
66. Malinovskaya SV, Dubrovskaya YV, Vitavetskaya LA (2005) Advanced quantum mechanical calculation of the beta decay probabilities. Energy approach to resonance states of compound superheavy nucleus and EPPP in heavy nuclei collisions. In: Grzonka D, Czyzykiewicz R, Oelert W, Rozek T, Winter P (eds) Low energy antiproton physics, vol 796. AIP, New York, pp 201–205 (AIP Conf. Proc.)

67. Malinovskaya SV, Glushkov AV, Khetselius OY (2008) New laser-electron nuclear effects in the nuclear γ transition spectra in atomic and molecular systems. In: Wilson S, Grout P, Maruani J, Delgado-Barrio G, Piecuch P (eds) Frontiers in quantum systems in chemistry and physics. Series: Progress in Theoretical Chemistry and Physics, vol 18. Springer, Dordrecht, pp 525–541
68. Glushkov AV, Khetselius OY, Malinovskaya SV (2008) Optics and spectroscopy of cooperative laser-electron nuclear processes in atomic and molecular systems—new trend in quantum optics. Eur Phys J ST 160:195–204
69. Glushkov AV, Malinovskaya SV, Gurnitskaya EP, Khetselius OY, Dubrovskaya YV (2006) Consistent quantum theory of recoil induced excitation and ionization in atoms during capture of neutron. J Phys: Conf Ser 35:425–430
70. Glushkov AV, Ambrosov SV, Lobod AV, Gurnitskaya EP, Prepelitsa GP (2005) Consistent QED approach to calculation of electron-collision excitation cross sections and strengths: Ne-like ions. Int J Quant Chem 104:562–569
71. Glushkov AV, Shpinareva IM, Ignatenko V, Gura VI (2006) Study of atomic systems in strong laser fields: spectral hierarchy, dynamical stabilisation and generation of ultra-short vuv and x-ray pulses. Sens Electr Microsyst Tech 3(1):29–35
72. Glushkov AV (2005) Energy approach to resonance states of compound superheavy nucleus and EPPP in heavy nuclei collisions. In: Grzonka D, Czyzykiewicz R, Oelert W, Rozek T, Winter P (eds) Low energy antiproton physics, vol 796. AIP, New York, pp 206–210 (AIP Conf. Proc.)
73. Glushkov AV (2012) Spectroscopy of cooperative muon-gamma-nuclear processes: energy and spectral parameters. J Phys: Conf Ser 397:012011
74. Glushkov AV (2014) Spectroscopy of atom and nucleus in a strong laser field: Stark effect and multiphoton resonances. J Phys: Conf Ser 548:012020
75. Khetselius OY (2012) Spectroscopy of cooperative electron-gamma-nuclear processes in heavy atoms: NEET effect. J Phys: Conf Ser 397:012012
76. Khetselius OY (2012) Relativistic energy approach to cooperative electron-γ-nuclear processes: NEET effect. In: Nishikawa K, Maruani J, Brändas E, Delgado-Barrio G, Piecuch P (eds) Quantum systems in chemistry and physics. series: progress in theoretical chemistry and physics, vol 26. Springer, Dordrecht, pp 217–229
77. Khetselius OY (2012) Quantum geometry: new approach to quantization of the quasistationary states of Dirac equation for super heavy ion and calculating hyper fine structure parameters. Proc Intern Geom Cent 5(3–4):39–45
78. Khetselius OY (2008) Hyperfine structure of atomic spectra. Astroprint, Odessa
79. Glushkov AV (1992) Oscillator strengths of Cs and Rb-like ions. J Appl Spectrosc 56(1):5–9
80. Glushkov AV (1992) Negative ions of inert gases. JETP Lett 55:97–100
81. Glushkov AV (1990) Relativistic polarization potential of a many-electron atom. Sov Phys J 33(1):1–4
82. Glushkov AV, Ivanov LN (1993) DC strong-field Stark effect: consistent quantum-mechanical approach. J Phys B: At Mol Opt Phys 26:L379–L386
83. Glushkov AV (2013) Operator perturbation theory for atomic systems in a strong DC electric field. In: Hotokka M, Brändas E, Maruani J, Delgado-Barrio G (eds) Advances in quantum methods and applications in chemistry, physics, and biology. Series: Progress in Theoretical Chemistry and Physics, vol 27. Springer, Cham, pp 161–177
84. Glushkov AV, Malinovskaya SV, Ambrosov SV, Shpinareva IM, Troitskaya OV (1997) Resonances in quantum systems in strong external fields consistent quantum approach. J Techn Phys 38(2):215–218
85. Glushkov AV, Khetselius OY, Malinovskaya SV (2008) Spectroscopy of cooperative laser-electron nuclear effects in multiatomic molecules. Mol Phys 106:1257–1260
86. Glushkov AV (2006) Relativistic and correlation effects in spectra of atomic systems. Astroprint, Odessa
87. Khetselius OY (2011) Quantum structure of electroweak interaction in heavy finite Fermi-systems. Astroprint, Odessa

88. Glushkov AV, Khetselius OY, Svinarenko AA (2012) Relativistic theory of cooperative muon-γ-nuclear processes: negative muon capture and metastable nucleus discharge. In: Hoggan P, Brändas E, Maruani J, Delgado-Barrio G, Piecuch P (eds) Advances in the theory of quantum systems in chemistry and physics. Series: Progress in Theoretical Chemistry and Physics, vol 22. Springer, Dordrecht, pp 51–68
89. Glushkov AV, Khetselius OY, Lovett L (2009) Electron-β-nuclear spectroscopy of atoms and molecules and chemical bond effect on the β-decay parameters. In: Piecuch P, Maruani J, Delgado-Barrio G, Wilson S (eds) Advances in the theory of atomic and molecular systems dynamics, spectroscopy, clusters, and nanostructures. Series: Progress in Theor. Chem. and Phys., vol 20. Springer, Dordrecht, pp 125–152
90. Khetselius OY, Glushkov AV, Dubrovskaya YV, Chernyakova YG, Ignatenko AV, Serga IN, Vitavetskaya LA (2018) Relativistic quantum chemistry and spectroscopy of exotic atomic systems with accounting for strong interaction effects. In: Wang YA, Thachuk M, Krems R, Maruani J (eds) Concepts, methods and applications of quantum systems in chemistry and physics. Series: Progress in Theoretical Chemistry and Physics, vol 31. Springer, Dordrecht, pp 71–91
91. Glushkov AV, Rusov VD, Ambrosov SV, Loboda AV (2003) Resonance states of compound super-heavy nucleus and EPPP in heavy nucleus collisions. In: Fazio G, Hanappe F (eds) New projects and new lines of research in nuclear physics. World Scientific, Singapore, pp 126–132
92. Glushkov AV, Khetselius OY, Gurnitskaya EP, Loboda AV, Sukharev DE (2009) Relativistic quantum chemistry of heavy ions and hadronic atomic systems: spectra and energy shifts. Theory and applications of computational chemistry. AIP Conf Proc 1102:168–171
93. Glushkov AV, Ambrosov SA, Loboda AV, Gurnitskaya EP, Khetselius OY (2006) QED calculation of heavy multicharged ions with account for correlation, radiative and nuclear effects. In: Julien P, Maruani J, Mayou D, Wilson S, Delgado-Barrio G (eds) Recent advances in the theory of chemical and physical systems, vol 15. Springer, Dordrecht, pp 285–299 (Series: Progress in Theoretical Chemistry and Physics)
94. Glushkov AV, Dan'kov SV, Prepelitsa G, Polischuk VN, Efimov AE (1997) Qed theory of nonlinear interaction of the complex atomic systems with laser field: Multiphoton resonances. J Tech Phys 38(2):219–222
95. Ambrosov S, Ignatenko V, Korchevsky D, Kozlovskaya V (2005) Sensing stochasticity of atomic systems in crossed electric and magnetic fields by analysis of level statistics for continuous energy spectra. Sens Electron Microsyst Techn N2: 19–23
96. Glushkov AV, Malinovskaya SV, Svinarenko AA, Vitavetskaya LA (2005) Sensing spectral hierarchy, quantum chaos, chaotic diffusion and dynamical stabilisation effects in a multiphoton atomic dynamics with intense laser field. Sens Electron Microsyst Tech 2(2):29–36
97. Glushkov AV, Buyadzhi VV, Kvasikova AS, Ignatenko AV, Kuznetsova AA, Pre-pelitsa GP, Ternovsky VB (2017) Non-Linear chaotic dynamics of quantum systems: Molecules in an electromagnetic field and laser systems. In: Tadjer A, Pavlov R, Maruani J, Brändas E, Delgado-Barrio G (eds) Quantum Systems in Physics, Chemistry, and Biology, Series: Progress in Theoretical Chemistry and Physics, vol 30. Springer, Cham, pp169–180
98. Khetselius OY, Glushkov AV, Gurskaya MY, Kuznetsova AA, Dubrovskaya YV, Serga IN, Vitavetskaya LA (2017) Computational modelling parity non-conservation and electroweak interaction effects in heavy atomic systems within the nuclear-relativistic many-body perturbation theory. J Phys: Conf Ser 905:012029
99. Ignatenko AV, Buyadzhi AA, Buyadzhi VV, Kuznetsova AA, Mashkantsev AA, Ternovsky EV (2019) Nonlinear chaotic dynamics of quantum systems: Molecules in an electromagnetic field. Adv. Quantum Chem, vol 78. Elsevier, pp 149–170
100. Glushkov AV, Khetselius OY, Svinarenko AA, Buyadzhi VV (2015) Methods of computational mathematics and mathematical physics. P.1. TES, Odessa

Hyperfine and Electroweak Interactions in Heavy Finite Fermi Systems and Parity Non-conservation Effect

Olga Yu. Khetselius, Alexander V. Glushkov, Eugeny V. Ternovsky, Vasily V. Buyadzhi and Oleksii L. Mykhailov

Abstract The consistent theoretical approach, namely, nuclear-relativistic many -body perturbation theory is applied to study of the hyperfine and electroweak interaction parameters, parity non-conservation effect in heavy atomic systems. In fact the nuclear-relativistic many-body perturbation theory is based on the combining ab initio perturbation theory formalism for electron subsystem, nuclear relativistic middle-field model for nuclear subsystem and an energy approach for computing radiation transition amplitude and allows to fulfil computing the finite Fermi systems (atomic systems) with taking into account the relativistic, correlation, nuclear, radiative effects. The important feature is the correct accounting for the inter electron correlations, nuclear, Breit and QED radiative corrections. All correlation corrections of the second order and dominated classes of the higher orders diagrams are taken into account. The results of accurate calculation of the hyperfine structure parameters for the caesium are listed. There are presented the values of the nuclear spin dependent corrections to the PNC ^{133}Cs: 6 s–7 s amplitude, calculated on the basis of different theoretical methods. The estimated values of a weak charge Q_W for different heavy atoms (^{133}Cs, ^{173}Yb and others) are presented and compared with alternative theoretical data.

Keywords Hyperfine structure · Parity non-conservation effect · Relativistic perturbation theory · Correlation · Nuclear · Radiative corrections

1 Introduction

The parity non-conservation (PNC) or violation experiments in atomic physics provide an important possibility to deduce information on the Standard Model independent of high-energy physics experiments [1–12]. The recent LEP experiments are fulfilled [1, 2], that yield extremely accurate values for Z-boson properties. In the last two decades a status of the Standard model has been strengthened by different experimental achievements of particle physics. It should be mentioned Higgs boson

O. Yu. Khetselius (✉) · A. V. Glushkov · E. V. Ternovsky · V. V. Buyadzhi · O. L. Mykhailov
Odessa State Environmental University, L'vovskaya str., bld. 15, Odessa 65016, Ukraine

L. Mammino et al. (eds.), *Advances in Quantum Systems in Chemistry, Physics, and Biology*, Progress in Theoretical Chemistry and Physics 32,
https://doi.org/10.1007/978-3-030-34941-7_4

discovery, measurement of CP violations in the K^-, B^- mesons, evidence of accelerated expansion of the universe, determination of the fraction of dark energy and dark matter in the universe, etc. From the other side, there are a number of the serious factors which clearly points to some new physics beyond the Standard model despite the desperate lack of direct experimental evidence. One could remind that the density of matter included into the Standard model is approximately 5% of the energy density of the universe; besides, neutrinos in the Standard model are massless, and there are no neutrino oscillations (not to mention gravity). As it is known, the Standard model can be divided into three sectors: the calibration sector, the fragrance sector, and the symmetry-breaking sector. While the first two sectors are being actively studied in accelerator experiments (LEP, SLD, BELLE, etc.), the sector of spontaneous symmetry breaking is now attracting close attention, as it may give clear hints of existence in New Physics experiments beyond the Standard model. The observation of a static electric dipole moment of a many-electron atom which violates parity, P, and time reversal, T, symmetry, represents a great fundamental interest in a search of these hints. The detailed review of these topics can be found in Refs. [1–89].

Atomic optical and Stark pumping PNC measurements have been fulfilled in a whole number of heavy atoms, namely, in caesium (0.35% accuracy [1]), thallium (1.7%), bismuth (2%), Pb (1.2%) etc. The atomic optical tests of the Standard model provide important constraints on possible extensions of the SM. A recent analysis [2] of parity-violating electron-nucleus scattering measurements combined with atomic PNC measurements placed tight constraints on the weak neutral-current lepton-quark interactions at low energy, improving the lower bound on the scale of relevant new physics to ~TeV. The precise measurement of the PNC amplitudes in Cs [1] led to an experimental value of the small contribution from the nuclear-spin dependent PNC accurate to 14%. So, form the one side there is very actual necessity of the further development and increasing of the theoretical approaches accuracy and carrying out new atomic optical and Stark pumping PNC experiments.

The different methods have been used in calculation of the hyperfine structure parameters, PNC effect. The most popular multiconfiguration Dirac-Fock (MCDF) method for calculating parity and time reversal symmetry violations in many-electron atoms is often, however it application requires some additional generalizations [3, 17, 26]. Among other well-known calculation methods, a relativistic many-body perturbation theory (RMBPT), namely, the PT with relativistic Hartree-Fock (RHF) and Dirac-Fock (DF) zeroth approximations, the relativistic all-order method, QED perturbation theory (PT) etc. should be mentioned (e.g. [9–69]).

In present paper we present the results of application of the consistent theoretical approach, namely, the nuclear-relativistic many-body perturbation theory (N-RMBPT), to study the hyperfine and electroweak interaction parameters in the heavy finite Fermi systems and PNC effect. The N-RMBPT formalism is based on the combining ab initio perturbation theory formalism for electron subsystem, nuclear relativistic middle-field model for nuclear subsystem and an energy approach for computing radiation transition amplitude. It allows to fulfil computing the PNC amplitudes in the finite Fermi systems (atomic systems) [3, 10, 40, 41, 48–53, 90–99]. The important feature is the correct accounting for the inter electron correlations, nuclear, Breit and QED radiative corrections. All correlation corrections of the second order and dominated classes of the higher orders diagrams are taken into account. The results of calculation of the hyperfine structure parameters, the PNC amplitudes, the nuclear spin dependent corrections to the PNC, a weak charge Q_W for different atomic systems are presented and compared with available data in the literature.

2 Relativistic Nuclear-RMBPT Formalism in Theory of Heavy Finite Fermi Systems

Here we present a brief description of the key moments of our approach (more details can be found in Refs. [3, 4, 10, 50, 63–96]). The wave electron functions zeroth basis is found from the Dirac equation solution with potential, which includes the self-consistent ab initio potential (in the Dirac-Kohn-Sham approximation), electric, polarization potentials of a nucleus. All correlation corrections of the second and high orders of PT (electrons screening, particle-hole interaction etc.) are accounted for.

The concrete model for nuclear subsystem is based on the relativistic mean-field model for the ground-state calculation of the nucleus, which was developed as a renormalizable meson-field theory for nuclear matter and finite nuclei. The realization of nonlinear self-interactions of the scalar meson led to a quantitative description of nuclear ground states. As a self-consistent mean-field model (for a comprehensive review see Ref. [36, 99]), its ansatz is a Lagrangian or Hamiltonian that incorporates the effective, in-medium nucleon-nucleon interaction. As a Kohn-Sham scheme, the relativistic mean-field model can incorporate certain ground-state correlations and yields a ground-state description beyond the literal mean-field picture. As indicated in Refs. [36, 37] the strong attractive scalar (*S*: −400 MeV) and repulsive vector (*V*: +350 MeV) fields provide both the binding mechanism (*S* + *V*: −50 MeV) and the strong spin-orbit force (*S*–*V*: −750 MeV) of both right sign and magnitude. In our opinion, the most preferable one for the class of problems under consideration is so called NL3-NLC version (see details in Refs. [3, 60, 68]), which are among the most successful parameterizations available.

Let us consider the procedure of computing the PNC transition amplitude. The dominative contribution to the PNC amplitude is provided by the spin-independent part of the operator for a weak interaction, which should be added to the atomic Hamiltonian [3]:

$$H = H_{at} + \mu \sum_j H_W(j), \tag{1}$$

$$H_W^1 = \frac{G}{2\sqrt{2}} Q_W \gamma_5 \rho(r). \tag{2}$$

Here $G_F = g^2/4\sqrt{2}m_W^2$ is the Fermi constant of the weak interaction, γ_5—is the Dirac matrix, $\rho(r)$ is a density of the charge distribution in a nucleus and Q_W is a weak charge of a nucleus, linked with number of neutrons N and protons Z and the Weinberg angle θ_W in the Standard model (c.f. [2, 5]):

$$Q_W = Z(1 - 4\sin^2\theta_W) - N, \tag{3}$$

with accounting for the radiative corrections, Eq. (2) can be rewritten as [1, 2]:

$$\begin{aligned} Q_W &= \{Z(1 - [4.012 \pm 0.010]\sin^2\theta_W) - N\} \cdot (0.9857 \pm 0.0004)(1 + 0.0078T) \\ \sin^2\theta_W &= 0.2323 + 0.00365S - 0.00261T). \end{aligned} \tag{4}$$

The parameters S, T parameterize the looped corrections in the terms of conservation (S) and violation (T) of an isospin.

The spin-dependent contribution to the PNC amplitude has a few distinct sources: the nuclear anapole moment (that is considered as an electromagnetic characteristics of system, where the PNC takes a place; generally speaking, speech is about the arisen spin structure and the magnetic field distribution is similar to the solenoid field), the Z-boson exchange interaction from nucleon axial-vector currents (A_nV_e), and the combined action of the hyperfine interaction and spin-independent Z-boson exchange from nucleon vector (V_nA_e) currents (e.g. [3, 10, 28]).

The above-mentioned interactions can be represented by the Hamiltonian

$$H_W^i = \frac{G}{\sqrt{2}}k_i(\alpha \cdot I)\rho(r), \tag{5}$$

where $k(i = a)$ is an anapole contribution, $k(i = 2) = k_{Z0}$—axial-vector contribution, $k(i = kh) = k_{Qw}$ is a contribution due to the combined action of the hyperfine interaction and spin-independent Z exchange. It is well known that the contribution into a PNC amplitude, provided by the anapole moment term, significantly dominates.

The estimate of the corresponding matrix elements is in fact reduced to the calculation of the following integrals [10]:

$$< i|H_W^1|j >= i\frac{G}{2\sqrt{2}}Q_W\delta_{k_i - k_j}\delta_{m_i m_j}\int_0^\infty dr[F_i(r)G_j(r) - G_i(r)F_j(r)]\rho(r). \tag{6}$$

The reduced matrix element is as follows:

$$< i\|H_W^1\|j >= i\frac{G}{2\sqrt{2}}Q_W\int_0^\infty dr[F_i(r)G_j(r) - G_i(r)F_j(r)]\rho(r). \tag{7}$$

Further the general expression for the corresponding PNC amplitude for a-b transition is written as follows:

$$< a|PNC|b >= -\sum_{n} [\frac{< b\left|e\alpha_{\nu} A^{\nu}\right|n >< n\left|H_{W}^{(1)}\right|a >}{\varepsilon_{a}-\varepsilon_{n}} + \frac{< b\left|H_{W}^{(1)}\right|n >< n\left|e\alpha_{\nu} A^{\nu}\right|a >}{\varepsilon_{b}-\varepsilon_{n}}. \tag{8}$$

The corresponding spin-dependent PNC contribution is:

$$< a|PNC|b >^{sd}= k_{a} < a|PNC|b >^{(a)} + k_{2} < a|PNC|b >^{(2)} + k_{hf} < a|PNC|b >^{(hf)}, \tag{9}$$

where

$$\begin{aligned}
< a|PNC|b >^{(hf)} = & \sum_{\substack{m\neq a\\ n\neq a}} \frac{< a\left|H_{W}^{(1)}\right|n >< n\left|H_{W}^{(hf)}\right|m >< m|e\alpha_{\nu}A^{\nu}|b >}{(\varepsilon_{a}-\varepsilon_{m})(\varepsilon_{a}-\varepsilon_{n})} \\
& + \sum_{\substack{m\neq a\\ n\neq a}} \frac{< a\left|H_{W}^{(hf)}\right|n >< n\left|H_{W}^{(1)}\right|m >< m|e\alpha_{\nu}A^{\nu}|b >}{(\varepsilon_{a}-\varepsilon_{m})(\varepsilon_{a}-\varepsilon_{n})} \\
& + \sum_{\substack{m\neq a\\ n\neq b}} \frac{< a\left|H_{W}^{(1)}\right|m >< m|e\alpha_{\nu}A^{\nu}|n >< n\left|H_{W}^{(hf)}\right|m >}{(\varepsilon_{a}-\varepsilon_{m})(\varepsilon_{b}-\varepsilon_{n})} \\
& + \sum_{\substack{m\neq a\\ n\neq b}} \frac{< a\left|H_{W}^{(hf)}\right|m >< m|e\alpha_{\nu}A^{\nu}|n >< n\left|H_{W}^{(hf)}\right|b >}{(\varepsilon_{a}-\varepsilon_{m})(\varepsilon_{b}-\varepsilon_{n})} \\
& + \sum_{\substack{m\neq b\\ n\neq b}} \frac{< a|e\alpha_{\nu}A^{\nu}|n >< n\left|H_{W}^{(1)}\right|m >< m\left|H_{W}^{(hf)}\right|b >}{(\varepsilon_{b}-\varepsilon_{m})(\varepsilon_{b}-\varepsilon_{n})} \\
& + \sum_{\substack{m\neq b\\ n\neq b}} \frac{< a|e\alpha_{\nu}A^{\nu}|n >< n\left|H_{W}^{(hf)}\right|m >< m\left|H_{W}^{(1)}\right|b >}{(\varepsilon_{b}-\varepsilon_{m})(\varepsilon_{b}-\varepsilon_{n})} \\
& - < a\left|H_{W}^{(hf)}\right|a > \sum_{m\neq a} \frac{< a\left|H_{W}^{(1)}\right|m >< m|e\alpha_{\nu}A^{\nu}|b >}{(\varepsilon_{a}-\varepsilon_{m})^{2}} \\
& - \sum_{n\neq b} \frac{< a|e\alpha_{\nu}A^{\nu}|n >< n\left|H_{W}^{(1)}\right|b >}{(\varepsilon_{b}-\varepsilon_{n})^{2}} < b\left|H_{W}^{(hf)}\right|b >.
\end{aligned} \tag{10}$$

Here the following notations are used: $|a >=|aIF_{F}M_{F} >$, $|b >=|bIF_{I}M_{I} >$, I spin of a nucleus, $F_{I,F}$ is a total momentum of an atom and M its z component (I, F are the initial and final states). It should be noted the expressions for the matrix elements $< a|PNC|b >^{(a)}$, $< a|PNC|b >^{(2)}$ are similar to Eq. (10). The full description of the corresponding matrix elements and other details of the general method and PC code are presented in Refs. [3, 4, 10, 50, 63–98].

The fundamentals of the RMBPT formalism are presented previously in details in Refs. [40–49] and here we mention only the key points. The RMBPT formalism includes the optimized Dirac-Kohn-Sham (DKS) zeroth approximation and allows to provide an effective taking the relativistic, exchange-correlation, nuclear, radiative effects into account. The relativistic electron wave functions are determined from solution of the relativistic Dirac equation with a general potential. The latter includes ab initio mean-field potential, electric, polarization potentials of a nucleus. There have been considered all correlation corrections of the second order and dominated classes of the higher orders diagrams (electrons screening, mass operator iterations etc.).

A multielectron system is described by the relativistic Dirac Hamiltonian (the atomic units are used) as follows [3, 4]:

$$H = \sum_{i} \{\alpha c p_i - \beta c^2 - Z/r_i\} + \sum_{i>j} \exp(i|\omega| r_{ij})(1 - \alpha_i \alpha_j)/r_{ij}, \qquad (11)$$

where Z is a charge of nucleus, α_i,α_j are the Dirac matrices, ω_{ij} is the transition frequency, c—the velocity of light. The interelectron interaction potential second term in (3)) takes into account the retarding effect and magnetic interaction in the lowest order on parameter of the fine structure constant α^2 (α is the fine structure constant). The mean-field self-consistent potential in the zeroth–order Hamiltonian is as follows:

$$V_{MF} = V^{DKS}(r) = [V^{D}_{Coul}(r) + V_X(r) + V_C(r|b)], \qquad (12)$$

with the standard Coulomb-like potential $V^{D}_{Coul}(r)$, is the Kohn-Sham exchange potential $V_X(r)$ [38]:

$$V_X[\rho(r), r] = V_X^{KS}(r) \cdot \left\{ \frac{3}{2} \ln \frac{[\beta + (\beta^2 + 1)^{1/2}]}{\beta(\beta^2 + 1)^{1/2}} - \frac{1}{2} \right\}, \qquad (13a)$$

$$\beta = [3\pi^2 \rho(r)]^{1/3}/c, \qquad (13b)$$

and a correlation functional $V_C(r|b)$, taken in the Lundqvist-Gunnarsson form [4] with ab intio optimization parameter b (for details, see below and Refs. [41, 54–59]). The approach includes a generalized procedure (based on an relativistic energy approach) of generating the optimal basis set of relativistic electron wave functions with performance of the gauge invariance principle. To reach the latter we focus on accurate consideration of the QED PT fourth order (a second order of the atomic perturbation theory) Feynman diagrams, whose contribution into imaginary part of radiation width Im δE for the multi-electron ions accounts for multi-body correlation effects. This value is considered to be representative for the correlation effects, whose minimization is a reasonable criterion in the searching for the optimal one-electron basis of the many-body PT. A minimization of the functional ImδE leads to integral-differential Dirac-Kohn-Sham-like density functional equations. The magnetic inter-electron interaction is accounted for in the lowest order on α^2 (α is the fine structure constant) parameter.

The Coulomb-like potential of a nucleus (for the spherically symmetric nuclear density $\rho(r|R)$) is determined as follows:

$$V_{nucl}(r|R) = -(1/r)\int_0^r dr' r'^2 \rho(r'|R) + \int_r^\infty dr' r' \rho(r'|R). \tag{14}$$

To take into account the radiation (QED) corrections, we used the procedures, described detail in Refs. [40–49, 81–86]. A method for calculation of the self-energy part of the Lamb shift is based on an idea by Ivanov-Ivanova et al. (e.g. [39]), which generalizes the known hydrogen-like method by Mohr [34] and radiation model potential method by Flambaum-Ginges [35] (look details in Refs. [3, 10, 60, 68, 81–86]). According to Ref. [39], the radiative shift and the relativistic part of energy in an atomic system are, in principle, defined by one and the same physical field. One could suppose that there exists some universal function that connects the self-energy correction and the relativistic energy. It is worth to note that the low-energy part of the Lamb shift is determined by the following expression:

$$E_H(\Lambda) = \mathrm{Re}\frac{1}{\pi Z}\int_0^\infty d\xi[E(\xi,0) - E(\xi,\Lambda)], \tag{15a}$$

$$E_H(\xi,\Lambda) = \iint dr_1 dr_2 \frac{1}{r_{12}} \exp\left[(E_0 - i\xi)^2 - \Lambda^2\right]^{1/2} \Psi^+(r_2)\alpha^\mu G(r_1 r_2)\alpha^\mu \Psi(r_1), \tag{15b}$$

(here $\Psi(r)$ is the Dirac function, an energy parameter $E = i\xi$ is imaginary, G is the complex Green's function) and calculated by means of the complex Green function method in version [89]. The important radiation contributions are given by the standard Uehling-Serber term and the Källen-Sabry and Wichmann-Kroll corrections of higher orders (such as $[\alpha(Z\alpha)]^n$ (n = 2, ...), $\alpha^2(\alpha Z)$, $\alpha(Z\alpha)^n$ (n = 3) etc.; α is the fine structure constant). In order to take into consideration the effect of the vacuum polarization in the first PT order the generalized Uehling-Serber potential is used and modified to account for the high-order radiative corrections according to the procedure [3]. It is written as follows:

$$U(r) = -\frac{2\alpha}{3\pi r}\int_1^\infty dt\, \exp(-2rt/\alpha Z)\left(1 + 1/2t^2\right)\frac{\sqrt{t^2-1}}{t^2} \equiv -\frac{2\alpha}{3\pi r}C(g), \tag{16}$$

where $g = r/(\alpha Z)$. A More correct and consistent approach is presented in Refs. [3, 10, 60, 68, 81–86]. Taking into account the nuclear finite size effect modifies the potential (16) as follows [3]:

$$U^{FS}(r) = -\frac{2\alpha^2}{3\pi}\int d^3r' \int_1^{\infty} dt \exp(-2t|r - r'|/\alpha Z) \times \left(1 + \frac{1}{2t^2}\right)\frac{\sqrt{t^2 - 1}}{t^2}\frac{\rho(r')}{|r - r'|}. \quad (17)$$

Other details of the general method and PC code are described in Refs. [3, 4, 10, 50, 53–102]. All calculations are performed with using the numeral codes SuperAtom (Nucleus) (modified versions 93).

3 Results and Conclusions

As the first illustration (the test), we consider ^{133}Cs and present the results (Table 1) of calculation of the hyperfine structure (hfs) parameters for Cs. In Table 1 the experimental (A^{Exp}) and our (A^{N-Qed}) data for magnetic dipole constant A (MHz) for valent states of ^{133}Cs ($I = 7/2$, $g_i = 0.7377208$) are presented. The calculation results within standard (A^{RHF}) RHF and RHF with accounting for the second and higher PT corrections, the MCDF approximation and QED formalism are given too (from Refs. [3, 22–33, 62, 68]). The following notations are used: A^{RCC}—calculation by relativistic cluster-coupled (RCC) method; A^{DF}—DF method; A^{RHF}—RHF method and A^{QED}—the QED calculation; A^{N-Qed} is the result of this work. The key quantitative factor of physically reasonable agreement between theory and experimental data is connected with the correct accounting for the inter electron correlations, nuclear, Breit and QED radiative corrections.

In Table 2 the PNC amplitudes (in units of 10^{-11}iea_B($-Q_W$)/N) are listed and calculated on the basis of the different methods (without the Breit corrections): DF, RHF, MCDF, MBPT and nuclear-RMBPT results (data from Refs. [3, 22–33, 62, 68]).

In Table 3 we present the Breit correction (in units of 10^{-11}($-Q_W$)/N) to the PNC amplitude, which are calculated by the different methods (without the Breit corrections): DF, RHF, MCDF, and nuclear-RMBPT (data from Refs. [3, 22–33, 62, 68]). Let us note that the radiative corrections to the PNC amplitude, provided by the

Table 1 The values (MHZ) of the hfs constant A for valent states of ^{133}Cs: A^{Exp}—experiment; A^{RHF}, dA^{RHF}—RHF calculation plus the second and higher PT orders contribution [A^{QED}—data]; A^{N-Qed}—this work

State	A^{MCDF}	A^{RHF}	$A^{RHF} + dA$	A^{Qed}	A^{N-Qed}	A^{Exp}
$6s_{1/2}$	1736.9	1426.81	2291.00	2294.45	2296.78	2298.16(13)
$6p_{1/2}$	209.6	161.09	292.67	292.102	292.118	291.90(13)

Table 2 PNC amplitudes (in units of 10^{-11}iea_B(-Q_W)/N), which are calculated by different methods (without the Breit corrections): DF, RHF, MCDF, MBPT and nuclear-QED PT

Atom trans.	Spin of Nucl.	Nucl. Moment μ_N	Radius (fm) of nucleus	DF	RHF	MCDF	MBPT	N-RMBPT
^{85}Rb 5s–6s	5/2	1.3534	4.246	−0.110	−0.138	−0.134	−0.135	−0.132
^{133}Cs 6s–7s	7/2	2.5826	4.837	−0.741	−0.926 −0.897	−0.904	−0.906 −0.908	−0.903
^{223}Fr 7s–8s	3/2	1.1703	5.640	−13.72	−16.63	−15.72	−15.56 −15.80	−15.54
^{211}Fr 7s–8s	9/2	4.0032	5.539	−12.51	−15.16	−14.34	–	−14.17

Table 3 The Breit correction (in units of 10^{-11}(−Q_W)/N) to the PNC amplitude, which are calculated by the different methods (without the Breit corrections): DF, RHF, MCDF, and nuclear-RMBPT results (other data from Refs. [25, 29, 33])

Atom: transition.	DF	RHF	MCDF	N-QED PT
^{133}Cs 6s–7s	0.0022	0.0018	0.0045	0.0049
^{223}Fr 7s–8s	0.0640	0.0650	0.1430	0.1703

vacuum-polarization (VP) effect and the self-energy (SE) part are as follows: E_{PNC} – ^{133}Cs – VP = 0.38%, SE = −0.74%; ^{223}Fr – VP = 1.025%, SE = −1.35%.

In Table 4 we list the nuclear spin dependent corrections to PNC (^{133}Cs: 6s−7s) amplitude, calculated by different theoretical methods (in units of the $k_{a,2,hf}$ coefficient): MBPT, DF–PT, the shell model, N-RMBPT (from Refs. [3, 22–33, 62, 68]).

In Table 5 we present the estimated values of the weak charge Q_W for different heavy atoms, predicted in different approaches and determined in the Standard model (SM) (from Refs. [1–7, 22–33, 62, 68]). The analysis of results shows that in

Table 4 The nuclear spin-dependent corrections to the PNC ^{133}Cs: 6s-7s amplitude E_{PNC}, calculated by different methods (in units of $k_{a,2,hf}$ coeff.): MBPT, DF-PT, shell model, N-RMBPT (see text)

Correction	MBPT	Shell model	DF	N-RMBPT
K (sum)	0.1169	0.1118	0.112	0.1159
k_2—the Z-boson exchange interaction from nucleon axial-vector currents ($A_n V_e$)	0.0140	0.0140	0.0111 0.0084	0.0138
k_{hf}—the combined action of the hyperfine interaction and spin-independent Z exchange	0.0049	0.0078	0.0071 0.0078	0.0067
k_a—anapole moment	0.0980	0.090	0.0920	0.0954

Table 5 The estimated values of the weak charge Q_W and final PNC amplitudes (in units $10^{-11} iea_B(-Q_W)/N$) for different heavy atoms, predicted in different approaches

Contribution	$E_{PNC}Q_W$	N-RMBPT	MCDF	MBPT-DF	MCDF-QED	RHF+Breit+Correlation	RCC
^{85}Rb 5s–6s	E_{PNC}	−0.1318	−0.135	–	–	−0.134	–
^{133}Cs 6s–7s	E_{PNC}	−0.8985	−0.935 −0.905	−0.897 −0.904	−0.8981 −0.9055	−0.898 −0.910 −0.902	−0.9054 −0.899
^{133}Cs 6s–7s	Q_W	−72.62	−69.78 −71.09	−72.69 −72.18	−72.65 −72.06	−72.66 −71.70 −72.42	−72.06 −72.58
$^{137}Ba^+$ 6s–5d$_{3/2}$	E_{PNC}	−2.385	–	−2.35	–	−2.34	−2.46
^{173}Yb	E_{PNC}	−97.07	–	–	–	–	–
$6s^{2\,1}S_0$–$5d6s^3D_1$	Q_W	−92.31	–	–	–	–	–
^{205}Tl 6p$_{1/2}$–6p$_{3/2}$	E_{PNC}	26.5114	−26.75	−26.5	–	–	–
^{205}Tl 6p$_{1/2}$–6p$_{3/2}$	Q_W	−116.55	−112.4	−116.2 −116.7	–	–	–

principle a majority of theoretical approaches provide physically reasonable agreement with the Standard model data, but the important question is how much exact this agreement is. Some received data on estimating these constants directly indicate the necessity of new adequate précised experiments. The rare-earth elements (and corresponding multicharged ions), in particular, ytterbium, are especially interesting as they have very complicated spectra of energy levels with very unusual behavior in relatively weak electric and laser fields. In our opinion, particular attention should be paid to the ^{173}Yb ytterbium atom, where the theoretical PNC values of the EPNC amplitude differ from similar values of all considered heavy alkaline atoms by almost two orders of magnitude, which makes this atom particularly important in terms of studying the weak electron-nuclear interaction, the PNC effect, and of course, the Standard model check. Excessive complexity of the ^{173}Yb calculation, where the correlation effects corrections (including quick "blurring" of the initial state over an infinite set of additional configurations and other effects) is very large, making it difficult to obtain data on the fundamental parameters of Yb. Using the experimental value $\Delta E_1^{PNC}/\beta$ [11]: $(8.7\pm1.4)10^{-10}ea_B$ $\left(\Delta E_1^{PNC}/\beta = 39\text{mV/cm}\right)$ and the calculated atomic constant value of $99.707\cdot10^{-10}ea_B$ (for 173Yb; Z = 70, N = 103) it is not difficult to determine the value of a weak charge $Q_W = -92.31$, which is different from Q_W (the Stanard model) = −95.44. This circumstance imposes unambiguous restrictions on the fundamental values of S, T. It is interesting to note that the estimate of difference [QW (theoretical)—QW (CM)] ~ 6 indicated in [11] in our opinion, is a little overestimated due to the neglect of the contribution of QED, neutron skin effects etc. Perhaps the increase in the PNC effect at ^{173}Yb can be explained qualitatively and quantitatively in terms of a quantum chaos theory and strong inter-electron correlations (e.g. [3, 98]). In any case it is worth noting the sensitivity of PNC experiments to New Physics at energies, which even today are difficult to reach on modern colliders, including the restrictions on the mass of the Z 'boson and the mixing angle in models beyond the Standard model. The analysis shows that the perspectives of the PNC experiments with Stark pumping of the individual states in the rare-earth atoms (and probably more effective multicharged ions of these elements) and simultaneously polarized laser field dressing (with a cold-atom fountain or interferometer) may provide comfortable conditions for precise observation of weak effects.

References

1. Grojean C (2007) New approaches to electroweak symmetry breaking. Physics-Uspekhi 50:3–42; Review of Particle Properties (1996) Particle Physics Booklet. AIP. July, 1996
2. Shabalin EP (2001) What future for CP- and T-violation studies and CPT-invariance tests? Phys Usp 171:951–976
3. Khetselius OY (2011) Quantum structure of electroweak interaction in heavy finite Fermi systems. Astroprint, Odessa
4. Glushkov AV (2008) Relativistic quantum theory. Quantum, mechanics of atomic systems. Astroprint, Odessa
5. Khriplovich IB (1991) Parity Nonconservation in Atomic Phenomena. Gordon and Breach, Philadelphia

6. Grojean C (2007) New approaches to electroweak symmetry breaking. Phys Usp 177:3–42
7. Khriplovich IB (1997) Discovery of anapole moment. Phys Usp 167:1214–1216
8. Auerbach N (2008) Search for electric dipole moments in atoms of radioactive nuclei. J Phys G: Nucl Part Phys 35:P014040
9. Safronova MS, Rupsi P, Jiang D et al (2009) New directions in atomic PNC. Nucl Phys A 827:411–413
10. Khetselius OY (2008) Hyperfine structure of atomic spectra. Astroprint, Odessa
11. Tsigutkin K, Dounas-Frazer D, Family A, Stalnaker JE, Yashchuk VV, Budker D (2009) Observation of a large atomic parity violation effect in Ytterbium. Phys Rev Lett 103:071601
12. Nagasawa T, Haga A, Nakano M (2004) Hyperfine splitting of hydrogenlike atoms based on relativistic mean field theory. Phys Rev C 69:034322
13. Bouchiat MA (2007) Linear stark effect in dressed atoms as a signal to measure a nuclear anapole moment with a cold-atom fountain or interferometer. Phys Rev Lett 98:043003
14. Tomaselli M, Kuhl T, Seelig P, Holbrow C, Kankeleit E (1998) Hyperfine splittings of hydrogenlike ions and the dynamic-correlation model for one-hole nuclei. Phys Rev C 58(3):1524–1534
15. Tomaselli M, Schneider SM, Kankeleit E, Kühl T (1995) Ground state magnetization of ^{209}Bi in a dynamic-correlation model. Phys Rev C 51(6):2989–2997
16. Labzowsky LN, Johnson WR, Soff G, Schneider SM (1995) Dynamic proton model for the hyperfine structure of H-like ion $^{209}Bi^{+82}$. Phys Rev A 51(6):4597–4602
17. Kozlov MG, Porsev SG (2010) Parity non-conservation in thallium arXiv:physics/0105090.v1
18. Porsev SG, Rakhlina YG, Kozlov MG (1995) Non-conservation of spatial parity in atomic ytterbium. JETP Lett 61:449–453
19. Dzuba VA, Flambaum VV, Sushkov OP (1997) Polarizabilities and parity nonconservation in the Cs atom and limits on the deviation from the standard electroweak model. Phys Rev A 56:R4357–4360
20. Safronova MS, Johnson WR, Derevianko A (1999) Relativistic many-body calculations of energy levels, hyperfine constants, electric-dipole matrix elements, and static polarizabilities for alkali-metal atoms. Phys Rev A 60:044103
21. Bennett SC, Roberts JL, Wieman CE (1999) Measurement of the dc Stark shift of the 6s–7s transition in atomic caesium. Phys Rev A 59:3254–3260
22. Dzuba VA, Flambaum VV (2000) Off-diagonal hyperfine interaction and parity nonconservation in cesium. Phys Rev A 62:052101
23. Johnson WR, Bednyakov I, Soff G (2001) Vacuum-polarization corrections to the parity-nonconserving 6s–7s transition amplitude in ^{133}Cs. Phys Rev Lett 87:233001
24. Dzuba VA, Harabati C, Johnson WR, Safronova MS (2001) Breit correction to the parity-nonconservation amplitude in cesium. Phys Rev A 63:044103
25. Kozlov MG, Porsev SG, Tupitsyn II (2001) High-Accuracy calculation of 6s–7s parity-nonconserving amplitude in Cs. Phys Rev Lett 86:3260–3263
26. Johnson WR, Sapistein J, Blundell SA (1993) Atomic structure calculations associated with PNC experiments in atomic caesium. Phys Scripta 46:184–192
27. Vasilyev AA, Savukov IM, Safronova MS, Berry HG (2002) Measurement of 6s–7p transition probabilities in atomic Cs and revised value for the weak charge Q_W. Phys Rev A 66:020101(R)
28. Johnson WR, Safronova MS, Safronova UI (2003) Combined effect of coherent Z exchange and hyperfine interaction in parity-nonconserving interaction. Phys Rev A 67:062106
29. Shabaev VM, Tupitsyn II, Pachucki K, Plunien G, Yerokhin V (2005) Radiative and correlation effects and parity-nonconserving transition amplitude in heavy alkali-metal atoms. Phys Rev A 72:062105
30. Safronova MS, Johnson WR, Safronova UI, Cowan TE (2006) Relativistic many-body theory calculation of the Stark-induced amplitude of the 6p–7p transition in thallium. Phys Rev A 74:022504
31. Dzuba VA, Flambaum VV, Safronova MS (2006) Breit interaction and parity-nonconservation in many-electron atoms. Phys Rev A 73:022112

32. Glushkov AV, Ambrosov SV, Loboda AV, Svinarenko AA, Chernyakova Yu, Khetselius OY (2004) QED calculation of the super heavy elements ions: energy levels, radiative corrections and hfs for different nuclear models. Nucl Phys A: Nucl and Hadr Phys 734:21–28
33. Kuchiev MY, Flambaum VV (2003) Radiative corrections to parity nonconservation in atoms and test of the standard model. J Phys B 36:R191–222
34. Mohr PJ (1982) Self-energy of the n = 2 states in a strong Coulomb field. Phys Rev A 26:2338–2354
35. Flambaum VV, Ginges JS (2005) Radiative potential and calculations of QED radiative corrections to energy levels and electromagnetic amplitudes in many-electron atoms. Phys Rev A 72:052115
36. Serot BD, Walecka JD (1986) Advances in nuclear physics. The relativistic nuclear many body problem, vol 16. Plenum Press, New York
37. Burvenich TJ, Evers J, Keitel CH (2006) Dynamic nuclear Stark shift in superintense laser fields. Phys Rev C 74:044601
38. Kohn W, Sham LJ (1965) Self-consistent equations including exchange and correlation effects. Phys Rev A 140:1133
39. Ivanov LN, Ivanova EP, Aglitsky EV (1988) Modern trends in the spectroscopy of multicharged ions. Phys Rep 166:315–388
40. Glushkov AV, Ambrosov SV, Loboda AV, Gurnitskaya EP, Khetselius OY (2005) QED calculation of heavy multicharged ions with account for correlation, radiative and nuclear effects. In: Julien J-P, Maruani J, Mayou D, Wilson S, Delgado-Barion G (eds) Recent Advances in the Theory of Chemical and Physical Systems. Series: Progress in Theoretical Chemistry and Physics, vol 15. Springer, Dordrecht, pp 285–299
41. Glushkov AV, Svinarenko AA, Ignatenko AV (2011) Spectroscopy of autoionization resonances in spectra of the lanthanides atoms. Photoelectronics 20:90–94
42. Glushkov AV, Khetselius OY, Dubrovskaya YV, Loboda AV (2006) Sensing the capture of negative muon by atoms: Energy approach. Sens Electron Microsyst Technol N3:31–35
43. Khetselius OY (2008) On possibility of sensing nuclei of the rare isotopes by means of laser spectroscopy of hyperfine structure. Sens Electron Microsyst Technol N3:28–33
44. Khetselius OY (2009) Relativistic calculating the hyperfine structure parameters for heavy-elements and laser detecting the isotopes and nuclear reaction products. Phys Scripta T 135:014023
45. Khetselius OY (2009) On sensing nuclei of the 207Bi & 207Pb isotopes by means of laser spectroscopy of hyperfine. Sens Electron Microsyst Technol N2:26–29
46. Khetselius OY (2009) Relativistic perturbation theory calculation of the hyperfine structure parameters for some heavy-element isotopes. Int J Quantum Chem 109:3330–3335
47. Khetselius OY (2012) Spectroscopy of cooperative electron-gamma-nuclear processes in heavy atoms: NEET effect. J Phys: Conf Ser 397:012012
48. Glushkov AV (2006) Relativistic and correlation effects in spectra of atomic systems. Astroprint, Odessa
49. Khetselius OY (2012) Relativistic energy approach to cooperative electron-γ-nuclear processes: NEET effect. In: Nishikawa K, Maruani J, Brändas E, Delgado-Barrio G, Piecuch P (eds) Quantum Systems in Chemistry and Physics: Progress in Methods and Applications. Series: Progress inTheoretical Chemistry and Physics, vol 26. Springer, Dordrecht, pp 217–229
50. Glushkov AV, Rusov VD, Ambrosov SV, Loboda AV (2003) Resonance states of compound super-heavy nucleus and EPPP in heavy nucleus collisions. In: Fazio G, Hanappe F (eds) New projects and new lines of research in nuclear physics. World Scientific, Singapore, pp 126–132
51. Glushkov AV (2005) Energy approach to resonance states of compound superheavy nucleus and EPPP in heavy nuclei collisions. In: Grzonka D, Czyzykiewicz R, Oelert W et al (eds) Low energy antiproton physics. AIP Conf Proc, New York, 796:206–210
52. Glushkov AV (2012) Spectroscopy of cooperative muon-gamma-nuclear processes: energy and spectral parameters. J Phys: Conf Ser 397:012011

53. Malinovskaya SV, Glushkov AV, Dubrovskaya YV, Vitavetskaya LA (2006) Quantum calculation of cooperative muon-nuclear processes: Discharge of metastable nuclei during negative muon capture. In: Julien J-P, Maruani J, Mayou D, Wilson S, Delgado-Barion G (eds) Recent Advances in the Theory of Chemical and Physical Systems. Series: Progress in Theoretical Chemistry and Physics, vol 15. Springer, Dordrecht, pp 301–307
54. Glushkov AV, Khetselius OY, Gurnitskaya EP, Loboda AV, Florko TA, Sukharev DE, Lovett L (2008) Gauge-invariant QED perturbation theory approach to calculating nuclear electric quadrupole moments, hyperfine structure constants for heavy atoms and ions. In: Wilson S, Grout PJ, Maruani J, Delgado-Barrio G, Piecuch P (eds) Frontiers in Quantum Systems in Chemistry and Physics. Series: Progress in Theoretical Chemistry and Physics, vol 18. Springer, Dordrecht, pp 507-524
55. Glushkov AV, Ivanov LN (1992) Radiation decay of atomic states: atomic residue polarization and gauge noninvariant contributions. Phys Lett A 170:33–36
56. Glushkov AV, Ivanov LN, Ivanova EP (1986) Radiation decay of atomic states. Generalized energy approach. In: Autoionization Phenomena in Atoms. Moscow State Univ., Moscow, p 58
57. Ivanova EP, Glushkov AV (1986) Theoretical investigation of spectra of multicharged ions of F-like and Ne-like isoelectronic sequences. J Quant Spectr Rad Transf 36:127–145
58. Khetselius OY (2015) Optimized Perturbation Theory for Calculating the Hyperfine Line Shift and Broadening of Heavy Atoms in a Buffer Gas. In: Nascimento M, Maruani J, Brändas E, Delgado-Barrio G (eds) Frontiers in Quantum Methods and Applications in Chemistry and Physics. Series: Progress in Theoretical Chemistry and Physics, vol 29. Springer, Cham, pp 55–76
59. Khetselius OY (2019) Optimized relativistic many-body perturbation theory calculation of wavelengths and oscillator strengths for Li-like multicharged ions. Advances in Quantum Chemistry, vol 78. Elsevier, Amsterdam, pp 223–251. https://doi.org/10.1016/bs.aiq.2018.06.001
60. Khetselius OY, Lopatkin YM, Dubrovskaya YV, Svinarenko AA (2010) Sensing hyperfine-structure, electroweak interaction and parity non-conservation effect in heavy atoms and nuclei: New nuclear-QED approach. Sens Electron Microsyst Technol 7(2):11–19
61. Khetselius OY (2009) Atomic parity non-conservation effect in heavy atoms and observing P and PT violation using NMR shift in a laser beam: to precise theory. J Phys: Conf Ser 194:022009
62. Khetselius OY (2010) Relativistic hyperfine structure spectral lines and atomic parity nonconservation effect in heavy atomic systems within QED theory. AIP Conf Proc 1290:29–33
63. Glushkov AV (1992) Oscillator strengths of Cs and Rb-like ions. J Appl Spectrosc 56(1):5–9
64. Glushkov AV (1990) Relativistic polarization potential of a many-electron atom. Sov Phys J 33(1):1–4
65. Khetselius OY, Zaichko PA, Smirnov AV, Buyadzhi VV, Ternovsky VB, Florko TA, Mansarliysky VF (2017) Relativistic many-body perturbation theory calculations of the hyperfine structure and oscillator strength parameters for some heavy element atoms and ions. In: Tadjer A, Pavlov R, Maruani J, Brändas E, Delgado-Barrio G (eds) Quantum Systems in Physics, Chemistry, and Biology. Advances in Concepts and Applications. Series: Progress in Theoretical Chemistry and Physics, vol 30. Springer, Cham, pp 271–281
66. Khetselius OY, Glushkov AV, Dubrovskaya YV, Chernyakova YG, Ignatenko AV, Serga IN, Vitavetskaya LA (2018) Relativistic quantum chemistry and spectroscopy of exotic atomic systems with accounting for strong interaction effects. In: Wang YA, Thachuk M, Krems R, Maruani J (eds) Concepts, Methods and Applications of Quantum Systems in Chemistry and Physics. Series: Progress in Theoretical Chemistry and Physics, vol 31. Springer, Cham, pp 71–91
67. Dubrovskaya YV, Khetselius OY, Vitavetskaya LA, Ternovsky VB, Serga IN (2019) Quantum chemistry and spectroscopy of pionic atomic systems with accounting for relativistic, radiative, and strong interaction effects. Advances in Quantum Chemistry, vol 78. Elsevier, Amsterdam, pp193–222. https://doi.org/10.1016/bs.aiq.2018.06.003

68. Khetselius OY, Glushkov AV, Gurskaya MY, Kuznetsova AA, Dubrovskaya YV, Serga IN, Vitavetskaya LA (2017) Computational modelling parity nonconservation and electroweak interaction effects in heavy atomic systems within the nuclear-relativistic many-body perturbation theory. J Phys: Conf Ser 905:012029
69. Glushkov AV, Khetselius OY, Svinarenko AA, Buyadzhi VV (2015) Spectroscopy of autoionization states of heavy atoms and multiply charged ions. TEC, Odessa
70. Glushkov AV, Khetselius OY, Svinarenko AA, Buyadzhi VV (2015) Methods of computational mathematics and mathematical physics. P.1. TES, Odessa
71. Glushkov AV, Malinovskaya SV, Filatov VV (1989) S-Matrix formalism calculation of atomic transition probabilities with inclusion of polarization effects. Sov Phys J 32(12):1010–1014
72. Glushkov AV, Butenko YV, Serbov NG, Ambrosov SV, Orlova VE, Orlov SV, Balan AK, Dormostuchenko GM (1996) Calculation and extrapolation of oscillator strengths in Rb-like, multiply charged ions. Russ Phys J 39(1):81–83
73. Glushkov AV (2012) Advanced relativistic energy approach to radiative decay processes in multielectron atoms and multicharged ions. In: Nishikawa K, Maruani J, Brändas E, Delgado-Barrio G, Piecuch P (eds) Quantum Systems in Chemistry and Physics: Progress in Methods and Applications. Series: Progress in Theoretical Chemistry and Physics, vol 26. Springer, Dordrecht, pp 231–252
74. Glushkov AV (2013) Operator perturbation theory for atomic systems in a strong DC electric field. In: Hotokka M, Brändas E, Maruani J, Delgado-Barrio G (eds) Advances in Quantum Methods and Applications in Chemistry, Physics, and Biology. Series: Progress in Theoretical Chemistry and Physics, vol 27 Springer, Cham, pp 161–177
75. Glushkov AV, Malinovskaya SV, Loboda AV, Shpinareva IM, Gurnitskaya EP, Korchevsky DA (2005) Diagnostics of the collisionally pumped plasma and search of the optimal plasma parameters of x-ray lasing: Calculation of electron-collision strengths and rate coefficients for Ne-like plasma. J Phys: Conf Ser 11:188–198
76. Glushkov AV, Malinovskaya SV, Chernyakova YuG, Svinarenko AA (2004) Cooperative Laser-electron-nuclear processes: QED calculation of electron satellites spectra for multi-charged ion in laser field. Int J Quant Chem 99(5):889–893
77. Glushkov AV, Ambrosov SV, Ignatenko AV, Korchevsky DA (2004) DC strong field stark effect for non-hydrogenic atoms: new consistent quantum mechanical approach. Int J Quant Chem 99(5):936–939
78. Glushkov AV, Malinovskaya SV, Loboda AV, Shpinareva IM, Prepelitsa GP (2006) Consistent quantum approach to new laser-electron-nuclear effects in diatomic molecules. J Phys: Conf 35:420–424
79. Malinovskaya SV, Dubrovskaya YV, Zelentzova TN (2004) The atomic chemical environment effect on the β decay probabilities: relativistic calculation. Herald of Kiev Nat Univ. Series: Phys Math 4:427–432
80. Glushkov AV (1994) Calculation of parameters of the interaction potential between excited alkali atoms and mercury atoms-the Cs-. Fr-Hg interaction. Opt. and Spectr 77(1):5–10
81. Glushkov AV, Gurskaya MYu, Ignatenko V, Smirnov AV, ISerga IN, Svinarenko AA, Ternovsky EV (2017) Computational code in atomic and nuclear quantum optics: advanced computing multiphoton resonance parameters for atoms in a strong laser field. J Phys: Conf Ser 905:012004
82. Svinarenko AA, Glushkov AV, Khetselius OY, Ternovsky VB, Dubrovskaya YuV, Kuznetsova AA, Buyadzhi VV (2017) Theoretical spectroscopy of rare-earth elements: spectra and autoionization resonance. In: Jose EA (ed) Rare earth element. InTech, Orjuela, pp 83–104. https://doi.org/10.5772/intechopen.69314
83. Glushkov AV, Khetselius OY, Svinarenko AA, Buyadzhi VV, Ternovsky VB, Kuznetsova AA, Bashkarev PG (2017) Relativistic perturbation theory formalism to computing spectra and radiation characteristics: application to heavy element. In: Dimo I (ed) Recent studies in perturbation theory. InTech, Uzunov, pp 131–150. https://doi.org/10.5772/intechopen.69102

84. Malinovskaya SV, Glushkov AV, Khetselius OY (2008) New Laser-Electron Nuclear Effects in the Nuclear γ Transition Spectra in Atomic and Molecular Systems. In: Wilson S, Grout P, Maruani J, Delgado-Barrio G, Piecuch P (eds) Frontiers in Quantum Systems in Chemistry and Physics. Series: Progress in Theoretical Chemistry and Physics, vol 18. Springer, Dordrecht, pp 525–541
85. Bystryantseva AN, Khetselius OY, Dubrovskaya YuV, Vitavetskaya LA, Berestenko AG (2016) Relativistic theory of spectra of heavy pionic atomic systems with account of strong pion-nuclear interaction effects: 93Nb, 173Yb, 181Ta, 197Au. Photoelectronics 25:56–61
86. Khetselius OY (2012) Quantum Geometry: New approach to quantization of quasistationary states of Dirac equation for superheavy ion and calculating hyperfine structure parameters. Proc Intern Geom Center 5(3–4):39–45
87. Glushkov AV, Khetselius OY, Svinarenko AA (2013) Theoretical spectroscopy of autoionization resonances in spectra of lanthanide atoms. Phys Scr T 153:014029
88. Malinovskaya SV, Glushkov AV, Khetselius OY, Svinarenko AA, Mischenko EV, Florko TA (2009) Optimized perturbation theory scheme for calculating the interatomic potentials and hyperfine lines shift for heavy atoms in the buffer inert gas. Int Journ Quant Chem 109:3325–3329
89. Glushkov AV, Malinovskaya SV, Khetselius OY, Loboda AV, Sukharev DE, Lovett L (2009) Green's function method in quantum chemistry: New numerical algorithm for the Dirac equation with complex energy and Fermi-model nuclear potential. Int Journ Quant Chem 109:1717–1727
90. Glushkov AV, Lovett L, Khetselius OY, Gurnitskaya EP, Dubrovskaya YuV, Loboda AV (2009) Generalized multiconfiguration model of decay of multipole giant resonances applied to analysis of reaction (μ-n) on the nucleus ^{40}Ca. Int Journ Modern Phys A 24(2–3):611–615
91. Glushkov AV, Khetselius OY, Svinarenko AA (2012) Relativistic theory of cooperative muon-γ -nuclear processes: Negative muon capture and metastable nucleus discharge. In: Hoggan P, Brändas E, Maruani J, Delgado-Barrio G, Piecuch P (eds) Advances in the Theory of Quantum Systems in Chemistry and Physics. Series: Progress in Theoretical Chemistry and Physics, vol 22. Springer, Dordrecht, pp 51–68
92. Glushkov A, Khetselius OY, Lovett L (2009) Electron-β-nuclear spectroscopy of atoms and molecules and chemical bond effect on the β-decay parameters. In: Piecuch P, Maruani J, Delgado-Barrio G, Wilson S (eds) Advances in the Theory of Atomic and Molecular Systems: Dynamics, Spectroscopy, Clusters, and Nnanostructures. Series: Progress in Theoretical Chemistry and Physics, vol 20. Springer, Dordrecht, pp 125–152
93. Glushkov AV, Khetselius OY, Lopatkin YM, Florko TA, Kovalenko OA, Mansarliysky VF (2014) Collisional shift of hyperfine line for rubidium in an atmosphere of the buffer inert gas. J of Phys: Conf Ser 548:012026
94. Malinovskaya SV, Dubrovskaya YuV, Vitavetskaya LA (2005) Advanced quantum mechanical calculation of the beta decay probabilities. Energy approach to resonance states of compound superheavy nucleus and EPPP in heavy nuclei collisions. In:Grzonka D, Czyzykiewicz R, Oelert W, Rozek T, Winter P (eds) Low energy antiproton physics. AIP Conference Proceedings, vol 796. AIP, New York, pp 201–205
95. Florko TA, Ambrosov SV, Svinarenko AA, Tkach TB (2012) Collisional shift of the heavy atoms hyperfine lines in an atmosphere of the inert gas. J Phys: Conf Ser 397:012037
96. Buyadzhi VV, Zaichko PA, Antoshkina OA, Kulakli TA, Prepelitsa GP, Ternovsky VB, Mansarliysky VF (2017) Computing of radiation parameters for atoms and multicharged ions within relativistic energy approach: Advanced Code. J Phys: Conf Ser 905(1): 012003
97. Svinarenko AA (2014) Study of spectra for lanthanides atoms with relativistic many- body perturbation theory: Rydberg resonances. J Phys: Conf Ser 548:012039
98. Glushkov AV (2012) Methods of a chaos theory. Astroprint, Odessa
99. Glushkov AV (2019) Multiphoton spectroscopy of atoms and nuclei in a laser field: Relativistic energy approach and radiation atomic lines moments method. Advances in Quantum Chemistry, vol 78. Elsevier, Amsterdam, pp 253-285. https://doi.org/10.1016/bs.aiq.2018.06.004

100. Chernyakova YuG, Vitavetskaya LA, Bashkaryov PG, Serga IN, Berestenko AG (2015) The radiative vacuum polarization contribution to the energy shift of some levels of the pionic hydrogen Photoelectronics 24:108–111
101. Kuznetsova AA, Vitavetskaya LA, Chernyakova YG, Korchevsky DA (2013) Calculating the radiative vacuum polarization contribution to the energy shift of 2p-2s transition in pionic deuterium. Photoelectronics 22:108–111
102. Furnstahl RJ (2004) Next generation relativistic models. In: Lalazissis GA, Ring P, Vretenar D (eds) Extended Density functionals in nuclear structure physics, vol 641. Springer, Berlin, pp 1–30

Clusters and Molecules Interactions

Quantum Study of Helium Clusters Doped with Electronically Excited Li, Na, K and Rb Atoms

David Dell'Angelo

Abstract I shall present a systematic study of the lowest electronically excited state of alkali (Li, Na, K, Rb) atoms in helium clusters with up to 15 helium atoms, for which a diatomics-in-molecule model combined with an Importance-Sampling Diffusion Monte Carlo (IS-DMC) technique to compute the vibrational ground state energies has been used. Solvation structures are examined via various density distributions. Results show a first solvation shell around the *n*p-orbital completed with $5 \leq n \leq 7$ depending on the alkali atom and in agreement with other theoretical works. Except for Li, a shift of one helium atom is observed between classical and quantum first filled shells. Dealing with the number of helium atoms in the second shell, for K and Rb a closure at eleven atoms can be easily concluded whereas number ranging from 9 to 11 for Li and 9 or 10 for Na are found. The planar structures of the $Ak^{\star}He_n$ obtained in the first shell are maintened up to the second shell and the radius of the second shells are twice as large as the inner ones.

Keywords Excited states · Quantum Monte Carlo · Alkali · Helium clusters · Doping · Liquid helium · Quantum dynamics · Molecular physics

1 Introduction

Helium clusters have become these last three decades [1, 2] the subject of extensive experimental and theoretical studies [3, 4]. The unique and particular properties of these cold (T = 370 mK for ^4He_N [5]) droplets permit one to study superfluidity [6] and chemical reaction at low temperatures [7, 8]. As they offer a weakly perturbing environment, helium nanodroplet isolation [1, 2, 9] (HENDI) has been established as a useful tool for spectroscopic studies of molecules and weakly bound complexes [1, 2, 10]. Dopants such as atoms and molecules can be added to these nanodroplets and most submerge into the liquid helium. Alkali metal atoms represent an exception.

D. Dell'Angelo (✉)
Department of Physics and Astronomy, University College of London,
Gower Street, London, WC1E 6BT, UK
e-mail: ucapdd0@ucl.ac.uk

L. Mammino et al. (eds.), *Advances in Quantum Systems in Chemistry, Physics, and Biology*, Progress in Theoretical Chemistry and Physics 32,
https://doi.org/10.1007/978-3-030-34941-7_5

In their electronic ground state are known to reside on the surface of the helium droplet [11]. Upon excitation, they can remain on the droplet, detach from it or go inside the droplet, depending on the helium interaction potential in the corresponding electronically excited state. Whitin the variety of possible dopants, alkali atoms appear to be one of the best candidates to study electronic excitations due to their relatively simple electronic configuration with a single electron in the valence orbitals and the optically accessible electronic transitions. Because of their simple, well known, absorption spectrum and because they introduce only weak perturbations, alkali atoms constitute a minimally perturbing probe capable of investigating the surface of a helium cluster. These alkalis features permit to address the question of solvation versus surface location for an impurity atom in liquid helium or how the finite size of a helium cluster affects the superfluidity [12]. Spectra of the D lines ($n\mathrm{P}_{(1/2,3/2)} \leftarrow n\mathrm{S}$,$n$ being the principal quantum number in the ground state) of alkali-metal atoms on the surface of superfluid helium nanodroplets obtained via Laser-induced-fluorescence (LIF) and beam-depletion (BD) spectroscopy have been measured for Li [13–16], Na [13–15, 17–19], K [13, 15, 18, 20, 21], Rb [15, 21–24], and Cs [21, 25–29]. Excitation spectra of alkalis on helium nanodroplets in the neighborhood of the free alkali D lines have been interpreted in terms of an excitation of the alkali s electrons into a p_σ or a p_π orbital, referring to the orbital angular momentum being perpendicular or parallel, respectively, to the helium nanodroplet surface. In fact, a peculiarity in the excitation of light alkali atoms on the surface of the helium droplet is the formation of alkali-helium exciplexes which form when one or more He atoms are attracted into the nodal plane of the excited p orbital [30]. Alkali-helium exciplexes are not formed when the p orbital is aligned perpendicularly to the helium surface (Σ excitation); aligning the p orbital parallel, as an increase in both attraction and repulsion leads to a bound-bound transition, mainly the Ak*He diatomic exciplex forms.

In bulk liquid helium, fluorescence of laser-excited light alkalis has not been found and the formation of exciplexes like Ak*He$_5$ has been proposed [20] as the cause of the quenching of fluorescence emission, because of the crossing between the excited and the ground state potential energy surface and the possibility of a decay via non-radiative transitions. In contrast to bulk helium, fluorescence spectra of light alkalis attached to helium droplets have been recorded upon $p \leftarrow s$ excitation [13]. When Ak$=$Rb ($5P_{(1/2,3/2)} \leftarrow 5S$ transition), as argued by Reho et al. [20] and experimentally verified through LIF excitation spectroscopy by Brühl et al. [22], a high barrier along the $5^2P_{1/2}$ Rb-He potential prohibiting the tunneling of a helium atom from the droplet toward the Rb atom prevents formation of the exciplex. For Rb on helium only the excitation of a He$_{n-1}$-HeRb $5^2P_{3/2}$ complex would yeld fluorescence emission from Rb*He while the barrier would prevent desorption of the diatomic in its $5^2P_{1/2}$ state within its fluorescence lifetime. The desorption of Rb* off the droplet [31] competes with the submersion of the Rb$^+$ ion into the droplet interior, because of the repulsive interaction of the droplet with the Rb atom in an excited state and the attractive interaction of the droplet with the Rb$^+$ [32]. The formation of Rb*He$_n$ exciplexes has been observed in pressurized liquid He [33] and explained as a consequence of the quenching of the Rb $5^2P_{1/2}$ state. Also in solid He, Rb atoms do not emit visible fluorescence whereas Cs atoms do. Moroshkin et al. [34] presented spectroscopic studies of $Cs^{\star}He_n$ exciplexes produced by laser excitation of Cs atoms isolated in a solid ^{4}He matrix. In these experiments the laser-excited $6P_{3/2}$ state is completely quenched and does not emit any fluorescence on the $6P_{3/2} \rightarrow 6S_{1/2}$ transition. The $Cs(6P_{3/2})He_{n=2}$ exciplex, formed directly from the $6P_{3/2}$ state, is quenched by a radiationless transformation into the larger $Cs(6P_{1/2})He_{n=6,7}$ complex through the attachment of additional He atoms. Exciplex formation from the $6P_{1/2}$ state proceeds via a strongly suppressed tunnelling transition.

Theoretical calculations for pure and atom- or molecule-doped helium clusters have been widely performed, using the quantum Monte Carlo (QMC) method [35, 36], the path integral Monte Carlo (PIMC) method [37–39], or density functional approaches [12, 32, 40–43]. They confirmed that alkali atoms in the ground electronic state are attached to the surface of helium clusters because of the weaker Ak-He interaction (0.7–1.5 cm^{-1}) with respect to the He-He bond. A deep description of the solvation process can be obtained by solving exactly the Schrödinger equation of nuclear motion using an explicit many-body algorithm. In systems where accurate interaction potentials between helium and the impurity are available, the quantum Monte Carlo QMC approach is probably the best suited, as the DFT approach does not take into account properly the discrete nature and the anisotropic deformation [44, 45] of the helium aggregates, and this might lead to overestimate the overall interaction energy with the impurity. Furthermore, we have to consider the strong dependence of the DFT on the choice of exchange-correlation functional which implies that the description of van der Waals bonding within DFT is unreliable, given the local nature of the approximations to the exchange-correlation functional. Quantum Monte Carlo (QMC) techniques are useful in this regard as they can provide a highly accurate description of electron correlation effects. Although QMC methods are computationally expensive, they can be applied to systems which are large enough to model condensed matter. This was shown for instance by Drummond and Needs on solid neon [46]. In their work, they showed that standard DFT methods do not describe van der Waals bonding accurately, but they might be expected to work quite well at high densities, where the short-range repulsion dominates. Alkali atoms are displaced by about 5–6 Å from the droplet surface and the helium surface near alkali atoms is slightly distorted. As reported by Ancilotto et al. [41], when an alkali atom is attached to a large He-cluster, the Ak atom deforms the surface of the cluster giving it a form of a dimple. The depth of the dimple increases with the mass of the alkali metal atom, and it is larger for ^{3}He than for ^{4}He. $Ak^{\star}He_n$ exciplexes with the alkali atom in the first excited state, consist of a ring of helium atoms located at the waist of the alkali atom p orbital. Theoretical studies predict different numbers of helium atoms in the ring. For Na, Dupont-Roc [47] found from five to six helium in the ring, Kanorsky et al. [48] five, while De-Toffol et al. [49] estimated at least eight helium atoms. Quantum Monte Carlo studies of $K^{\star}He_n$ [50] and of $Rb^{\star}He_n$ [51, 52] led to rings formed by six or seven, and seven helium atoms for K and Rb, respectively. Anyway, to the best of my knowledge, no detailed theoretical studies after this first ring-shaped structure have been conducted so far. As the number of helium atoms in these ring structures varies depending on the alkali species, I investigated helium solvation structures up to the second shell of $Ak^{\star}He_n$ clusters using Importance Sampling Diffusion Monte Carlo (IS-DMC) technique. Though DMC simulations cannot recover the temporal evolution of the sytem, they provide many important quantities that are hardly accessible experimentally, such as radial and angular distribution functions, solvation energies, excitation spectra, as well as their dependence on the size of the helium aggregate.

The inspection permitted to observe how the structures and the geometries of these clusters change moving down the alkali periodic group. The model used for the interaction potential of the full cluster is described in Sect. 2 together with the technical details of the quantum Monte Carlo approach. Detailed analyses of $Ak^{\star}He_n$ energetics and structural properties at the classical and quantum levels are presented in Sect. 3. Finally, Sect. 4 concludes.

2 Classical Analysis

2.1 The Dimer Potential Energy Curves

The interaction potential of alkali atoms in their ground electronic state attached to the surface of helium clusters [41] can be modeled quite accurately within the pair potential approximation. By contrast, for *p*-electronically excited alkali atoms the pair approximation is no longer valid due to the anisotropic character of the interaction. Like in previous works [51, 52], in which the description of the system has been detailed, I rely on the Diatomics-In-Molecule (DIM) [53] approximation to account for the anisotropy of $Ak^{\star}$ interacting with multiple helium atoms. After including the spin-orbit interaction assumed to be independent of the intermolecular distance, three global $Ak^{\star}He_n$ potential energy surfaces have been obtained from the two first excited adiabatic Ak-He one-dimensional potential energy curves by diagonalization of a 6×6 complex matrix. Pascale's curves [54, 55] have been used. The choice is motivated by the classical analysis extended by IS-DMC computations of energetics and structural properties of $Li^{\star}He_{n\leq5}$ and $Na^{\star}He_{n\leq5}$ clusters in the first electronic excited state carried out in our previous work [56]. In particular, it was shown that globally the results were independent of the diatomic curves used and the physical conclusions obtained with Pascale's curves were quite reasonable. Figure 1 shows the $^2\Pi_{1/2}$ adiabatic curve for $Ak^{\star}He$ dimers when Pascale's interaction potential together with the spin-orbit terms, listed in Table 1, are added. Most recently, curves for $Rb^{\star}He$ [57–59] and spin-orbit terms [60] have been calculated.

The shape of these curves results from the mixing of the attractive Π and repulsive Σ curves by the spin-orbit coupling term. All curves present two wells separated by a barrier. A systematic reduction of the depth of the well occuring at short distance is observed along the periodic table. This is due to the depth of the Π curves as well as to the increased mixing with the repulsive Σ state induced by growing the spin-orbit value. The cesium represents an extreme case in which the first well is a local

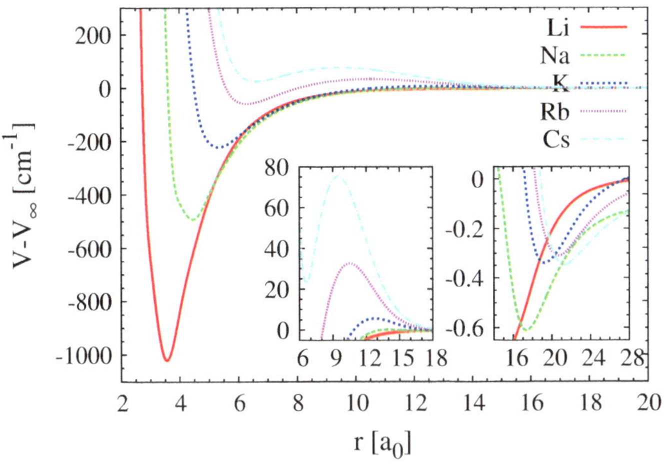

Fig. 1 [color online] $^2\Pi_{1/2}$ potential curves for Ak*He dimers using the curves of Pascale [54, 55]. The different alkali atoms are presented using color code. Two insets focussing on the barriers (left) and on the local minima (right) are also shown

Table 1 Atomic mass (most abundant isotope) and spin-orbit constant for the alkali atoms [61]. Masses are in amu. Positions [P] are in atomic units, spin-orbit costants and potential values in cm^{-1}

Element	Mass	Δ_{SO}	Barrier [P]	Global minimum [P]	Local minimum [P]
Li	7.016003	0.340	1.5×10^{-4} [47.0]	1022.8 [3.5]	0.2 [57.5]
Na	22.989760	17.196	0.2 [14.0]	494.1 [4.4]	0.5 [17.3]
K	38.963707	57.706	6.7 [12.7]	223.3 [5.4]	0.3 [19.3]
Rb	84.911794	237.595	32.4 [10.5]	60.7 [6.3]	0.3 [20.7]
Cs	132.905429	554.039	74.5 [9.6]	0.2 [21.3]	51.2 [6.6]

minimum. As pointed out by Dupont-Roc [47] and reported by Nettels et al. [62], the height of the potential barrier, which increases from Li to Cs, is determined by the strength of the spin-orbit interaction of the P state. If the spin-orbit coupling is weak compared to the alkali-helium interaction as in the case of the light alkalis, it can be neglected and the electronic configuration can be approximated by $P_{x,y,z}$ orbitals that allows the formation of dumbbell-shaped exciplexes with $n > 2$. If on the other hand the spin-orbit interaction dominates, as for Rb or Cs, one has to consider the electron distribution of the $L-S$-coupled $P_{1/2}$ state, which is spherical and hence repulsive. In the extreme case represented by Li, this barrier lays below the asymptote value. For Na, K and Rb the barrier is above the asymptote value and in order to form the quasibound complex, helium atom has to tunnel under the potential barrier. For Cs, the global minimum occurs at very long distance and is very shallow. No bound states are expected from such diatomic curve.

2.2 Larger Clusters

Given the complexity of the systems as n increases, I considered first the geometries corresponding to rings, expected structures for the exciplex. Theoretical works dealing with excited alkali atoms in liquid helium [47, 49], in solid helium [62–65], in ^{4}He cold gas [28, 66] as well as in clusters [50] agree on considering larger $Ak^{\star}He_{n\geq3}$ structures made up of a ring of n He atoms around the p-like electron density of the excited alkali atom. I studied these particular ring geometries as a function of the number n of helium atoms for Li, Na, K and Rb. The PES of $Ak^{\star}He_2$ presents in the lowest $\Pi_{1/2}$ excited state a global and a local minimum, at least for reasonable Ak-He distances. Figure 2 presents on the top panel the global minimum values for the five alkalis and for $n = 2$. The lower panel gives the energy difference between the global and the local minimum. The corresponding geometries (linear or bent) are pointed out by the symbols. The global minimum corresponds to identical Ak-He distances indicated as abscissae on the figure. For the lighter (Li, Na)$^{\star}He_{n=2}$ alkali clusters it corresponds to a bent and symmetric structure. For Na the energy difference between bent and linear geometries is small. For heavy alkalis the global minimum occurs at linear geometry with the Ak atom halfway between the He atoms.

The difference between global and local minimum energy increases from K to Cs. As argued by Nettels et al. [62], when some helium atoms are positioned next to the cesium atom the electronic wave function of $6P_{1/2}$ state is deformed and looses its spherical symmetry. In the case of more than two helium atoms, it becomes 'dumbbell' shaped. Helium atoms are then attracted by van der Waals force along its nodal plane and may form an exciplex with $n \geq 3$. With an increasing number of helium atoms around the waist of the dumbbell the repulsive potential between those atoms increases, which puts a natural limit on the maximum number n_{max} that can be accommodated. The rearrangement of the second He atom proceeds more easily than in the case of the formation of $Ak^{\star}He_3$, because He atoms in the potential well can move along the circle more freely than those in the depressions of the applelike

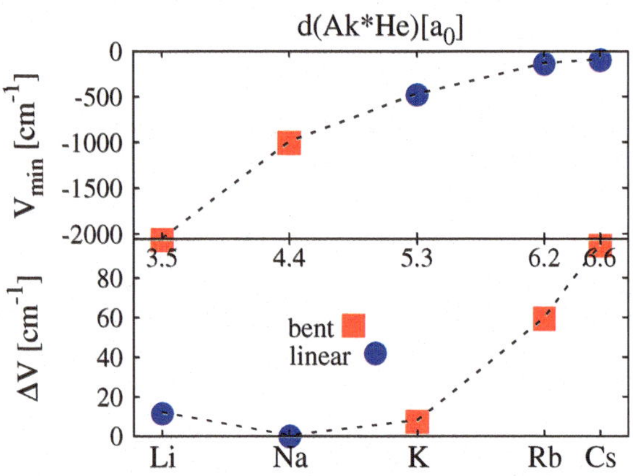

Fig. 2 Top panel: global potential energy minima for the $Ak^{\star}He_2$ systems as a function of the Ak-He distance. Bottom panel: energy difference between the local and the global potential minima

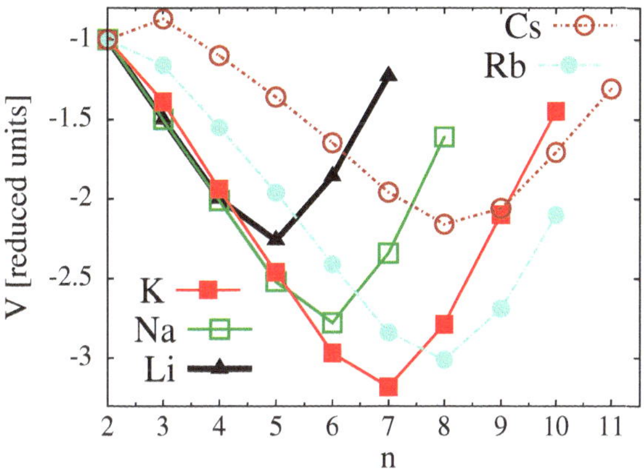

Fig. 3 Evolution of the minimum of the potential as a function of the number of helium atoms within the regular ring constraint for the Li, Na, K, Rb and Cs alkali atoms

density distribution for $Ak^{\star}He_2$ [27]. By contrast, when the third He atom approaches the alkali, the two potential minima are already occupied by the two He atoms. When we add the fourth helium atom, the ring structure obtained for this n value is stable with respect to $n = 2$, 3 helium atoms. Results for the regular ring are presented in Fig. 3, which shows the evolution of the minimum of potential as a function of n for these particular geometries, defined by $\widehat{HeAkHe}$ angles equal to $2\pi/n$ and identical Ak-He distances. The minimum for each curve corresponds to the largest number of helium atoms that the potential surface can accommodate in a regular ring. Thus at this classical level, ring with $n = 5$ for Li, $n = 6$ for Na, $n = 7$ for K and $n = 8$ for Rb and Cs are obtained. The $Cs^{\star}He_n$ system presents for $n = 3$ a local minimum of $-379.4\,\text{cm}^{-1}$ within the ring constraint. This value is larger than for $n = 2$, which corresponds to $-435\,\text{cm}^{-1}$. This means that $Cs^{\star}He_3$ is unstable with respect to the evaporation of one helium atom. However, other geometries are to be considered in order to confirm or infirm the above statement. Then I fixed the $Cs^{\star}$ at $(y, z) = (0, 0)$ and the two helium atoms of $Cs^{\star}He_2$ at equilibrium position $(0, \pm r_0$, with $r_0 = 6.6$ a.u.). I found that with the third He atom located on the yz plane at (4.5, 4.8 a.u.) or at (11.9, 0 a.u.) the molecule reaches a potential minimum, respectively, of $-424\,\text{cm}^{-1}$ and $-452.3\,\text{cm}^{-1}$. Unlike the former, the latter is stable with respect to $Cs^{\star}He_2$.

3 Dynamical Results

3.1 Method

I use the importance sampling diffusion Monte Carlo (IS-DMC) algorithm to obtain energies and structural properties of $Ak^{\star}He_n$. Since DMC methods are described in numerous publications [3, 67–69], I only give here the details specific to $Ak^{\star}He_n$ systems. The product

$$\Psi_T = \prod_{i \in He} \chi^{He}(r_{iA})\, \Phi^{HeAk}(r_{iA}) \prod_{i,j \in He} \Phi^{HeHe}(r_{ij}) \tag{1}$$

is used for the trial function. An explicit reference to the alkali atom is made in the trial function, Eq. (1), in which r_{iA} are the distances between the He atoms and the alkali atom, and r_{ij} the distances between two He atoms. The product in Eq. (1) contains Fermi-type and Jastrow-type functions,

$$\chi(r) = \{1 + \exp(r - r_0)\}^{-1}, \tag{2}$$

$$\Phi(r) = \exp\left(-ar^{-5}\right), \tag{3}$$

which are parameterized by r_0 and a given in Table 2. The Jastrow-type functions avoid the sampling of too short helium-helium and alkali-helium distances for which the potential is strongly repulsive. The Fermi-type function confines the helium atoms inside a sphere centered at the alkali position, preferred location for the dopant in this first electronic state as shown for the $Rb^{\star}He_n$ system [51]. Fermi parameter has been tailored to the largest $Rb^{\star}He_n$ systems and then used to study the other alkali systems, so that they depend only on the cluster size. Computed energy carries a bias due to the use of a trial function. IS-DMC yelds the exact ground state energy in the limit of infinite ensemble size and vanishing time step $\Delta\tau$. After equilibration of the ensemble distribution, about 8000 blocks of 100,000 atomic unit of imaginary time using time step of 50 down to 10 a.u. were performed. The bias from the finite ensemble size is corrected by extrapolation, using simulations from 2000 to 8000 walkers, which has been found adequate. The density distributions are computed using a descendant-weighted scheme similar to the one described in Ref. [70]. In the present work, a single accumulation step is used (cf Eq. (15) of Ref. [70]) and a relax time from 4000 to 6000 a_0 depending on the cluster size.

Table 2 Parameters (in a.u.) of the trial wave function for the $Ak^{\star}He_n$ clusters as defined in Eq. (1) and Eqs. (2)–(3)

n	≤ 7	≤ 11	≤ 15	≤ 20
r_0	16	18	21	26
	Li	Na	K	Rb
$a_{(HeHe)}$	500	500	500	500
$a_{(HeAk)}$	50	200	400	900

3.2 Energetics

Given the large zero point energy effects in helium clusters, the classical analysis obtained for $Ak^{\star}He_n$ needs to be corrected by dynamical studies. In particular, the ring structure at the quantum level is expected to be affected as it was previously found for $Rb^{\star}He_n$ [51].

I report in Table 3 the exact vibrational ground state energies computed using IS-DMC for $Ak^{\star}He_n$ with n up to 15. The reported values correspond to a double extrapolation to infinite ensemble size and zero time step of the growth energy. The $Li^{\star}He_n$ and $Na^{\star}He_n$, $n = 1$–5 are reproduced from my previous paper [56]. Note that in Ref. [51], the $Rb^{\star}He_n$ carries an ensemble size bias not properly corrected as explained in Ref. [52]. In addition to the ground state energies, the chemical potential $\mu(n) = E(n) - E(n-1)$ is also provided. It measures the binding energy for the nth helium, $\mu(1)$ being computed taking into account the spin-orbit coupling of the bare alkali atoms. For all sizes studied, I found negative chemical potential which indicates the stability of the clusters with respect to helium evaporation. Analysis of $\mu(n)$ shows that helium atoms are strongly bound for small n. The abrupt variation of $\mu(n)$ clearly indicates the first shell closing with 5, 6, and 7 helium atoms for Li, Na, K and Rb respectively. The number of helium in the first shell was found to be consistent with the Ak-He dimer interaction curves used for both Li and Na alkali

Table 3 Vibrational ground state energies and chemical potential (in cm^{-1}) of the $Ak^{\star}He_n$ systems as a function of n. Horizontal lines indicate the first and (possible) second shells closing. $Li^{\star}He_n$ and $Na^{\star}He_n$ with $n = 1, \ldots, 5$ are taken from Ref. [56]

	$Li^{\star}He_n$		$Na^{\star}He_n$		$K^{\star}He_n$		$Rb^{\star}He_n$	
n	-E	-μ	-E	-μ	-E	-μ	-E	-μ
1	862.033±0.002	861.80	428.593±0.003	417.13	216.395±0.003	177.92	189.269±0.002	30.87
2	1725.641±0.007	863.61	846.675±0.003	418.08	395.214±0.005	178.81	257.795±0.004	68.53
3	2430.69±0.02	705.05	1201.452±0.004	354.78	553.161±0.008	157.95	291.396±0.006	33.60
4	3081.51±0.02	650.82	1552.63±0.01	351.18	711.53±0.01	158.37	366.852±0.008	75.46
5	3259.16±0.02	177.65	1829.42±0.02	276.79	870.98±0.01	159.45	442.18±0.01	75.32
6	3277.40±0.03	18.24	1841.80±0.03	12.38	961.48±0.01	90.50	508.69±0.01	66.51
7	3290.56±0.04	13.16	1855.40±0.07	13.60	968.27±0.03	6.79	542.64 ±0.02	33.95
8	3301.32±0.04	10.76	1869.80±0.08	14.40	976.20±0.08	7.93	546.68±0.06	4.04
9	3312.7±0.10	11.38	1883.0±0.1	13.2	982.8±0.1	6.6	550.43±0.08	3.75
10	3317.6±0.1	4.9	1897.6±0.1	14.6	990.1±0.1	7.3	555.7±0.1	5.27
11	3319.8±0.1	2.2	1905.1±0.1	7.5	997.5±0.1	7.4	559.4±0.1	3.7
12	3322.6±0.1	2.8	1912.1±0.1	7.0	998.2±0.1	0.7	559.7±0.1	0.3
15	3330.1±0.1		1928.0±0.1		1001.5±0.1		560.9±0.1	

atoms [56]. In the first shell systematic increase of the chemical potential is obtained for Li and Na when n increases from 2 to 5. For K and Rb, the energy gained adding the fourth He atom is greater than the one for adding the third one. The relative energy difference between the two last He atoms which fill the first shell decreases moving down the alkali series: we can infer that the escape probability of one He atom from the filled shell decreases in the same way. For larger n, He atoms are much more weakly bound. A second drop in the absolute value of the chemical potential can suggest the closing of the second shell. Anyway, this drop is less pronounced for Li and Na. Thus at this level one can observe that a closure of the second shell possibly corresponds to 9 atoms for Li, 10 for Na, 11 for K and Rb. Yet, if we look at the $\mu(n)$ of Li, we may also suppose a closure to 10 atoms: density distributions introduced below will add structural informations to better analyze this issue. With the small number of helium atoms considered in this work, the chemical potential values are still far from the pure cluster limit.

3.3 Densities

One of the attractive features of the Monte Carlo method is that it provides geometrical information on the system. At the cost of descendent weighting scheme [70], positional functions such as radial and angular distributions can be calculated. Figure 4 shows the radial distributions, Ak-He and He-He, obtained for $Ak^{\star}He_2$. In order to emphasize linear geometries, $\rho_{Ak-He}(x)$ and $\rho_{HeHe}(2x)$ are presented on the same figure with x indicated on the lower horizontal axis and 2x on the upper one. A linear

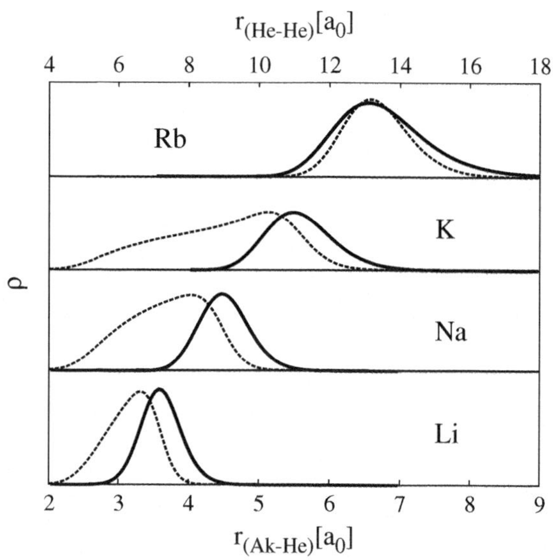

Fig. 4 Radial $Ak^{\star}$-He (full line) and He-He (dashed line) density distributions as a function of $r_{(AkHe)}$ (lower axis) and $r_{(HeHe)}$ (upper axis)

geometry implies an overlap of the two distributions ($r_{HeHe} = \frac{1}{2} r_{AkHe}$). As expected from the potential energy surface study, the maximum of ρ_{AkHe} shifts towards larger r_{iAk} values going from Li to Rb. At the same time the width of peak increases. For Rb, the ρ_{HeHe} is strictly overlapping the ρ_{AkHe} which is coherent with a linear He-Rb*-He system as previously shown [51]. From Li to K, the comparison of ρ_{AkHe} and ρ_{HeHe} indicates that both linear and bent geometries are present. Moreover, the ρ_{HeHe} broadens as we go from Li to K which implies more floppy $K^\star He_2$ systems. All of this is coherent with the double well picture I discussed in Sec. II. For Li, Na and K, the wave function has thus no negligeable amplitudes in both the local and global minima while for $Rb^\star He_2$, the local bent minimum is too high in energy. The angular distribution (not shown) confirms a mostly linear configuration for Rb and a wide range from $\pi/4$ to π non-zero probability for $\widehat{HeAkHe}$ angle for Li, Na and K. Figure 5 shows the density distributions for $Ak^\star He_{n\geq 3}$. In particular, I consider three cluster sizes: $n = 3$, 4 and 12. The use of red and black lines aims at pointing out the subtle difference between the density profiles obtained for light (Li, Na) and those obtained for heavier (K, Rb) alkali systems.

$n = 12$ emphasizes how the density distribution for larger sizes show the same grouping Li, Na versus K, Rb. $Li^\star He_3$ shows a He-He density distribution with a single peak at ~6 a.u. which indicates an equilateral-like triangle shape. For heavier alkalis this peak broadens more and more, indicating that now He-He distances are different. Examining He-He density distributions for $n = 4$, one can state, assuming a planar geometry [51], that the shape of the $Li^\star He_4$ system is quite squarish, because we have two peaks and one is higher than the other. With Na as dopant this squarish-like shape is lost, as pointed out by the same intensity of the two peaks. A perfect

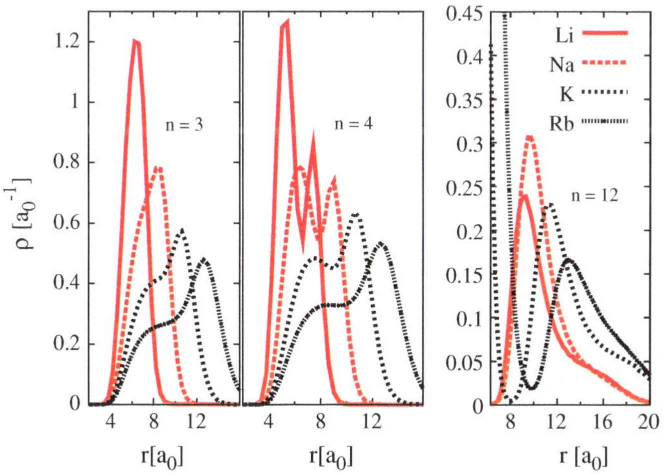

Fig. 5 Helium-helium radial density distributions of $Ak^\star He_3$ (left panel) and $Ak^\star He_4$ systems (center). Right panel: radial distributions when $n = 12$. In particular I display only the peak of the second solvation shell of the Ak-He radial distributions

agreement between density distributions and chemical potentials has been found up to the first solvatation shell, which for Li and Na occurs with $n = 5$, $n = 6$ for K and $n = 7$ for Rb.

Except for Li, a shift of one helium atom between the supposed "classical" first filled shells as shown in Fig. 3 and those obtained by IS-DMC computations is present. The closure of the first shell is showed in Fig. 6. Figure 7 displays the flat geometries dealing with these closures. If we look at the angular density distributions for Li and Na, we can notice that the rigidity, evaluated like a measure of the minimum reached in the distribution, increases from 4 to 5. Moreover, when $n = 6$, the density does not vanish anymore and this means that a new angle is present. If we focus on $K^{\star}He_n$ systems, when $n \leq 6$ a single peak occurs in the K-He radial density distributions. When $n = 6$ the He-He density shows two consistent distributions peaked at 5.8 and

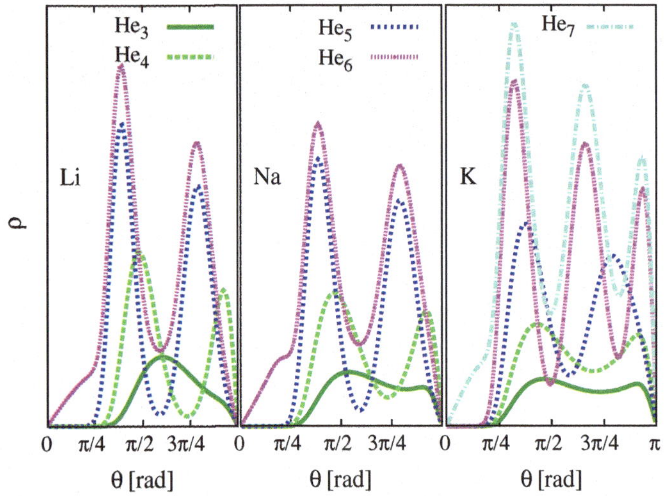

Fig. 6 Angular density distributions of $Ak^{\star}He_n$ systems up to one He atom after the first shell, as a function of $\widehat{HeAkHe}$ angle θ

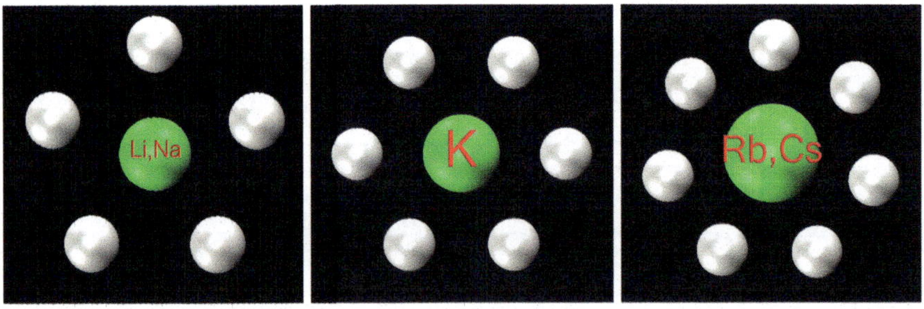

Fig. 7 Ring geometries dealing with the first shell closures of helium clusters: $Li^{\star}He_5$ and $Na^{\star}He_5$ (left); $K^{\star}He_6$ (center); $Rb^{\star}He_7$ and $Cs^{\star}He_7$ (right)

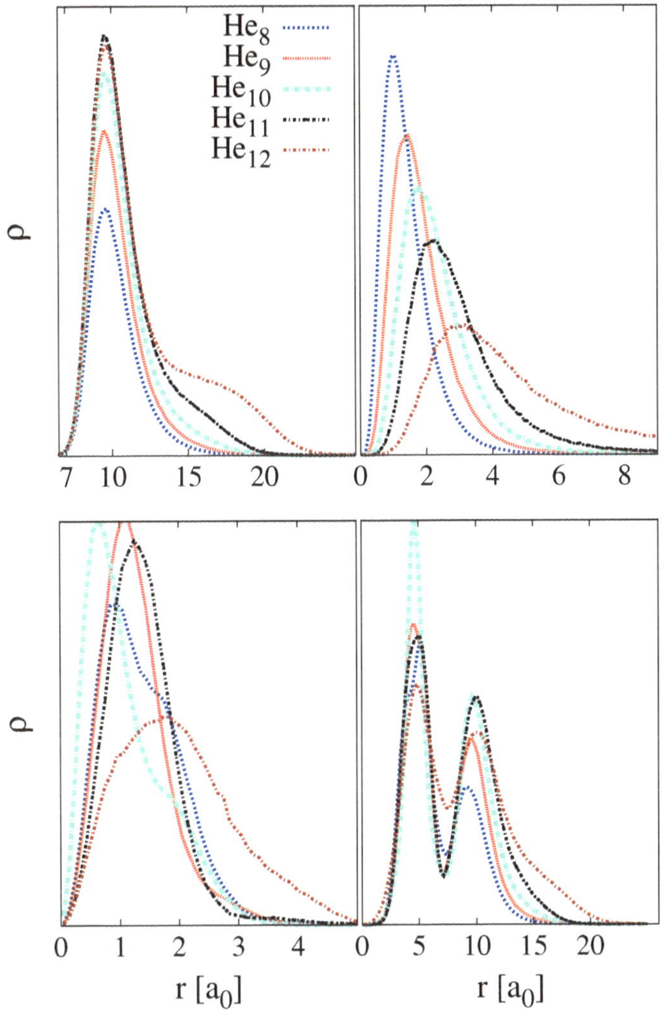

Fig. 8 Radial density distributions of $Ak^{\star}He_n$ systems, when $8 \leq n \leq 12$. Top panel: alkali-helium density profiles (left) and average distance distributions of He atoms from the mean plane (right). Bottom panel: density distributions concerning the distance between Ak and the center of mass of the molecule (left) and density distributions of the distance between helium atoms and the helium center of mass (right). The cyan profile emphasizes the number of helium atoms for which a symmetrical structure has been obtained

10.6 a_0. The angular density distribution $\rho(\theta)$ shows three peaks at 60°, 120°, and 180°. The structure is more pronounced than with $n = 7$, consistent with the closing of a first solvation shell. If we add one helium atom, all density distributions change, showing that the 7th helium floats around the first shell. With $n \geq 7$ a third broadened peak appears at $\sim$17 a_0 in the He-He density distribution whereas the K-He density distribution shows that one helium is far $\simeq$5 a_0 from the first shell. These findings are in agreement with the solvation structures obtained by Takayanagi and Shiga [50] who investigated helium solvation structures of $K^{\star}He_n$ exciplexes by path integral molecular dynamics simulations. Considering the radial density distributions of the clusters shown in Fig. 9, one can conclude that the second solvation shells around the np-orbital are likely completed with $n = 11$, because the first and the second peak do not increase after this value. However, if we look at the symmetry of the cluster which can be evaluated looking at the relative position of its center of mass, light alkali helium clusters present such a symmetry for $n = 9$ (Li) and $n = 10$ (Na). Since these values of n correspond also to the closure of the second shell as argued by examining the chemical potentials, we can estimate that the shells are able to host one (Na) or two (Li) additional atoms. With Fig. 8 I want to discuss the case of Na more in details. Beyond the first shell, inspection via Ak-He density distributions becomes more difficult as peaks broaden. Anyway, if we examine the density distribution of the distance between He atoms and the He-He center of mass, we can notice two more pronounced peaks in the He-$Na^{\star}He_{10}$ density profiles with respect to those of the other cluster sizes. Moreover, the density between peaks decreases and this indicates that a symmetric structure with Na as dopant is obtained for $n = 10$. Also the density distribution of the distance between Na and $NaHe_{10}$ center of mass is consistent with a second shell of 5 helium atoms. The right panel of Fig. 5 shows the profiles of Ak-He radial distributions when $n = 12$, that is one atom after the second shell. Each minimum corresponds to the separation between two peaks in the Ak-He density distribution. These minima become less and less pronounced from Li to Rb. This could mean that the possibility of exchange between He atoms of different shells should increase moving down the alkali periodic group. However, an alternative QMC approach by path integral simulations could measure the exchange of He atoms. By examining the least squares plane of all the $n + 1$ atoms [51], alkali clusters remained almost flat up to the second shell, after which tridimensional structures become more relevant, as shown on the top panel of Fig. 8. In agreement with Takayanagi and Shiga [50], these clusters have nearly planar structures due to the repulsive interaction of He and the electron in the p-orbital of the Ak atom. Table 4 summarizes some structural informations concerning the external shell. Its radius, found by considering the maximum of the second peak of He-He_n center of mass distribution, is roughly twice that of the inner shell. Obviously, the rigidity of the cluster decreases from the second to the first shell. In fact, the height of the peaks of radial density distributions decreases as a function of n. Although the density profile suggests an interpretation in terms of shells, the ongoing density increase in the nearest neighbor density maximum shows that the concept of shells, irrespective of the dopant considered, becomes inappropriate for $n \gtrsim 11$. In the fact that the density between peaks remains very high and particles exchange positions between

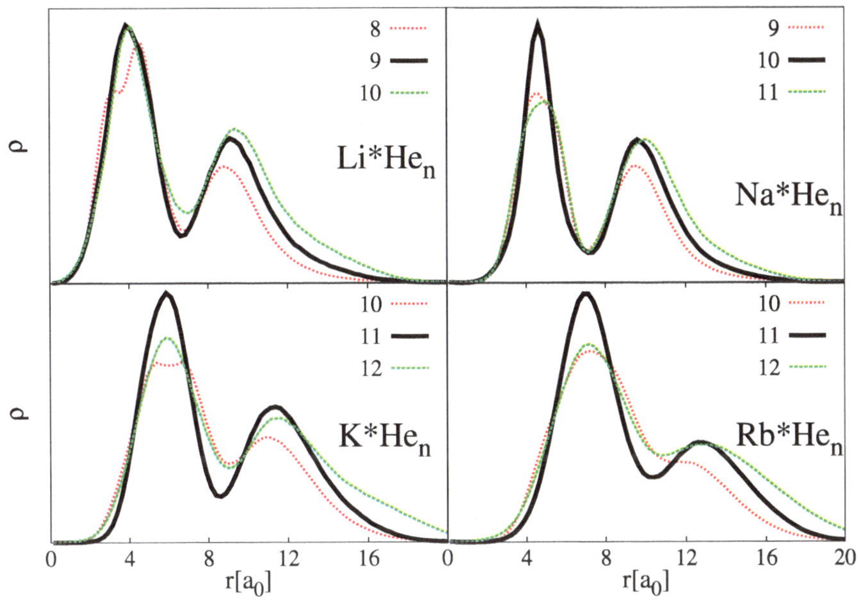

Fig. 9 Radial distributions of the distance between He atoms and He_n center of mass of the $Ak^{\star}He_n$ systems. The black line emphasizes the number of He atoms for which a symmetrical structure has been obtained

the inner and outer regions, one can see a typical signature of a quantum liquid. After the second shell it is difficult to obtain a sort of symmetric, organized structure. However, for some sizes a few organized structures have been noticed. For instance, as shown in the right panel of Fig. 10, for $n = 20$ the radial density distribution concerning the distance between He atoms and $Li^{\star}He_n$ center of mass shows that a third peak occurs.

4 Conclusions

Alkali-metal atoms excited to the first P states in a cold He environment produce alkali-metal-He_n exciplexes. In this work my contribution was to examine the stability for different sizes of these clusters using quantum Monte Carlo Methods. A Diatomics-In-Molecule approach including the spin-orbit coupling term was used to model the interaction potential of $Ak^{\star}He_n$ systems involving up to 15 helium atoms. Given the large zero point energy effects in helium clusters, the classical scenario of $Ak^{\star}He_n$ has to be corrected by dynamical studies. For this reason vibrational ground state energies and densities were computed using IS-DMC method. The comparative study of the alkali-helium clusters going down the periodic group revealed a structural difference between light (Li, Na) and heavy (K, Rb) alkali clusters. The stronger

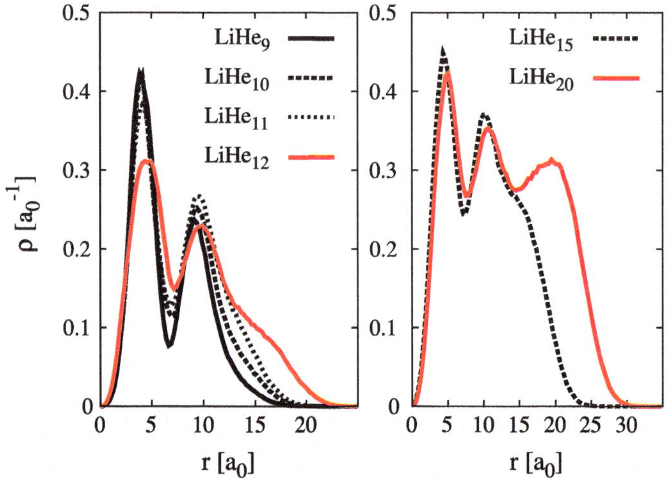

Fig. 10 Radial distributions of the distance between He atoms and Hen center of mass for Li*He_n systems when $9 \leq n \leq 20$

Table 4 Structural estimations for the Ak*He_n systems by examinating the density distribution profiles. (a) Distance between the shell and the He_n center of mass. (b) Distance between the nearest neighbour He atoms. (c) Average distance between He atoms and the mean plane. (d) Distance between the alkali atom and the center of mass of Ak*He_n system. Values are given in atomic units

First shell	Li*He_5	Na*He_6	K*He_6	Rb*He_7
a	3.8	4.5	5.7	6.7
b	4.5	5.3	5.6	5.9
c	0.3	0.5	0.7	1.1
Second shell	Li*He_9	Na*He_{10}	K*He_{11}	Rb*He_{11}
a	9.1	9.6	11.3	12.7
b	10.1	11.2	9.8	10.7
c	1.4	1.7	2.5	2.8
d	1.2	0.6	1.2	1.3

binding energy for small n is obviously due to the dopant, especially for Li and Na. The shells of the light alkalis look more rigid if compared to those of K and Rb. The energy to remove a helium atom drops off quite smoothly for higher n values. Anyway, with the small number of helium atoms considered in this work, the chemical potential values are still far from the pure cluster limit. Cesium shows some distinctive features if compared to the other alkali examined here. Because of the stronger Cs*He interaction, the probability of exciplex formation for the system Cs*He_n increases with the density of He, i.e., with rising He pressure, or when going from the liquid to the solid phase. Experimental [34, 62] as well as theoretical [71] works dout about the possibility of $Cs^{\star}He_n$ exciplexes with $n = 2$ or 3. Moreover, Moroshkin et al. [34] estimated that in solid helium the $Cs(6P_{3/2})^{\star}He_{n=2}$ exciplex is quenched by a radiationless transformation into the larger $Cs(6P_{1/2})^{\star}He_{n=6,7}$ complex through the attachment of additional He atoms. For these reasons a more meticulous study of first and second excited states of Cs in helium clusters is now in progress [72]. At the moment I can observe that both triatomic $Cs(6P_{1/2,3/2})^{\star}He_{n=2}$ exciplexes have a linear structure with the Cs atom halfway between the He atoms and a closure at seven He atoms for the first shell of the $Cs(6P_{1/2})^{\star}He_{n=2}$ system. Both findings are in agreement with the experiments [34].

References

1. Toennies J, Vilesov A (1998) Annu Rev Phys Chem 49:1
2. Lehmann K, Scoles G (1998) Science 279:2065
3. Reynolds PJ, Ceperley DM, Alder BJ, Lester WA (1982) J Chem Phys 77:5593
4. Germer EHTA, Stephenson JC, Cavanagh R (1994) Int Rev Phys Chem 13:41
5. Hartmann M, Miller RE, Toennies JP, Vilesov A (1995) Phys Rev Lett 75:1566
6. Hartmann M, Mielke F, Toennies JP, Vilesov AF, Benedek G (1996) Phys Rev Lett 76:4560
7. Levy D (2001) Special topic on helium nanodroplets: a novel medium for chemistry and physics. J Chem Phys 115
8. Chatterley AS, Schouder C, Christiansen L, Shepperson B, Rasmussen MH, Stapelfeldt H (2019) Nat Commun 10:133
9. Toennies JP, Vilesov AF, Whaley KB (2001) Phys Today 54:31
10. Higgins J, Callegari C, Reho J, Stienkemeier F, Ernst W, Lehmann K, Gutowski M, Scoles G (1996) Science 273:629
11. Callegari C, Ernst WE (2011) In: Quack M, Merkt F (eds) Handbook of high-resolution spectroscopy. Wiley, Chichester
12. Pototschnig JV, Lackner F, Hauser AW, Ernst WE (2017) Phys Chem Chem Phys 19:14718
13. Stienkemeier F, Higgins J, Callegari C, Kanorsky S, Ernst W, Scoles G (1996) Z Phys D: At Mol Clusters 38:253
14. Callegari C, Higgins J, Stienkemeier F, Scoles G (1998) J Phys Chem A 102:95
15. Bünermann O, Droppelmann G, Hernando A, Mayol R, Stienkemeier F (2007) J Phys Chem A 111:12684
16. Lackner F, Poms J, Krois G, Pototschnig JV, Ernst WE (2013) J Phys Chem A 117:11866
17. Reho J, Callegari C, Higgins J, Lehmann KK, Scoles G, Ernst W (1997) Faraday Discuss 108:161
18. Reho J, Higgins J, Lehmann KK, Scoles G (2000) J Chem Phys 113:9686
19. Loginov E, Callegari C, Ancilotto F, Drabbels M (2011) J Phys Chem A 115:6779
20. Reho J, Higgins J, Lehmann KK, Scoles G (2000) J Chem Phys 113:9694
21. Pifrader A, Allard O, Auböck G, Callegari C, Ernst WE, Huber R, Ancilotto F (2010) J Chem Phys 133:164502
22. Brühl F, Trasca R, Ernst W (2001) J Chem Phys 115:10220
23. Nagl J, Auböck G, Callegari C, Ernst WE (2007) Phys Rev Lett 98:075301
24. Auböck G, Nagl J, Callegari C, Ernst WE (2008) Phys Rev Lett 101:035301
25. Bünermann O, Mudrich M, Weidemüller M, Stienkemeier F (2004) J Chem Phys 121:8880
26. Kinoshita T, Takahashi Y, Yabuzaki T (1994) Phys Rev B 49:3648
27. Hirano K, Enomoto K, Kumakura M, Takahashi Y, Yabuzaki T (2003) Phys Rev A 68:012722
28. Enomoto K, Hirano K, Kumakura M, Takahashi Y, Yabuzaki T (2002) Phys Rev A 66:042505
29. Mudrich M, Stienkemeier F (2014) Int Rev Phys Chem 33:301

30. Schulz CP, Claas P, Stienkemeier F (2001) Phys Rev Lett 87:153401
31. Dozmorov NV, Baklanov AV, von Vangerow J, Stienkemeier F, Fordyce JAM, Mudrich M (2018) Phys Rev A 98:043403
32. von Vangerow J, Coppens F, Leal A, Pi M, Barranco M, Halberstadt N, Stienkemeier F, Mudrich M (2017) J Chem Phys 8:307
33. Kinoshita T, Fukuda K, Yabuzaki T (1996) Phys Rev B 54:6600
34. Moroshkin P, Hofer A, Lebedev V, Weis A (2008) Phys Rev A 78:032501
35. Barnett R, Whaley K (1993) J Chem Phys 99:9730
36. Chin S, Krotscheck E (1992) Phys Rev B 45:852
37. Sindzingre P, Klein M, Ceperley D (1989) Phys Rev Lett 63:1601
38. Kwon Y, Ceperley D, Whaley K (1996) J Chem Phys 104:2341
39. Nakayama A, Yamashita K (2000) J Chem Phys 112:10966
40. Dalfovo F, Lastri A, Pricaupenko L, Stringari S, Treiner J (1995) Phys Rev B 52:1193
41. Ancilotto F, DeToffol G, Toigo F (1995) Phys Rev B 52:16125
42. Barranco M, Guardiola R, Hernández S, Mayol R, Navarro J, Pi M (2006) J Low Temp Phys
43. Hernando A, Barranco M, Mayol R, Pi M, Ancilotto F, Bünermann O, Stienkemeier F (2009) J Low Temp Phys
44. Ogata S (1999) J Phys Soc Jpn 68:2153
45. Mella M, Colombo MC, Morosi G (2002) J Chem Phys 117:9695
46. Drummond ND, Needs RJ (2006) Phys Rev B 73:024107
47. Dupont-Roc J (1995) Z Phys B 98:383
48. Kanorsky S, Weis A, Arndt M, Dziewior R, Hansch TW (1995) Z Phys B 98:371
49. DeToffol G, Ancilotto F, Toigo F (1996) J Low Temp Phys 102:381
50. Takayanagi T, Shiga M (2004) Phys Chem Chem Phys 6:3241
51. Leino M, Viel A, Zillich R (2008) J Chem Phys 129:184308
52. Leino M, Viel A, Zillich RE (2011) J Chem Phys 134:024316
53. Ellison FO (1963) J Am Chem Soc 85:3540
54. Pascale J (1983) Phys Rev A 28:632
55. Pascale J (1983) Technical report, Service de Physique des Atomes et des Surfaces (C.E.N. Saclay), Gif sur Yvette, France
56. Dell'Angelo D, Guillon G, Viel A (2012) J Chem Phys 136:114308
57. Dhiflaoui J, Bejaoui M, Berriche H (2017) Eur Phys J D 71:331
58. Sharma AR, Weeks DE (2018) Phys Chem Chem Phys 20:29274
59. Blank LA, Sharma AR, Weeks DE (2018) Phys Rev A 97:032705
60. Bormotova EA, Kozlov SV, Pazyuk EA, Stolyarov AV, Skomorowski W, Majewska I, Moszynski R (2019) Phys Rev A 99:012507
61. Taken from the NIST web site
62. Nettels D, Hofer A, Moroshkin P, Müller-Siebert R, Ulzega S, Weis A (2005) Phys Rev Lett 94:063001
63. Hofer A, Moroshkin P, Nettels D, Ulzega S, Weis A (2006) Phys Rev A 74:032509
64. Moroshkin P, Hofer A, Nettels D, Ulzega S, Weis A (2006) J Chem Phys 124:024511
65. Moroshkin P, Hofer A, Ulzega S, Weis A (2006b) Low Temp Phys 32:981
66. Enomoto K, Hirano K, Kumakura M, Takahashi Y, Yabuzaki T (2004) Phys Rev A 69:012501
67. Suhm MA, Watts RO (1991) Phys Rep 204:293
68. Umrigar CJ, Nightingale MP, Runge KJ (1993) J Chem Phys 99:2865
69. Hammond BL, Lester WA, Reynolds PJ (1994) Monte Carlo methods in ab initio quantum chemistry. World Scientific, Singapore
70. Casulleras J, Boronat J (1995) Phys Rev B 52:3654
71. Kunz CF, Hess BA (2000) J Chem Phys 112:1383
72. Dell'Angelo D (2019) Electronically excited cesium atom in helium clusters: I and II excited states (in progress)

A Quantum Chemical Approach for the Characterization of the Interaction Potential of Propylene Oxide with Rare-Gas Atoms (He, Ne, Ar)

Patricia R. P. Barreto, Ana Claudia P. S. Cruz, Henrique O. Euclides, Alessandra F. Albernaz, Federico Palazzetti and Fernando Pirani

Abstract Propylene oxide is one of the simplest organic chiral molecules and has attracted considerable interest from the scientific community a few years ago, when it was discovered in the interstellar medium. Here, we report a preliminary study on the interaction between propylene oxide and rare-gas atoms, specifically He, Ne, and Ar. The interaction potentials as a function of the distance between the center-of-mass of propylene oxide and the rare-gas-atom are calculated for fourteen leading configurations at CCSD(T)/aug-cc-pVDZ level of theory. Symmetry Adapted Perturbation Theory has been employed for the analysis of the intermolecular potential, revealing that most of the contribution is given by dispersion and exchange forces.

Keywords Van der Waals clusters · Interaction potential · Quantum chemistry · Ab initio calculations

1 Introduction

Propylene oxide is one of the simplest organic chiral molecules, making it an ideal candidate for studies concerning chiral selectivity mechanisms [1, 2]. It is characterized by a high vapor pressure and stability that makes this molecule suitable for scattering experiments, where high beam intensity and duty cycle are important. Its commercial availability is also remarkable for both the racemic mixture and the two enantiomers. It has been largely studied both experimentally, for example by synchrotron radiation [3] and molecular beams [4, 5], and from a computational point of view. For this latter aspect, see for example the representation of the potential

P. R. P. Barreto (✉) · A. C. P. S. Cruz · H. O. Euclides
Instituto Nacional de Pesquisas Espaciais, São José dos Campos, São Paulo, Brazil

A. F. Albernaz
Instituto de Física, Universidade Nacional de Brasília, Brasília, Brazil

F. Palazzetti (✉) · F. Pirani
Dipartimento di Chimica, Biologia e Biotecnologie, Università degli Studi di Perugia, via Elce di Sotto, 8, Perugia, Italy
e-mail: federico.palazzetti@unipg.it

L. Mammino et al. (eds.), *Advances in Quantum Systems in Chemistry, Physics, and Biology*, Progress in Theoretical Chemistry and Physics 32,
https://doi.org/10.1007/978-3-030-34941-7_6

energy surfaces of propylene oxide with helium [6, 7] and the study of the structural isomers manifold of C_3H_6O [8]. Recent discovery in the interstellar medium of this molecule [9] determined unusual interest, considering that numerous organic, even complex, molecules are continuously discovered [10–12].

In this work, we present the study by a quantum mechanical approach of the binary interaction between propylene oxide and the rare gas atoms He, Ne and Ar. Ab initio calculations have been carried out at different levels of theory to determine the most stable geometry of the molecule and the energy profiles of some significant configurations, called leading configurations (see also Refs. [13, 14]), chosen upon physical and geometrical considerations. The ab initio points have been fitted with a fifth-order Rydberg potential function [15]. The Symmetry Adapted Perturbation Theory (SAPT) method has been employed to calculate electric properties of the systems propylene oxide–rare-gas atom, such as dipole moment, polarizability, quadrupole moment, ionization potential, electron affinity and proton affinity. These properties are employed for a preliminary analysis of electrostatic, exchange, induction and dispersion contributions to the intermolecular forces.

The article is structured as follows: in Sect. 2, we give a background of the quantum mechanical methods we employed and describe the leading configurations; in Sect. 3, we discuss the results; in Sect. 4, final remarks end the paper.

2 Background

2.1 Geometry Optimization and Calculation of Potential Energy Profile of the Leading Configurations

The structure of propylene oxide has been optimized by the Gaussian package [16] at various levels of theory (Table 1). One hundred energy points have been calculated at CCSD(T)/aug-cc-pVDZ by using MOLPRO [17] for a certain number of representative configurations, including the counterpoise correction to the basis set superposition error. As mentioned in Sect. 1, the choice of the representative configurations, namely leading configurations, relay on geometric and physical considerations of the system. The leading configurations of propylene oxide–He were discussed in Ref. [6] (see also [18]) and the same choice is applied to the systems propylene oxide–Ne and propylene oxide–Ar. Propylene oxide molecule is considered as built in a distorted tetrahedron (Fig. 1), where the carbon atoms of CH_3 and CH_2 fragments, referred as C3 and C2, the oxygen, O, and the hydrogen, H, atom bound to the asymmetric carbon are the vertices. Fourteen configurations have been identified and indicated by the letters V, E and F, that stand for vertex, edge and face of the tetrahedron, respectively (Fig. 2). In the four V configurations, the rare-gas-atom lies through the vertex–center-of-mass of propylene oxide direction. V1 is directed toward C3, V2 in correspondence of C2, V3 toward O, and V4 along the H–center-of-mass direction H. For the E configurations, the direction of the rare-gas-atom is referred to the line

Table 1 Interatomic distances, r (Å), angles among three atoms, A (in degrees), and dihedral angles, D (in degrees), in the optimized geometries of propylene oxide, calculated at various levels of theory and compared with experimental reference data

Methods	r(C1C2)	r(C2C3)	r(C1O)	r(C2O)	r(C2H)	r(C1H)	r(C3H)
CBS-G3	1.47	1.51	1.43	1.43	1.09	1.09	1.09
G2	1.46	1.50	1.44	1.44	1.09	1.09	1.09
G3B3	1.47	1.51	1.43	1.43	1.09	1.09	1.10
G3	1.46	1.50	1.44	1.44	1.09	1.09	1.09
Ref. [25]	1.471	1.513		1.436	1.082		1.09

Methods	A(C2C1C3)	A(OC1C3)	A(HC1C2)	A(HC2H)	A(HC3H)	D(HC1C2H)	D(HC2C1C3)	D(OC1C3H)	D(HC1C3H)
CBS-G3	122.4	116.5	117.2	115.3	108.7	−154.1	152.8	−43.7	−60.8
G2	121.9	115.9	117.3	115.4	108.9	−154.3	153.4	−43.9	−61.0
G3B3	122.4	116.6	117.2	115.0	108.7	−154.1	152.7	−44.1	−61.2
G3	121.9	115.9	117.3	115.4	108.9	−154.3	153.4	−43.9	−61.0
Ref. [25]	120.95			116.25	109.5				

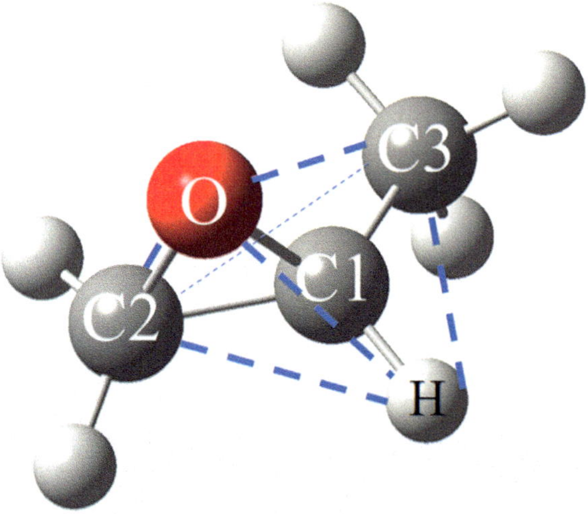

Fig. 1 The propylene oxide molecule: in white the hydrogen atoms; in grey the carbons C1, C2 and C3; in red the oxygen

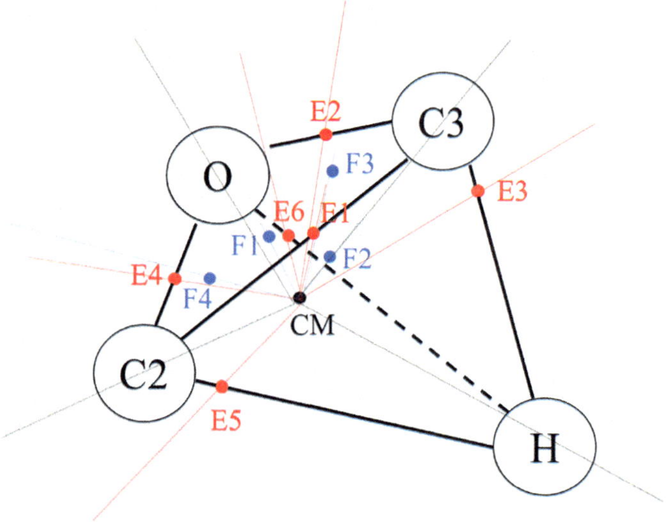

Fig. 2 The propylene oxide is represented as a distorted tetrahedron, whose vertices are the H, O, C2, and C3; the edges are indicated by thick black lines. The leading configurations are classified as V, E and F. The four V configurations correspond to the rare-gas atom directed through the vertices of the tetrahedron and the center-of-mass of the molecule (black dot); in V1 the rare-gas atom is directed through C3, in V2 through C2, in V3 through O, and in V4 through H (the direction of the V configurations are indicated by a black line). The six leading configurations E correspond to the direction (red line) connecting the center-of-mass of the molecule and the centers-of-mass of the six edges of the tetrahedron (red dots). The four leading con-figurations F correspond to the direction (blue line) connecting the center-of-mass of the molecule and the centers-of-mass of the faces of the tetrahedron (blue dots)

that connects the center-of-mass of the molecule and the center-of-mass of the edge of the tetrahedron, defined by the center-of-mass of two vertices. In the determination of the centers-of-mass of the edges involving C3 and C2, the whole masses of the groups CH_3 and CH_2 have been considered. Thus, E1 represents the configuration directed in correspondence of the center-of-mass of the edge C3-C2; E2 corresponds to the center-of-mass C3-O, E3 goes through C3-H, E4 through C2-O, E5 through C2-H, finally, and E6 is in correspondence of the center-of-mass of the edge O-H. The F configurations are defined as the directions that connect the rare-gas-atom to the centers-of-mass of the faces of the distorted tetrahedron and the center-of-mass of the molecule: in F1, the rare-gas is directed toward the center-of-mass of C3-C2-O, in F2 through C3-C2-H, in F3 through C3-O-H, and in F4 through the center of mass of C2-O-H.

The analytical form of the potential energy surface for the fourteen leading configurations is built by fitting a fifth degree generalized Rydberg function, V(R), to the calculated ab initio points:

$$V(R) = D_e \sum_{k=1}^{5} (1 + a_k (R - R_{eq})^k) \exp[-a_1 (R - R_{eq})] + E_{ref} \tag{1}$$

where D_e, a_i, R_{eq} and E_{ref} are adjustable parameters. The adjustable parameters have been obtained by a non-linear least-square procedure that minimized the differences between the analytic energies obtained by the potential function reported above and the ab initio energy points.

2.2 Symmetry Adapted Perturbation Theory Calculations

The SAPT method has been performed using the PSI4 code [19] at the CCSD(T)/aug-cc-pVDZ level of theory, on the configuration of minimum energy, among those considered in this work. In SAPT, the total Hamiltonian for the molecule–atom system is partitioned as $\mathbf{H} = \mathbf{F} + \mathbf{V} + \mathbf{W}$, where $\mathbf{F} = \mathbf{F_A} + \mathbf{F_B}$ is the sum of the Fock operators for species **A** and **B**, **V** is the intermolecular interaction operator, and $\mathbf{W} = \mathbf{W_A} + \mathbf{W_B}$ is the sum of the Møller-Plesset operators. The Fock operators are treated as zero-order Hamiltonian, while the interaction energy is evaluated through a perturbative expansion of V given by:

$$E_{int} = \sum_{n=1}^{\infty} \sum_{j=1}^{\infty} \left(E_{pol}^{(nj)} + E_{exch}^{(nj)} \right) \tag{2}$$

where n and j indicate the perturbation order. The polarization energies $\mathrm{E}_{\mathrm{pol}}^{(\mathrm{nj})}$ are identical to the corrections obtained in a regular Rayleigh-Schrödinger perturbation theory and the exchange corrections $E^{(nj)}_{exch}$ arise from the use of a global antisymmetrizer to force the correct permutation symmetry of the dimer wave function in each order, hence the name "symmetry adaptation". In this way, the interaction energy given by SAPT can be written as:

$$E_{SAPT} = E_{elst} + E_{exch} + E_{ind} + E_{disp} \tag{3}$$

Using higher-order SAPT, as SAPT2+3, the electrostatic part is given by:

$$E_{elst} = E^{(10)}_{elst} + E^{(12)}_{elst,resp} + E^{(13)}_{elst,resp} \tag{4}$$

The superscript defines the order in V, while the subscript, *resp*, means that orbital relaxation effects are included. The exchange part is given by:

$$E_{exch} = E^{(10)}_{exch} + E^{(11)}_{exch} + E^{(12)}_{exch} \tag{5}$$

The induction part is formulated as:

$$E_{ind} = E^{(20)}_{ind,r} + E^{(30)}_{ind,r} + E^{(22)}_{ind} + E^{(20)}_{exch-ind,r} + E^{(22)}_{exch-ind} - \delta^{(2)}_{HP} + \delta^{(3)}_{HH} \tag{6}$$

The $\delta^{(2)}_{HF}$ and $\delta^{(3)}_{HF}$ terms take into account higher-order induction effects, not included in MP2 correlation. Finally, the dispersion energy is given by:

$$E_{disp} = E^{(20)}_{disp} + E^{(21)}_{disp} + E^{(30)}_{disp} + E^{(20)}_{exch-disp} + E^{(30)}_{exch-disp} + E^{(30)}_{ind-disp} + E^{(30)}_{exch-ind-disp} + \Delta CCD \tag{7}$$

where ΔCCD is an improved version of the CCD (couple-cluster doubles) treatment of dispersion [20, 21].

3 Results and Discussion

In Table 1, we report the structural properties of the optimized geometry of propylene oxide, calculated at various levels of theory and compared with reference data. The CBS-QB3 method [22] presents overall the best agreement with reference data, for both distances and angles. Thus, the geometry optimized by CBS-QB3 is chosen for the calculation of the interaction potentials of the leading configurations as a function of the distance and for the analysis of the contributions to the intermolecular forces.

Potential energy profiles of the leading configurations of propylene oxide–He have been already presented and discussed in Ref. [6], where they have been compared with the Pirani potential function, also known as Improved Lennard-Jones. Here, we use a fifth order Rydberg potential function, as discussed in Sect. 2.2 for the three systems. The single point energies of propylene oxide with the He, Ne and Ar are reported in Figs. 3, 4 and 5, respectively. For propylene oxide–He, the minimum energy configuration is F1, at ca. 45 cm^{-1} and 3.7 Å followed by E5, at a similar value of energy, but with an equilibrium distance of ca. 4.2 Å. The stability of F1 is probably due to hindrance steric reasons and to the polarizability effect of the groups CH_3, CH_2 and O atom. This is confirmed by the comparison with the systems propylene oxide–Ne and propylene oxide–Ar, where F1 is also in these cases the leading configuration with the minimum energy. Among the V configurations, V1 is the most stable for the three systems; V1, V2 and V3 present also similar values of

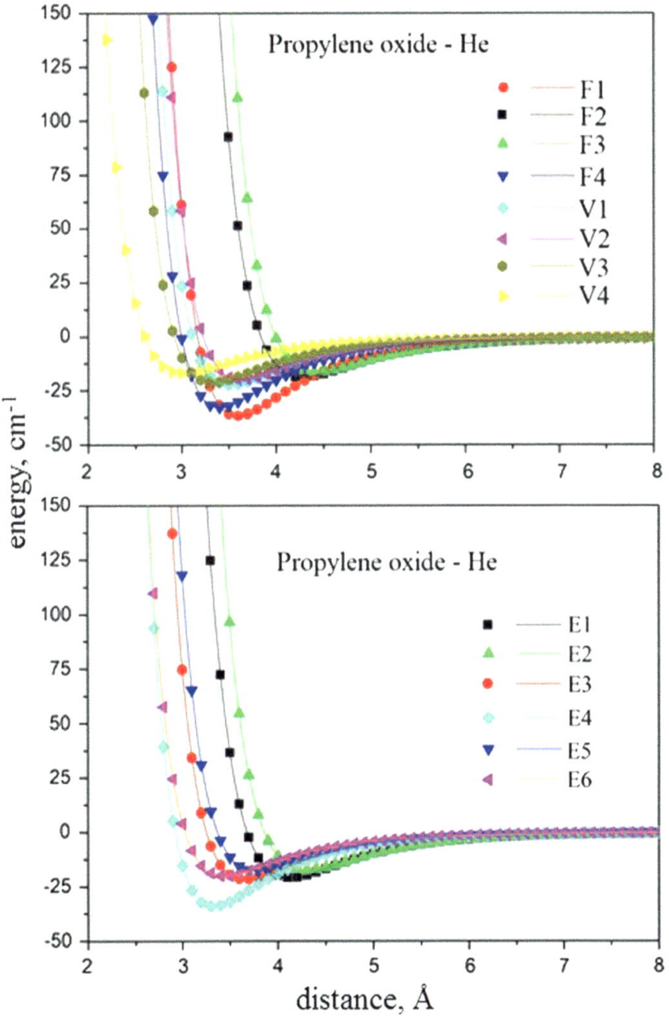

Fig. 3 Potential energy curves of propylene oxide–He of the fourteen leading configurations. The ab initio points are indicated by symbols and the Rydberg fitting by lines

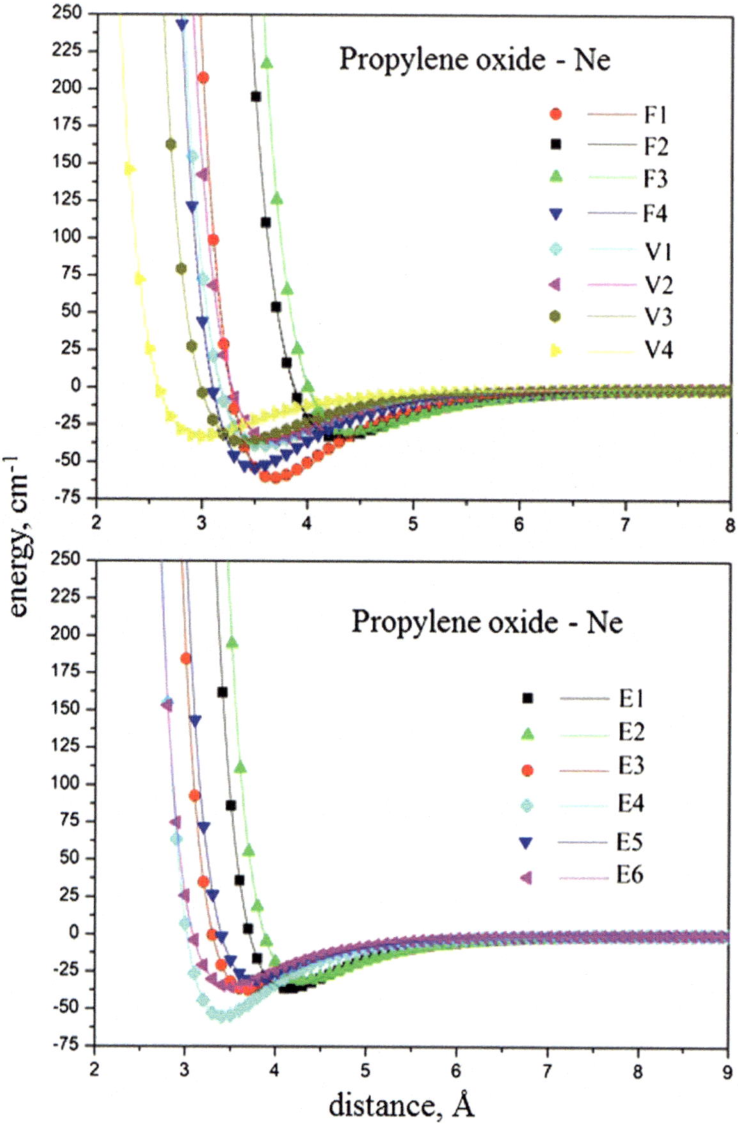

Fig. 4 Potential energy curves of propylene oxide–Ne of the fourteen leading configurations. The ab initio points are indicated by symbols and the Rydberg fitting by lines

the equilibrium, while V4, where the rare-gas atom is directed along the H, is less stable and at higher values of the equilibrium distance. Configurations E3 and E5, i.e. the "edges" CH_3–H and CH_2–H, also present higher values of the minimum energy and equilibrium distance, with respect to the other E configurations. Concerning F configurations, F2 and F3 present the highest repulsive character. It is important

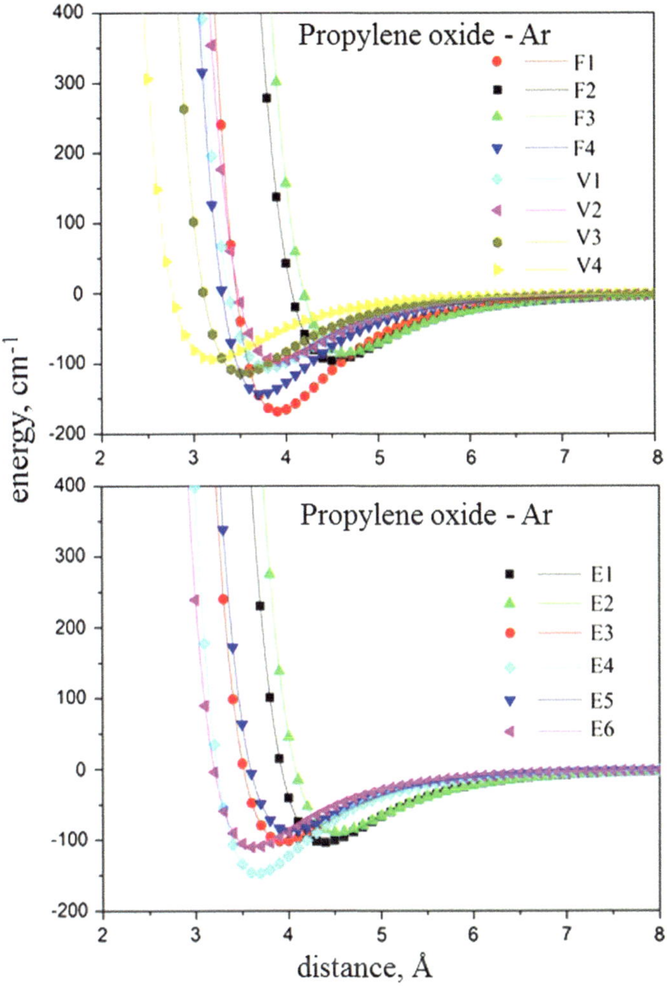

Fig. 5 Potential energy curves of propylene oxide–Ar of the fourteen leading configurations. The ab initio points are indicated by symbols and the Rydberg fitting by lines

to point out that the potential energy profiles of the leading configurations have a similar trend for the three systems. As mentioned above, F1 is the configuration of minimum energy for the three systems. For propylene oxide–He, F1 presents a minimum energy of 36.9 cm^{-1} at 3.6 Å; for propylene oxide–Ne, the depth of the potential well is ca. 60.7 cm^{-1} at 3.7 Å; finally, for propylene oxide–Ar, we have a minimum of 169.3 cm^{-1} at 3.9 Å.

In Table 2, we show the minimum energy and equilibrium distances for the three molecule–rare-gas-atom systems calculated at the CCSD(T)/aug-cc-pVDZ level of theory in the F1 configuration, comparing the results with and without counterpoise

Table 2 Equilibrium distance R_{eq} and minimum energy D_e of the F1 configuration for the three systems, propylene oxide–He, –Ne and –Ar, calculated ab initio with and without counterpoise correction of the basis set superposition error, and by SAPT calculation

	Ab initio calculation with CC		Ab initio calculation without CC		SAPT	
	R_{eq} (Å)	D_e (cm^{-1})	R_{eq} (Å)	D_e (cm^{-1})	R_{eq} (Å)	D_e (cm^{-1})
He	3.591	36.880	3.377	79.557	3.306	36.772
Ne	3.675	60.749	3.453	175.265	3.403	60.531
Ar	3.903	169.310	3.665	405.280	3.611	167.084

correction for the basis set superposition error. The result shows that the counterpoise correction tends to increase the equilibrium distance of about 0.2 Å for all the systems. However, the most significant effect is found in the minimum energy, where the counterpoise correction considerably decreases the attractive component of the potential. For the propylene oxide–He systems, the potential calculated without counterpoise correction is more attractive by a factor two than that calculated with the counterpoise correction, while for propylene oxide–Ne and –Ar systems this difference is even higher. We must note that the lower values of energy are probably the result of an overcorrection effect, especially considering the limitation of the basis set. Table 3 reports the difference between the minimum energy and equilibrium distance calculated ab initio with and without counterpoise correction, confirming the trend already observed for the F1 configuration.

Table 4 reports the electric properties dipole moment, polarizability, quadrupole moment, ionization potential, electron affinity and proton affinity, for the three systems investigated. For propylene oxide the zero-point energy is also presented. These properties have been then employed to analyze the contributions to the interaction potentials [23–26].

Figure 6 compares the energy profiles at different SAPT correction levels for the F1 configuration. Except for the Hartree-Fock (HF), that presents repulsive character, all the corrections present similar energy profiles. For propylene oxide–He, the minimum has been obtained at SAPT2+(CCD) level, for propylene oxide–Ne at SAPT2+, and for propylene oxide–Ar at SAPT0.

Figure 7 shows the interaction potential and the related contributions of the three systems for the F1 configuration calculated at SAPT2+(3)(CCD) level of theory. As expected, the equilibrium distance and the minimum energy increases with the atomic mass of the rare gas atom. The plots show that a fundamental role is played by dispersion forces, whose attractive character is already present in the long-range distance and increases by passing from He, to Ne and Ar. The exchange contribution, which presents a purely repulsive character, also depends on the atomic mass of the rare gas atom and its effect, as well as that of dispersion forces, is present at increasing distances moving from He to Ar. Induction and electrostatic contributions have been also considered in this treatment; their contribution is negligible in the long- and medium-range of distances. For what concerns the electrostatic contribution,

Table 3 Equilibrium distance R_{eq} and minimum energy D_e of thirteen leading configurations (F1 is given in Table 2) for the three systems, propylene oxide–He, –Ne and –Ar, calculated ab initio with and without counterpoise correction (with CC and without CC, respectively) of the basis set superposition error

	He				Ne				Ar			
	With CC		Without CC		With CC		Without CC		With CC		Without CC	
	R_{eq} (Å)	D_e (cm^{-1})	R_{eq} (Å)	D_e (cm^{-1})	R_{eq} (Å)	D_e (cm^{-1})	R_{eq} (Å)	D_e (cm^{-1})	R_{eq} (Å)	D_e (cm^{-1})	R_{eq} (Å)	D_e (cm^{-1})
E1	4.6728	20.5548	4.4812	49.3347	4.6920	36.2098	4.4526	126.8794	4.9331	102.9272	4.6579	272.3694
E2	4.8326	17.0660	4.6050	45.6085	4.8398	30.4258	4.5700	118.8607	5.0628	88.8797	4.7674	257.6504
E3	5.1837	21.4981	4.9765	49.8263	5.2183	37.5646	4.9868	130.1674	5.4649	101.8159	5.1821	284.7835
E4	4.2348	34.0712	4.0274	72.7400	4.3221	55.5152	4.1075	160.7506	4.5722	147.9199	4.3316	347.5941
E5	4.9297	17.6074	4.6595	54.6252	4.9492	31.0759	4.6546	132.9460	5.1859	87.1221	4.8703	282.9066
E6	4.4696	20.0215	4.3132	50.8045	4.5254	35.2318	4.3175	114.4462	4.6507	107.9729	4.4673	240.8238
F2	4.8025	18.5614	4.5929	47.3507	4.8147	32.8479	4.5510	120.9613	5.0438	95.1472	4.7599	264.5207
F3	4.9392	16.0346	4.6758	45.3633	4.9453	28.7276	4.6394	113.7882	5.1552	85.2261	4.8375	261.4505
F4	4.2593	32.9372	4.0719	69.6199	4.3555	53.6852	4.1428	156.0898	4.6011	143.2986	4.3613	339.3510
V1	5.1420	22.5037	4.9153	53.9265	5.1809	39.1942	4.9480	135.3902	5.4322	106.0917	5.1479	297.2911
V2	4.8899	20.1249	4.6520	56.3232	4.9244	34.8735	4.6597	137.8608	5.1638	96.4503	4.8716	291.7076
V3	4.4346	21.1676	4.2819	52.9715	4.4997	36.5377	4.2895	117.5765	4.6291	113.9263	4.4625	254.9733
V4	4.5998	17.1921	4.2587	58.1925	4.5716	32.6033	4.2503	145.1977	4.7842	94.6309	4.4090	343.4104

Table 4 The electric properties dipole moment (μ), polarizability (α), quadrupole moment (θ), ionization potential, electron affinity and proton affinity, calculated for the propylene oxide–rare-gas atoms systems, and zero point energy (ZPE) for the propylene oxide molecule, compared with reference data

	μ (D)	α (a. u.)	θ (a. u.)	Ionization potential (eV)	Electron affinity (eV)	Proton affinity (kcal/mol)	ZPE (kcal/mol)
He		1.38		24.56	2.63	37.1	
		1.40		24.59		42.5	
		[26]		[27]		[28]	
Ne		2.59		21.53	5.28	42.3	
		2.57		21.57		47.5	
		[26]		[27]		[28]	
Ar		11.08		15.74	2.76	84.9	
		11.23		15.76		88.2	
		[26]		[27]		[28]	
C_3H_6O	2.32	36.11	43.50	9.07	2.15	175.3	53.3823
	2.01			10.22 ± 0.02		192.0	
	[29]			[30]		[28]	

it becomes attractive in the vicinity of the potential-well, where the dispersion and exchange forces play the most important role, thus the contribution of the electrostatic interactions is almost negligible. Even lower is the effect of induction forces, whose attractive character becomes relevant in the short range of distances, where the repulsive contribution of the exchange forces conceals all the other contributions. Table 4 compares the minimum obtained from ab initio and SAPT calculation for the configuration F1. The energies obtained by the SAPT and ab initio method with counterpoise correction present a good agreement, with an error lower than 1% in energy.

4 Final Remarks

The interaction potential of fourteen configurations of the systems propylene oxide–rare-gas-atoms (He, Ne and Ar) have been calculated at the CCSDT(t) level of theory, using the aug-cc-pVDZ basis set. The geometry of propylene oxide, modeled as rigid, has been optimized at various levels of theory and CBS-Q3 gave the best agreement with reference data. The analytical form of the potential energy surface is generated by fitting the single point energies with a fifth-order generalize Rydberg potential function. The potential energy profiles of the three systems present a similar trend and F1 is the most stable leading configuration for all the three rare-gases. As expected,

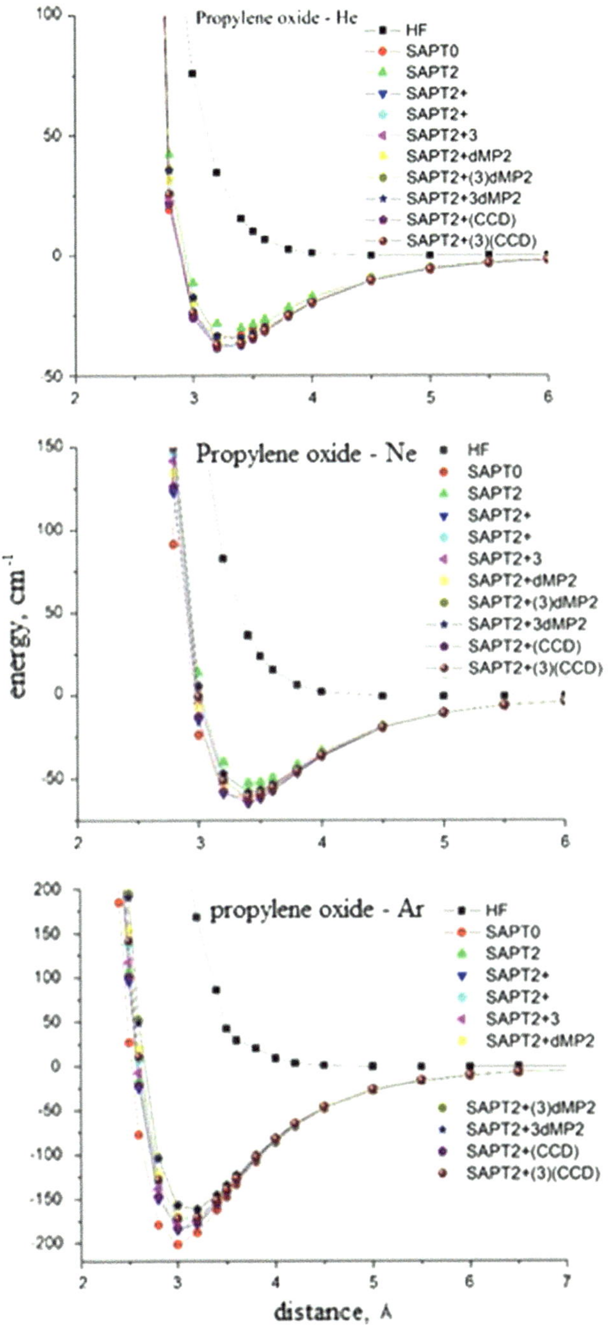

Fig. 6 Interaction potential as a function of the distance, at different SAPT corrections, up to SAPT2+(3) (CCD), for the F1 configuration. Propylene oxide–helium (upper panel), neon (medium panel) and argon (lower panel)

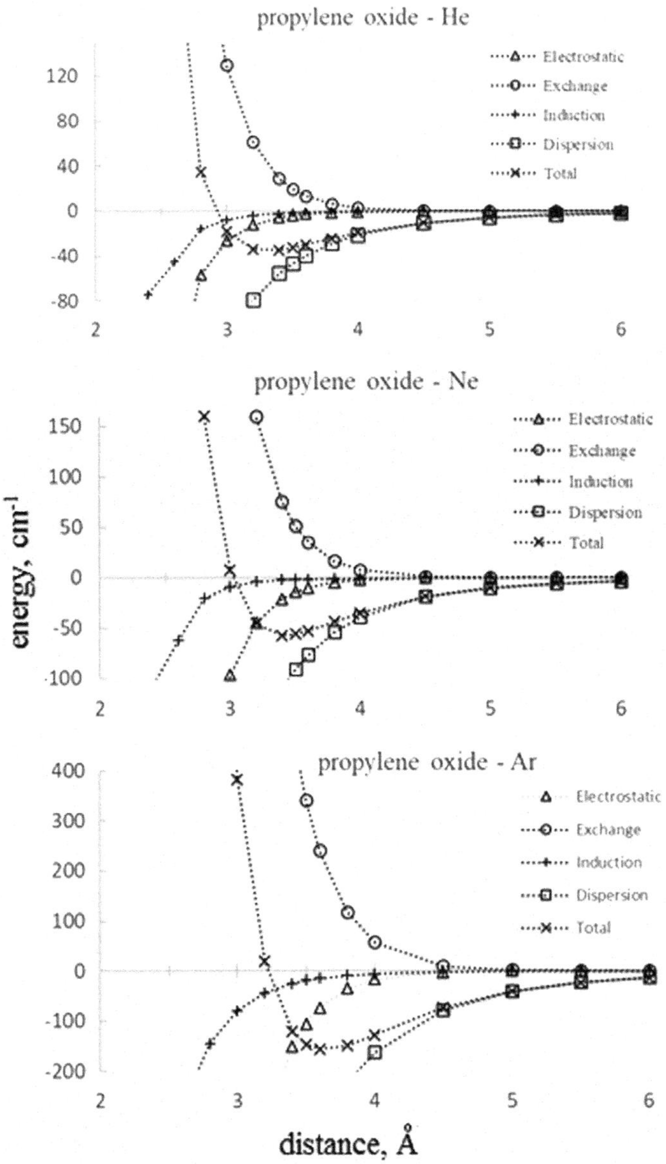

Fig. 7 Contributions to the interaction potential determined at SAPT2+(3)(CCD) level for the F3 configuration: helium (upper panel), neon (middle panel), argon (lower panel)

the minimum energy of the leading configuration become more negative passing from the He to the Ne ad the Ar. Similarly, the equilibrium distances become larger.

The SAPT method has been applied for a preliminary analysis of the contributions to the interaction potentials from the calculated electric properties. The three systems present a similar behavior; the repulsive contribution is given exclusively the exchange interaction, while the dispersion forces play the most important role for what concerns the attractive character in the portion of distances we have considered.

Future development requires experimental investigations on propylene oxide–rare gas collision processes for an accurate analysis on the nature of the interaction potentials (see for example [27–29]) and through quantum mechanical calculations that make use of more accurate basis set functions.

Acknowledgements Federico Palazzetti acknowledges the Italian Ministry for Education, University and Research, MIUR, for financial supporting through SIR 2014 "Scientific Independence for young Researchers" (RBSI14U3VF).

References

1. Aquilanti V, Grossi G, Lombardi A et al (2008) The origin of chiral discrimination: supersonic molecular beam experiments and molecular dynamics simulations of collisional mechanisms. Phys Scr 78:058119. https://doi.org/10.1088/0031-8949/78/05/058119
2. Lombardi A, Maciel GS, Palazzetti F et al (2010) Alignment and chirality in gaseous flows. J Vac Soc Jpn 53:645–653. https://doi.org/10.3131/jvsj2.53.645
3. Falcinelli S, Vecchiocattivi F, Alagia M et al (2018) Double photoionization of propylene oxide: a coincidence study of the ejection of a pair of valence-shell electrons. J Chem Phys 148:114302. https://doi.org/10.1063/1.5024408
4. Che D-C, Palazzetti F, Okuno Y et al (2010) Electrostatic hexapole state-selection of the asymmetric-top molecule propylene oxide. J Phys Chem A 114:3280–3286. https://doi.org/10.1021/jp909553t
5. Che D-C, Kanda K, Palazzetti F et al (2012) Electrostatic hexapole state-selection of the asymmetric-top molecule propylene oxide: rotational and orientational distributions. Chem Phys 399:180–192. https://doi.org/10.1016/j.chemphys.2011.11.020
6. Barreto PRP, Albernaz AF, Aquilanti V et al (2018) Potential energy surface for the interaction of helium with the chiral molecule propylene oxide. Lect Notes Comput Sci 10964:593–604. https://doi.org/10.1007/978-3-319-95174-4_46
7. Faure A, Dagdigian PJ, Rist C et al (2019) Interaction of chiral propylene oxide (CH_3CHCH_2O) with helium: potential energy surface and scattering calculations. ACS Earth Space Chem 3:964–972
8. Elango M, Maciel GS, Palazzetti F et al (2010) Quantum chemistry of C_3H_6O molecules: structure and stability, isomerization pathways, and chirality changing mechanisms. J Phys Chem A 114:9864–9874. https://doi.org/10.1021/jp1034618
9. McGuire BA, Carroll PB, Loomis RA, Finneran IA, Jewell PR, Remijan AJ (2016) Discovery of the interstellar chiral molecule propylene oxide (CH_3CHCH_2O). Science (80-) 352:1449–1452
10. Palazzetti F, Maciel GS, Lombardi A et al (2012) The astrochemical observatory: molecules in the laboratory and in the cosmos. J Chin Chem Soc 59:1045–1052. https://doi.org/10.1002/jccs.201200242
11. Lombardi A, Palazzetti F, Aquilanti V et al (2017) The astrochemical observatory: experimental and computational focus on the chiral molecule propylene oxide as a case study. Lect Notes Comput Sci 10408:267–280. https://doi.org/10.1007/978-3-319-62404-4_20

12. Aquilanti V, Caglioti C, Casavecchia P et al (2017) The astrochemical observatory: computational and theoretical focus on molecular chirality changing torsions around O–O and S–S bonds. AIP Conf Proc 1906:030010. https://doi.org/10.1063/1.5012289
13. Barreto PRP, Vilela AFA, Lombardi A et al (2007) The hydrogen peroxide-rare gas systems: quantum chemical calculations and hyperspherical harmonic representation of the potential energy surface for atom-floppy molecule interactions. J Phys Chem A 111:12754–12762. https://doi.org/10.1021/jp076268v
14. Maciel GS, Barreto PRP, Palazzetti F et al (2008) A quantum chemical study of H2S2: intramolecular torsional mode and intermolecular interactions with rare gases. J Chem Phys 129:164302. https://doi.org/10.1063/1.2994732
15. Barreto PRP, Ribas VW, Palazzetti F (2009) Potential energy surface for the H_2O-H_2 system. J Phys Chem A 113:15047–15054. https://doi.org/10.1021/jp9051819
16. Frisch MJ, Trucks GW, Schlegel HB, et al Gaussian 09 Revision E.01
17. Werner H-J, Knowles PJ, Knizia G et al (2012) Molpro: a general-purpose quantum chemistry program package. WIREs Comput Mol Sci 2:242–253
18. Albernaz AF, Barreto PRP, Aquilanti V et al (2018) The astrochemical observatory: the interaction between helium and the chiral molecule propylene oxide. AIP Conf Proc 2040:020018. https://doi.org/10.1063/1.5079060
19. Parrish RM, Burns LA, Smith DGA, Simmonett AC, DePrince AE III, Hohenstein EG, Bozkaya U, Sokolov AYu, Di Remigio R, Richard RM, Gonthier JF, James AM, McAlexander HR, Kumar A, Saitow M, Wang X, Pritchard BP, CDS P (2017) An open-source electronic structure program emphasizing automation, advanced libraries, and interoperability. J Chem Theory Comput 13:3185–3197
20. Parker TM, Burns LA, Parrish RM, Ryno AGSCD (2014) J Chem Phys 140:094106
21. Jeziorski B, Moszynski RSK (1994) Chem Rev 94:1887
22. Montgomery JAJ, Frisch MJ, Ochterski JW (2000) A complete basis set model chemistry. VII. Use of the minimum population localization method. J Chem Phys 112:6532
23. Aquilanti V, Cappelletti D, Pirani F (1996) Range and strength of interatomic forces: dispersion and induction contributions to the bonds of dications and of ionic molecules. Chem Phys 209:299–311
24. Cambi R, Cappelletti D, Liuti G, Pirani F (1991) Generalized correlations in terms of polarizability for van der Waals interaction potential parameter calculations. J Chem Phys 95:1852
25. Barreto PRP, Palazzetti F, Grossi G et al (2010) Range and strength of intermolecular forces for van der Waals complexes of the type H_2X_n-Rg, with $X = O$, S and $n = 1, 2$. Int J Quantum Chem 110:777–786. https://doi.org/10.1002/qua.22127
26. Pirani F, Maciel GS, Cappelletti D, Aquilanti V (2006) Experimental benchmarks and phenomenology of interatomic forces: open-shell and electronic anisotropy effects. Int Rev Phys Chem 25:165–199
27. Lombardi A, Faginas-Lago N, Grossi G et al (2016) Collisional energy exchange in CO_2–N_2 gaseous mixtures. Lect Notes Comput Sci (including Subser Lect Notes Artif Intell Lect Notes Bioinform) 9786:246–257. https://doi.org/10.1007/978-3-319-42085-1_19
28. Lombardi A, Palazzetti F (2008) A comparison of interatomic potentials for rare gas nanoaggregates. J Mol Struct THEOCHEM 852:22–29. https://doi.org/10.1016/j.theochem.2007.12.011
29. Albernaz AF, Aquilanti V, Barreto PRP et al (2016) Interactions of hydrogen molecules with halogen-containing diatomics from ab initio calculations: spherical-harmonics representation and characterization of the intermolecular potentials. J Phys Chem A 120:5315–5324. https://doi.org/10.1021/acs.jpca.6b01718
30. Olney TN, Cann NM, Cooper G (1976) Atomic and molecular physics, vol 7: structure data of free polyatomic molecules

A Theoretical Study of the Preferred Reaction Mechanism Between Chloroacetic Acid and Thiourea

Mwadham M. Kabanda and Kgalaletso P. Otukile

Abstract 2-iminothiazolidin-4-one derivatives are known to possess various biological activities such as anticancer, antiviral, antibacterial and antifungal, making them interesting lead compounds in the design of pharmaceutically useful drugs. For instance, recent studies have shown that 2-imino-thiazolidin-4-one derivatives are vital scaffold for the design of potent, orally active lysophospholipid receptor agonists selected for clinical development. Despite their potential application in drug development, the reaction mechanism for the formation of 2-iminothiazolidin-4-one is still a challenge. The current theoretical work attempts to study the mechanistic pathways for the formation of 2-iminothiazolidin-4-one from chloroacetic acid and thiourea. Calculations were performed with MP2 and DFT methods (with M06-2X and M11-L functionals) using the 6-31+G(d,p) and the 6-311++G(3df,2p) basis sets. The study was performed in vacuo and in water solution. The influence of water was assessed by predicting its role as bulk solvent. The results of the study show that the reaction between chloroacetic acid and thiourea may involve the iminothiol form of thiourea rather than the thione form. The reaction is spontaneous and prefers to occur in vacuo to occurring in water solution.

Keywords DFT calculations · Nucleophilic substitution · Tautomerism · Hydrogen bonding

M. M. Kabanda (✉) · K. P. Otukile
Department of Chemistry, Faculty of Natural and Agricultural Sciences, North-West University, Private Bag X2046, Mmabatho 2735, South Africa
e-mail: mwadham.kabanda@nwu.ac.za

Material Science Innovation and Modelling (MaSIM) Research Focus Area, Faculty of Natural and Agricultural Sciences, North-West University, Private Bag X 2046, Mmabatho 2735, South Africa

L. Mammino et al. (eds.), *Advances in Quantum Systems in Chemistry, Physics, and Biology*, Progress in Theoretical Chemistry and Physics 32,
https://doi.org/10.1007/978-3-030-34941-7_7

1 Introduction

Thiazolidin-4-one derivatives are known for their biological activities such as anti-diabetic [1], anti-cancer [2], anti-HIV and platelet activating factor antagonist [3, 4]. In particular, functionalised 2-iminothiazolidin-4-one heterocyclic derivatives exhibit several biological activities such as anti-tuberculosis [5], anti-convulsant [6], anti-fungal [7], antibacterial [8], antimicrobial [9], antiviral [10] and immunodulation properties [11]. Recent studies have shown that 2-iminothiazolidin-4-one derivatives are already considered as potential prodrug compounds because of their potent, orally active Sphingosine-1-phosphate ($S1P_1$) receptor agonist [12]. The potentiality of their applications in the design of pharmaceutical drugs has been the driving force for the large number of synthetic protocols leading to the synthesis of 2-iminothiazolidin-4-ones [13–15]. The traditional approach for the synthesis of 2-iminothiazolidin-4-ones involved cyclization from thioureas with chloroacetic acid derivatives via using various catalysts, such as sodium acetate, sodium hydride, and pyridine [10, 11, 16–19]. Several studies have recently reported on the possible green pathways in the preparation of 2-iminothiazolidin-4-ones [14, 15].

To our knowledge, the mechanism for the reaction of chloroacetic acid ($ClCH_2COOH$) with thiourea to form 2-iminothiazolidin-4-one has been studied through experimental techniques [13, 14] and has not yet been studied through theoretical methods. Therefore, to shed light on this point, this study presents a detailed computational investigation into the mechanism of formation of 2-iminothiazolidin-4-one. The specific objectives of the study are to identify the intermediate and transition structures and to characterise their geometric features. In experimental conditions, the synthesis of 2-iminothiazolidin-4-ones has been reported to occur preferably in water solution [15]. To simulate such conditions, the study reported here is performed in vacuo and in water solution in order to analyze the specific influences of the solvent on the process as a medium, but also its possible interaction with the reactants and its role on the mechanistic pathway.

To model the mechanism of the process in the gas phase and in aqueous solution, two density functional theory (DFT) methods and an MP2 method were selected. DFT with hybrid GGA functionals, such as MPWB1K, and meta-hybrid functionals (e.g. M06-2X) are increasingly utilised for the study of reaction mechanisms for organic compounds and for characterising the kinetic features of chemical reactions [20–27]. Moreover, because of their computational affordability, DFT with hybrid meta-GGA functionals are also increasingly providing benchmark results for systems where MP2 and its variations or CCSD(T) methods are too expensive to afford [23–26]. In this work, two functionals, the M06-2X (a global-hybrid meta-GGA) and M11-L (a meta-nonseparable gradient approximations functional) are utilised throughout the study.

The schematic representation of the starting reactant molecules, chloroacetic acid and thiourea, and the atom numbering utilised in this work are shown in Fig. 1. The reaction mechanism hypothesised and investigated in the present study is illustrated in Fig. 2. Thiourea can exist in two tautomeric forms, the thione and iminothiol, as

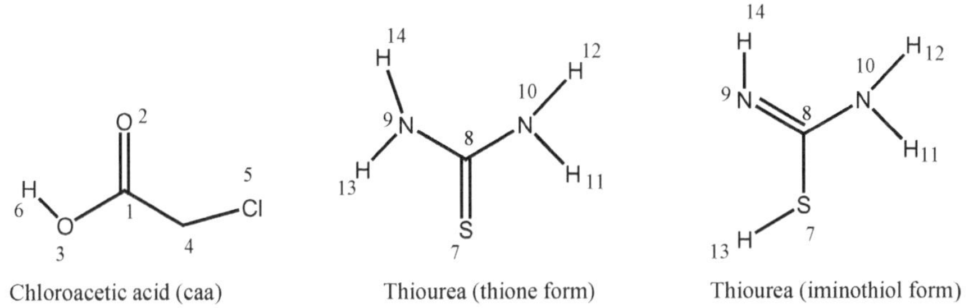

Fig. 1 The structures of the starting reactant molecules and the atom numbering utilised in the discussion

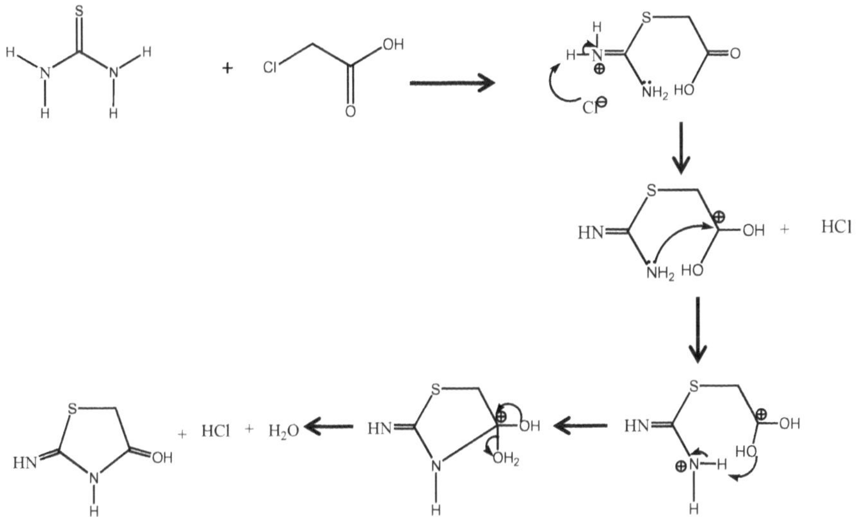

Fig. 2 Schematic representation of the plausible mechanism for the reaction between chloroacetic acid and thiourea

shown in Fig. 1. In a situation in which the keno-enol (amido-iminol) tautomerism occurs, theoretical studies suggest that the thione form (NH_2CSNH_2) is the most stable and indeed many active drugs exist in this stable form [28, 29]. However, in other studies, both experimental studies (mass spectra) and theoretical studies indicate that the iminothiol form ($NHCSHNH_2$) has a higher relative tendency to occur than the thione form [29, 30]. Moreover, although $NHCSHNH_2$ are less stable (and therefore extremely difficult to isolate experimentally) and are present to a small extent at equilibrium, they are nevertheless extremely important because they are highly reactive. In order to investigate the preferred form reacting with chloroacetic acid, both tautomers of thiourea are utilised. Throughout the rest of the text, the thione form is referred to as thiourea and the iminothiol form is called isothiourea.

2 Computational Details

The geometry optimisations of the reactants, transition states and products were performed with the DFT method and the MP2 method. DFT calculations were performed utilising the M06-2X and the M11-L functionals and using the 6-31+G(d,p) and 6-311++(3df,2p) basis set. These basis sets are extensively utilised in the study of chemical reactions and specifically in the characterisation of transition state geometries [20–35]. The reaction transition states (TSs) were located using the synchronous transit-guided quasi-newton technique for the saddle point search (QST3 module).

Frequency calculations were performed, at the same level of calculations as the optimised geometry procedure, on fully optimized conformers to determine the nature of the stationary points. For ground state geometries no imaginary frequencies were observed, whereas there was only one imaginary frequency for the transition state geometries. Zero-point and thermal enthalpies corrections, computed at T = 298 K and pressure of 1 atm in the rigid rotor harmonic oscillator approximation, were performed to obtain free energies. Solvent effects on geometries and relative conformational stabilities were taken into consideration using the continuum solvation model density (SMD) [23], in which, unlike other continuum models, the full solute electron density is used without defining partial atomic charges.

All calculations were performed with Gaussian09 [36]. The schematic representations were drawn using the ChemOffice package in the UltraChem 2010 [37] version and conformers were drawn using GaussView5 program [38].

The Quantum Mechanics Atoms in Molecule (QMAIM) was performed using the AIMAll [39]. The wavefunctions necessary for utilisation in the QMAIM program were calculated on the basis of fully optimised geometries using the same level of accuracy as in the optimisation. The number of critical points (CPs) found for all of the analysed systems are in agreement with the Poincare-Hopf rule. The electron density (ρ) and its Laplacian ($\nabla^2\rho$) at the bond critical point (BCP) were calculated in order to characterise the various structures. According to the topological analysis of electronic charge density, in the theory of the atoms in molecules (AIM [40]), electron density () and Laplacian of the electron density, (∇^2), are used to describe the strength and the characteristic of the bond, respectively. The Laplacian (∇^2) is the sum of λ_1, λ_2, and λ_3, where λ_i is the th eigenvalue of the Hessian matrix of the electronic density. In general, when $\nabla^2 < 0$ the bond is covalent, but when $\nabla^2 > 0$ the bond belongs to the electrostatic interaction.

3 Results and Discussions

3.1 $NHCSHNH_2$ + $ClCH_2COOH$ Reaction Mechanism in Vacuo

The optimised geometries, associated to the stationary points located on the potential energy surface, are shown in Fig. 3. The relative energies of the isolated reactant molecules (**R**$_{\mathbf{iso}}$), pre-association complexes (**RC**), transition state (**TS**) structures, intermediates (**IM**) and products of the reaction of $ClCH_2COOH$ with $NHCSHNH_2$ are shown in Table 1. The relative energies refer to the sum of the energies of the separate reactants. The selected bond distances of the optimised structures, necessary for the discussion, are reported in Table 2 as well as indicated in Fig. 3. The mechanism for the cyclisation is better shown through the potential energy diagram, which is shown in Fig. 4. Unless otherwise specified, relative energies and bond distances referred to in the discussion correspond to M06-2X/6-311++G(3df,2p) results. The reaction is assumed to take place in several steps. In the first step, the pre-association

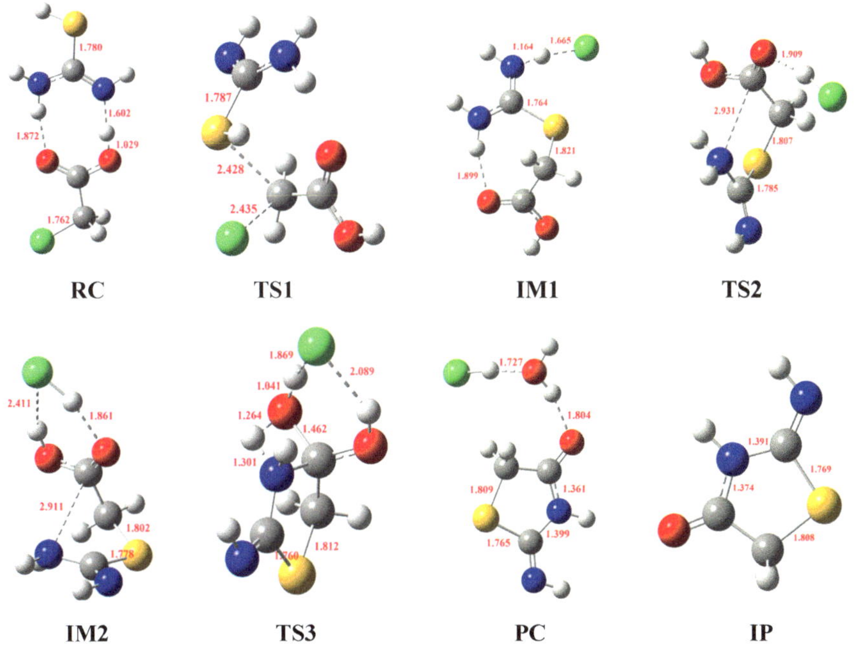

Fig. 3 In vacuo geometries (reactant complex, transition states and product complexes) computed at the M06-2X/6-311++G(3df,2p) level for the reaction between chloroacetic acid and the iminothiol form of thiourea. The sulfur atom is depicted in yellow, oxygen in red, chlorine in green, nitrogen in blue, carbon in grey and hydrogen in grey. The relevant interatomic distances are given in angstroms; covalent bonds are indicated with solid lines while non-covalent bonds (including hydrogen bonds) are indicated with dotted lines

Table 1 Relative energies (ΔE, kcal/mol) and relative free energies (ΔG, kcal/mol) for the geometries of the species involved in the reaction between thiourea and chloroacetic acid, results with different methods

Results of the study of the reaction of the iminothiol form of thiourea with chloroacetic acid						
Iminothiol form	M06-2x/6-31+G(d,p)		M11-L/6-31+G(d,p)		MP2/6-31+G(d,p)	
	ΔE	ΔG	ΔE	ΔG	ΔE	ΔG
R_{iso}	0.000	0.000	0.000	0.000	0.000	0.000
RC-t	−18.706	−7.702	−14.881	−4.088	−17.668	−6.943
TS1-t	60.368	72.580	50.635	62.407	66.808	79.256
IM1	−17.081	−5.320	−14.516	−2.692	−16.806	−6.196
TS2	−24.271	−12.303	−6.907	3.735	−10.280	0.702
IM2	−8.993	1.938	−7.598	3.171	−10.138	0.714
TS3	25.321	39.073	30.474	44.171	30.097	44.027
PC	−19.345	−9.272	−19.728	−9.780	−22.897	−12.351
P_{iso}	−5.032	−11.142	−7.818	−14.173	−9.111	−15.579
Iminothiol form						
R_{iso}	0.000	0.000	0.000	0.000	0.000	0.000
RC-t	−17.400	−6.474	−15.390	−4.862	−14.689	−4.263
TS1-t	60.776	72.945	58.211	69.921	53.959	65.796
IM1	−15.227	−3.374	−29.624	−18.230	−13.019	−1.217
TS2	−7.532	4.337	−7.983	2.511	−5.826	5.057
IM2	−8.906	1.810	−8.794	1.868	−7.291	3.520
TS3	28.706	42.341	23.695	37.146	32.387	45.958
PC	−18.237	−8.346	−22.105	−9.893	−18.808	−9.207
P_{iso}	−6.029	−12.240	6.456	1.479	−8.6213	−14.993
Results of the study of the reaction of the thione form of thiourea with chloroacetic acid						
Thione form						
R iso	0.000	0.000	0.000	0.000	0.000	0.000
RC	−24.581	−15.353	−20.243	−10.121	−22.295	−11.731
TS1	21.626	32.678	18.493	29.356	26.422	37.919
IM1	−11.928	−0.548	−8.980	2.641	−13.434	−3.024
TS2	−19.118	−7.531	−1.371	9.068	−6.908	3.874
IM2	−3.840	6.709	−2.062	8.504	−6.766	3.886
TS3	30.474	43.845	36.010	49.504	33.469	47.199
PC	−14.192	−4.501	−14.192	−4.447	−19.526	−9.179
P iso	0.121	−6.371	−2.282	−8.840	−5.739	−12.407

(continued)

Table 1 (continued)

Thione form						
R_{iso}	0.000	0.000	0.000	0.000	0.000	0.000
RC	−14.526	−4.881	−6.771	3.460	−12.148	−1.929
TS1	30.880	42.455	18.231	29.522	28.623	39.568
IM1	−2.243	9.615	−11.351	0.023	−1.149	10.633
TS2	5.4512	17.326	10.291	20.764	6.044	16.906
IM2	4.077	14.799	9.479	20.120	4.579	15.370
TS3	41.689	55.330	41.968	55.399	44.257	57.807
PC	−5.254	4.643	−3.832	8.360	−6.938	2.642
P_{iso}	6.954	0.749	24.730	19.732	3.249	−3.144

complex is formed from $ClCH_2COOH$ and $NHCSHNH_2$. Several starting geometries were considered, but the structure in Fig. 3 was determined to be the global minimum of those studied. This is a barrier-less weakly bound pre-association reactant complex (**RC-t**) stabilised by two simultaneous intermolecular hydrogen bonds (H-bonds) involving the COOH group of chloroacetic acid and the amino (NH_2) and the imino (N=C) group of isothiourea. The OH group of the acid is the H-bond donor to the N atom of the C=N group and the C=O group is the H-bond acceptor for the NH_2. The O3–H6···N9 intermolecular H-bond is shorter (1.602 Å) than the O2···H11–N10 (1.872 Å) because the sp^2 N atom has greater tendency to donate electrons than the sp^2 O atom as a result of the difference in their electronegativity values. **RC-t** is 17.400 kcal/mol more stable than the sum of the energies of the starting reactant molecules.

The second step of the reaction is the formation of the intermediate **IM1** from the reactant complex **RC-t** by passing through the first transition state (**TS1-t**), which possesses an imaginary frequency of −472.70i cm^{-1}. **TS1-t** is characterised by the attack of the C4 atom by the S atom and the subsequent removal of the Cl5 atom. The imaginary vibrational mode of **TS1-t** corresponds to the formation of a C4–S7 single bond and a simultaneous elongation of the C4–Cl5 bond. The bond lengths of the essentially broken C4–Cl5 bond (2.435 Å) and the concurrently forming C4–S7 bond (2.428 Å) and the small absolute value of the imaginary frequency indicate that this is a loose SN_2 transition state structure. Loose TS structures have also been identified for a number of other concerted SN_2 reactions such as those of glycosides [41–44]. The N10–H11···O2 intermolecular H-bond that is present in **RC-t** is maintained in **TS1-t**. **TS1-t** is found to be 60.776 kcal/mol higher than the starting reactant molecules and ≈76 kcal/mol above the complex reactant molecule.

IM1 is characterised by the formation of the much shorter C4–S7 bond (1.782 Å) than in **TS1-t** and also by the presence of the N–H···O intramolecular H-bond (1.899 Å) involving the N–H and the C=O group. **IM1** is also characterised by the concomitant formation of the HCl (H13···Cl5) molecule as a result of the Cl5 anion abstracting the H13 proton, thereby permitting the rebalance of electronic charge

Table 2 Representative bond lengths (Å) and bond distances (Å) for the isolated reactants complex, transition states intermediates and products of the reaction of chloroacetic acid (acc) and isothiourea

Structure	C1–O2	C4–Cl5	C1–O3	C1–C4	C8–N10	C8–S7	S7–C4	N10–C1	N10–H11	N9–H13
M06-2X/6-31+G(d,p)										
RC	1.213	1.771	1.318	1.518	1.336	1.695	–	–	1.020	1.008
Trans1	1.204	2.430	1.340	1.499	1.336	1.710	2.178	–	1.012	1.009
Inter1	1.214	5.171	1.330	1.508	1.346	1.775	1.828	3.230	1.020	1.188
Tran1b	1.207	3.808	1.338	1.508	1.387	1.791	1.814	2.918	1.012	1.022
Inte1b	1.210	–	1.336	1.514	1.392	1.787	1.813	2.895	1.012	1.021
Trans1c	1.337	–	1.464	1.525	1.441	1.769	1.822	1.514	1.303	1.020
Prod-2	1.220	–	3.301	1.515	1.404	1.774	1.820	1.364	3.877	1.020
Prod-iso	1.208	–	–	1.522	1.396	1.779	1.819	1.378	–	1.020
M06-2X/6-311++G(3df,2p)										
RC	1.204	1.761	1.314	1.516	1.332	1.685	–	–	1.017	1.005
Trans1	1.195	2.408	1.337	1.496	1.332	1.700	2.164	–	1.009	1.006
Inter1	1.205	5.075	1.328	1.504	1.342	1.764	1.821	3.196	1.016	1.164
Tran1b	1.203	3.801	1.337	1.507	1.385	1.785	1.807	2.931	1.010	1.020
Inte1b	1.201	–	1.332	1.513	1.385	1.778	1.803	2.911	1.009	1.018
Trans1c	1.332	–	1.462	1.524	1.434	1.760	1.812	1.512	1.301	1.018
Prod-2	1.211	–	3.315	1.513	1.399	1.765	1.809	1.361	3.867	1.017
Prod-iso	1.199	–	–	1.520	1.391	1.769	1.808	1.374	–	1.017

(continued)

Table 2 (continued)

Structure	C1–O2	C4–Cl5	C1–O3	C1–C4	C8–N10	C8–S7	S7–C4	N10–C1	N10–H11	N9–H13
M11-L/6-31+G(d,p)										
RC	1.198	1.745	1.304	1.508	1.331	1.664	–	–	1.015	1.007
Trans1	1.192	2.433	1.322	1.488	1.329	1.676	2.250	–	1.008	1.007
Inter1	1.204		1.311	1.496	1.331	1.745	1.821	3.154	1.019	1.082
Tran1b	1.201	–	1.317	1.500	1.371	1.773	1.799	3.014	1.011	1.023
Inte1b	1.198	–	1.316	1.505	1.378	1.764	1.785	2.880	1.012	1.021
Trans1c	1.309	–	1.473	1.512	1.422	1.748	1.792	1.481	1.306	1.019
Prod-2		–								
Prod-iso	1.195		–	1.509	1.379	1.753	1.791	1.361	–	1.019
M11-L/6-311++G(3df,2p)										
RC	1.187	1.732	1.294	1.502	1.320	1.654	–	–	1.013	1.006
Trans1	1.181	2.382	1.313	1.481	1.319	1.665	2.261	–	1.004	1.003
Inter1	1.192		1.303	1.490	1.322	1.732	1.804	3.122	1.016	1.082
Tran1b	1.190	–	1.309	1.494	1.363	1.757	1.781	3.018	1.008	1.020
Inte1b	1.187	–	1.306	1.500	1.367	1.751	1.770	2.884	1.009	1.019
Trans1c	1.301	–	1.462	1.507	1.410	1.735	1.778	1.476	1.301	1.017
Prod-2	1.193	–	3.419	1.495	1.343	1.736	1.776	1.343	3.899	1.017
Prod-iso	1.183	–	–	1.503	1.369	1.739	1.775	1.353		1.017

(continued)

Table 2 (continued)

Structure	C1–O2	C4–Cl5	C1–O3	C1–C4	C8–N10	C8–S7	S7–C4	N10–C1	N10–H11	N9–H13
MP2/6-31+G(d,p)										
RC	1.227	1.766	1.334	1.515	1.348	1.677	–	–	1.018	1.009
Trans1	1.220	2.417	1.357	1.489	1.344	1.693	2.174	–	1.009	1.008
Inter1	1.227	5.498	1.350	1.506	1.370	1.778	1.821	3.288	1.016	1.775
Tran1b	1.222	–	1.362	1.501	1.387	1.784	1.826	3.504	1.012	1.022
Inte1b	1.223	–	1.352	1.511	1.399	1.778	1.810	2.939	1.012	1.023
Trans1c	1.343	–	1.491	1.518	1.446	1.764	1.819	1.510	1.306	1.021
Prod-2	1.234	–	3.270	1.516	1.406	1.768	1.817	1.368	3.829	1.021

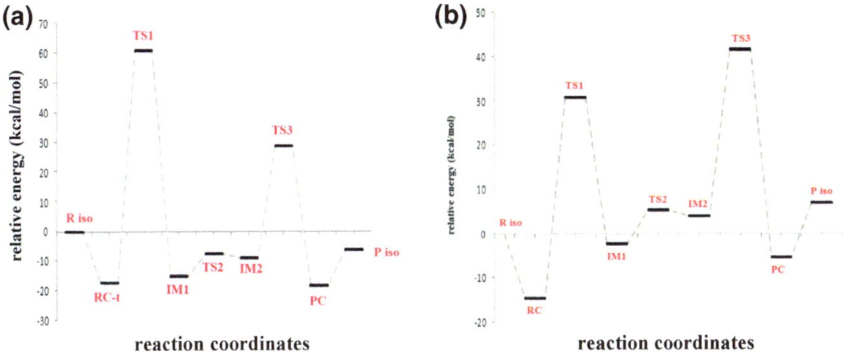

Fig. 4 Energy diagram of the gas phase mechanism for the unassisted reaction between chloroacetic acid and the two taumers of thiourea: **a** iminothiol + chloroaceti acid and **b** thione + chloroacetic acid, M06-2X/6-311+G(3df,2p) results in vacuo. The acronyms RC, TS and IM, PC and P_{iso} denote reactant complex, transition state, intermediate, product complex and isolated product respectively

throughout the complex. With all the methods, ΔE and ΔG values (Table 1) suggest that the conversion of **RC-t** to **IM1** is unfavourable. The second intermediate (**IM2**) is formed from **IM1** by going through the second transition state (**TS2**), which possess an imaginary frequency of $-17.50i\ cm^{-1}$. The constituents of **TS2** molecular entity are arranged in such a way that N10 is in nearby proximity to C1 and the N10···C1 bond distance (Å) is 2.931; the O2 atom forms an intermolecular H-bond (1.909 Å) with the HCl group. In **IM2**, the N10···C1 bond length is slightly shorter than in **TS2**. Therefore, the difference between **IM1** and **IM2** is that in **IM2**, the C1 atom and the N10 are oriented towards each other and the HCl molecule is bonded to the O2 atom.

The third step of the mechanism is the formation of the product from **IM2** by going through the third transition state (**TS3**), whose imaginary vibrational mode (−1647.14i) corresponds to the breaking of the N10–H11 bond and the simultaneous formation of the O3–H11 bond. **TS3** is characterised by the formation of the N10–C1 bond (1.512 Å), resulting in the tetrahedral character of the C1 atom. This step is possible because the proximity of the N10 and C1 atoms in **IM2** allows the lone pair of electrons on the amine group (hard nucleophile) to attack the C atom of the carbonyl group (hard electrophile). The attack on the carbonyl carbon atom is concomitantly followed by the conversion of the C=O group to a single bond, with the O2 atom acquiring a negative charge. Of the two negative charges, $O2^-$ and $Cl5^-$, the $O2^-$ has greater tendency to attract the proton and therefore it becomes protonised by accepting the proton from the nearby H13···Cl5 group. The $Cl5^-$ ion is then stabilised by weakly interacting with both the neighbouring OH groups. The transition structure is also characterised by the concomitant movement of the H11 proton from N10 to O3; the N10–H11 bond length is 1.301 Å, which is an indication of the elongation of the N10–H11 bond and the O3–H11 bond length is 1.264 Å. In this way, the proton transfer from the N10 atom is nearly complete, while the protonation of the

O3 atom is not yet complete (because the 1.264 Å O–H bond distance is quite large in comparison to the standard 0.97 Å O–H bond distance). When the transfer of the H11 proton from N10 to O3 is complete, it results in the formation of the H_2O group, which is an easy leaving group. The IRC results indicates that the protonation of the carbonyl O2, the formation of the oxocyclic C1–N10, coupled with the release of the H11 from N10 and its subsequent attachment to O3 (resulting in the formation of the H_2O bonded to C1) may occur as a concerted reaction step involving asynchronous bond-formation and bond-breaking processes.

The final stage in the process is the formation of the product from **TS3** by the elimination of the leaving group (the water molecule) and the simultaneous restoration of the sp^2 character of the C1 atom (i.e., the simultaneous conversion of the C–O back to C=O) as a result of the concomitant transfer of the proton from O2H13 to the chloride ($Cl5^-$) ion. The formation of the product from **IM2** has an activation energy barrier of 37.612 kcal/mol; however, the product is only 6.029 kcal/mol lower than the starting isolated reactants.

An examination of the relative energies obtained with all the methods also suggest that the products have lower energy than the starting reactant molecules, which implies that the process is exothermic. The ΔG values suggest that the reaction is exergonic, which implies that it is thermodynamically feasible to occur. The energetic values (reported in Table 1 and represented graphically in Fig. 4) also show that the step connecting **RC-t** with **TS1-t** is the rate-determining step of the investigated reaction, with energy lying about 60.776 kcal/mol above the energy of the free reactants.

To obtain more information on the transition structures, intermediates and the resulting complex product, NBO and QTAIM analyses were performed, which provide deeper insight into their charge and electron distributions. Figure 5 shows the molecular graphs along with the NBO charges on the structures, while Table 3 reports properties related to bond critical point data (i.e., the electron density (ρ) and the Laplacian of the electron density, $\nabla^2\rho$). NBO charges on **TS1** show a regular charge distribution, in which the leaving hydrogen atom (H13) of the S atom has a positive charge of 0.035 a.u. and the leaving chloride ion has a negative charge of −0.532 a.u. (which is higher than in **RC-t**). The S atom also attains a higher positive charge than in the **RC-t** structure. In **TS2**, the electrostatic stabilisation contributes to the lowering of the negative charge on the chloride ion and an increase in the positive charge on the S atom.

In **RC-t**, the Laplacians of the electron density for all the bonds, with the exception of the O2···H11 and H6···N9 bonds, are covalent in nature. The Laplacians of the electronic charge density of the O2···H11 and H6···N9 bonds are 0.1045 and 0.0795 a.u, respectively. The low and positive values of $\nabla^2\rho$ for both the O2···H11 and H6···N9 bonds of the **RC-t** complex show that there is a weak intermolecular hydrogen bond interaction between the carbonyl O atom of chloroacetic acid and the hydrogen atom of the amino group and between the hydroxyl H atom of chloroacetic acid and the imine N atom [45–47]. The electronic charge density (ρ) corresponding to both O2···H11 and H6···N9 bonds are 0.0288 and 0.0669 au, respectively. These

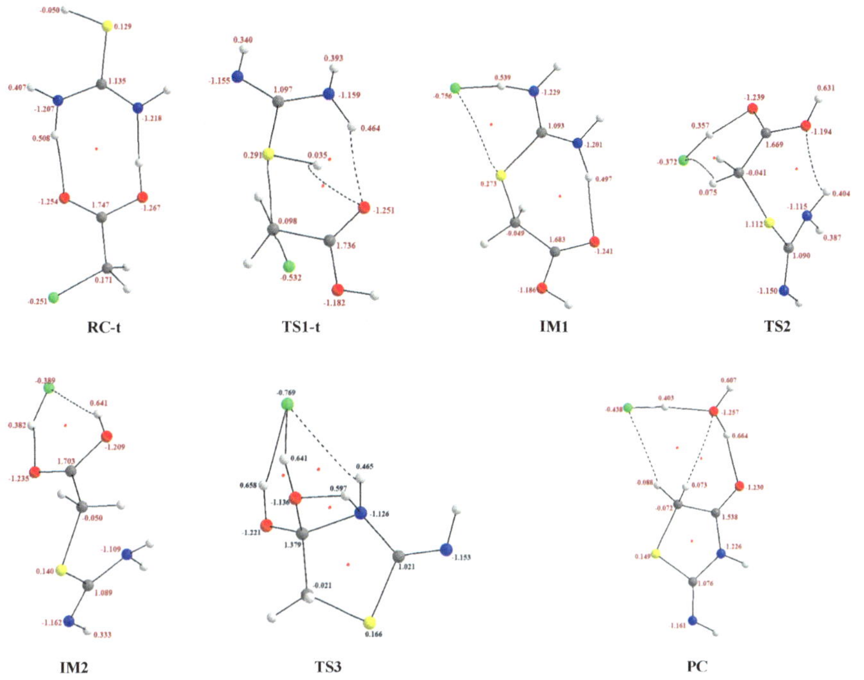

Fig. 5 Molecular graphs with selected NBO charges for the studied geometries (reactant complex, transition states and product complexes) for the reaction between chloroacetic acid and the iminothiol form of thiourea

values suggest that the O2···H11 H-bond is weaker than the H6···N9 H-bond, which is also confirmed by the bond length parameters discussed earlier.

In **TS1-t**, the value of the charge density (ρ) for the C4–Cl5 bond is relatively small (0.050), while $\nabla^2\rho$ has a value of 0.070. These values characterise local charge depletion on the C4–Cl5 bond (as seen from the decrease in the $\nabla^2\rho$ values between reactant complex and **TS1-t**). An analysis of the $\nabla^2\rho$ value for the C4–S7 bond for the **TS1-t** structure suggests closed-shell interactions. However, the C4–S7 bond becomes a covalent bond in nature in all other structures (i.e., **IM1**, **TS2**, **IM2**, **TS3** and **PC**), as seen from the negative value of $\nabla^2\rho$. Each of the two C–O groups in **TS3** has a bond length of 1.362 Å, the C–N bond is 1.512 Å and the C1–C4 bond is 1.524 Å. These geometric parameters are characteristic of a tetrahedral C atom. The computed results are in qualitative agreement with experimental results [14] both concerning the existence of intermediate complexes and adducts and the fact that the thiol group is the most probable tautomer for interaction with chloroacetic acid.

Table 3 Topological properties of the electronic Density $\rho(r)$ at relevant bond critical points (BCPs) for the Species Involved in the cyclization reaction between thiourea and chloroacetic acid

Results for the reaction involving the imidothiol form of thiourea											
Bond distance	Critical point data	caa	tu-t	RC-t	TS1-t	IM1	TS2	IM2	TS3	PC	P iso
C1–O2	ρ	0.444		0.426	0.430	0.430	0.431	0.432	0.329	0.423	0.435
	$\nabla^2\rho$	−0.202		−0.315	−0.344	−0.328	−0.303	−0.290	−0.856	−0.357	−0.282
C4–Cl5	ρ	0.194		0.192	0.050						
	$\nabla^2\rho$	−0.274		−0.267	0.070						
C1–O3	ρ	0.307		0.336	0.321	0.319	0.312	0.316	0.248		
	$\nabla^2\rho$	−0.583		−0.593	−0.606	−0.569	−0.575	−0.579	−0.591		
C1–C4	ρ	0.259		0.258	0.272	0.263	0.259	0.258	0.255	0.255	0.252
	$\nabla^2\rho$	−0.633		−0.628	−0.723	−0.653	−0.636	−0.624	−0.605	−0.600	−0.582
C8–N10	ρ		0.316	0.391	0.321	0.338	0.309	0.308	0.278	0.295	0.301
	$\nabla^2\rho$		−0.991	−1.266	−0.998	−1.084	−0.938	−0.934	−0.764	−0.856	−0.883
C8–S7	ρ		0.192	0.196	0.197	0.201	0.196	0.198	0.203	0.202	0.201
	$\nabla^2\rho$		−0.328	−0.351	−0.360	−0.387	−0.354	−0.362	−0.384	−0.379	−0.372
S7–C4	ρ				0.054	0.177	0.182	0.184	0.182	0.183	0.184
	$\nabla^2\rho$				0.047	−0.262	−0.285	−0.298	−0.285	−0.291	−0.291
N10–C1	ρ								0.245	0.322	0.313
	$\nabla^2\rho$								−0.558	−0.989	−0.944
N10–H11	ρ		0.346	0.067		0.335	0.343		0.150		
	$\nabla^2\rho$		−1.778	+.080		−1.957	−1.776		−0.218		
O3–H11	ρ								0.153		

(continued)

Table 3 (continued)

Results for the reaction involving the imidothiol form of thiourea											
Bond distance	Critical point data	caa	tu-t	RC-t	TS1-t	IM1	TS2	IM2	TS3	PC	P iso
	$\nabla^2\rho$								−0.181		
Results involving the thione form of thiourea											
Bond distance	Critical point data	caa	tu-t	RC-t	TS1-t	IM1	TS2	IM2	TS3	PC	P iso
C1–O2	ρ	0.444		0.429	0.441	0.430	0.431	0.432	0.329	0.423	0.435
	$\nabla^2\rho$	−0.202		−0.291	−0.269	−0.328	−0.303	−0.290	−0.856	−0.357	−0.282
C4–Cl5	ρ	0.194		0.193	0.049						
	$\nabla^2\rho$	−0.274		−0.269	0.085						
C1–O3	ρ	0.307		0.330	0.313	0.319	0.312	0.316	0.248		
	$\nabla^2\rho$	−0.583		−0.587	−0.605	−0.569	−0.575	−0.579	−0.591		
C1–C4	ρ	0.259		0.259	0.270	0.263	0.259	0.258	0.255	0.255	0.252
	$\nabla^2\rho$	−0.633		−0.633	−0.719	−0.653	−0.636	−0.624	−0.605	−0.600	−0.582
C8–N10	ρ		0.329	0.344	0.343	0.338	0.309	0.308	0.278	0.295	0.301
	$\nabla^2\rho$		−1.027	−1.095	−1.085	−1.084	−0.938	−0.934	−0.764	−0.856	−0.883
C8–S7	ρ		0.218	0.214	0.213	0.201	0.196	0.198	0.203	0.202	0.201
	$\nabla^2\rho$		−0.069	−0.186	−0.300	−0.387	−0.354	−0.362	−0.384	−0.379	−0.372
S7–C4	ρ				0.086	0.177	0.182	0.184	0.182	0.183	0.184
	$\nabla^2\rho$				0.029	−0.262	−0.285	−0.298	−0.285	−0.291	−0.291
N10–C1	ρ								0.245	0.322	0.313

(continued)

Table 3 (continued)

Results involving the thione form of thiourea

Bond distance	Critical point data	caa	tu-t	RC-t	TS1-t	IM1	TS2	IM2	TS3	PC	P iso
	$\nabla^2\rho$								−0.558	−0.989	−0.944
N10–H11	ρ		0.346	0.333	0.342	0.335	0.343		0.150		
	$\nabla^2\rho$		−1.800	−2.024	−2.005	−1.957	−1.776		−0.218		
O3–H11	ρ								0.153		
	$\nabla^2\rho$								−0.181		

3.2 $NHCSHNH_2 + ClCH_2COOH$ Reaction Mechanism in Water Solution

The optimised geometries for the study of the $NHCSHNH_2 + ClCH_2COOH$ reaction mechanism in the presence of water solvent are shown in Fig. 6. The geometries optimised in vacuo and in water tend to be significantly different. For instance, the geometry of IM1 in vacuo has the chloride ion abstracting H13 atom from N9 (the N9–H13 bond distance is 1.164 Å). In solution, however, H13 is not abstracted from N9 (the N9–H13 bond distance is 1.037 Å), which may be a result of the fact that the chloride ion prefers to interact with the solvent molecules and is therefore stabilised by the bulk solvent molecules. Another major geometry difference between the results in vacuo and the result in water solution is on the product complex, where it is observed that while in vacuo the product complex is composed of 2-iminothiazolidin-4-one, a water molecule and a hydrogen chloride molecule, the result in water solution indicates that the products are 2-iminothiazolidin-4-one, the hydronium ion and the chloride ion. The difference in the geometry of the product complex may also be explained on the basis of the stabilisation of the chloride ion by the continuum medium. Some of the **TSs** and **IMs** present H-bond like interactions, which in general

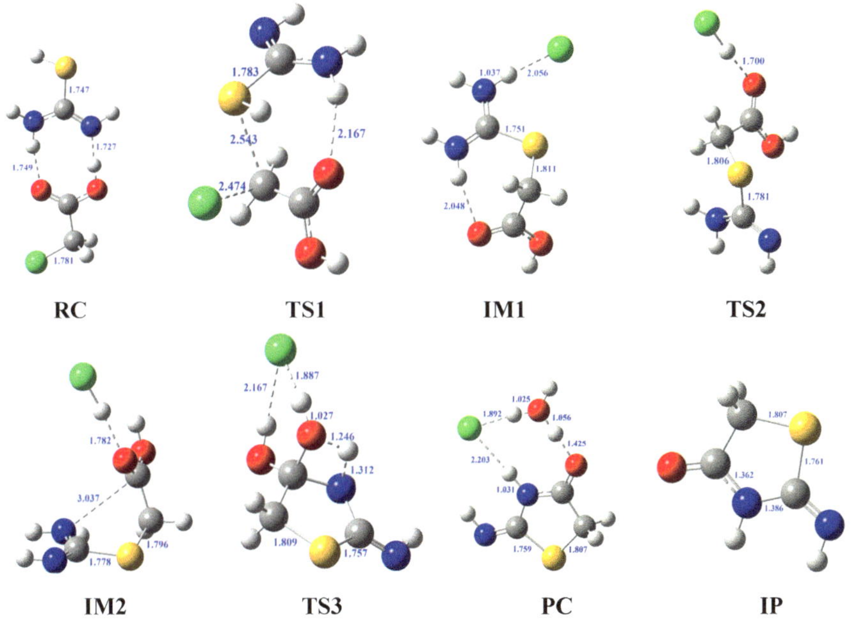

Fig. 6 Geometries (reactant complex, transition states and product complexes) computed at the M06 2X/6 311++G(3df,2p) level in water solution. The sulfur atom is depicted in yellow, oxygen in red, chlorine in green, nitrogen in blue, carbon in grey and hydrogen in grey. The relevant interatomic distances are given in angstroms; covalent bonds are indicated with the solid line while the non-covalent (including hydrogen bonds) are indicated with dotted lines

are weaker in water solution compared to *vacuum*. The exceptions are **TS2** and **IM2**, where the interaction distances are slightly shorter for the geometries optimized in aqueous solution. Among the **TSs** and **IMs** presenting H-bond like interactions, the shortest distance correspond to the N–H···Cl5 bond in **IM1**, corresponding to the process of abstraction of the H13 proton by the Cl ion.

The ΔE and ΔG values suggest that the reaction is endothermic and endergonic. In addition, the calculated energy barriers for the formation of **IM1** from **RC**-t and the formation of the product complex from **TS3** are significantly reduced in solution. Although the reaction appears to have lower barrier heights in water solution as compared to the results in vacuo, it is still overall not a spontaneous reaction in water solution, as evidenced by its ΔG value.

3.3 $NH_2CSNH_2 + ClCH_2COOH$ Reaction Mechanism in Vacuo

The relative energies of the isolated reactant molecules, pre-association complexes, transition state (TS) structures, intermediates and products of the reaction of $ClCH_2COOH$ with NH_2CSNH_2 are shown in Table 1b and the corresponding optimised geometries are shown in Fig. 7. The relative energies refer to the sum of the energies of the separate reactants (R). Selected bond distances necessary for the discussion are indicated in Fig. 7. The potential energy diagram depicting the reaction mechanism for the cyclisation is shown in Fig. 4b. As in the previous case, relative energies and bond distances referred to in the discussion correspond to M06-2X/6-311++G(3df,2p) results.

In the first step, the pre-association complex is formed from $ClCH_2COOH$ and NH_2CSNH_2. This is a barrier-less weakly bound pre-association reactant complex (**RC**) stabilised by two simultaneous intermolecular H-bonds involving the COOH group of chloroacetic acid and the amino (NH_2) group of NH_2CSNH_2. In this geometry, the OH group of $ClCH_2COOH$ is the H-bond donor to the S atom and the C=O group is the H-bond acceptor to the NH_2 in thiourea. The O2···H11–N10 intermolecular H-bond is shorter than the S7···H6–O3. **RC** is 14.526 kcal/mol more stable than the sum of the energies of the starting reactant molecules.

The cyclisation reaction starts when the most nucleophilic centre in one of the starting molecules attacks the most electrophilic centre in the second molecule. In the two starting molecules, there are four possible nucleophilic centres, which are O, S, N and Cl. Among these possible centres, S is the most nucleophilic because it has lower electronegativity value (2.5) than both O (3.5) and N (3.0) and because it is a largest atom (atomic size effect) so that its lone pair of electrons are farthest from the nucleus in comparison with O, N and Cl (i.e., it is the most polarisable atom among the possible nucleophiles). The most electrophilic centre is the α-position C4 atom of $ClCH_2COOH$. Therefore, an attack by the soft nucleophilic S atom on the soft electrophilic C atom leads to an SN_2 nucleophilic reaction (with S replacing

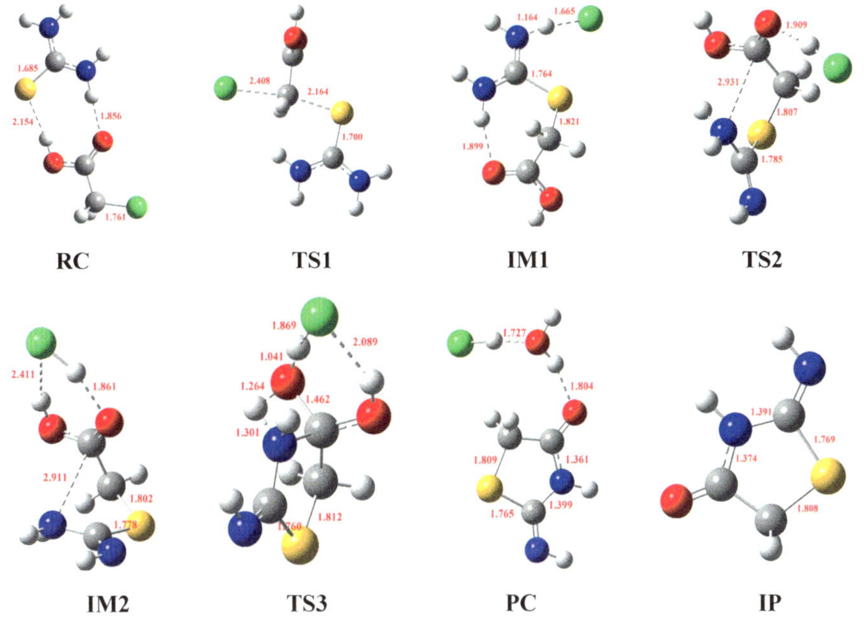

Fig. 7 Geometries (reactant complex, transition states and product complexes) computed at the M06-2X/6-311++G(3df,2p) level for the reaction between chloroacetic acid and the thione form of thiourea. The sulfur atom is depicted in yellow, oxygen in red, chlorine in green, nitrogen in blue, carbon in grey and hydrogen in grey. The relevant interatomic distances are given in angstroms; covalent bonds are indicated with the solid line while the non-covalent (including hydrogen bonds) are indicated with dotted lines

the Cl group), in which the first intermediate (**IM1**) is formed from **RC**. **IM1** is formed from **RC** by passing through the first transition state (**TS1**), whose imaginary vibrational mode (-468.01i) corresponds to the formation of the S7–C4 single bond and a simultaneous breaking of the C4–Cl5 bond. This step is characterised by an increasing degree of charge delocalisation while proceeding from **RC** to **IM1** along the reaction coordinate, as it can be noted by the progressive shortening of the N10–C1 and S7–C4 bond distances (Table 2). **TS1** is found to be 30.880 kcal/mol higher than the starting reactant molecules and 45.406 kcal/mol above the complex reactant molecule. In comparison to the reactant molecules, the geometric features of the **TS1** are such that the C8–S7 bond is elongated by 0.015 Å and the C4–Cl5 bond is elongated by 0.647 Å. The geometry features of **IM1** obtained here is essentially the same as that obtained for the reaction between $NHCSHNH_2$ and $ClCH_2COOH$. Consequently, all the subsequent reaction steps and geometries (second transition state, second intermediate, third transition state and products) are structurally and energetically similar for the two mechanisms. For this reason, these the geometries of this reaction mechanism are not discussed further in this article. However, it is interesting to discuss and compare the energy barriers and the nature of the reaction because these properties are strongly determined by the starting reactant molecules and the subsequent structures. The overall energetic values are however significantly different; the **RC** to **TS1** barrier is much lower than the energy required for the formation of **TS1-t** from **RC-t**. In this reaction, the rate determining step corresponds to **TS3**, which lies 41.689 kcal/mol above the free reactants. An analysis of the ΔE and ΔG values suggests that the reaction is always endothermic and endergonic. Therefore, the large barrier height and the endergenicity of the reaction implies that it may be both kinetically and thermodynamically unfavourable pathway.

3.4 *NH_2CSNH_2 + $ClCH_2COOH$ Reaction Mechanism in Water Solution*

Geometry optimisations in solution were performed to investigate the geometric and energetic differences that might be due to solute solvent interaction. The geometry of the **RC** complex is water solution is similar to that in vacuo, however, the H-bond lengths are such that the N10–H11···O2 is longer in water solution (1.960 Å) than in vacuo (1.856 Å) and the O3–H6···S7 H-bond is shorter in in water solution (2.143 Å) than in vacuo (2.154). The **TS1** structure is also similar between the results in water solution and in vacuo, characterised by the breaking of the C4–Cl5 bond and the concomitant formation of the C4–S7 bond. The C4–Cl5 bond distance is shorter in water solution (2.275 Å) than in vacuo (2.408 Å). The C4–S7 bond distance is longer in water solution (2.379 Å) than in vacuo (2.164 Å). These results suggest that in solution the transition state is closer to the reactant geometry than in vacuo. As with the results in vacuo, the geometry features of **IM1** obtained here are essentially the same as those obtained for the reaction between $NHCSHNH_2$ and $ClCH_2COOH$. Consequently, all the subsequent reaction steps and geometries (second transition state, second intermediate, third transition state and products) are structurally and energetically similar for the two mechanisms in water solution and therefore are not discussed further in this work.

In the presence of the bulk solvent (Table 1), the reaction barrier is reduced from 30.880 to 18.231 kcal/mol. The determining factor for barrier reduction could be the formation of H-bond networks involving the transition state structure and the solvent molecules [48–53]. Interestingly, the relative stability of the third transition state structure does not change significantly between the results in vacuo (41.689 kcal/mol) and the results in water solution (41.968 kcal/mol). Moreover, the result suggests that, as it is in vacuo, the step **IM2** → **TS3** is the energetically most costly step, and it involves the H11 proton transfer from N10 to O3 and the subsequent formation of the tetrahedral C2 carbon atom. The overall reaction is endothermic and endergonic, which means that the reaction of thiourea and chloroacetic acid may not be thermodynamically preferred in the presence of bulk water solvent.

4 Conclusions

The reaction of thiourea and chloroacetic acid has been investigated to determine the preferred geometries of the reactants, the possible transition states, intermediate and products. The two tautomeric forms of thiourea have been utilised in the study to determine the tautormer which preferably reacts with chloroacetic acid to form 2-iminothiazolidin-4-one. The study has been performed with both MP2 and DFT methods and considering both *vacuum* and water solvent media. The results of the investigation show that the reaction of $ClCH_2COOH$ with NH_2CSNH_2, in vacuo and in water solution, is thermodynamically disfavoured because it is both endothermic and endergonic. The reaction of $ClCH_2COOH$ and $NHCSHNH_2$ is favoured in vacuo and disfavoured in water solution. Therefore, is reasonable to infer that the reaction of thiourea and chloroacetic acid involves the imidothiol form of thiourea. Since the in vacuo medium (with dielectric constant, $\varepsilon = 1$) is the most ideal representation of non-polar media, it is also reasonable to conclude that the reaction between thiourea and chloroacetic acid is best carried out in non-polar media. A rationalisation for the difference in the behaviour of the two tautomers can be explained as follows; in practice, $NHCSHNH_2$ is likely to behave more like a zwitterionic form. In such a case, the S atom in $NHCSHNH_2$ has a negative charge and is therefore a better nucleophile than the S atom in NH_2CSNH_2.

In the reaction between thiourea and chloroacetic acid, the formation of the **TS3** products is a result of the concerted reaction step involving asynchronous bond-formation and bond-breaking processes. However, the formation of the product from **TS3** is both concerted and synchronous bond breaking (during the removal of the water molecule) and bond formation (the restoration of the carbonyl O2 atom).

Investigation of the reaction of $ClCH_2COOH$ with either of the tautomer of thiourea in water solution indicates that the reaction is endothermic and therefore not thermodynamically favoured. Moreover, while in vacuo the by-products for the reaction include water and hydrogen chloride, in water solution the by-products are the hydronium ion and the chloride ion. The difference in the by-products may be related to the interaction between the solvent molecules and the chloride ions.

A comparison of the results across methods (MP2 and DFT) suggests similar trends are obtained, in terms of energetic and electronic properties, for the species involved in the investigated mechanisms; this is true in all the media.

References

1. Skyler JS (2004) Diabetes mellitus: pathogenesi and treatment strategies. J Med Chem 47:4113–4117
2. Kaminskyy O, Bednarczyk-Cwynar B, Vasylenko O, Kazakova O, Zimenkovsky B, Zaprutko L, Lesyk R (2012) Synthesis of new potential anticancer agents based on 4-thiazolidinone and oleanane scaffolds. Med Chem Res 21:3568–3580

3. Tanabe Y, Suzukamo G, Komuro Y, Imanishi N, Morooka S, Enomoto M, Kojima A, Sanemitsu Y, Mizutani M (1991) Structure-activity relationship of optically active 2-(3-pyridyl)thiazolidin-4-ones as a PAF antagonists. Tetrahedron Lett 32:379–382
4. Rawal RK, Prabhakar YS, Katti SB, De Clercq E (2005) 2-(Aryl)-3-furan-2-ylmethyl-thiazolidin-4-ones as selective HIV-RT Inhibitors. Bioorg Med Chem 13:6771–6776
5. Srivastava T, Gaikwad AK, Haq W (2005) Synthesis and biological evaluation of 4-thiazolidinone derivatives as potential antimycobacterial agents. Arkivoc ii: 120–130
6. Rajopadhye M, Popp FD (1987) Synthesis and antileukemic activity of spiro[indoline-3,2′-thiazolidine]-2,4′-diones. J Heterocyclic Chem 24:1637–1642
7. Metwally NH, Radwan IT, El-Serwy WS, Mohamed MA (2019) Design, synthesis, DNA assessment and molecular docking study of novel 2-(pyridin-2-ylimino)thiazolidin-4-one derivatives as potent antifungal agents. Bioorg Chem 84:456–467
8. Pan B, Huang R-Z, Han S-Q (2010) Design, synthesis, and antibiofilm activity of 2-arylimino-3-aryl-thiazolidine-4-ones. Bioorg Med Chem Lett 20:2461–2464
9. El-Gazzar A-RBA, Gaffar MA-E-D, Ali SA (2008) Design, synthesis, and preliminary evaluation as antimicrobial activity of novel *spiro*-1, 3-thiazolidine C-acyclic nucleoside analogs. J Sulfur Chem 29:549–558
10. Nagalakshmi G, Maity TK, Maiti BC (2013) Synthesis, characterization and antiviral evaluation of some novel 2-[(substitutedphenyl/heteroaryl)imino]-3-phenyl-1,3- thiazolidin-4-ones. Der Pharmacia Lett 5:177–188
11. Bolli MH, Abele S, Binkert C, Bravo R, Buchmann S, Bur D, Gatfield J, Hess P, Kohl C, Mangold C, Mathys B, Menyhart K, Müller C, Nayler O, Scherz M, Schmidt G, Sippel V, Steiner B, Strasser D, Treiber A, Weller T (2010) 2-imino-thiazolidin-4-one derivatives as potent, orally active S1P1 receptor agonists. J Med Chem 53:4198–4211
12. Akerblom E (1967) 2-aminothiazolidin-4-one and 2-iminothiazolidin-4-one. Part I. The reaction of chloroacetic acid with N-monoalkylthiurea. Acta Chim Scand 21:843–848
13. Kasmi-Mir S, Djafri A, Paquin L, Hamelin J, Rahmouni M (2006) One-pot synthesis of 5-arylidene-2-imino-4-thiazolidinones under microwave irradiation. Molecules 11:597–602
14. Sathishkumar M, Nagarajan S, Shanmugavelan P, Dinesh M, Ponnuswamy A (2013) A facile, rapid, one-pot regio/stereoselective synthesis of 2-iminothiazolidin-4-ones under solvent/scavenger-free conditions. Beilstain J Org Chem 9:689–697
15. Meng G, Zheng M, Dong M, Qu Q (1998) An eco-friendly preparation of 2-iminothiazolidin-4-ones derivatives. Org Prep Proced Int 44:184–186
16. Kasmi S, Hamelin J, Benhaoua H (1998) Microwave-assisted solvent-free synthesis of iminothiazolines. Tetrahedron Lett 39:8093–8096
17. Murru S, Singh CB, Kavala V (2008) A convenient one-pot synthesis of thiazol-2-imines: application in the construction of pifithrin analogues. Tetrahedron 64:1931–1942
18. Meng G, Li ZY, Zheng ML (2008) An efficient one-step method for the large-scale synthesis of 2,4-thiazolidinedione. Org Prep Proced Int 40:572–574
19. Meng G, Gao Y, Zheng ML (2011) Improved preparation of 2,4-thiazolidinedione. Org Prep Proced Int 43:312–313
20. Lynch BJ, Truhlar DG (2001) How well can hybrid density functional methods predict transition sate geometries and barrier heights? J Phys Chem A 105:2936–2941
21. Zhao Y, Schultz NE, Truhlar DG (2006) Design of density functionals by combining the method of constraint satisfaction with parametrization for thermochemistry, thermochemical kinetics, and noncovalent interactions. J Chem Theory Comput 2:364–382
22. Zhao Y, Truhlar DG (2008) Exploring the limit of accuracy of the global hybrid meta density functional for main-group thermochemistry, kinetics, and noncovalent interactions. J Chem Theory Comput 4:1849–1868
23. Marenich AV, Cramer CJ, Truhlar DG (2009) Universal solvation model based on solute electron density and a continuum model of the solvent defined by the bulk dielectric constant and atomic surface tensions. J Phys Chem B 113:6378–6396
24. Peverati R, Truhlar DG (2014) The quest for a universal density functional: the accuracy of density functionals across a broad spectrum of databases in chemistry and physics. Phil Trans R Soc A 372: 20120476/1-51

25. Peverati R, Truhlar DG (2012) M11-L: a local density functional that provides improved accuracy for electronic structure calculations in chemistry and physics. J Phys Chem Lett 3:117–124
26. Chan B, Gilbert ATB, Gill PMW, Radom L (2014) Performance of density functional theory procedures for the calculation of proton-exchange barriers: unusual behavior of M06-type functionals. J Chem Theory Comput 10:3777–3783
27. Linder M, Brinck T (2012) Stepwise diels-alder: more than just an oddity? A computational mechanistic study. J Org Chem 77:6563–6573
28. Jayaram PN, Roy G, Mugesh G (2008) Effect of thione–thiol tautomerism on the inhibition of lactoperoxidase by anti-thyroid drugs and their analogues. J Chem Sci 120:143–154
29. Allegretti PE, Castro EA, Furlong JJP (2000) Tautomeric equilibrium of amides and related compounds: theoretical and spectral evidences. J Mol Struct (Theochem) 499:121–126
30. Allegretti PE, de las Mercedes MS, Castro EA, Furlong JJP (2007) Tautomeric equilibria studies by mass spectrometry. World J Chem 2:25–62
31. Gao JY, Zhang CH, Luo MM, Kim CK, Chu W, Xue Y (2012) Mechanism for the reaction of 2-naphthol with N-methyl-N-phenyl-hydrazine suggested by the density functional theory investigations. J Comp Chem 33:220–230
32. Nowacki A, Sikora K, Dmochowska B, Wiśniewski A (2012) Studies of the forrmation of N-substituted pyridinium mesylates: a theoretical approach. Comp Theoret Chem 1000:33–41
33. Nowacki B, Dmochowska K, Sikora J, Madaj A, Wiśniewski (2012) Theoretical studies of the formation of quanternary pyridinium mesylates. Comp Theoret Chem 986:85–92
34. Velez E, Quijano J, Notario R, Pabón E, Murillo J, Leal J, Zapata E, Alarcón G (2009) A computational study of stereospecifity in the thermal elimination reaction of menthyl benzoate in the gas phase. J Phys Org Chem 22:971–977
35. Sandhiya L, Kolandaivel P, Senthikumar K (2012) Theoretical studies on the reaction mechanism and kinetics of the atmospheric reactions of 1,4-thioxane with OH radical. Struct Chem 23:1475–1488
36. Gaussian 09, Revision *D.01*, Frisch MJ, Trucks GW, Schlegel HB, Scuseria GE, Robb MA, Cheeseman JR, Scalmani G, Barone V, Mennucci B, Petersson GA, Nakatsuji H, Caricato M, Li X, Hratchian HP, Izmaylov AF, Bloino J, Zheng G, Sonnenberg JL, Hada M, Ehara M, Toyota K, Fukuda R, Hasegawa J, Ishida M, Nakajima T, Honda Y, Kitao O, Nakai H, Vreven T, Montgomery, Jr. JA, Peralta JE, Ogliaro F, Bearpark M, Heyd JJ, Brothers E, Kudin KN, Staroverov VN, Kobayashi R, Normand J, Raghavachari K, Rendell A, Burant JC, Iyengar SS, Tomasi J, Cossi M, Rega N, Millam JM, Klene M, Knox JE, Cross JB, Bakken V, Adamo C, Jaramillo J, Gomperts R, Stratmann RE, Yazyev O, Austin AJ, Cammi R, Pomelli C, Ochterski JW, Martin RL, Morokuma K, Zakrzewski VG, Voth GA, Salvador P, Dannenberg JJ, Dapprich S, Daniels AD, Farkas O, Foresman JB, Ortiz JV, Cioslowski J, Fox DJ, Gaussian, Inc., Wallingford CT, 2009
37. Chem. Office 12.0, Perkin Elmer Inc., USA
38. Dennington R, Keith T, Millam J (2009) Gauss view, version 5. Semichem Inc., Shawnee Mission
39. Keith TA (2014) AIMAll (version 14.06.21). TK Gristmill Software, Overland Park KS, USA
40. Bader RFW (1990) Atoms in molecules. A quantum theory. Clarendon Press, Oxford
41. Banait NS, Jencks WP (1991) Reactions of anionic nucleophiles with cu-D-glucopyranosyl fluoride in aqueous solution through a concerted, A_ND_N (s_N2) mechanism. J Am Chem Soc 113:7951–7958
42. Lee JK, Bain AD, Berti PJ (2004) Probing the transition states of four glucoside hydrolyses with ^{13}C kinetic isotope effects measured at natural abundance by NMR spectroscopy. J Am Chem Soc 126:3769–3776
43. Banait NS, Jencks WP (1991) General-acid and general-base catalysis of the cleavage of a-D-glucopyranosyl fluoride. J Am Chem Soc 113:7958–7963
44. Tanaka Y, Tao W, Blanchard JS, Hehre EJ (1994) Transition state structures for the hydrolysis of alpha-D-glucopyranosyll fluoride by retaining and inverting reactions by glycolases. J Biol Chem 269:32306–32312

45. Kabanda MM, Tran VT, Seema KM, Serobatse KRN, Tsiepe TJ, Tran QT, Ebenso EE (2015) Conformational, electronic and antioxidant properties of lucidone, linderone and methyllinderone: DFT, QTAIM and NBO studies. Mol Phys 113:683–697
46. Kabanda MM, Tran VT, Tran QT, Ebenso EE (1046) A computational study of pyrazinamide: tautomerism, acid–base properties, micro-solvation effects and acid hydrolysis mechanism. Comp Theor Chem 1046: 30–41
47. Kolandaivel P, Nirmala VJ (2004) Study of proper and improper hydrogen bonding using Bader's atoms in molecules (AIM) theory and NBO analysis. Mol Struct 694:33–38
48. D'Cunha C, Morozov AN, Chatfield DC (2013) Theoretical study of HOCl-catalyzed keto–enol tautomerization of β-cyclopentanedione in an explicit water environment. J Phys Chem A 117:8437–8448
49. Otukile KP, Kabanda MM (2019) A DFT mechanistic and kinetic study on the reaction of phloroglucinol with OH in different media: hydrogen atom transfer versus oxidation. J Theor Comput Chem 18:1950017
50. Otukile KP, Kabanda MM (2019) A DFT mechanistic, thermodynamic and kinetic study on the reaction of 1, 3, 5-trihydroxybenzene and 2, 4, 6-trihydroxyacetophenone with •OOH in different media. J Theor Comput Chem 18:1950023
51. Otukile KP, Mammino L, Kabanda MM (2018) A Theoretical study on the degradation of 2-mercaptobenzothiazole and 2-mercaptobenzimidazole by _OH in vacuo and aqueous media. Comp Theor Chem 1125:112–127
52. Kabanda MM, Serobatse KRN (2018) A DFT study on the addition and abstraction reactions of thiourea with hydroxyl radical. J Sulfur Chem 39:23–46
53. Serobatse KRN, Kabanda MM (2017) An appraisal of the hydrogen atom transfer mechanism for the reaction between thiourea derivatives and OH radical: a case-study of dimethylthiourea and diethylthiourea. Comp Theor Chem 1101:83–95

Density Functional Theory Studies of Ruthenium Dye (N3) Adsorbed on a TiO_2 Brookite Cluster for Application in Dye Sensitized Solar Cells

I. F. Elegbeleye, N. E. Maluta and R. R. Maphanga

Abstract Titanium dioxide has been a subject of increasing interest due to its application in pigments, photocatalysis and semiconductor materials in dye sensitized solar cells (DSSC). Recent studies suggest that TiO_2 brookite exhibits good photocatalytic properties. Efficiency in excess of 11% has been achieved with the use of ruthenium (N3) dyes as DSSC sensitizer. The optical properties, energy level alignment and electronic state energy of the ruthenium (N3) sensitizer to TiO_2 cluster were studied to gain insight into the electron injection kinetics and electron injection efficiency of the dye/TiO_2 complex. The simulated absorption spectra show absorption peaks at 311, 388 and 480 nm. The HOMO of (N3) lies at −5.03 eV and is centred on the NCS moiety where the donor group is situated, while the LUMO lies at −3.01 eV and is centred on the 4,4-dicarboxy-2,2-bipyridine moiety where the acceptor group is situated. Upon absorption on a brookite cluster, the light absorption maximum red shifted to higher wavelength; this results in the distribution of the LUMO shifting from the dye to the TiO_2 cluster. The results suggest favourable electron injection from the dye excited state into TiO_2 semiconductor. The results suggest that TiO_2 brookite is a promising entrant for DSSC semiconductor.

Keywords Density functional theory · Dye sensitized solar cell · Ruthenium dye · Red shift on cluster formation · TiO_2 brookite

1 Introduction

In recent years, TiO_2 has been a subject of increasing interest due to its applications in pigments, photocatalysis and semiconductor materials in dye sensitized solar cells

I. F. Elegbeleye · N. E. Maluta
Department of Physics, University of Venda, Thohoyandou, South Africa

N. E. Maluta (✉) · R. R. Maphanga
National Institute for Theoretical Physics (NITheP), Gauteng, South Africa
e-mail: Eric.Maluta@univen.ac.za

R. R. Maphanga
Council for Science and Industrial Research, P.O. Box 395, Pretoria, South Africa

L. Mammino et al. (eds.), *Advances in Quantum Systems in Chemistry, Physics, and Biology*, Progress in Theoretical Chemistry and Physics 32,
https://doi.org/10.1007/978-3-030-34941-7_8

(DSSCs) [1, 2]. The interest in TiO_2 as semiconductor for dye sensitized solar cell is motivated by its non-toxicity and excellent stability upon illumination [1, 3–5]. However, the major drawback limiting its usage is its inability to absorb visible and infra-red photons of the solar spectrum because of its wide conduction band gap (3.0–3.2 eV) [5–7]. In DSSCs, dye molecules chemisorbed on the surface of TiO_2 are used to harness light, and the consequent photoexcited electrons are injected into the conduction band of TiO_2 [8–10]. An efficient photosensitizer must have good excited state properties and intense absorption in the visible and near infra-red region of the solar spectrum. Several dye molecules have been employed as sensitizer for DSSC; the highest efficiency in excess of 13% has been achieved in cells with nanostructured TiO_2 semiconductor sensitized by Ruthenium (II) polypyridyl complex(N3) [11–15].

A significant number of studies has been done towards the surface modification of TiO_2 crystals with atoms/sensitizing dye molecules to step-wise reduce the band gap and further enhance their activities in the visible and near infra-red region of the solar spectrum. Polymorphs of TiO_2 have been a model for such studies, aimed at improving photocurrent yield and light harvesting in DSSCS [3, 16]. Energy band modulation by elemental doping, monodoping, codoping with nonmetals and transition metals, and adsorption of dye molecules on TiO_2 surfaces, have been tested; the results showed improved spectral response and enhanced photocatalytic performances of TiO_2 [7, 17]. Surfaces of rutile and anatase polymorphs have been greatly exploited and have been a prototypical model for basic studies on TiO_2 oxide [18, 19]. Relatively limited work has been done on the brookite form of TiO_2, in contrast to rutile and anatase polymorphs that has been greatly exploited [4].

A recent study on TiO_2 brookite suggested that it is a good photocatalyst and may exhibit higher photocatalytic activity than both rutile and anatase [4]. The absorption edge of brookite observed in a prior study was also reported to be broad and to extend to the visible region of the solar spectrum, in contrast to steep edges in the visible region observed for rutile and anatase polymorphs of TiO_2 [19]. Since brookite surfaces have not been studied and are reported to have better photocatalytic properties, it is of keen interest to study the interactions of promising ruthenium (N3) dye molecules with surfaces of brookite TiO_2 for optimization of photon current density in dye sensitized solar cells.

2 Computational Procedures

Optimization of the ground state geometries of the TiO_2 brookite/ruthenium complex in the gas phase was performed using DFT with the hybrid B3LYP exchange correlation functional and the LANL2DZ pseudopotential for Ti atom; the 6–311+G (d, p) basis set was used for the C, N, O and H atoms. All DFT/TD-DFT calculations on the ruthenium (N3) complex were done using Gaussian 03 quantum chemical package [20]. The UV-Vis simulated absorption spectrum of the TiO_2/dye complex was computed by Time dependent (TD) DFT *in vacuo* with the same functional and basis set. 80 singlet to singlet electronic transitions were considered for the excitations, to account for the whole absorption spectrum [21]. The UV-Vis spectra, maximum excitation wavelength, excitation energies, oscillator strength and light harvesting efficiency of the dyes were computed. Their matching with the solar spectrum is discussed. The light harvesting efficiency (LHE) at the maximum spectrum wavelength (λ_{max}) was computed for the ruthenium (N3) complex using Eq. 1,

$$LHE = 1 - 10^{-f} \tag{1}$$

where “f” is the absorption strength of the dye associated with the maximum absorption of the dye (also called the oscillator strength) [15, 22, 23].

The absolute values of the adsorption energies of ruthenium (N3) complex on TiO_2 were computed using Eq. 2.

$$E_{ads} = E_{dye} + E_{TiO_2} - \left(E_{dye+TiO_2}\right) \tag{2}$$

where E_{ads} is the adsorption energy, E_{dye} is the energy of the dye, $E_{(TiO_2)}$ is the energy of the TiO_2 slab and $E_{(dye+TiO_2)}$ is the total energy of the dye-TiO_2 complex. A positive value of E_{ads} indicates stable adsorption [14, 24, 25].

The energies of the HOMO and LUMO, the HOMO-LUMO energy gap and the isodensity surfaces of the molecular orbitals involved in the excitation of the ruthenium (N3) complex were identified from the fchk file obtained from the simulations.

The bulk structure of brookite TiO_2 that was used for this study was optimized using the CASTEP module in Materials Studio BIOVIA [26] to obtain the ground state structure of the TiO_2 brookite semiconductor. The convergence energies and k-points were 650 eV and 4 × 7 ×7 respectively where k-points is the radius of convergence in the brillouin zone. The optimized structure of the ruthenium (N3) complex was exported into an atomic simulation environment via Avogadro Software [27]. DFT with the PBE functional was used through GPAW [28] and AVOGADRO [27] computational software within the atomic simulation environment to explore the optical properties of two modelled TiO_2 brookite clusters, that is Ti_8O_{16} and $Ti_{68}O_{136}$. The optical properties of the interplay of ruthenium (N3) dye molecules with brookite Ti_8O_{16} and $Ti_{68}O_{136}$ clusters were investigated using DFT, in order to optimize photon current densities in DSSCs.

3 Results and Discussions

3.1 Geometric Properties of the Ruthenium (N3) Complex

The molecular structure of the ruthenium (N3) complex studied in this work is shown in Fig. 1 and its optimized geometry in Fig. 2. Figure 2 shows that the ruthenium atom is octahedral coordinated to six nitrogen atoms pertaining to the two thiocyanate

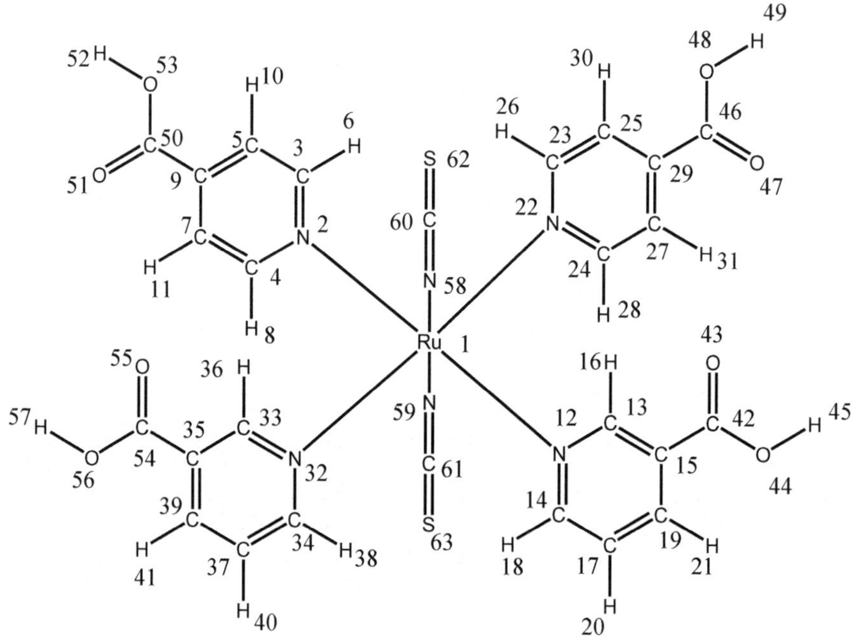

Fig. 1 Molecular structure of ruthenium (N3) complex and atom numbering utilized in this work

and the four carboxylic acids of the two bipyridyl ligands. Selected bond lengths and bond angles of the optimized geometry of the ruthenium (N3) complex are reported in Table 1. The geometry optimization of ruthenium (N3) complex converged when the internal forces acting on all the atoms were less than 4.9×10^{-5} eV and a threshold value of 4.5×10^{-4} eV/atom. The dipole moments obtained after optimization is 2.8633.

3.2 UV/VIS Absorption Spectrum of the Calculated Ruthenium (N3) Complex

As already mentioned, 80 singlets to singlet transitions were considered for this complex in order to account for the whole spectrum. The absorption spectrum in gas phase is presented in Fig. 3. The ruthenium sensitizer shows good absorption in the UV and visible region of the solar spectrum with intense absorption around 311 nm, 388 nm and 480 nm, but the peak (λ_{max}) is more notable at 480 nm. This calculated λ_{max} peak of the ruthenium (N3) sensitizer is in close agreement with experimental and computed values reported in literature [23, 29]. Hence, the DFT/B3LYP level of theory with the selected basis set can be considered adequate for the simulation of the absorption spectra of the ruthenium (N3) complex dye.

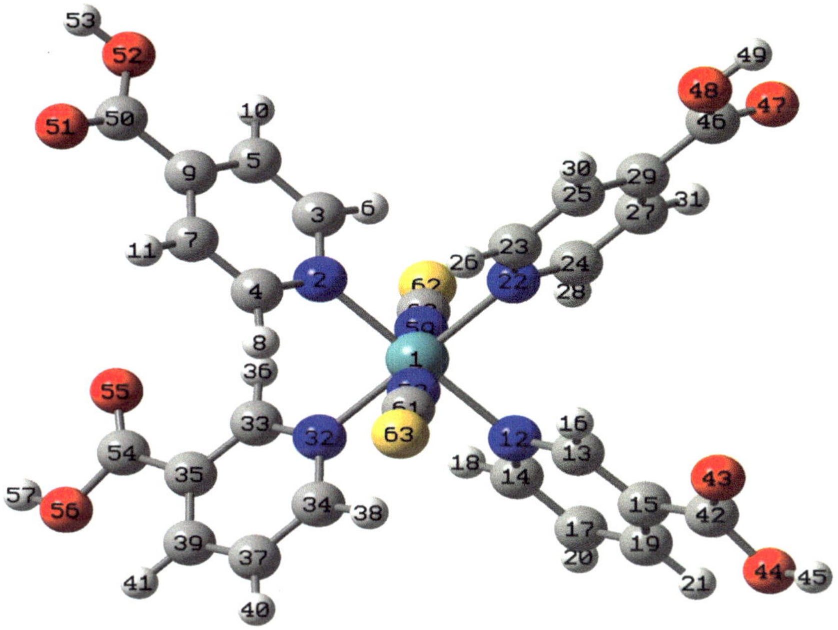

Fig. 2 Optimized geometry of the ruthenium (N3) complex ([cis-di (thiocyanate) bis (2,2-bipyridine-4,4-dicarboxylate) ruthenium]) Ru(dcbpy)(NCS)2 and colours of atoms utilized in this work (Olive green for ruthenium atoms, blue for nitrogen atoms, yellow for sulphur atoms, grey for carbon atoms and red for oxygen atoms)

Table 1 Selected bond lengths (Å) and bond angles (degrees) of ruthenium (N3) complex

Bond length (Å)		Bond angles (°)	
Length considered	Value of length	Angles considered	Value of angles
Ru-N2	2.096	N2-Ru-N12	89.8
Ru-N12	2.109	N2-Ru-N12	89.8
Ru-N22	2.096	N32-Ru-N59	89.8
Ru-N32	2.109	N12-Ru-N59	89.2
Ru-N58	2.065	N32-Ru-N12	90.3
Ru-N59	2.065	N58-Ru-12	89.8

3.2.1 Light Harvesting Efficiency of Ruthenium (N3) Complex

The light harvesting efficiencies of the dye molecules were calculated theoretically using Eq. 1.

The highest oscillator strength of the ruthenium (N3) complex is 0.15, yielding a corresponding calculated light harvesting efficiency of 0.29 (Table 2). Considering

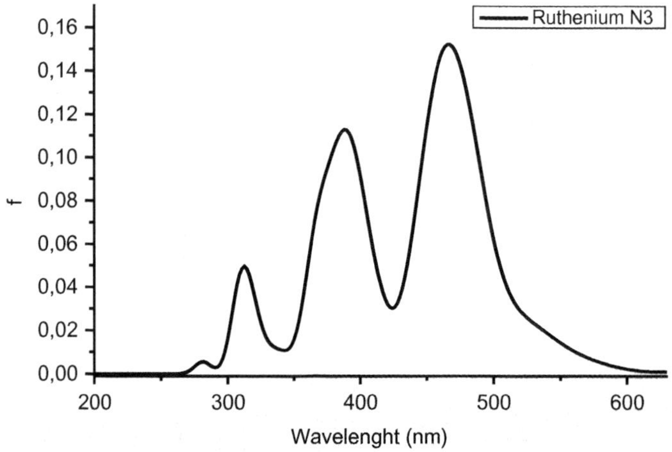

Fig. 3 Simulated absorption spectra (oscillator strength, f, versus wavelength) of ruthenium (N3) complex

Table 2 Computed light harvesting efficiencies of ruthenium (N3) complex at the three absorption peaks

Wavelength (nm)	Oscillator strength	LHE	LHE (%)	Wavelength (nm)
311	0.05	0.10	10	311
388	0.11	0.22	22	388
480	0.15	0.29	29	480

all the three absorption peaks, the ruthenium complex has an average LHE value of 20.1%.

3.2.2 The Computed Energies of HOMO and LUMO, HOMO-LUMO Energy Gap and Isodensity Surfaces of Ruthenium (N3) Complex

The computed energies of HOMO and LUMO and the HOMO-LUMO energy gap of the ruthenium (N3) complex are presented in Table 3 and compared with the results of

Table 3 Energies of HOMO, LUMO and HOMO-LUMO energy gap of the ruthenium (N3) complex

Dye (N3)	HOMO (eV)	LUMO (eV)	H-L gap (eV)
Experimental work [24, 29]	−5.36	−3.47	1.89
Computed [23]	−5.67	−3.63	1.96
This work	−5.03	−3.01	2.02

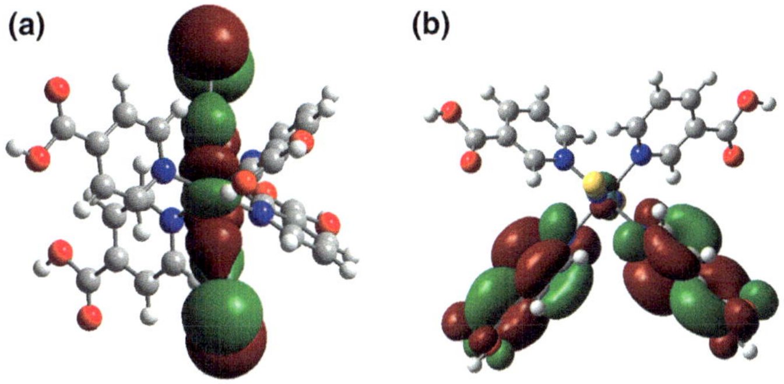

Fig. 4 Isodensity surfaces of ruthenium (N3) complex **a** HOMO **b** LUMO. Isovalue = 0.02

another works [23, 26], and with the experimental values [29]. The reported experimental values of the energies of the HOMO, LUMO and HOMO-LUMO energy gap of the ruthenium (N3) complex are −5.36 eV, −3.47 eV and 1.89 respectively. The results obtained in this work compares favourably with the simulated and experimental result [23, 26, 29].

The isodensity surfaces of the HOMO and LUMO of the ruthenium (N3) complex are presented in Fig. 4. The HOMO is distributed on the thiocyanate ligands, while the LUMO is distributed on the two dcbpy 4,4-dicarboxy-2,2-bipyridine ligands. The different distributions of the HOMO and LUMO indicates good electron injection properties of the ruthenium (N3) complex, due to various donor and acceptor levels.

3.3 Adsorption of Ruthenium (N3) Complex Dye on Brookite TiO_2 Nanocluster

The brookite clusters studied in this work are $(TiO_2)_n$ with n = 8 and n = 68. Figure 5 shows the brookite $(TiO_2)_8$ model comprises of eight titanium and sixteen oxygen atoms, the structure of brookite $(TiO_2)_8$ was imported from CASTEP materials studio [29] via the crystallographic mode (cif) without periodicity into Avogadro visualizing interface. Figure 6 reveals a periodic brookite $(TiO_2)_{68}$ supercell 2 × 2 × 2 Å comprising of sixty-eight titanium atoms and one hundred and thirty six oxygen atom. The ruthenium (N3) dye molecules were adsorbed on $(TiO_2)_8$ and $(TiO_2)_{68}$ brookite cluster by bidentate adsorption mode in which each of the oxygen of the carboxylic group binds to a two-fold coordinated and four-fold coordinated titanium atom on $(TiO_2)_8$ and $(TiO_2)_{68}$ brookite cluster, as shown in Fig. 7. This adsorption mode was found to be the most energetically favourable, especially for ruthenium complexes with two bipyridine ligands having carboxylic acid functional groups. The results were presented on relaxation energies, optical excitation spectrum and isodensity surfaces of the key molecular orbital involved in excitation.

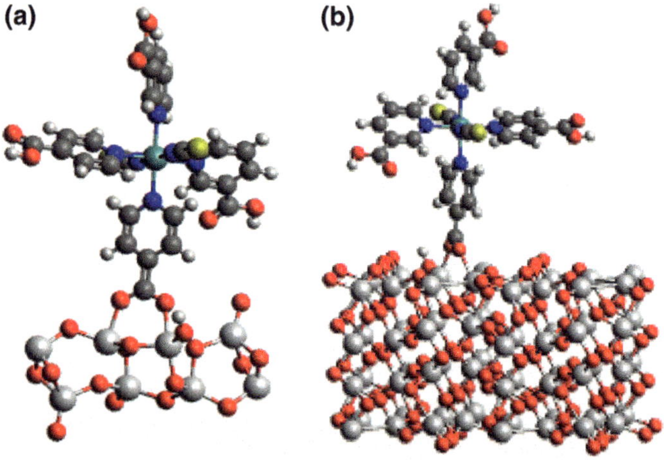

Fig. 5 Ruthenium (N3) complex absorbed on **a** $(TiO_2)_8$ brookite cluster **b** $(TiO_2)_{68}$ brookite cluster

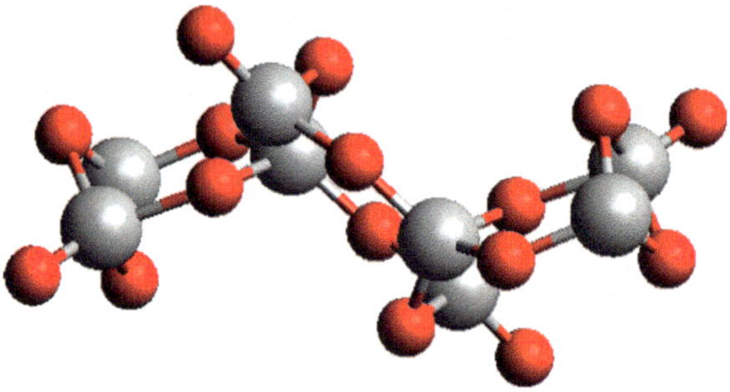

Fig. 6 $(TiO_2)_8$ brookite cluster

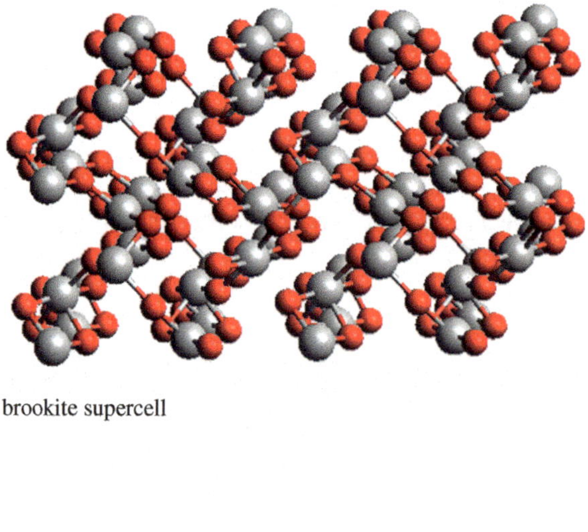

Fig. 7 $(TiO_2)_{68}$ brookite supercell

Table 4 Relaxation energies of ruthenium (N3) dye molecule absorbed on $(TiO_2)_8$ and $(TiO_2)_{68}$ brookite cluster

System	Relaxation energy (eV)	Energy $(TiO_2)_n$ cluster $_{(slab)}$ (eV)	Energy$_{(slab\ +molecule)}$ Dyes@$(TiO_2)_n$(eV)
Ruthenium (N3)	−432.919		
$(TiO_2)_8$		−198.308	
$(TiO_2)_{68}$		−1635.558	
Ruthenium@$(TiO_2)_8$			−632.953
Ruthenium@$(TiO_2)_{68}$			−2072.267

3.4 Adsorption Energies of Ruthenium (N3) Dye Molecule Absorbed on $(TiO_2)_n$, n = 8, 68 Brookite Complex

The relaxation energy for ruthenium (N3) dye is −432.91 eV. The energy of the clusters are −198.31 eV, −1635.55 eV for $(TiO_2)_8$ and $(TiO_2)_{68}$ respectively (Table 4). The adsorption energies of ruthenium (N3) dye on $(TiO_2)_8$ and $(TiO_2)_{68}$ brookite cluster were computed using Eq. 2.

The adsorption energy of ruthenium (N3) dye @ $(TiO_2)_8$ is 1.73 eV and of ruthenium (N3) dye @ $(TiO_2)_{68}$ is 3.76 eV. These positive adsorption energies denote the ability of the dye molecules to bind to the surface of $(TiO_2)_8$ and $(TiO_2)_{68}$ clusters. The adsorption energy of ruthenium (N3) dye on $(TiO_2)_{68}$ and $(TiO_2)_8$ brookite cluster are 3.84 eV and 1.73 eV respectively. The results shows that the ruthenium (N3) dye binds more strongly to the surface of the larger cluster $(TiO_2)_{68}$ brookite than the corresponding $(TiO_2)_8$ brookite cluster.

3.4.1 Absorption Spectrum of Ruthenium (N3) Dyes Absorbed on $(TiO_2)_8$ and $(TiO_2)_{68}$ Brookite Cluster

The optical spectra of the ruthenium (N3) dye absorbed on $(TiO_2)_8$ is presented in Fig. 8, in which the absorption corresponds to the optical excitation of the ruthenium (N3) dye absorbed on $(TiO_2)_8$. Comparing the absorption spectra of the $(TiO_2)_8$ cluster (Fig. 8) with the absorption spectra of the ruthenium (N3) dye absorbed on the $(TiO_2)_8$ cluster (Fig. 9), shows that the absorption spectra of $(TiO_2)_8$ cluster shows excitation in the UV region around 200–400 nm, while the absorption spectra of the ruthenium@ $(TiO_2)_8$ shows absorption in the visible region around 400–600 nm, with a near infra-red region where a maximum occur around 1100 nm. The results suggest that the absorption of the ruthenium (N3) dye on $(TiO_2)_8$ brookite gives rise to a bathocromatic shift of the absorption maxima to higher wavelengths. Furthermore, absorption is observed over a wide range from the visible to the far infrared regions. The red shift observed upon absorption on ruthenium (N3) dye on $(TiO_2)_8$ brookite clusters suggests good optical properties of the ruthenium (N3) dye molecules and agrees with earlier reports on the good photocatalytic properties of brookite TiO_2 [17, 30, 31].

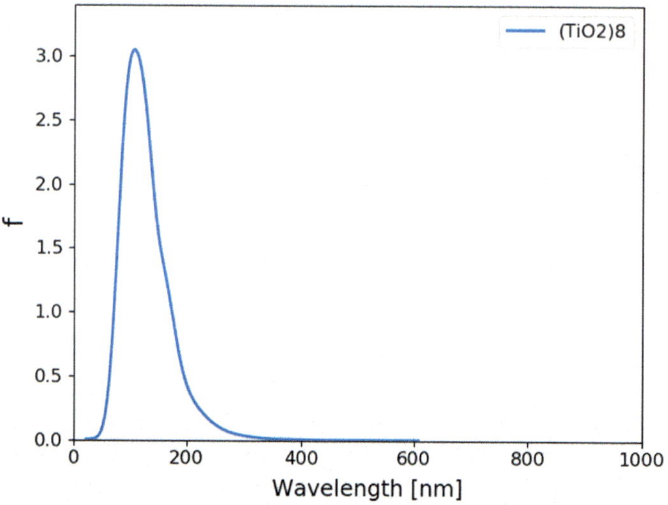

Fig. 8 UV/Vis absorption spectrum (oscillator strength, f, versus wavelength) of $(TiO_2)_8$ brookite cluster

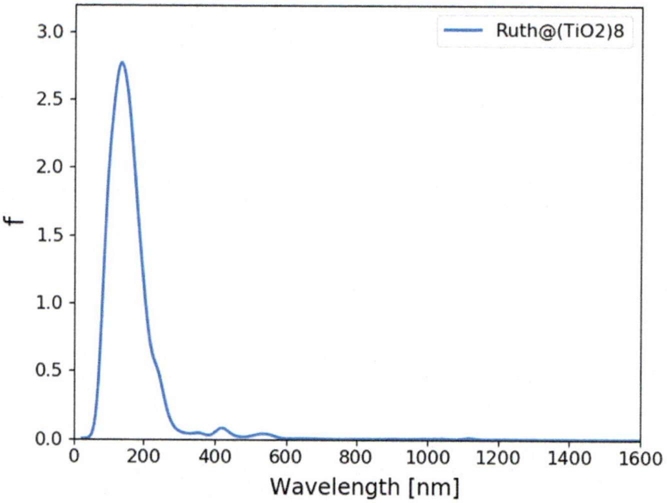

Fig. 9 UV/Vis absorption spectrum (oscillator strength, f, versus wavelength) of ruthenium (N3) complex absorbed on $(TiO_2)_8$ brookite cluster

3.4.2 Isodensity Surfaces of the Ruthenium (N3) Dye Absorbed on $(TiO_2)_n$, n = 8, 68 Brookite Cluster

The isodensity surfaces of the frontier molecular orbitals (HOMO and LUMO) involved in the photoexcitation of the ruthenium (N3) complex absorbed on $(TiO_2)_8$

and $(TiO_2)_{68}$ clusters are shown in Figs. 10 and 11 respectively. The results show that the HOMO is mainly concentrated on the dye donor group where the occupied electronic orbital are located, while the LUMO is distributed over the TiO_2. The distribution of the LUMO very close to the surface of TiO_2 is due to the nature of the ruthenium dye and also suggest that the dyes are energetically in proximity to bind to the surface of TiO_2 upon adsorption, the position of the LUMO agrees with other literature report on ruthenium N3 complex [32]. This suggests electron transfer between the occupied excited state of the dye and the unoccupied acceptor levels of the semiconductor conduction band. The localization of the HOMO electronic level

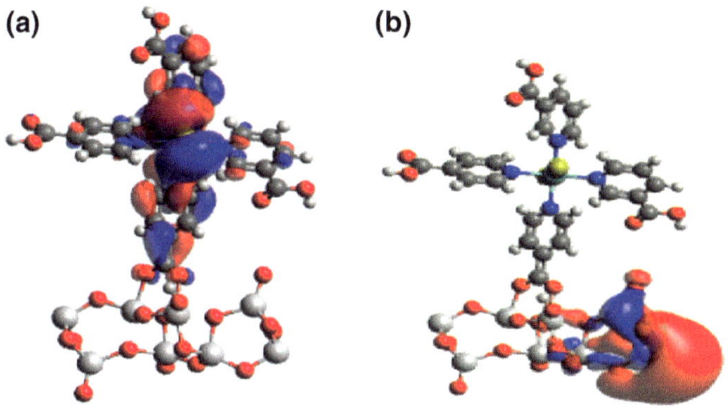

Fig. 10 Isodensity surfaces of **a** HOMO of ruthenium (N3) @ $(TiO_2)_8$ **b** LUMO of ruthenium (N3) @ $(TiO_2)_8$ brookite cluster

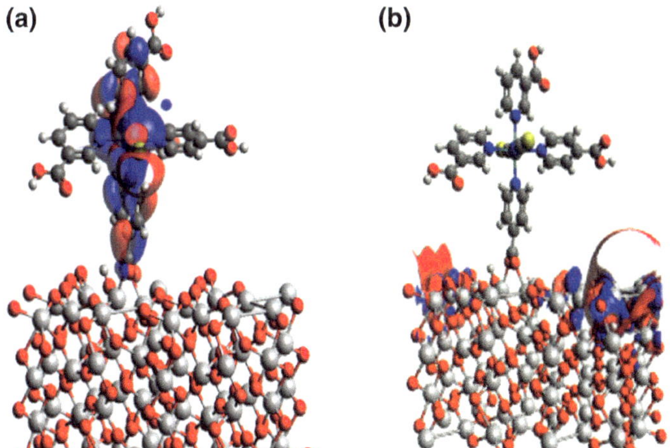

Fig. 11 Isodensity surfaces of **a** HOMO of ruthenium (N3) @ $(TiO_2)_{68}$ brookite cluster **b** LUMO of ruthenium (N3) @ $(TiO_2)_{68}$ brookite cluster

on the dye molecules and the LUMO electronic level on the TiO_2 clusters in Figs. 10 and 11 implies efficient separation of charge upon adsorption and electron injection from the dye excited state into the TiO_2 semiconductor conduction band.

4 Conclusions

A DFT study of a ruthenium (N3) complex was performed successfully for application in dye sensitized solar cells. The optical absorption, UV-Vis spectrum and light harvesting efficiency of the dye molecule were investigated. The results show that the light harvesting efficiency depends on the absorption strength of the dyes. The absorption of the ruthenium (N3) complex on the TiO_2 brookite semiconductors showed that the dye has a stable grafting to the surface of TiO_2. A red shifting of the absorption maxima to the near infra-red region also suggests good optical properties of the dye and the brookite polymorph. The location of the HOMO and LUMO which is visible on the isodensity surfaces of the dye-TiO_2 implies efficient separation of charges upon adsorption and electron injection from the dye excited state into the TiO_2 semiconductor conduction band. Our findings generally suggest that the ruthenium (N3) complex dye molecule and brookite semiconductors exhibit promising features for application in DSSCs.

References

1. Beltran A, Gracia L, Andres J (2006) J Phys Chem B 110:23417–23423
2. Tong Z, Shang PG (2014) Amer Chem Soc 118:11385–11396
3. Hao Y, Jia L, Gang Z, Sum WC, Hongda D, Lin G, Chengjun FK, Wenhui D (2015) Roy Soc Chem Advan 5:60230–60236
4. Monique MR, Xihong P, Lianjun L, Ying L, Jean MA (2012) J Phys Chem 116:19755–19764
5. Yaqin W, Ruirui Z, Jianbao L, Liangliang L, Shiwei L (2014) Nano Res Lett 9:1–8
6. Jun Z, Lisha Q, Wei F, Junhua X, Zhenguo J (2014) Amer Cer Soc 97:2615–2622
7. Hou XG, Huang M, Wu XL, Liu A (2009) Science Chin Ser G-Phys Mech Astr 52:838–842
8. Mohammad NK, De Angelis F, Fantacci S, Selloni A, Viscardi G, Liska P, Ito S, Bessho M, Gratzel T (2005) J Amer Chem Soc 127:16835–16846
9. Jinxia L, Chun Z, Zexing C (2013) Phys Chem Chem Phys 15:13844–13851
10. Chiara A, Edoardo M, Mariachiara P, Enrico R, Fillipo A (2012) Phys Chem Chem Phys 14:15963–15974
11. Yella A, Lee HW, Tsao HN, Yi C, Chandiran AK, Nazeeruddin MK, Diau EWG, Yeh CY, Zakeeruddin SM, Gratzel M (2011). *Sci* 334, 629–634
12. Cai RZ, Zi Jiang L, Yu HC, Hong SC, You ZW, Wang JF, Dao BW (2010) DFT Curr Appl Phys 10:77–83
13. Jungsuttiwong S, Tarsang R, Pansay S, Yakhantip T, Promarak V, Sudyoadsuk T, Kaewin T, Saengsuwan S, Namuangrak S (1999) Int J Chem Mol Nuc Mater and Metallur Eng 77:561–567
14. Avinnash P, Ramesh KCH, Bhanuprakash K (2012) J Chem Sci 124:301–310
15. Ramesh KC, Manho L, Xingfa G, Joonkyung J (2015) J Mol Mod 21(297):1–8
16. Barbara V, Folarin W, Thomas B, Dominic L (2012) ACS PubAmer Chem Soc 28:11354–11363

17. Schaub R, Wahlstrom E, Ronnau A, Laegsgaard E, Stensgaard I, Besenbacher F (2003) Sci, 299:377–379
18. Ken O, Bin L, Jin Z, Kenneth DJ, Jinlong Y, Hrvoje P (2005) Sci 308:1154–1158
19. Zallen R, Moret MP (2006) Sol Stat Comm 137:154–157
20. Frisch MJ, Trucks GW, Schlegel HB, Scuseria GE, Robb MA, Cheeseman JR, Montgomery JA, Jr Vreven T, Kudin KN, Burant JC, Millam JM, Iyengar SS, Tomasi J, Barone V, Mennucci B, Cossi M, Scalmani G, Rega N, Petersson GA, Nakatsuji H, Hada M, Ehara M, Toyota K, Fukuda R, Hasegawa J, Ishida M, Nakajima T, Honda Y, Kitao O, Nakai H, Klene M, Li X, Knox JE, Hratchian HP, Cross JB, Adamo C, Jaramillo J, Gomperts R, Stratmann RE, Yazyev O, Austin AJ, Cammi R, Pomelli C, Ochterski JW, Ayala PY, Morokuma K, Voth GA, Salvador P, Dannenberg JJ, Zakrzewski VG, Dapprich S, Daniels AD, Strain MC, Farkas O, Malick DK, Rabuck AD, Raghavachari K, Foresman JB, Ortiz JV, Cui Q, Baboul AG, Clifford S, Cioslowski J, Stefanov BB, Liu G, Liashenko A, Piskorz P, Komaromi I, Martin RL, Fox DJ, Keith T, Al-Laham MA, Peng CY, Nanayakkara A, Challacombe M, Gill PMW, Johnson B, Chen W, Wong MW, Gonzalez C, Pople JA (2004) Gaussian 03, revision D.01. Gaussian, Inc., Wallingford, CT
21. Corneliu I, Opre PP, Fanica C, Marilena F, Mihai AG (2013) Mat 6:2372–2392
22. Asif M, Salah UK, Usman AR, Mudassir HT (2014) Arab J Chem 1–7
23. Umer M, Ibnelwaleed A, Hussein KH, Shakeel A (2015) J Advan Mater Sci Eng 2015:1–8
24. Emildo M (2017) Phys Sci Rev 6:1–9
25. Prajongtat P, Suramitr S, Nokbin S, Nakajima K, Misuke K, Hannongbua S (2017) J mol Graph and Model 76:551–561
26. Clark SJ, Segall MD, Pickard CJ, Hasnip PJ, Probert MJ, Refson K, Yates JR, Payne MC (2005) 220:576–570
27. Hanwell MD, Curtis DE, Lonie DC, Vandermeersch T, Zurek E, Hutchison GR (2012) J Cheminform 4
28. Mortensen JJ, Hansen LB, Jacobsen KW (2005) Phys Rev B 71:1–11
29. Anthonysamy A, Lee Y, Karunagaran B (2015) J Mat Chem 21:12389–12397
30. Agatino DP, Marianna B, Leonardo P (2013) Cat 3:36–73
31. Gong X, Selloni A (2007) Phys Rev B 76:235307
32. Ramesh KC, Joonkyung J (2018) Sol Ener 383–290

Biochemistry and Biophysics

Complexes of Furonewguinone B with a Cu^{2+} Ion. A DFT Study

Liliana Mammino

Abstract Furonewguinone B is a prenylated acylphloroglucinol (ACPL) isolated from *Hypericum* species and exhibiting antioxidant activity. Its molecule contains a substituted furanoid ring fused to the benzene ring. The molecule has four OH groups, two sp^2 O atoms and one ether O, which results in a high number of possible coordination sites for a metal ion—the seven O atoms and the π bond in the prenyl chain. Complexes of furonewguinone B with a Cu^{2+} ion were calculated considering all the conformers of the molecule and all the sites to which Cu^{2+} might bind, coordinating either to individual sites or simultaneously to two geometrically suitable sites. Calculations were performed at the DFT/B3LYP level, using the 6-31+G(d,p) basis set for the C, O and H atoms and the LANL2DZ pseudopotential for the Cu^{2+} ion. The complexes' relative energies, molecule-ion affinities and other computed molecular properties are analysed in terms of the binding sites of the ion. The effects of complexation on the intramolecular hydrogen bonds and on the other characteristics of the conformers are also analysed. The results highlight the preferred binding sites of the Cu^{2+} ion and show that its charge is reduced effectively in the complexes. Comparison with previously-calculated complexes of other antioxidant ACPLs highlight similarities and differences related to the structural features of the molecules.

Keywords Acylphloroglucinols · Antioxidants · Effects of complexation with a metal ion on molecular properties · Intramolecular hydrogen bonding · Molecule-metal ion affinity in complexes · Phenolic compounds · Reducing ability

---

Electronic supplementary material The online version of this chapter (https://doi.org/10.1007/978-3-030-34941-7_9) contains supplementary material, which is available to authorized users.

L. Mammino (✉)
School of Mathematical and Natural Sciences, University of Venda, Thohoyandou, South Africa

L. Mammino et al. (eds.), *Advances in Quantum Systems in Chemistry, Physics, and Biology*, Progress in Theoretical Chemistry and Physics 32,
https://doi.org/10.1007/978-3-030-34941-7_9

1 Introduction

Acylphloroglucinols (ACPLs, Fig. 1) are phloroglucinol (1,3,5-trihydroxybenzene) derivatives characterised by the presence of a CRO group, where R is more often an alkyl group [1]. Many ACPLs are of natural origin and exhibit a variety of biological activities [1]. Furonewguinones are prenylated ACPLs isolated from *Hypericum* species. They exhibit antioxidant activity in polymorphonuclear cells, and are capable of strongly reducing oxygen production by PMNs after stimulation with *N*-formyl-methionyl-leucylphenylalanine or opsonised zymosan [1–5].

The current work focuses on furonewguinone-B (Fig. 2, hereafter concisely denoted as FNGB). The work pertains to an ongoing computational study of ACPLs

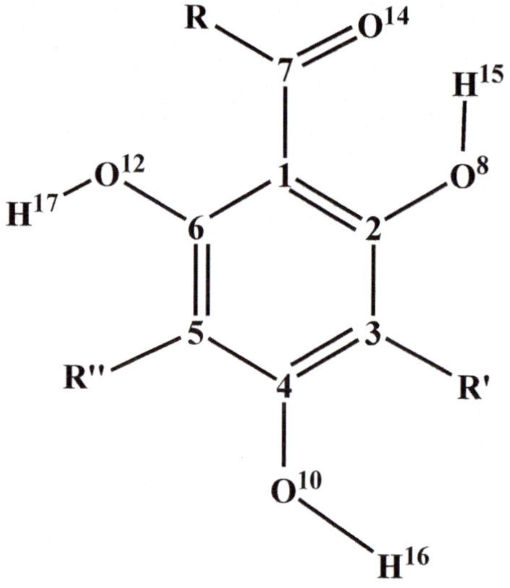

Fig. 1 General structure of acylphloroglucinols and atom numbering of the acylphloroglucinol moiety utilized in the previous works on acylphloroglucinols

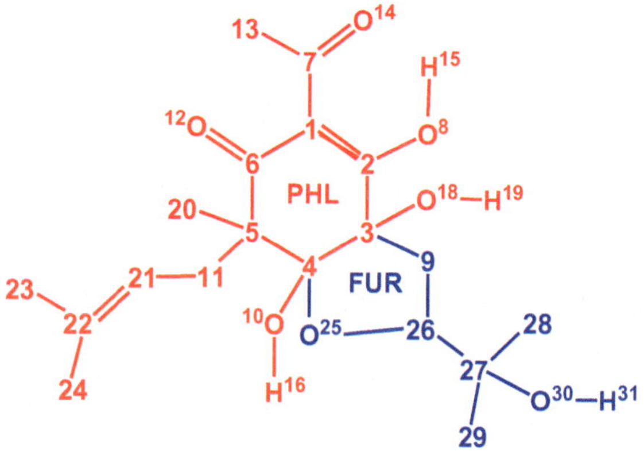

Fig. 2 Structure of furonewguinone B and atom numbering utilized in this work. The C atoms are denoted only by their numbers. The H atoms not pertaining to OH groups are not shown

with antioxidant properties [6–8]. Antioxidant compounds help protect the organism against reactive oxygen species, whose excessive presence appears to increase the risk of degenerative diseases such as ischemia, Alzheimer disease, Parkinson disease and schizophrenia [9–11]. Biologically active compounds of natural origin are viewed as particularly promising for drug development because of their proven compatibility with a living organism and ability to reach their biological target. Understanding the molecular origin of the activity of a compound is relevant for the design of compounds with more potent activity [12]; the initial step is the identification of the features influencing the properties of a molecule.

Several phenolic compounds (commonly present in plants as secondary metabolites [13]) exhibit antioxidant activity, which can be ascribed to the presence of phenol OHs. Possible mechanisms for this activity have been objects of intensive studies [13–24]. It has been established that the activity is enhanced by the presence of ortho OHs (whose intramolecular hydrogen bond contributes to stabilize the molecular radical formed at some stage of the antioxidant action) and by the presence of a C=C double bond in a suitable position in a substituent [14, 16].

Because of containing three phenol OHs, ACPLs could be expected to have good antioxidant properties, and they are actually viewed as potentially interesting lead compounds for the development of drugs against degenerative diseases [25]. Their phenol OHs cannot form intramolecular hydrogen bonds (IHB) with each other, because of being mutually *meta*. On the other hand, ACPLs are characterised by a comparatively strong IHB between the sp^2 O in the CRO group and an *ortho* OH ([26–28]), which may have a role in their antioxidant action. ACPLs of natural origin reported to have interesting antioxidant activity [1–5, 9, 29–31] contain substituents with C=C double bonds, such as a prenyl or a geranyl chain, or additional OH groups, or both.

The FNGB molecule contains a furanoid ring fused to the phloroglucinol ring between C3 and C4, and a prenyl chain at C5. It contains four OH groups and two sp^2 O atoms. It differs from hyperguinone B (HPGB [8]) by the nature of the ring fused along the C3–C4 bond and by the presence of three additional OHs, at C2, C4 and C27. The conformers of FNGB are characterised by different IHB patterns. The H15···O14 IHB (here termed “first IHB” for analogy with previous studies on ACPLs) is the strongest, as the acceptor is an sp^2 O. Additional O–H···O IHBs are possible and have stabilising effects [32]; O–H···π IHBs are also possible and also have stabilising effects [33].

After a preliminary conformational study of the uncomplexed FNGB molecule, the current work focuses on the ability of FNGB to coordinate a Cu^{2+} ion and reduce it to Cu^{+}. This is a way of modelling the molecule’s antioxidant ability [12, 34]. Metal chelation is also viewed as a possible pathway in the mechanisms of the antioxidant activity of polyhydroxybenzenes [13], which contributes to the significance of this type of modelling.

The FNGB molecule contains eight sites apt to bind Cu^{2+}: the seven O atoms (four in the four OH groups, two sp^2 O and the O heteroatom in the furanoid ring) and the C21=C22 π bond in the prenyl chain (π1). Complexes were calculated for all the significantly different conformers of FNGB, considering each binding site individually and also simultaneous binding to two or three geometrically suitable sites. Particular attention was given to including all the predictably stable complexes. The 58 complexes thus calculated can be considered a sufficiently representative (though not exhaustive) ensemble.

The relative energies and the molecule-ion affinity (MIA) of the complexes provide indications about the Cu^{2+} binding preferences for the different sites, showing greater preference for Cu^{2+} to bind simultaneously to O25 and O30, followed by simultaneous binding to O25, O30 and O12, simultaneous binding to O8 and O18, and binding to O30 alone (the preferred site when the ion binds to only one site). Complexes in which the ion binds to these four sites account for relative energies up to 22.65 kcal/mol and for the best MIA values. In all the calculated complexes, the charge on the copper ion is reduced to less than +1, proving the ability of FNGB to reduce the ion by transferring an electron to it and becoming a molecular ion with a positive charge and an unpaired electron. Spin density maps show that this electron is distributed in regions of the molecule far away from the ion.

Single point calculations in solution were added to evaluate the changes in the relative energies and other properties of the complexes, whose changes can realistically be identified using a continuum solvent model with a single-point approach.

Tables reporting the values of relevant properties for all the calculated complexes (relative energy, MIA, natural and Mulliken charge and Mulliken spin density on the ion, distances of the ion from its binding site/s, comparisons of IHB parameters in the uncomplexed FNGB molecule and in the complexes, etc.) and figures showing the geometries of the calculated uncomplexed conformers and of the calculated complexes and the spin density maps in the complexes, as well as graphs illustrating the trends of relevant quantities, are included in the Electronic Supporting Material (ESM). Their numbering is independent of the numbering of the tables and figures included in the text; their numbers are preceded by the letter S to clearly distinguish them when they are referred to in the text.

2 Computational Details

Calculations were performed at the same level of theory utilized in the study of the complexes of other antioxidant ACPLs with a Cu^{2+} ion [6–8], i.e., Density Functional Theory (DFT) with the B3LYP functional [35–37]. The conformational study of the uncomplexed molecule utilised the 6-31+G(d,p) basis set, as the inclusion of diffuse and polarisation functions had proved particularly relevant for DFT descriptions of IHBs in ACPLs [26–28, 32, 33]). The calculation of complexes utilised the 6-31+G(d,p) basis set for the C, O and H atoms and the LANL2DZ pseudopotential [38] for Cu^{2+}, as this combination is known to enable better estimation of the molecule-ion interaction for transition metals complexes [39, 40]. The reasons for selecting the B3LYP functional have been explained in detail in [6] and are not repeated here; it suffices to recall that the comparability of values (such as relative stabilities or binding energies) remains good and, therefore, B3LYP is suitable and widely utilized [15] when comparisons are of interest, as in the current work (comparison of complexes with the same binding site/s, of the binding abilities of different sites, of the ion charge in different complexes, etc.).

Natural Bond Orbital [41–45] analysis was utilised to get detailed information about the electronic structure and realistic values of the charge on the ion in the complexes (the Mulliken charges are often too small to be realistic). A realistic estimation of the charge on the ion is particularly important in studies like the current one, because its value provides an indication of the molecule's ability to reduce an oxidant species.

The molecule-ion affinity (MIA) was taken as the energy difference between the final situation (complex) and the initial situation (the molecule and the ion at infinite distance), and was thus calculated using the equation [12]

$$MIA = E_{complex} - E_{ligand} - E_{ion} \quad (1)$$

where $E_{complex}$ is the energy of the complex, E_{ligand} is the energy of the corresponding uncomplexed conformer of the molecule and E_{ion} is the energy of the isolated Cu^{2+} ion. Basis set superposition error (BSSE) corrections were not included in the MIA calculation because the counterpoise correction is explicitly meant [46] for weak interactions (such as H-bonds), whereas the molecule-ion interaction in the complex is strong; furthermore, the BSSE error is usually small for DFT methods when the basis set expansion is sufficiently flexible.

Harmonic vibrational frequencies were calculated for both the uncomplexed conformers and the complexes, to ascertain the true-minima nature of the identified stationary points, to obtain the zero point energy (ZPE) corrections and to compare the strengths of IHBs through the red shifts that they cause in the vibrational frequency of the donor. The computed frequency values were scaled by 0.9648, as recommended for DFT/B3LYP/6-31+G(d,p) calculations [47].

Calculations in solution utilized the Polarisable Continuum Model (PCM [48–50], which considers the solute as embedded in a cavity surrounded by the continuum solvent), and were performed with the default settings of Gaussian-03 [51] for PCM. While calculations in vacuo were performed with fully relaxed geometry, calculations in solution were performed as single point calculations on the in-vacuo-optimized geometries (at the same level of theory) because the computational demands of PCM optimization for molecular systems of this size make re-optimisation in solution too costly; previous studies of ACPLs [26–28] and ACPL complexes [6–8] had shown fair consistency between the results of full reoptimization and single point PCM calculations, above all for lower energy conformers and for trends-identification. The same solvents utilized in previous studies [6–8, 26–28] were considered (chloroform, acetonitrile and water), as they cover the polarity and H-bonding ability ranges of the media in which a molecule may preferably be present within a living organism. A quick evaluation [52] of the octanol/water partition coefficient of FNGB yields 0.627749; this does not exclude non-negligible presence of the molecule from any medium, thus making the consideration of those three solvents significant also in view of the possible distribution of FNGB molecules in the different media in an organism.

All the calculations were performed on desktop PCs using Gaussian 03, Revision D 01 [51]. Visualization utilised GaussView [53] and Chem3D [52].

All the reported energy values are in kcal/mol and all the distances are in Å.

3 Results

3.1 *Naming of Conformers and Complexes*

The study of ACPLs [26–28, 32, 33] has shown the importance of keeping track of the relevant features characterising the conformers, in a way that enables quick comparisons of specific characteristics. This is done through acronyms in which different letters are utilised to denote individual characteristics. For what could be called the "canonical structure" of ACPLs (the pattern shown in Fig. 1), the acronym system remains basically the same [e.g., 6, 7, 26–28, 32, 33]. For ACPLs with more complex structures, the symbols to be used in the acronyms are adapted to the characteristics of the given structure. In the case of FNGB, the main characterising features and stabilising factors are the IHBs present in a conformer; therefore, each of the possible IHBs is denoted by a symbol, with the letter d denoting the H15···O14 first IHB as in all previous works, and Greek letters denoting the other IHBs (H219···O10, H19···O8, H30···O16, H30···O25, H31···O25, H15···O19 and the O–H···π IHBs). All the symbols are listed in Table 1. Figure 3 shows the possible IHBs through selected conformers, thus also illustrating the meaning of the corresponding symbols. Since the orientation of the prenyl chain has been found to be energy-influencing in some prenylated ACPLs, it is also assigned symbols, in the same way as for hyperguinones [7]. Thus, for instance, d-β-γ-ε-b denotes a conformer having the H15···O14 IHB (d), the H19···O8 IHB (β), the H16···O18 IHB (γ), and the H31···O25 IHB (ε), and the prenyl chain oriented 'upwards' and bending towards the benzene ring (b); d-α-ε-η-e denotes a conformer having the H15···O14 IHB (d), the H19···O25 IHB (α), the H31···O25 IHB (ε), the H16···π1 IHB (η) and the prenyl chain oriented 'downwards' and bending towards the benzene ring (e); and so on.

The complexes are also denoted by acronyms, specifying both the type of conformer and the binding site/s of the Cu^{2+} ion (by writing "Cu" followed by the binding sites). For instance, d*-β-γ-a-Cu-O25-O30 denotes a complex where the H15···O14 IHB is present (d) and H15 has transferred to O14 on complexation (*), the H19···O8 and H16···O18 IHBs are also present (β and γ), the prenyl chain is oriented "upwards" (a) and the Cu^{2+} ion binds simultaneously to O25 and O30.

While the IHBs present in the conformers of the isolated molecule are always easily recognisable as such, because their bond length is shorter than the sum of the van der Waals radii of the O and H atoms (2.7 Å) and their directionality is sufficiently

Table 1 Symbols utilized in the acronyms denoting the conformers and complexes of furonewguinone B

Symbol	Meaning
d	The H15···O14 first IHB is present
α	The H19···O25 IHB is present
β	The H19···O8 IHB is present
γ	The H16···O18 IHB is present
δ	The H16···O25 IHB is present
ε	The H31···O25 IHB is present
τ	The H15···O18 IHB is present
η	The H16···π1 interaction is present
ξ	The H19···π1 interaction is present
^	When added after the symbol denoting an IHB, informs that the length of that IHB is between 2.7 and 3.0 Å
*	When added after the symbol denoting an IHB, informs that a proton transfer has occurred on complexation
a	The prenyl chain is oriented 'upwards' and extended
b	The prenyl chain is oriented 'upwards' and bends towards or over the benzene ring
c	The prenyl chain is oriented 'downwards' and is extended
e	The prenyl chain is oriented 'downwards' and bends towards or over the benzene ring
π1	Indicates the C21=C22 π bond
′	The conformer or the complex is similar to a lower energy one denoted with the same acronym.

clear, some IHBs may get distorted on complexation. In some cases, it becomes difficult to establish whether a certain O–H···O pattern in a complex can be considered an IHB. O–H···O patterns with an H···O distance greater than 2.7 Å show some stabilising effect and may be considered particularly weak IHBs. As a compromise option, the IHB concept is here expanded to the cases when 2.7 Å ≤ H···O ≤ 3.0 Å; such IHBs are denoted with the same Greek letter as the analogous IHB with length ≤2.7 Å, followed by the ^ symbol. Visualization can also help discriminate borderline cases; for instance, in d*-γ-ε-a-Cu-O30, what could have been an H19···O8 IHB has a bond length of 3.037 Å (just slightly greater than the selected 3.0 Å cut-off value), but space-filling visualization shows that the directionality is too poor to consider it an IHB; therefore, it is not included in the acronym denoting the complex.

The atom numbering utilized in this work (Fig. 2) pursues maximum consistency with the one utilized in other studies of ACPLs (Fig. 1), to facilitate comparisons and enable analysis in consistent terms. The atoms of the acylphloroglucinol moiety maintain the same numbering as in previous works [e.g., 6, 7, 26–28, 32, 33]; the numbering of the atoms of the prenyl chain is the same as for hyperguinones [8]; the numbering of the other atoms is specific for FNGB. To facilitate conciseness when expedient, the two moieties are denoted as PHL (phloroglucinol moiety

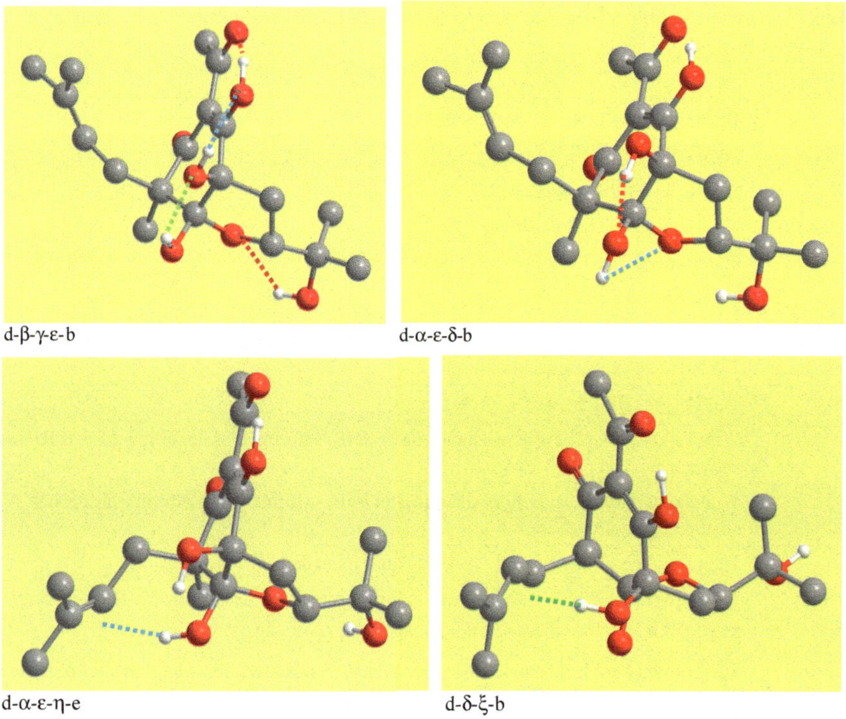

Fig. 3 The possible intramolecular hydrogen bonds (IHBs) in furonewguinone B. The IHBs are represented by coloured dotted segments joining the H atom and the acceptor atom. The image of conformer d-β-γ-ε-b shows H15···O14 (red), H19···O8 (blue), H16···O18 (green) and H31···O25 (dark red). The segments in the other images show only the IHBs not present in d-β-γ-ε-b. Thus, the segments in the image of conformer d-α-ε-δ-b show H19···O10 (red) and H16···O25 (blue). The images in the second row show the O–H···π1 interactions: H16···π1 in d-α-ε-η-e and H19···π1 in d-δ-ξ-b; in this case, the dotted segment points to the double bond, but does not reach any atom, because the acceptor is a π electron cloud

and its substituents) and FUR (the furanoid ring and its substituents), as indicated by the different colours in Fig. 2 (although O18–H19 is not part of the canonical phloroglucinol moiety, it appears convenient to ascribe it to it in this case).

3.2 Conformational Preferences and Hydrogen Bonding of the Uncomplexed Molecule. Results in Vacuo

Figure 4 shows the geometries of the FNGB conformers with relative energy ≤3.5 kcal/mol (selected as a cautious threshold value, below which conformers

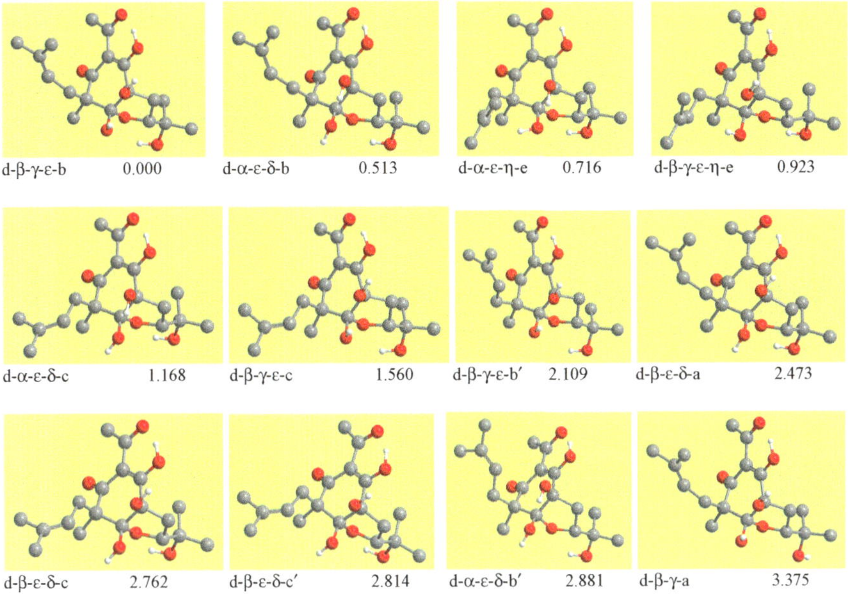

Fig. 4 Conformers of furonewguinone B with relative energy lower than 3.5 kcal/mol. The acronym denoting the conformer is reported under each image, followed by its relative energy. The H atoms not pertaining to OH groups are hidden to better highlight the structural skeleton. The meanings of the symbols in the acronyms are explained in Table 1

might be investigated as possible responsible for the biological activity of the compound). Fig. S1 shows the geometries of all the calculated conformers and table S1 lists their relative energies.

IHBs have the main stabilising roles and the relative energy of the conformers is predominantly determined by their IHB patterns. Besides the first IHB, other possible O–H···O IHBs are H19···O10, H19···O8, H16···O18, H16···O25, H31···O25, and H15···O19. Not all of them can be present at the same time, because the same H atom engages only one acceptor at a time (bifurcation on the donor H has not been noted in these conformers); thus, for instance, H15···O14 and H15···O19 are mutually excluding, and the same is true for H19···O10 and H19···O8, H16···O18 and H16···O25, etc.

Table S2 reports the parameters of the IHBs in the calculated conformers; it also reports the dihedral angles of the donor H atoms, to enable comparison of the changes in the orientation of this atom brought about by complexation. Table 2 summarises the ranges of the parameters of the different IHBs (three out-of-range values have been excluded from the evaluation of the ranges, namely the values for H16···O18 in d-β-γ-η-e and d-β-γ-ε-η-e and for H31···O25 in d-α-ε-δ-b′). The first IHB has the parameters of a moderate-bordering-on-strong IHB [27] and is the strongest IHB present. The stabilising effect of the O–H···O IHBs in which the acceptor is an sp^3 O is generally weaker than when the acceptor is an sp^2 O [32]. In the case

Table 2 Ranges of the parameters of the intramolecular hydrogen bonds, and of the red shifts they cause on the vibrational frequency of the donor OH, in the calculated conformers of furonewguinone B in vacuo

IHB considered	Ranges of parameters			Ranges of red shifts (cm^{-1})
	H···O (Å)	O···O (Å)	OĤO (°)	
H15···O14	1.427–1.472	2.412–2.434	151.8–152.7	1204–1427
H16···O18	1.905–1.964	2.519–2.585	118.6–121.6	63–85
H19···O10	2.018–2.093	2.610–2.654	113.2–117.8	52–113
H19···O8	2.226–2.461	2.703–2.815	101.3–109.2	22–35
H31···O25	2.318–2.366	2.773–2.815	105.6–108.5	14–24
H16···O25	2.362–2.564	2.326–2.331	65.2–73.0	5–23

of FNGB, the directionality of the additional O–H···O IHBs is mostly non-optimal or straightforwardly poor, which makes them weaker than O–H···O IHBs with good directionality. In Table 2, these IHBs are listed in order of increasing bond length, which also corresponds to decreasing bond angle, i.e., poorer directionality; thus, in the first column, the IHBs are likely listed in order of decreasing strength. H16···O25 has very poor directionality (very small bond angle) and is the weakest O–H···O IHB.

O–H···π IHBs are present in a number of conformers, more frequently with O10–16 as donor. Table S3 reports the distances of the donor H atom from the C atoms forming the π1 bond in the prenyl chain (C21 and C22). The distances suggest that this IHB may be comparatively strong in FNGB.

C–H···O IHBs have some stabilising effect [33] and play important roles in determining the orientation of the methyl groups, which mostly orient their H atoms in such a way as to maximise the number of C–H···O IHBs. There are 13–15 C–H···O IHBs in each conformer of FNGB, in relation to the high number of O atoms. It is here opted not to take them into individual account, because their high number would increase the complexity of the acronyms, if each of them is assigned an individual identifying letter. However, some attention is given to the effects of complexation on their length for sampling conformer/complexes pairs.

Table S4 reports the stretching vibrational frequencies of the OH groups (harmonic approximation). Table S5 reports the red shifts caused by the various IHBs on the vibrational frequencies of the OHs when these act as IHB donors, and Table 2 reports the red shift ranges. The selection of a suitable reference for the evaluation of the red shifts is not easy for FNGB, because the reference should be a case in which a given OH is free (not engaged in IHBs), whereas, in the conformers of FNGB, the OHs are mostly engaged in one or another IHB. It was opted to evaluate the red shifts for O8–H15, O10–H16 and O18–H19 with respect to the frequency of the OH groups in the C_{3v} conformer of phloroglucinol (3692.08 cm^{-1}); the red shifts for O30–H31 are evaluated with respect to the average of the frequencies it has in the conformers in which H31 is oriented away from any other O atom (3683.60 cm^{-1}). The red shifts provide indications about the IHB relative strengths, according to the rule-of-thumb criterion that greater red shifts correspond to stronger IHBs. H15···O14 is by far the strongest IHB. It may be recalled that, in FNGB, the molecular context has all the features which enhance the strength of the H15···O14 first IHB in ACPLs, as it forms on the same side as a bulky substituent (FUR) and the OH at C6 is replaced by a keto O [27]. As in all ACPLs, the first IHB is also strengthened by its closing a 6-member ring bordering with a benzene ring and comprising the C7=O14 double bond, which responds to what used to be termed "resonance assisted H-bond" [54–56]. The other clearly identifiable red shifts are those associated with the O–H···π IHBs: 80–157 cm^{-1} for H16···π1 and 192 cm^{-1} for the only case in which H19···π1 is present. The red shifts for the other O–H···O IHBs are much smaller—actually so small that they may be comparable with frequency fluctuations related to other features of the molecular context. Nevertheless, they compare among themselves in the same way in which the IHB lengths compare, i.e., smaller red shifts correspond to greater bond lengths (Table 2). The fact that the red shifts are so small may be related to the poor directionality of these IHBs, for which the vibration of the donor does not bring the H close to the acceptor O along the O···O direction, but occurs towards a rather different or largely different direction. This also suggests that these IHBs are considerably weaker than what could be inferred from their bond lengths, and that the poor directionality has a dominant weakening role.

The orientation of the phenol OHs has non-negligible influence on the energy of the conformers of ACPLs having free OHs [26, 57]. This is not the case for FNGB, because its phenol OHs are always engaged in some IHBs; therefore, the analysis of the conformers' energy in terms of IHB patterns automatically entails the consideration of the orientations of the phenol OHs. The orientation of the prenyl chain does not appear to influence the conformers' energy significantly, likely because the IHB patterns have dominant roles.

3.3 *Complexes of Furonewguinone B with a Cu^{2+} Ion. Results in Vacuo*

Complexes were calculated considering all the possible sites of FNGB to which the ion can bind: simultaneously to three O atoms, or to two O atoms, or to an O atom and π1, and individually to each O atom or to π1. In a number of cases, different inputs optimised to the same complex, which supports the reliability of the identified binding and geometry preferences. Figure 5 shows the geometries of representative complexes and fig. S2 shows those of all the calculated complexes. Table 3 reports the relative energy, the MIA, and the charge and spin density on the ion, for representative complexes (including the ten lowest energy ones); table S6 reports these quantities for all the calculated complexes and table S7 analyses them in terms of the binding sites of the ion. Fig. S3 shows diagrams highlighting the trends of these quantities for all the complexes (part A) and in terms of the binding sites of the ion (part B). Fig. S4 shows diagrams illustrating the ranges of these quantities for different binding sites of the ion.

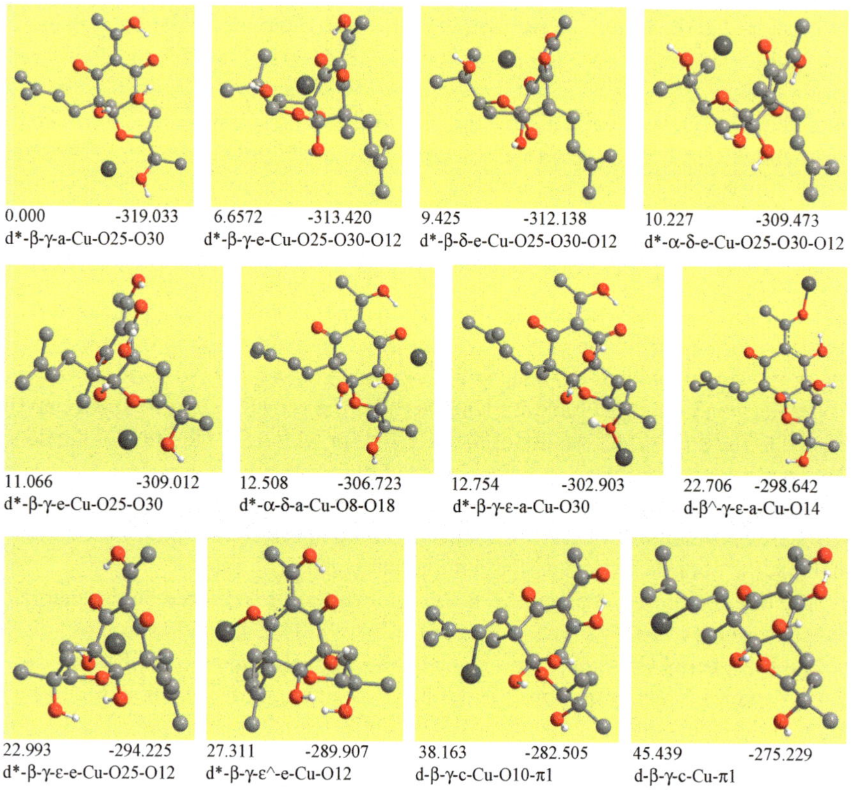

Fig. 5 Representative complexes of furonewguinone B with a Cu^{2+} ion. The figure shows the four lowest energy complexes and a representative complex for each of the other possible binding site/s of the Cu^{2+} ion. The complex relative energy (kcal/mol, not corrected for ZPE) is reported under each image on the left, and the molecule-ion affinity (kcal/mol, not corrected for ZPE or BSSE) on the right. The H atoms not pertaining to OH groups are not shown, to better visualize the skeleton structure

The relative energies not corrected or corrected for ZPE (table S6) have close values and show similar trends; the same is true for the non-corrected and corrected MIAs.

The relative energy and MIA values highlight the binding-site preferences of the ion. In order of decreasing preference, they are as follows: O25 and O30 simultaneously; O25, O30 and O12 simultaneously; O8 and O18 simultaneously; O30 alone; O14 alone; other options. Differently from the complexes of ACPLs considered previously, where π bonds are among the most preferred or the most preferred binding sites [6–8, 58], complexes of FNGB in which the ion binds to π1 are the least stable; only few such inputs optimise to the expected complexes, and their relative energies

Table 3 Relevant quantities for representative complexes of furonewguinone B with a Cu^{2+} cation: complex relative energy, molecule-metal ion affinity (MIA), charge on the Cu ion (both from natural population analysis and Mulliken charge) and Cu atomic spin density. The table considers the ten lowest energy complexes and, after these, the two lowest energy complexes for each of the binding site/s not included in the ten lowest energy complexes

Complex	Complex relative energy (kcal/mol)		MIA (kcal/mol)		Charge on Cu (a.u.)		Cu atomic spin density (Mulliken)
	Not corrected for ZPE	Corrected for ZPE	Not corrected for ZPE	Corrected for ZPE	Natural	Mulliken	
d*-β-γ-a-Cu-O25-O30	0.000	0.000	−319.033	−317.074	0.94293	0.578418	−0.000137
d*-β-γ-e-Cu-O25-O30-O12	6.657	5.048	−313.420	−313.100	0.92174	0.335682	0.002460
d*-β-δ-e-Cu-O25-O30-O12	9.425	7.806	−312.138	−311.895	0.92731	0.364715	0.010866
d*-α-δ-e-Cu-O25-O30-O12	10.227	8.252	−309.473	−309.837	0.91836	0.367389	0.001080
d*-β-γ-e-Cu-O25-O30	11.066	9.464	−309.012	−308.684	0.94545	0.474545	0.000549
d*-β-γ-b-Cu-O25-O30-O12	11.824	10.031	−309.337	−309.114	0.95095	0.361051	0.032917
d*-α-δ-b-Cu-O25-O30-O12	12.240	9.793	−308.898	−309.688	0.91968	0.365929	0.001446
d*-β-γ-b-Cu-O25-O30	12.468	10.466	−308.693	−308.679	0.94857	0.533202	0.000638
d*-α-δ-a-Cu-O8-O18	12.508	11.391	−306.723	−306.047	0.94693	0.625115	0.000222
d*-γ-ε-a-Cu-O30	12.754	11.895	−302.903	−302.049	0.94095	0.752014	−0.000001
d*-α-δ-ε^-a-Cu-O8-O18	14.016	12.844	−302.154	−301.703	0.94733	0.626723	0.000235
d*-β-γ-b-Cu-O25-O30	15.647	13.290	−304.431	−304.858	0.94303	0.479851	0.000558
d-β^-γ-ε-a-Cu-O14	22.706	21.184	−294.180	−298.690	0.93319	0.674985	0.000567
d*-β-γ-ε-e-Cu-O25-O12	22.993	20.826	−298.642	−294.888	0.92164	0.139134	0.001116
d-α-δ-ε^-a-Cu-O14	23.865	22.422	−294.096	−292.126	0.93400	0.686358	0.000630
d*-β-δ-a-Cu-O18-O10	24.167	22.825	−292.305	−296.168	0.95538	0.648891	−0.000156
d*-β-δ-e-Cu-O18-O10-π1	25.252	23.073	−296.929	−296.629	1.00400	0.239105	0.111432

(continued)

Table 3 (continued)

Complex	Complex relative energy (kcal/mol)		MIA (kcal/mol)		Charge on Cu (a.u.)		Cu atomic spin density (Mulliken)
	Not corrected for ZPE	Corrected for ZPE	Not corrected for ZPE	Corrected for ZPE	Natural	Mulliken	
d*-β-γ-ε^-e-Cu-O12	27.311	25.707	−289.907	−289.251	0.94623	0.539069	−0.000140
d-β-γ-ε-b-Cu-O25-O12	30.413	26.023	−287.353	−289.901	0.92826	0.191444	0.001058
d*-β-δ-ε-e-Cu-O12	30.513	28.452	−287.959	−288.378	0.94533	0.586944	0.000099
d-β-γ-c-Cu-O10-π1	38.163	34.912	−282.505	−283.964	0.93322	0.131212	0.029883
d-β-γ-c-Cu-π1	45.439	41.627	−275.229	−277.249	0.88792	0.152611	0.000275
d-β-γ-ε-c-Cu-π1	47.478	44.189	−269.740	−271.525	0.86998	0.315646	0.001614

Table 4 Ranges of the distance of the Cu ion from the atoms to which it is bonded, in the complexes of furonewguinone B with a Cu^{2+} cation. For the cases when the Cu ion binds to a C=C π bond, the distances from the two C atoms are reported

Binding site/s	Distance considered	Distance range (Å)	Binding site/s	Distance considered	Distance range (Å)
O25–O30–O12	Cu···O25	2.328–2.429	O18–O10	Cu···O18	2.030–2.061
	Cu···O30	2.009–2.018		Cu···O10	2.073–2.099
	Cu···O12	2.009–2.028	O14	Cu···O14	1.942–1.949
O8–O18	Cu···O8	2.039–2.072	O30	Cu···O30	1.941–1.948
	Cu···O18	2.091–2.132	O12	Cu···O12	1.922–1.948
O25–O12	Cu···O25	2.314–2.493	π1	Cu···C21	2.100, 2.106
	Cu···O12	1.977–2.020		Cu···C22	2.317, 2.274
O25–O30	Cu···O25	2.080–2.113	–	–	–
	Cu···O30	2.029–2.049	–	–	–

are among the highest; in most cases, inputs in which the ion binds to π1, or simultaneously to π1 and O12, optimise to situations in which the C26–C27 bond cleavages completely; the phenomenon is discussed more in detail in the last paragraph of this section.

Table 4 reports the ranges of the distances of the ion from the atom/s to which it binds and fig. S5 shows diagrams illustrating these ranges; table S8 reports the distances individually for all the calculated complexes. The ion approaches the binding site more closely when it binds to only one atom. When it binds to more than one O atom, Cu···O25 is the longest distance, likely because of steric hindrances by the neighbouring groups.

The calculated complexes of other ACPLs [6–8, 58] showed that a proton transfer from the donor to the acceptor atom of an IHB may occur on complexation. In the case of NFGB, it concerns only the first IHB and it occurs in the majority of complexes; like for the other ACPLs, it never occurs when the ion binds to O14. In the tables reporting IHB lengths, the values for H15···O14 are reported when no proton transfer occurs and the values for H15···O8 when it occurs; the distinction is clear from the complexes' acronyms, as the proton transfer is indicated by an asterisk following the letter d. The evaluation of changes occurring on complexation is always referred to the characteristics of H15···O14 in the uncomplexed conformers, because there is no equivalent of H15···O8 in the uncomplexed conformers.

Table S9 reports the parameters of the IHBs in the complexes. Table S10 reports the parameters' changes occurring on complexation, analysing them according to the ion binding sites. The length of the first IHB nearly always increases, except when Cu binds to O30, when it may also slightly decrease. Like in the other complexes of ACPLs [6–8, 58], it increases considerably (by 0.18–0.23 Å) when the ion binds to O14. It also increases when the ion binds to O8•O18, O25•O30, and O10•O18, with 0.15–0.19, 0.09–0.17 and 0.14–0.22 Å respective increases. The H19···O8 length may increase or decrease—depending on the conformer type—when the ion binds to O25•O30, O25•O30•O12, or O25•O12; it decreases considerably when the ion binds to O18•O10, and increases considerably (by 0.34–0.49 Å) when it binds to O14. The H19···O10 length decreases by 0.16–0.28 Å when the ion binds to O8•O18 and increases in the other cases. The H16···O18 length may increase or decrease, depending on the ion binding site and on the conformer type; the largest decreases (0.46–0.55 Å) occur with d*-β-γ-ε-e conformers. The H16···O25 length mostly decreases, with the greatest decreases (0.19–0.38 Å) when the ion binds to O8•O18. The H31···O25 length decreases when the ion binds to O30 and increases considerably in the other cases, with the greatest increase (0.45–0.50 Å) when the ion binds to O8•O18.

Table S11 reports the vibrational frequencies (harmonic approximation) of the OH groups in the calculated complexes and table S12 analyses them in terms of the binding sites of the ion. Table S13 reports the red shifts in the vibrational frequencies of the OH groups caused by IHBs, evaluated with respect to the same references as for the uncomplexed conformers. The changes in the red shifts induced by complexation are analysed both in terms of conformer types (table S14) and in terms of the ion binding sites (table S15). Comparison of tables S14 and S15 shows that the binding site of the ion is the major factor determining how the red shift changes on complexation, although the influence of the conformer-type also appears to be significant. The changes are discussed in some details in the next paragraphs, because they provide indications as to whether a certain IHB is weakened or strengthened by complexation.

Like for the complexes of other ACPLs [6–8, 58], the greatest decreases in the red shift (741–876 cm^{-1}) concern O8–H15 when the ion binds to O14. Together with the previously-analysed increases in the bond length, this is clear indication of considerable weakening of the first IHB when the ion binds to its acceptor O. The red shift of O8–H15 also decreases—often considerably—for most of the other binding sites. In complexes with the ion binding to O30, the red shift changes depend markedly on whether H15 transfers to O14 or not, as it decreases considerably when the proton transfer occurs and increases considerably when it does not occur; this is consistent with the observation that the IHB length decreases for all the complexes where no proton transfer occurs (table S16). The red shift also increases slightly for some complexes in which the ion binds to O25•O12 and no proton transfer occurs.

Like for the uncomplexed conformers, the red shifts for the other O–H···O IHBs are much smaller, and their changes are often comparable with frequency fluctuations related to other features of the molecular context. Therefore, they cannot be considered to offer unequivocal indications about the weakening or strengthening of an IHB on complexation; better indications are provided by the changes in the IHB parameters (table S10).

The O–H···π IHBs are practically lost on complexation: inputs in which they are present optimise to outputs in which they are not present.

The C–H···O IHBs are mostly maintained on complexation, although their length may vary significantly. Table S17 reports the lengths of the C–H···O IHBs for the representative complexes considered in Table 3, comparing them with the lengths of the corresponding IHBs in the uncomplexed conformer. The changes in the lengths of the IHBs are related to the geometry changes occurring in the molecule on complexation. The lengths of the IHBs involving the terminal C28 and C29 methyls and O30 do not change significantly, because both the methyls and the OH are attached to the same atom (C27). In the other cases, the donor and the acceptor are sufficiently distant that geometry changes can bring them significantly closer or farther away. In few cases, an IHB that was present in the conformer disappears on complexation, or a new IHB appears.

An electron transfer from the molecule to the ion reduces the charge on the ion from +2 to +1; in this way, the molecule acquires a positive charge and one of its electrons is unpaired. The Mulliken spin density values on the ion (Tables 3 and S6) confirm the absence of unpaired electrons. It is also important to recall that a verification of the reliability of the Mulliken spin densities calculated with Gaussian03 (as in the current work) had been performed for hyperguinones [8] and, therefore, the spin densities reported here can be considered reliable. Figure 6 shows

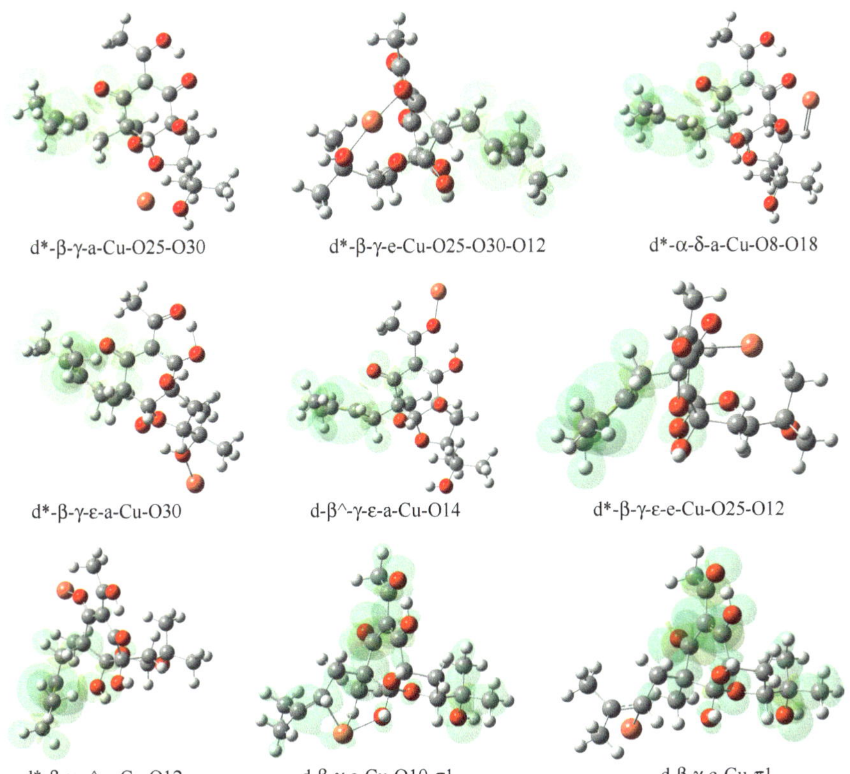

Fig. 6 Spin density maps for representative complexes of furonewguinone B. The lowest energy complex for each binding site of the Cu^{2+} ion is selected as representative. For each image, the conformer model is oriented in such a way as to better highlight both the spin distribution and the position of the Cu ion. The spin densities are mapped with isovalue 0.0004

the spin density maps for representative complexes of FNGB, selected to portrait the situation for complexes with different binding sites of the ion; Fig. 7 shows the similarities in the spin distribution for complexes with the same binding sites and different FNGB conformers; and fig. S6 shows the spin density maps for all the calculated complexes. Like in the considered complexes of other ACPLs [6–8, 58], the spin density is distributed in regions away from the copper ion.

An intriguing phenomenon—not encountered in the study of the complexes of other ACPLs with a Cu^{2+} ion—appears on the optimisation of several inputs in which the ion is located near π1 or between π1 and O12. The C26–C27 bond cleavages completely and the resulting fragment separates. Fig. S7 shows representative steps

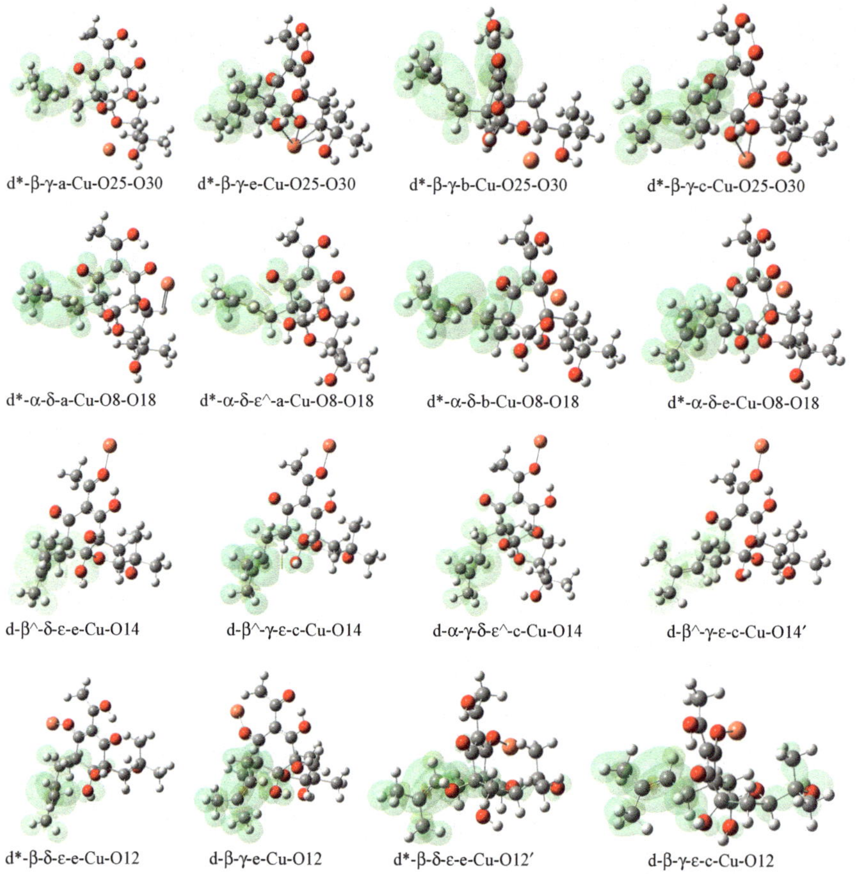

Fig. 7 Comparisons of spin density maps for the same binding sites of the Cu ion and different geometries of the molecule, for selected complexes of furonewguinone B. The figure compares the situations in which Cu binds to both O25 and O30 (first raw), to both O8 and O18 (second raw), to O14 (third raw) and to O12 (fourth raw). The spin densities are mapped with isovalue 0.0004

in the optimisation of one of these inputs. Some of the inputs undergoing this phenomenon were changed by removing O30 from the molecule, and they optimised without any bond cleavage. Fig. S8 compares the outputs of some inputs differing only by the presence or absence of O30. In the latter case, changes in the binding sites may occasionally occur; an example is shown in the second row of fig. S8, with the ion binds to the benzene ring and to π1 in the structure without O30. Although many more cases should be considered to be able to make generalised inferences, it may be suggested that the presence of O30 plays a relevant role for the cleavage of the C26–C27 bond when the ion is close to π1.

3.4 Results in Solution

Calculations in solution were performed for all the uncomplexed conformers and for the complexes with in-vacuo relative energy less than 26 kcal/mol, to ensure representativeness of binding sites and complexation effects. Table S18 reports the values of relevant quantities (relative energies, energy lowering in solution, dipole moment, free energy of solvation and its electrostatic component) in different media for the uncomplexed conformers and the diagrams in fig. S9 highlight their trends. Table S19 reports the values of the same quantities in different media for the complexes, adding the quantities specifically relevant to the complexes, such as the charge and the spin density on the ion; the diagrams in fig. S10 highlight their trends and the diagrams in fig. S11 highlight their ranges for complexes with the binding sites largely associated with lower relative energies (O25•O30, O25•O30•O12, O8•O18 and O30). The quantities considered in these tables and figures are those that can be evaluated realistically through PCM calculations, and provide information on the complexes' stability and binding sites preferences in different media (through the relative energies) and on the molecule's ability to reduce the ion (through the charge on the ion). The MIA is not considered because its evaluation would involve the estimation of the energy aspects of the partial desolvation of the molecule and the ion in the region where they come into contact on complexation, and this evaluation is not easy using a continuum solvent model [6].

Although acetonitrile has greater dipole moment than water, its effect on the properties of ACPLs is intermediate between that of chloroform and that of water (likely because the greater dipole moment is related to greater distance between the charges, not to the values of the charges). Therefore, on analysing results, the expression «as the solvent polarity increases» is referred to the chloroform-acetonitrile-water sequence and the expression «as the medium polarity increases» is referred to the vacuum-chloroform-acetonitrile-water sequence.

The relative energies of the uncomplexed conformers in different media show trends analogous to those of many ACPLs [26, 28], such as slight decrease of the energy gaps in chloroform and acetonitrile with respect to in vacuo and greater decrease in water, and some changes in the relative stabilities of the conformers (more frequent and remarkable in water). The dipole moment increases as the medium polarity increases, as generally noted for ACPLs [26]. The free energy of solvation (ΔG_{solv}) is negative in water, with small magnitudes, and positive in the other two solvents, with greater values in acetonitrile with respect to chloroform.

The relative energies of the complexes show similar trends: the energy of a given complex decreases as the medium polarity increases, with greater decreases for water solution. The energy decreases are not regular, leading to different relative-energy sequences for some complexes, above all in water solution. The relative energy ranges (fig. S11) show clear tendency to lower values in water.

The natural charge on the ion increases slightly as the medium polarity increases, but it always remains less than 1 a.u., except for the high energy d*-β-δ-e-Cu-O18-O10-π1 complex; the ranges of its values depend significantly on the ion binding site. The spin density on Cu may increase or decrease slightly with the medium polarity; the values corresponding to the O25♦O30♦O12 binding site are considerably greater than the others. The dipole moment of the complexes increases as the medium polarity increases, and the ranges of its values depend on the conformer type and on the ion binding site.

Differently from the uncomplexed conformers, ΔG_{solv} is negative in all the solvents, with large magnitudes. The magnitude is smaller in chloroform, greater in acetonitrile and greatest in water; it dominantly depends on the solvent, whereas the dependence on the ion binding site is comparatively minor. The difference with respect to the uncomplexed conformers is due to the fact that the entire molecule is an ion with a positive charge ($\approx$+1), and the Cu ion attached to it also has a $\approx$+1 charge; the charge greatly increases the strength of the interactions between the solute and the solvent.

4 Discussion and Conclusions

The study of 58 complexes of FNGB, with different geometries of the molecule and different binding sites of the ion, shows that the ion is effectively reduced in all the complexes and in all the media considered. This is consistent with the experimentally-proven antioxidant activity of FNGB. The obtained detailed information about the energetics and properties of the complexes highlights considerable dependence on the binding sites of the ion.

FNGB is the fourth antioxidant ACPLs whose complexes with a Cu^{2+} ion have been studied extensively within the ongoing research to which the current work pertains—the other ones being hyperjovinol A (HPJA [6]), arzanol (ARZ [7]) and hyperguinones A and B [8] (HPGB being taken as representative of the two hyperguinones for the comparisons in this section). The structures of these molecules differ greatly. R′ is a geranyl-type chain with an additional OH in HPJA, a substituted pyranoid ring linked to the phloroglucinol moiety through a methylene bridge in ARZ, a substituted pyranoid ring fused to the benzene ring at the C3–C4 bond in HPGB, and a substituted furanoid ring fused to the benzene ring at the C3–C4 bond in FNGB. ARZ, HPGB and FNGB also have a prenyl chain at C5. In terms of the OH groups and O atoms present, HPJA has the three phenol OHs and the keto O14 typical of ACPLs (as shown in Fig. 1) and an additional OH in R′; ARZ has the three phenol OHs and O14 as in Fig. 1, an additional OH and an additional keto O attached to the pyranoid ring, and the O heteroatom in the pyranoid ring; HPGB has only O14 and O8–H15 corresponding to Fig. 1, because O10 is incorporated in the fused pyranoid ring and O12 is a keto O; FNGB has O8–H15, O10–H16 and O14 corresponding to Fig. 1, an additional OH at C3 and an additional OH attached to the furanoid ring, and O12 is a keto O. In terms of C=C π bonds (besides the π system of the benzene ring), HPJA has a π bond in R′ (π1); ARZ has a π bond in the prenyl chain at C5 (π1) and two π bonds in the pyranoid ring (π2 and π3); HPGB has a π bond in the prenyl chain at C5 (π1) and another in the pyranoid ring (π2); and FNGB has one in the prenyl chain at C5 (π1). The structures of HPJA, ARZ and HPGB are shown after the table of values in table S20.

It is interesting to compare relevant features for the complexes of these molecules with a Cu^{2+} ion. A thorough comparison of the effects of complexation on IHBs has been carried out in a separate work [59], because of the importance of IHBs for biologically active molecules and for analysing conformational preferences in general. Table S20 attempts a comparison of other relevant features. The terms of the comparison are not straightforward because the different structural features of the molecules entail different binding possibilities and binding site preferences for the ion and, consequently, a variety of other differences. It was opted to consider partially-qualitative criteria for the comparison, and to include also those applicable to types of complexes possible for all these molecules, such as the characteristics of complexes in which the ion binds to O14 or to O12.

The number of calculated complexes is different for the different molecules, because it depends on the variety of conformers and of possible binding sites. In all cases, the calculated complexes comprise not only the lowest energy ones, but a number of high energy (totally unpopulated) ones; the latter cannot be related to the behaviour or to the biological activity of the complex, but are interesting to investigate the effects of specific changes (changes in the binding site, in the conformer geometry, and the like).

The most preferred binding option for HPJA, ARZ and HPGB is simultaneous binding to an O atom and a π bond; the second and third preferred binding options entail either simultaneous binding to another O atom and a π bond or simultaneous binding to two O atoms. For FNGB, the three preferred binding sites involve simultaneous binding to two or three O atoms; the phenomenon described at the end of Sect. 3.3. suggests that the presence of O30–H31 somehow makes binding to π1 unfavourable. The lowest energy complex in which the ion binds to O14 has high relative energy for all these molecules, with comparable values for HPJA, ARZ and HPGB ($\approx$15 kcal/mol) and higher value for FNGB.

HPJA has the best MIA values, followed by ARZ, which is closely followed by HPGB, while FNGB has the poorest MIA. The trend is highlighted by the comparison of the MIA ranges, and also by the comparison of the MIA of corresponding complexes. The comparison of the complexes with Cu binding to O14 is the most suitable option to highlight differences depending on the entire molecular structure, because the ion does not come close to other parts of the molecule.

The lowest natural charges are observed when the ion binds only to a π bond, although this is the third preferred binding site only for HPJA, while it falls in the high-relative-energy region for the other molecules. This points to great reducing ability of the π bond and is consistent with the importance of the presence of a π bond in a substituent for the antioxidant activity of polyhydroxybenzenes [14, 16].

The distances of the ion from its binding site/s in the lowest energy complex of the four molecules are comparable (somewhat longer for FNGB). The distances in the other complexes selected for comparison are fairly similar.

The length of the first IHB increases considerably when the ion binds to O14 and the red shift in the vibrational frequency of the donor OH decreases sharply (both phenomena indicating considerable weakening of the first IHB); the decrease for HPGB and FNGB is more than double with respect to the decrease for HPJA and ARZ. As for the other IHBs, they often weaken, although to a much less extent than the first IHB. There are also complexes where the length of some IHBs decreases and the corresponding red shift increases (indicating strengthening of the given IHB). The greatest noted increases in the red shift concern O8–H15 for all these molecules, but correspond to different binding sites of the ion in different molecules.

The ranges of the free energy of solvation are basically comparable for the complexes of all these molecules in the same solvent.

Table S21 summarises the same features considered in table S20, for a molecule (1-[3-geranyl-2,4,6-trihydroxyphenyl]-2-methylpropan-1-one, GTM [58]), that does not have additional OH groups or O atoms, and differs from HPJA only because R′ has the two π bonds typical of a geranyl chain, does not have the additional OH present in HPJA, but maintains the methyl at C18. The values of the considered features are comparable with those of the complexes of HPJA, ARZ, HPGB and FNGB. The main differences relate to the absence of the additional OH and the presence of two π bonds in R′; for instance, the three lowest energy complexes involve simultaneous binding to both π bonds (the two lowest energy ones also entailing additional simultaneous binding to either O8 or O10).

The results for the complexes of FNGB with a Cu^{2+} ion presented here add to those obtained for the complexes of other ACPLs with substantially different molecular structures [6–8, 58, 59]. Altogether, the results highlight a number of similarities in the behaviour of ACPLs on complexation with a Cu^{2+} ion, thus contributing information both about ACPLs and about the effects of complexation with a metal ion in general.

References

1. Singh IP, Bharate SB (2006) Nat Prod Rep 23:558–591
2. Heilmann J, Winkelmann K, Sticher O (2003) Planta Med 69:202–206
3. Winkelmann K, Heilmann J, Zerbe O, Rali T, Sticher O (2000) J Nat Prod 63:104–108

4. Winkelmann K, Heilmann J, Zerbe O, Rali T, Sticher O (2001) J Nat Prod 64:701–706
5. Winkelmann K, Heilmann J, Zerbe O, Rali T, Sticher O (2001) Helv Chim Acta 84:3380–3392
6. Mammino L (2013) J Molec Model 19:2127–2142
7. Mammino L (2017) J Mol Model. https://doi.org/10.1007/s00894-017-3443-4
8. Mammino L (2019) Adv Quantum Chem 78:83–108
9. Athanasas K, Magiatis P, Fokialakis N, Skaltsounis AL, Pratsinis H, Kletsas D (2004) J Nat Prod 67:973–977
10. Peuchen S, Bolanos JP, Heales SJR, Almeida A, Duchen MR, Clark JB (1997) Prog Neurobiol 52:261–281
11. Facchinetti F, Dawson VL, Dawson TM (1998) Cell Mol Neurobiol 18:667–677
12. Alagona G, Ghio C (2009) Phys Chem Chem Phys 11:776–790
13. Chiodo SG, Leopoldini M, Russo N, Toscano M (2010) Phys Chem Chem Phys 12:7662–7670
14. Leopoldini M, Prieto Pitarch I, Russo N, Toscano M (2004) J Phys Chem A 108:92–96
15. Leopoldini M, Marino T, Russo N, Toscano M (2004) J Phys Chem A 108:4916–4922
16. Leopoldini M, Marino T, Russo N, Toscano M (2004) Theor Chem Acc 111:210–216
17. Leopoldini M, Russo N, Toscano M (2006) J Agric Food Chem 54:3078–3085
18. Leopoldini M, Russo N, Toscano M (2007) J Agric Food Chem 55:7944–7949
19. Leopoldini M, Russo N, Toscano M (2011) Food Chem 125:288–306
20. Leopoldini M, Chiodo SG, Russo N, Toscano M (2011) J Chem Theory Comput 7:4218–4233
21. Bentes ALA, Borges RS, Monteiro WR, Luiz GM, de Macedo LGM (2011) Alves CN 16:1749–1760
22. Iuga C, Alvarez-Idaboy JR, Russo N (2012) J Org Chem 77:3868–3877
23. Mazzone G, Malaj N, Galano A, Russo N, Toscano M (2015) RSC Advances 5:565–575
24. Galano A, Mazzone G, Alvarez-Diduk R, Marino T, Alvarez-Idaboy JR, Russo N (2016) Annu Rev Food Sci Technol 7:335–352
25. Verotta L (2003) Phytochem Rev 1:389–407
26. Kabanda MM, Mammino L (2012) Int J Quant Chem 112:3691–3702
27. Mammino L, Kabanda MM (2009) J Mol Struct (Theochem) 901:210–219
28. Mammino L, Kabanda MM (2009) J Phys Chem A 113(52):15064–15077
29. Appendino G, Ottino M, Marquez N, Bianchi F, Giana A, Ballero M, Sterner O, Fiebich BL, Munoz E (2007) J Nat Prod 70:608–612
30. Bauer J, Koeberle A, Dehm F, Pollastro F, Appendino G, Northoff H, Rossi A, Sautebin L, Werz O (2011) Biochem Pharmacol 81:259–268
31. Rosa A, Pollastro F, Atzeri A, Appendino G, Melis MP, Deiana M, Incani A, Loru D, Dessì MA (2011) Chem Phys Lipids 164:24–32
32. Mammino L, Kabanda MM (2013) Molec Simul 39(1):1–13
33. Mammino L, Kabanda MM (2012) Int J Quant Chem 112:2650–2658
34. Alagona G, Ghio C (2009) J Phys Chem A 113:15206–15216
35. Becke AD (1992) J Chem Phys 96:9489
36. Becke AD (1993) J Chem Phys 98:5648–5652
37. Lee C, Yang W, Parr RG (1998) Phys Rev B 37:785–789
38. Hay J, Wadt WR (1985) J Chem Phys 82(270):284–299
39. Siegbahn PEM (2003) Q Rev Biophys 36(1):91–145
40. Siegbahn PE (2006) J Biol Inorg Chem 11(6):695–701
41. Reed AE, Weinhold F (1983) J Chem Phys 78(6):4066–4074
42. Reed AE, Weinstock RB, Weinhold F (1985) J Chem Phys 83(2):735–747
43. Reed AE, Weinhold F (1985) J Chem Phys 83(4):1736–1741
44. Reed AE, Curtiss LA, Weinhold F (1988) Chem Rev 88(6):899–926
45. Carpenter JE, Weinhold F (1988) J Mol Struct (Theochem) 169:41–62
46. Boys SF, Bernardi F (1970) Mol Phys 19:553–566
47. Merrick JP, Moran D, Radom L (2007) J Phys Chem A 111:11683–11700
48. Tomasi J, Persico M (1994) Chem Rev 94:2027–2094
49. Tomasi J, Mennucci B, Cammi R (2005) Chem Rev 105:2999–3093
50. Mennucci B (2010) J Phys Chem Lett 1:1666–1674

51. Frisch MJ, Trucks GW, Schlegel HB, Scuseria GE, Robb MA, Cheeseman JR, Montgomery JA, Vreven T, Kudin KN, Burant JC, Millam JM, Iyengar SS, Tomasi J, Barone V, Mennucci B, Cossi M, Scalmani G, Rega N, Petersson GA, Nakatsuji H, Hada M, Ehara M, Toyota K, Fukuda R, Hasegawa J, Ishida M, Nakajima T, Honda Y, Kitao O, Nakai H, Klene M, Li X, Knox JE, Hratchian HP, Cross JB, Adamo C, Jaramillo J, Gomperts R, Stratmann RE, Yazyev O, Austin AJ, Cammi R, Pomelli C, Ochterski JW, Ayala PY, Morokuma K, Voth GA, Salvador P, Dannenberg JJ, Zakrzewski VG, Dapprich S, Daniels AD, Strain MC, Farkas O, Malick DK, Rabuck AD, Raghavachari K, Foresman JB, Ortiz JV, Cui Q, Baboul AG, Clifford S, Cioslowski J, Stefanov BB, Liu G, Liashenko A, Piskorz P, Komaromi I, Martin RL, Fox DJ, Keith T, Al-Laham MA, Peng CY, Nanayakkara A, Challacombe M, Gill PMW, Johnson B, Chen W, Wong MW, Gonzalez C, Pople JA (2003) Gaussian 03. Gaussian Inc, Pittsburgh
52. Chem3D, version 8.0.3 (2003) Chemoffice, Cambridge Software
53. GaussView 4.1, Gaussian Inc, Pittsburgh
54. Gilli G, Bellucci F, Ferretti V, Bertolasi V (1989) J Am Chem Soc 111:1023–1028
55. Bertolasi V, Gilli P, Ferretti V, Gilli G (1991) J Am Chem Soc 113:4017–4925
56. Gilli P, Bertolasi V, Ferretti V, Gilli P (1994) J Am Chem Soc 116:909–915
57. Mammino L, Kabanda MM (2007) J Mol Struct (Theochem) 805:39–52
58. Mammino L (2019) Theor Chem Acc 138:15. https://doi.org/10.1007/s00214-018-2381-2
59. Mammino L, Theor Chem Acc accepted manuscript

Computational Study of Shuangancistrotectorine A: A Naphthylisoquinoline Alkaloid with Antimalarial Activity

Mireille K. Bilonda and Liliana Mammino

Abstract Shuangancistrotectorines A, B, C, D and E are naphthylisoquinoline alkaloids isolated from the twigs of *Ancistrocladus tectorius,* an indigenous plant in China and South East Asia. Their molecules are all C_2-symmetric, i.e., they consist of two identical units. Each unit contains a naphthalene moiety and an isoquinoline moiety. Shuangancistrotectorine B and E are atropo-diastereomers of shuangancistrotectorine A and D respectively. Shuangancistrotectorine A, B and D exhibit very good and specific antimalarial activity, with shuangancistrotectorine A being the most active. The current work presents the results of a detailed conformational study of shuangancistrotectorine A, performed *in vacuo* and in three solvents with different polarities and different H-bonding abilities (chloroform, acetonitrile and water), using two levels of theory, HF/6-31G(d,p) and DFT/B3LYP/6-31+G(d,p). Particular attention is given to intramolecular hydrogen bonds' patterns. The results show that intramolecular hydrogen bonds are the dominant factor influencing conformational preferences and energies, and also the other computable molecular properties. The mutual orientation of the moieties is also an energy-influencing factor, and the results show that all the moieties prefer to be perpendicular to each other. Comparisons with the results of other previously-investigated dimeric naphthylisoquinoline alkaloids are also included.

Keywords Alkaloids · Antimalarials · Intramolecular hydrogen bonding · Mutual orientation of aromatic rings · Shuangancistrotectorine A · Naphthylisoquinoline alkaloids · Solute-solvent interactions

---

Electronic supplementary material The online version of this chapter (https://doi.org/10.1007/978-3-030-34941-7_10) contains supplementary material, which is available to authorized users.

M. K. Bilonda · L. Mammino (✉)
University of Venda, Thohoyandou, South Africa

L. Mammino et al. (eds.), *Advances in Quantum Systems in Chemistry, Physics, and Biology*, Progress in Theoretical Chemistry and Physics 32,
https://doi.org/10.1007/978-3-030-34941-7_10

1 Introduction

Malaria is a transmittable disease caused by plasmodia, among which *Plasmodium falciparum* is the most dangerous and responsible for most deaths. According to WHO 2018 reports [1], 219 million cases were recorded in 2017, and the disease still causes more than 435 000 deaths each year. African countries still have the highest burden of malaria cases, and children under five years are the most vulnerable [1]. In terms of drugs viability, the major problem is the high rate with which *Plasmodium falciparum* develops resistance few years after a new drug enters into clinical use; this implies the need that new drugs, with different molecular structures and different modes of actions, become available to replace the currently-in-use ones as soon as the parasite's resistance lowers their efficacy.

Natural products represent a potentially immense source of new biologically active compounds, with diverse molecular structures and pharmacophores and, therefore, diverse mechanisms of action. For this reason, they are considered important for the identification of new lead compounds for drug development.

Shuangancistrotectorines A, B, C, D and E are naphthylisoquinoline alkaloids isolated from the twigs of *Ancistrocladus tectorius*, a plant indigenous in China and South East Asia [2]. Their molecules have a dimeric structure, i.e., they consist of two units, with each unit consisting of a naphthalene moiety and an isoquinoline moiety. The two units are identical, which make the molecules C_2-symmetric dimers. Shuangancistrotectorine B and E are atropo-diastereomers of shuangancistrotectorine A and D respectively [2]. Shuangancistrotectorine A, B and D show very good antimalarial activity, with shuangancistrotectorine A being the most active and having the lowest IC_{50}, 0.052 μM [2]. (The IC_{50} indicates the amount of a particular drug that is needed to inhibit 50% of a given biological process, or component of a process, such as an enzyme, cell, cell receptor or microorganism).

The current work is part of an ongoing computational study of naphthylisoquinoline (NIQ) alkaloids [3–6], with major focus on those having antimalarial activity. It investigates the conformational preferences of shuangancistrotectorine A (Fig. 1, hereafter denoted by the acronym SHA) in the gas phase and in three solvents with different polarities and different H-bonding abilities (chloroform, acetonitrile and water). SHA is selected because it has the strongest antimalarial activity among shuangancistrotectorines. Calculations were performed at two levels of theory: Hartree-Fock (HF/6-31G(d,p)) and Density Functional Theory with the B3LYP functional (DFT/B3LYP/6-31+G(d,p)). Calculations in solution used the Polarizable Continuum Model (PCM).

The results show that the conformational preferences are primarily influenced by the presence of intramolecular interactions, chiefly intramolecular hydrogen bonds (IHBs), followed by the mutual orientations of the four moieties. O–H···O IHBs have the dominant stabilizing effect. Other H-bond-type intramolecular interactions, such as O–H···π interactions, also have significant stabilizing effect. All the moieties prefer to be perpendicular to each other.

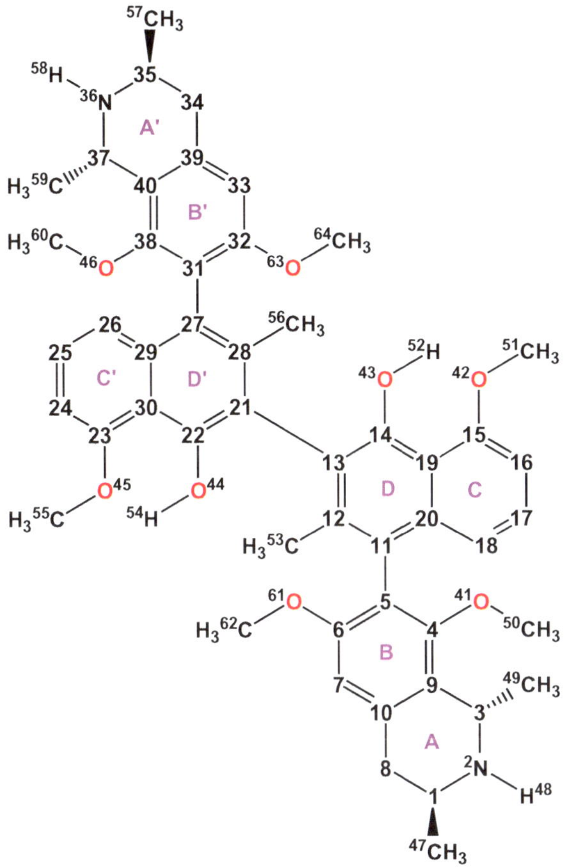

Fig. 1 Structure of the shuangancistrotectorine A molecule and atom numbering utilized in this work. The C atoms in the rings are denoted by the numbers showing their positions. Only the H atoms attached to O or N are numbered separately, while the H atoms attached to a C atom are given the same number as that C atom

Besides the conformational preferences, energetics and relevant geometric characteristics such as the parameters of the IHBs, all the other computable molecular properties (dipole moments, HOMO-LUMO gaps, solvent effect, etc.) are also considered and analysed. In addition, comparisons are made with the results of the studies of other dimeric NIQ alkaloids: jozimine A_2 [3] (JZM, another C_2-symmetric NIQ alkaloid with antimalarial activity [7]) and michellamine A [5] (MCA, a naphthylisoquinoline alkaloid with anti-HIV activity [8]). The three molecules differ by the substituents in the moieties and by the fact that the two naphthalene moieties in MCA are not identical, while they are identical in JZM and SHA (fig. S1). Comparison of the stabilising factors of the three molecules is given particular attention.

2 Computational Details

Calculations were performed using the same levels of theory as in previous works on NIQ alkaloids [3–6], to enable meaningful comparisons. These are Hartree-Fock (HF) with the 6-31G(d,p) basis set, and Density Functional Theory (DFT) with the B3LYP functional [9, 10] and the 6-31+G(d,p) basis set. The reasons for selecting these two levels of theory have been explained in [3–6]. The results obtained in [3–6], as well as studies on other molecules [11, 12] showed that HF can successfully handle intramolecular H-bonding at a comparatively low cost and yields HOMO-LUMO energy gaps approaching those of experiments. DFT calculations are important because DFT takes into account part of the electron correlation. Utilising at least two calculation methods of different natures provides more complete information and validates trends-verification. Although the molecule is symmetric, no symmetry conditions were imposed, to verify how closely the symmetry present in the input is maintained on optimization.

Vibrational frequencies (harmonic approximation) were calculated in the gas phase at the HF/6-31G(d,p) level to verify that the stationary points from optimization results correspond to true minima, to obtain the zero-point energy (ZPE) corrections, and to evaluate the red shifts caused by IHBs, which, in turn, enable an approximate comparison of the IHB strengths. The frequency values were scaled by 0.9024, as recommended for the HF/6-31G(d,p) level [13]. Frequency calculations at the DFT/B3LYP/6-31+G(d,p) level proved unaffordable (did not complete), likely because of the high number of atoms in the molecule.

Calculations in solution considered the same three solvents (chloroform, acetonitrile and water) as in the previous studies of NIQ alkaloids [3–6] and utilised the same computational approach. They were performed as single point (SP) calculations on the in-vacuo-optimised geometries, at the HF/6-31G(d,p) level and using the Polarizable Continuum Model (PCM [14–19]). In this model, the solvent is considered infinite and is modelled by a continuous isotropic dielectric, and the solute molecule is considered inserted in a cavity within the continuum solvent. The geometry of the cavity follows the geometry of the solvent accessible surface of the solute molecule. The default settings of Gaussian03 [20] for PCM were used, namely, the Integral Equation Formalism model (IEF [15–19]) and Gepol model [21–23]. The SCFVAC option was selected to obtain more thermodynamic data. These calculations were performed as SP because the size of the SHA molecule makes re-optimisation in solution computationally exceedingly expensive. While SP calculations cannot provide information on the geometry changes caused by the solvent (including changes in the IHB parameters), they can provide reasonable information on the energetics, such as the conformers' relative energies in solution and the free energy of solvation, and also information about the changes in the molecule's dipole moment caused by the solvent's polarizing effects. PCM calculations at the DFT/B3LYP/6-31+G(d,p) level exhibited frequent convergence problems, which prevented the obtainment of enough results to identify patterns.

A number of calculations were initially performed using GAUSSIAN 03, Revision D 01 [20], on desktop computers. At a later stage, the study utilised Gaussian 09, revision E01 [24] using the CHPC platform.

All the energy values reported in this work are in kcal/mol and all the distances are in Å. Acronyms are used for the calculation methods and for the media, for conciseness sake on reporting values: HF for HF/6-31G(d,p), DFT for DFT/B3LYP/6-31+G(d,p), 'vac' for vacuum, 'chlrf' for chloroform, 'actn' for acetonitrile and 'aq' for water. Tables with all the numerical values of the properties of the calculated conformers (relative energies, parameters of the IHBs, dipole moments, free energy of solvation in the solvents considered, HOMO-LUMO energy gaps, etc.) in the results of both calculation methods, figures showing all the calculated conformers, and figures showing comparisons of the SHA results with those of JZM and MCA, are included in the Electronic Supplementary Information (ESI). The ESI tables and figures are numbered independently from those included in the text; they may be cited in the text, and their numbers are preceded by the letter S.

3 Results

3.1 Naming of Conformers

When analysing the results, it is important to recognise the geometric characteristics of each conformer in an easy way. To this purpose, each conformer is denoted by an acronym built using symbols (letters) for each of its characterising features. The O–H···O IHBs and other IHB-type interactions such as O–H···π and C–H···O are the most important factors stabilizing the conformers, and different conformers have different IHB patterns. Each IHB or IHB-type interaction is denoted by a specific lowercase letter in the acronyms. The letters and their meanings are listed in Table 1. The letters a-h (except 'e' and 'f') maintain the same meanings as in previous studies [4–6], i.e., they denote analogous IHBs. The letters 'i' and 'u' are used to denote types of C–H···O interactions present in SHA but not in the previously-studied NIQs. The letters 'e' and 'f' are excluded from the list in the current study because the IHB-types that they denoted in previous works [3–6] are not present in SHA.

Each acronym starts with SHA, followed by a number giving a quick idea about the conformer's position in the increasing relative energy sequence. The number is followed by the letters denoting the IHBs present, organised according to the most common strength-scale of the corresponding IHBs, from stronger to weaker: O–H···O, O–H···π and C–H···O. Since the mutual orientations of the four moieties also have significant influence on the energy of the conformers, they are also denoted by specific letters, listed after those denoting the IHBs; these letters are the same as in previous works [4–6] and their meaning is explained in Table 1.

Table 1 Letters used in the acronyms denoting the conformers of shuangancistrotectorine A in this work, and their meanings

Letter	Meaning	Letter	Meaning
a	Presence of the O43–H52···O42 IHB	u	Presence of the C–H56···O43 IHB
b	Presence of the O44–H54···O45 IHB	t	C14–C13–C21–C28 torsion angle close to +90°
c	Presence of the O43–H52···π IHB	v	C14–C13–C21–C28 torsion angle close to −90°
d	Presence of the O44–H54···π IHB	p	C4–C5–C11–C12 torsion angle close to +90°
g	Presence of the C–H49···O41 IHB	q	C4–C5–C11–C12 torsion angle close to −90°
h	Presence of the C–H59···O46 IHB	x	C22–C23–C31–C38 torsion angle close to 90°
i	Presence of the C–H53···O44 IHB	y	C22–C25–C31–C38 torsion angle close to −90°

The following example illustrates how the acronyms describe the characteristics of individual conformers. In the SHA-1-a-b-g-h-i-u-q-v-x acronym, '1' represent the first position in order of increasing relative energy (i.e., the lowest energy conformer); 'a' and 'b' represent the O–H···O IHBs present (O43–H52···O44 and O44–H54···O45); 'g,h,i,u' represent the C–H···O interactions present (C–H49···O41, C–H59···O46, C–H53···O44 and C–H56···O43); and 'q-v-x' inform that the C4–C5–C11–C12 torsion angle is close to −90° (q), the C14–C13–C21–C28 torsion angle is close to −90° (v) and the C22–C23–C31–C38 torsion angle is close to 90° (x). In the text, the acronyms denoting the conformers start with 'SHA', whereas in the tables and figures 'SHA' is not reported for the sake of space-saving and the conformers are indicated only by the number and series of letters denoting their characteristics.

Since it is convenient to be able to mention each ring individually and concisely, the rings are denoted by uppercase letters. Since the two units are identical, identical rings are denoted by the same letters, priming those of one of the units (A, B, C, D, A′, B′, C′ and D′, Fig. 1). The whole unit comprising the A, B, C and D rings is called S and the whole unit comprising the A′, B′, C′ and D′ rings is called S′. This option to denote rings and units had initially been designed for JZM [4].

3.2 *Results in Vacuo*

16 conformers were identified and the frequency calculations confirm that they are true minima. Figure 2 shows their geometries and Table 2 reports their relative energies from HF and DFT results. Although it is known that the molecule is symmetric, some pairs of symmetric conformers were calculated independently to verify how

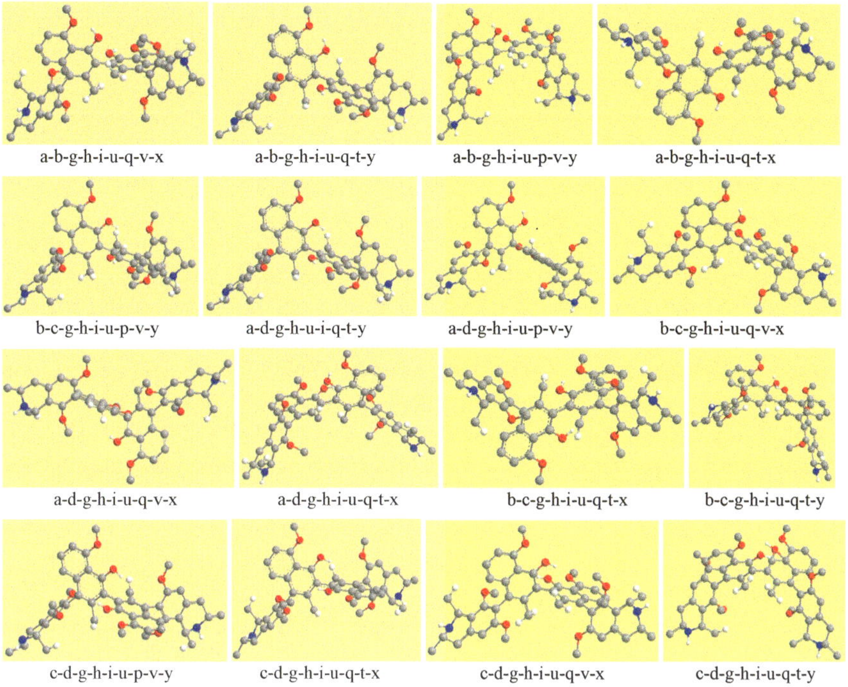

Fig. 2 Optimized geometries of the calculated conformers of the shuangancistrotectorine A molecule and acronyms denoting them. Geometries from HF/6-31G(d,p) results. The H atoms are hidden to better highlight the structures, except for the H atoms of the OH groups (which are usually engaged in IHBs). The initial part of the acronyms (SHA) is not reported under the images, because of space reasons

Table 2 Relative energies of the calculated conformers of shuangancistrotectorine A HF/6-31G(d,p) and DFT/B3LYP/6-31+G(d,p) results *in vacuo*, respectively denoted as HF and DFT in the columns' headings

Conformers	Relative energy (kcal/mol)		Conformers	Relative energy (kcal/mol)	
	HF	DFT		HF	DFT
1-a-b-g-h-i-u-q-v-x	0.000	0.000	9-a-d-g-h-i-u-q-v-x	4.928	5.141
2-a-b-g-h-i-u-q-t-y	1.194	0.833	10-a-d-g-h-i-u-q-t-x	5.620	5.665
3-a-b-g-h-i-u-p-v-y	1.218	0.828	11-b-c-g-h-i-u-q-t-x	5.620	5.665
4-a-b-g-h-i-u-q-t-x	1.904	1.501	12-b-c-g-h-i-u-q-t-y	5.736	5.652
5-b-c-g-h-i-u-p-v-y	4.503	4.583	13-c-d-g-h-i-u-p-v-y	8.008	8.252
6-a-d-g-h-i-u-q-t-y	4.503	4.583	14-c-d-g-h-i-u-q-t-x	8.753	9.076
7-a-d-g-h-i-u-p-v-y	4.866	4.903	15-c-d-g-h-i-u-q-v-x	9.368	9.550
8-b-c-g-h-i-u-q-v-x	4.928	5.141	16-c-d-g-h-i-u-q-t-y	8.411	8.619

The absolute energies of the lowest energy conformer (hartree) are −2633.861716/HF and −2650.476101/DFT closely the results correspond to the symmetry. These are the SHA-5-b-c-g-h-i-u-p-v-y and SHA-6-a-d-g-h-i-u-q-t-y, SHA-8-b-c-g-h-i-u-q-v-x and SHA-9-a-d-g-h-i-u-q-v-x, and SHA-10-a-d-g-h-i-u-q-t-x and SHA-11-b-c-g-h-i-u-q-t-x pairs. In the analysis hereafter, these conformers will be called 'conformers of a symmetric pair'. It has also to be noted that, for such pairs, the different numbers assigned to the two conformers do not correspond to different relative energies, but only to the fact that they are the outcomes of separate calculations.

The conformational preferences are influenced by the IHB patterns (numbers and types of IHBs present in a conformer) and, to a lesser extent, by the mutual orientations of the four moieties. As already mentioned, three types of IHBs may be present in the SHA molecule: O–H···O IHBs between the OH and the ether O attached to a naphthalene moiety (O43–H52···O42 and O44–H54···O45); O–H···π interactions between the OH attached to the naphthalene moiety of one unit and the closest π system of the naphthalene moiety of the other unit (O43–H52···π, O44–H54···π); C–H···O interactions between a methyl and the OH within the same isoquinoline moiety (C–H49···O41, C–H59···O46); and C–H···O interactions between the methyl attached to the naphthalene moiety of one unit and the OH attached to the naphthalene moiety of the other unit (C–H53···O44 and C–H56···O43). The O–H···O IHBs are the dominant stabilizing factors.

The lowest energy conformers are conformers in which all the possible O–H···O IHBs and C–H···O interactions are present simultaneously. The first four energy conformers have relative energy below 2 kcal/mol in both the HF and DFT results; they all contain the O43–H52···O44 and O44–H54···O45 IHBs and the C–H49···O41, C–H59···O46, C–H53···O44 and C–H56···O43 interactions, and differ only by the orientation of the moieties.

Table 3 reports the ranges of the parameters of the IHBs present in SHA and Table S1 reports the values of the parameters for all the conformers. On the basis of their bond lengths, O···O distances and bond angles, the O–H···O IHBs present in SHA can be considered moderate H-bonds [25]. The parameters of the O43–H52···O42 and O44–H54···O45 IHBs in the same conformer are very close. The two conformers

Table 3 Ranges of the parameters of the IHBs in the calculated conformers of shuangancistrotectorine A HF/6-31G(d, p) and DFT/B3LYP/6-31+G(d,p) results *in vacuo*, respectively denoted as HF and DFT in the columns' headings

IHB	IHB length range (Å)		O···O distance range (Å)		OĤO/CĤO range (°)	
	HF	DFT	HF	DFT	HF	DFT
O43–H52···O42	1.733–1.764	1.723–1.734	2.583–2.591	2.585–2.593	142.9–143.5	145.1–145.7
O44–H54···O45	1.733–1.764	1.723–1.734	2.583–2.591	2.585–2.593	142.9–143.5	145.1–145.7
C–H49···O41	2.510–2.529	2.502–2.528	3.114–3.130	3.120–3.314	114.1–114.2	114.4–114.7
C–H59···O46	2.510–2.529	2.502–2.528	3.114–3.130	3.120–3.314	114.1–114.2	114.4–114.7
C–H53···O44	2.726–2.955	2.737–3.031	3.372–3.645	3.362–3.729	112.3–122.4	115.0–122.3
C–H56···O43	2.726–2.939	2.737–3.029	3.372–3.645	3.362–3.727	114.0–122.5	113.5–122.1

The ranges are separated in two blocks, for the O–H···O and C–H···O IHBs respectively of a symmetric pair have the same IHBs in either S or S' and the parameters of these IHBs are identical.

The removal of the O43–H52···O42 IHB brings about the O43–H52···π interaction and the removal of the O44–H54···O45 IHB brings about the O44–H54···π interaction. Each of these removals causes an increase in the conformer's energy. For example, the removal of O43–H52···O42 IHB in SHA-1-a-b-g-h-i-u-q-v-x leads to the formation of O43–H52···π (yielding SHA-8-b-c-g-h-i-u-q-v-x); the corresponding energy increase (≈5 kcal/mol) can be considered an indication of the energy difference between the O43–H52···O42 and O43–H52···π IHBs. On the other hand, since the removal of an O–H···O IHB brings about an O–H···π IHB and vice versa, it is not possible to evaluate the energy of an individual IHB by comparison with a conformer in which it is removed by 180° rotation of the donor [26–35].

No bond length can be defined for the O–H···π IHB, as the acceptor is a whole π electron distribution and not an individual atom. It may be convenient to consider the distance between the H atom of the donor OH and the closest C atom in the acceptor aromatic system, in order to compare the O–H···π IHBs in different conformers. The H···C distances for the O43–H52···π and O44–H54···π IHBs are very close. The ranges of these distances (Å) are 2.216–2.228/HF and 2.184–2.198/DFT for both O43–H54···π and O44–H55···π. The values for all the conformers are reported in Table S2.

The ranges of the H···O distance for the C–H···O interactions/IHBs are reported in Table 3 and the values for all the conformers are reported in Table S3. This distance is considerably longer than the H···O distance for O–H···O IHBs, consistently with the fact that the C–H···O IHBs are considerably weaker H-bonds. The H···O distance is slightly shorter when the IHB involves a C–H and the O within the same isoquinoline moiety (C–H49···O41 and C–H59···O46) than when it involves a C–H in one naphthalene moiety and an O atom in the other naphthalene moiety (C–H53···O44 and C–H56···O43). Corresponding trends are also observed for the donor-acceptor C···O distance and the $\hat{CHO}$ bond angle. Thus, the C–H···O IHBs between naphthalene moieties are somewhat weaker than C–H···O IHBs within an isoquinoline moiety.

Table 4 reports the calculated harmonic vibrational frequencies of the O–H bonds present in SHA. For molecules containing IHBs, it is interesting to consider the red-shift (lowering of the vibrational frequency of the donor OH) caused by the IHBs. The red shift is evaluated with respect to the frequency of a free OH of the same type (an OH in the same position in the molecule, but not engaged in IHBs). Since the OHs in SHA are never free (as is also the case with other dimeric NIQ alkaloids [4, 6]), it is necessary to resort to a model structure to evaluate the frequency of a free OH in a similar or comparable molecular context. The model structure shown in Fig. 3 was utilised here to this purpose. It contains only a naphthalene moiety because OH groups are attached only to the naphthalene moieties in SHA; the OH is free, the CH_3 and OCH_3 substituents are in the same positions as in SHA, and the presence of the other moieties is mimicked by CH_3 groups. The vibrational frequency of the OH in the model structure is 3733.29 cm^{-1}, and it is used as a reference to calculate the

Table 4 Vibrational frequencies (harmonic approximation) of the O–H bonds in the calculated conformers of shuangancistrotectorine A HF/6-31G(d,p) results *in vacuo*

Conformers	Vibrational frequency (cm^{-1})		Conformers	Vibrational frequency (cm^{-1})	
	O43–H52	O44–H54		O43–H52	O44–H54
1-a-b-g-h-i-u-q-v-x	3729.36 3727.41	3729.36 3727.41	10-a-d-g-h-i-u-q-t-x 11-b-c-g-h-i-u-q-t-x	3723.53 3744.89	3744.89 3723.53
2-a-b-g-h-i-u-q-t-y	3733.28	3738.63	12-b-c-g-h-i-u-q-t-y	3751.12	3732.73
3-a-b-g-h-i-u-p-v-y	3740.48 3738.52	3740.48 3738.52	13-c-d-g-h-i-u-p-v-y	3728.99 3723.11	3728.99 3723.11
4-a-b-g-h-i-u-q-t-x	3735.05 3733.12	3735.05 3733.12	14-b-c-g-h-i-u-q-t-x	3730.13 3724.54	3730.13 3724.54
5-b-c-g-h-i-u-p-v-y 6-a-d-g-h-i-u-q-t-y	3745.25 3722.89	3728.32 3743.03	15-c-d-g-h-i-u-q-v-x	3731.23 3730.66	3731.23 3730.66
7-a-d-g-h-i-u-p-v-y 8-b-c-g-h-i-u-q-v-x	3728.43 3747.96	3745.36 3718.66	16-c-d-g-h-i-u-q-t-y	3734.85 3724.29	3734.85 3724.29
9-a-d-g-h-i-u-q-v-x	3718.75	3748.06			

The frequency values have been scaled by 0.9024, recommended for HF/6-31G(d,p) calculations [13]. When two values are reported for the same conformers, the vibrations of the two OHs are coupled, and one frequency corresponds to symmetric coupling and the other to antisymmetric coupling

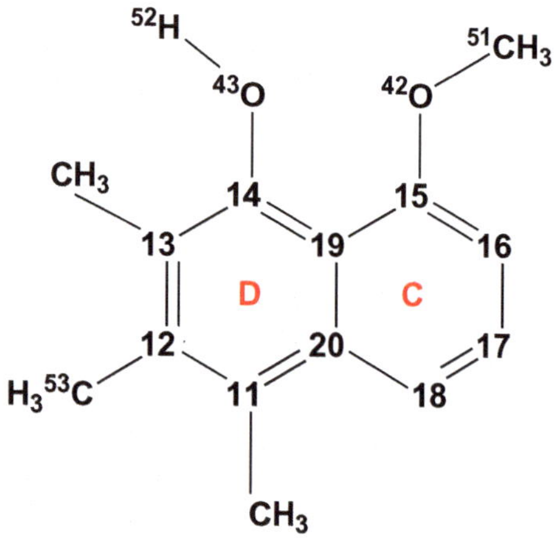

Fig. 3 Model structure used to calculate a reference frequency for the vibrations of the OHs attached to the naphthalene moieties of SHA (O43–H52 and O44–H54). The atom numbering of the D,C naphthalene moiety is selected for this model

red shifts of the OHs in SHA. The red shifts of the vibrations of O43–H52 and O44–H54 when they are engaged in O–H···O IHBs are reported in Table 5, while table S4 reports the red shifts when these OHs are engaged in O–H···π IHBs. The values of the red shifts for O–H···O IHBs and O–H···π IHBs are basically comparable, although the latter are slightly smaller. It is not easy to make straightforward inferences on the strength of the IHBs in SHA from the red shifts. The strength of an IHB and,

Table 5 Red shifts in the vibrational frequency of the O43–H52 and O44–H54 groups in the calculated conformers of shuangancistectorine A, when these OHs are engaged in O–H···O IHBs HF/6-31G(d,p) results *in vacuo*

Conformers	Red shift (cm^{-1})		Conformers	Red shift (cm^{-1})	
	O43–H52	O44–H54		O43–H52	O44–H54
1-a-b-g-h-i-u-q-v-x	91.54	91.54	6-a-d-g-h-i-u-q-t-y	98.01	
	93.49	93.49	7-a-d-g-h-i-u-p-v-y	92.47	
2-a-b-g-h-i-u-q-t-y	87.62	82.27	8-b-c-g-h-i-u-q-v-x		102.24
3-a-b-g-h-i-u-p-v-y	80.42	80.42	9-a-d-g-h-i-u-q-v-x	102.15	
	82.38	82.38	10-a-d-g-h-i-u-q-t-x	97.37	
4-a-b-g-h-i-u-q-t-x	85.85	85.85	11-b-c-g-h-i-u-q-t-x		97.37
	87.78	87.78	12-b-c-g-h-i-u-q-t-y		88.17
5-b-c-g-h-i-u-p-v-y		92.58			

therefore, also its red-shift, depend significantly on the molecular contexts of the donor and the acceptor, as shown, e.g., by the substantially different red shifts of the different O–H···O IHBs (with both donor and acceptor being sp^3 O) in arzanol [36], where the weaker IHBs between the two moieties are those in which the donor OH is attached to the aromatic benzene ring and the acceptor O to the substituent α-pyrone ring. The red shifts for the O–H···O IHBs in SHA are greater than those of the O–H···O IHBs in arzanol in which the donor is a phenol OH—consistently with the fact that they have better parameters. For the sake of comparison, the red shift of the O–H···O IHB in 1,2,4-tryhydroxybenzene, calculated at the same HF/6-31G(d,p) level as for SHA, is 23.62 cm^{-1}, and the parameters of this IHB (2.195 Å bond length, 2.687 Å O···O distance and 119.4° bond angle) indicate that it is substantially weaker than the IHBs in SHA. Studies on the relationships between red shifts and IHB strength (e.g. [37, 38]) have considered simpler molecular systems than SHA, thus not including effects relevant for SHA, such as effects related to the molecular size, or to its symmetry, or possible effects from the interactions of multiple aromatic moieties. All the same, the comparisons with arzanol and with 1,2,4-tryhydroxybenzene support the inference that the O–H···O IHBs in SHA are moderate IHBs, as also indicated by the IHB parameters. It has also to be recalled that HF underestimates the strength of H-bonds, thus yielding smaller red shifts than, e.g., DFT calculations; as already mentioned, DFT frequency calculations were not viable for SHA, due to its size.

Table 6 reports the relative energies corrected for ZPE and the ZPE corrections for the calculated conformers. The ZPE corrections are very close for all the conformers. Their range is 652.83–653.18 kcal/mol, with the greater values corresponding to lower energy conformers. The relative energies corrected for ZPE have the same trends as the uncorrected ones.

Table 7 reports the values of the dipole moments of the calculated conformers *in vacuo*. Their ranges (Debye) are 0.05–5.33/HF and 0.23–5.11/DFT. Both the magnitude and the direction of the dipole moment are largely influenced by the

Table 6 Relative energy ($\Delta E_{correct}$, kcal/mol) corrected for ZPE and ZPE corrections (kcal/mol) for the conformers of shuangancistrotectorine A Results from HF/6-31G(d,p) frequency calculations

Conformers	$\Delta E_{correct}$	ZPE correction	Conformers	$\Delta E_{correct}$	ZPE correction
1-a-b-g-h-i-u-q-v-x	0.000	653.177	9-a-d-g-h-i-u-q-v-x	4.745	652.993
2-a-b-g-h-i-u-q-t-y	1.068	653.027	10-a-d-g-h-i-u-q-t-x	5.429	652.986
3-a-b-g-h-i-u-p-v-y	1.055	653.038	11-b-c-g-h-i-u-q-t-x	5.429	652.986
4-a-b-g-h-i-u-q-t-x	1.800	653.073	12-b-c-g-h-i-u-q-t-y	5.457	652.897
5-b-c-g-h-i-u-p-v-y	4.604	652.914	13-c-d-g-h-i-u-p-v-y	7.657	652.826
6-a-d-g-h-i-u-q-t-y	4.275	652.948	14-c-d-g-h-i-u-q-t-x	8.498	652.922
7-a-d-g-h-i-u-p-v-y	4.604	652.914	15-c-d-g-h-i-u-q-v-x	9.078	652.885
8-b-c-g-h-i-u-q-v-x	4.744	652.993	16-c-d-g-h-i-u-q-t-y	8.080	652.845

The absolute energy of the lowest energy conformer (hartree) is −2633.861716/non-corrected and −2632.820812/corrected

Table 7 Dipole moment of the calculated conformers of shuangancistrotectorine A HF/6-31G(d,p) and (DFT/B3LYP/6-31+G(d,p) results *in vacuo*, respectively denoted as HF and DFT in the column headings

Conformers	Dipole moment (D)		Conformers	Dipole moment (D)	
	HF	DFT		HF	DFT
1-a-b-g-h-i-u-q-v-x	0.054	0.250	9-a-d-g-h-i-u-q-v-x	4.235	4.160
2-a-b-g-h-i-u-q-t-y	2.270	2.350	10-a-d-g-h-i-u-q-t-x	3.970	4.131
3-a-b-g-h-i-u-p-v-y	2.777	3.137	11-b-c-g-h-i-u-q-t-x	3.970	4.130
4-a-b-g-h-i-u-q-t-x	2.301	2.874	12-b-c-g-h-i-u-q-t-y	5.333	5.105
5-b-c-g-h-i-u-p-v-y	2.733	3.083	13-c-d-g-h-i-u-p-v-y	2.733	0.255
6-a-d-g-h-i-u-q-t-y	2.733	3.083	14-c-d-g-h-i-u-q-t-x	0.278	0.236
7-a-d-g-h-i-u-p-v-y	4.149	4.255	15-c-d-g-h-i-u-q-v-x	2.820	2.541
8-b-c-g-h-i-u-q-v-x	4.235	4.160	16-c-d-g-h-i-u-q-t-y	2.000	1.807

orientation of the OH groups. Conformers with the two O–H···O IHBs and all the C–H···O IHBs (the four lowest energy conformers) and conformers with the two O–H···π IHBs and all the C–H···O IHBs (the four highest energy conformers) have lower dipole moments, with the first lowest energy conformer having the lowest value. Most of the conformers with one O–H···O IHB, one O–H···π IHB and all the C–H···O IHBs have high dipole moments (with few exceptions). The mutual orientation of the moieties also has considerable influence on the dipole moment. Different orientations of the S and S' units may cause 2-4 D difference in the dipole moments, and different orientations of the isoquinoline and naphthalene moieties within each unit may cause ≈2 D difference. The two conformers of a symmetric pair have the same dipole moment.

Table 8 HOMO-LUMO energy gap of the calculated conformers of shuangancistrotectorine A *in vacuo* HF/6-31G(d,p) and (DFT/B3LYP/6-31+G(d,p) results, respectively denoted as HF and DFT in the columns' headings

Conformers	HOMO-LUMO energy gap (kcal/mol)		Conformers	HOMO-LUMO energy gap (kcal/mol)	
	HF	DFT		HF	DFT
1-a-b-g-h-i-u-q-v-x	234.732	99.090	9-a-d-g-h-i-u-q-v-x	225.269	89.075
2-a-b-g-h-i-u-q-t-y	231.714	96.787	10-a-d-g-h-i-u-q-t-x	224.291	88.429
3-a-b-g-h-i-u-p-v-y	234.469	98.695	11-b-c-g-h-i-u-q-t-x	224.291	88.429
4-a-b-g-h-i-u-q-t-x	233.076	97.603	12-b-c-g-h-i-u-q-t-y	221.617	86.333
5-b-c-g-h-i-u-p-v-y	226.324	90.380	13-c-d-g-h-i-u-p-v-y	235.202	98.996
6-a-d-g-h-i-u-q-t-y	226.324	90.380	14-c-d-g-h-i-u-q-t-x	234.073	99.140
7-a-d-g-h-i-u-p-v-y	224.215	88.391	15-c-d-g-h-i-u-q-v-x	236.138	100.050
8-b-c-g-h-i-u-q-v-x	225.263	89.075	16-c-d-g-h-i-u-q-t-y	233.157	97.992

Table 8 reports the HOMO-LUMO energy gaps of the calculated conformers of SHA. The gap is greater for conformers with the two O–H···O IHBs and all the C–H···O IHBs (the four lowest energy conformers) and conformers with the two O–H···π and all the C–H···O IHBs (the four highest energy conformers); it is smaller for conformers with one O–H···O IHB, one O–H···π IHB and all the C–H···O IHBs. SHA-15-c-d-g-h-i-u-q-v-x has the greatest gap and SHA-12-b-c-g-h-i-u-q-t-y has the smallest one. Figure 4 shows the shapes of the HOMO and LUMO orbitals for the four lowest energy conformers, some representative higher energy ones and some symmetric pairs; figure S2 shows the shapes of these orbitals for all the calculated conformers. The shapes indicate greater presence of electron density in the naphthalene moieties than in the isoquinoline moieties. The two conformers of a symmetric pair have the same HOMO-LUMO energy gap, and the electron density concentrates on one or the other of the naphthalene moieties.

3.3 Results in Solution

The dielectric constants of the three solvents considered are 4.81 (chloroform), 36.64 (acetonitrile) and 78.54 (water). It is important to note that acetonitrile has greater dipole moment than water, but its effect on solutes is often intermediate between that of chloroform and that of water; therefore, in the paragraphs that follow, "increasing solvent polarity" refers to the chloroform-acetonitrile-water sequence and "increasing medium polarity" refers to the vacuum-chloroform-acetonitrile-water sequence.

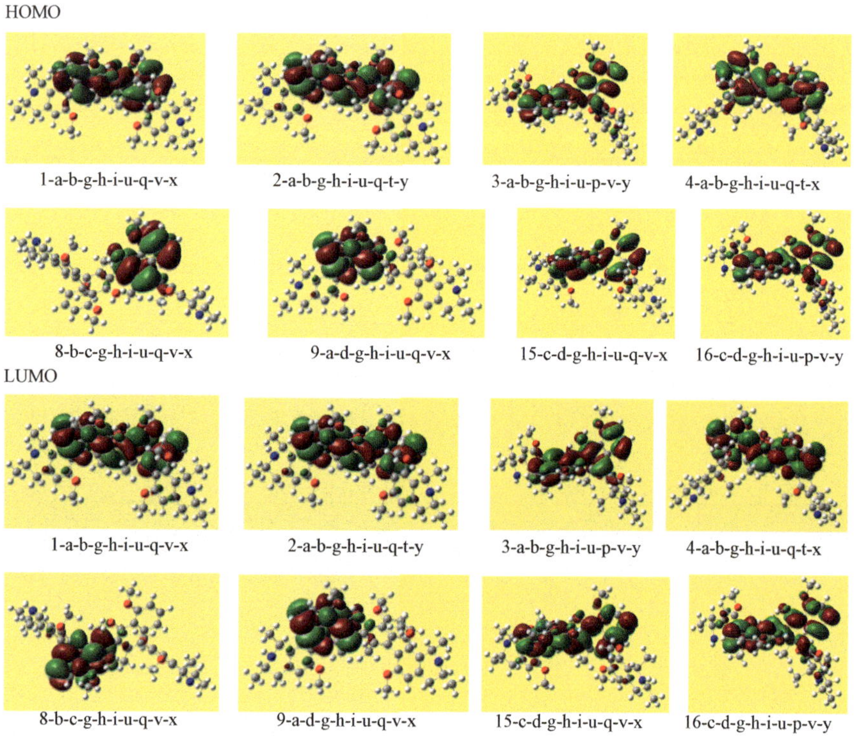

Fig. 4 Shapes of the HOMO and LUMO frontier orbitals of selected conformers of shuangancistrotectorine A

Table 9 compares the relative energies of the conformers in the four media considered (gas phase, chloroform, acetonitrile and water), in the HF results. The lowest energy conformer is the same in all the media. The increasing relative energy sequence is not always the same in different media. The relative energy in water solution is always smaller (often considerably smaller) than the relative energy in vacuo, except for the highest energy conformer. It decreases as the medium polarity increases only for SHA-4-a-b-g-h-i-u-q-t-x, SHA-10-a-d-g-h-i-u-q-t-x and SHA-11-b-c-g-h-i-u-q-t-x (the latter pertaining to a symmetric pair). For the other conformers, it may be greater than in vacuo for both chloroform and acetonitrile or only for one of them. No clear patterns are identifiable. For instance, SHA-4-a-b-g-h-i-u-q-t-x has the two O–H···O IHBs and all possible C–H···O IHBs and its relative energy decreases with increasing solvent polarity, but this does not happen with the other two conformers having the same IHBs (2-a-b-g-h-i-u-q-t-y and 3-a-b-g-h-i-u-p-v-y). It is also interesting to note that the relative energies of the two conformers of a symmetric pair may not remain the same in some of the solvents, suggesting that the presence of the solvent might somewhat affect their symmetry.

If 3.5 kcal/mol is taken as a cautious upper threshold value for the energy of conformers which might be responsible for the biological activity, the results in water

Table 9 Relative energies of the calculated conformers of shuangancistrotectorine A in the media considered HF/6-31G(d,p) results

Conformers	Relative energy (kcal/mol)				Conformers	Relative energy (kcal/mol)			
	vac	chlrf	actn	aq		vac	chlrf	actn	aq
1-a-b-g-h-i-u-q-v-x	0.000	0.000	0.000	0.000	9-a-d-g-h-i-u-q-v-x	4.928	4.967	5.136	3.746
2-a-b-g-h-i-u-q-t-y	1.194	1.397	1.260	0.871	10-a-d-g-h-i-u-q-t-x	5.620	5.526	5.448	3.908
3-a-b-g-h-i-u-p-v-y	1.218	1.037	1.630	0.645	11-b-c-g-h-i-u-q-t-x	5.620	5.486	5.065	3.544
4-a-b-g-h-i-u-q-t-x	1.904	1.435	1.321	0.571	12-b-c-g-h-i-u-q-t-y	5.736	5.380	5.545	4.823
5-b-c-g-h-i-u-p-v-y	4.503	5.023	5.347	4.389	13-c-d-g-h-i-u-p-v-y	8.008	8.875	9.243	7.994
6-a-d-g-h-i-u-q-t-y	4.503	5.023	5.347	4.389	14-c-d-g-h-i-u-q-t-x	8.753	9.277	9.279	7.677
7-a-d-g-h-i-u-p-v-y	4.866	4.910	5.179	4.060	15-c-d-g-h-i-u-q-v-x	9.368	9.770	10.168	8.816
8-b-c-g-h-i-u-q-v-x	4.928	4.938	4.777	3.688	16-c-d-g-h-i-u-q-t-y	8.411	9.231	9.765	8.949

The results in vacuo are from fully-relaxed-geometry calculations, the results in solution are from single point PCM calculations on the *in-vacuo*-optimized geometries

Table 10 Solvent effect (free energy of solvation, ΔG_{solv}) of the calculated conformers of shuangancistrotectorine A in the three solvents considered Results from HF/6-31G(d,p) single point PCM calculations on the *in-vacuo*-optimized geometries

Conformers	ΔG_{solv} (kcal/mol)			Conformers	ΔG_{solv} (kcal/mol)		
	chlrf	actn	aq		chlrf	actn	aq
1-a-b-g-h-i-u-q-v-x	11.79	23.39	1.35	9-a-d-g-h-i-u-q-v-x	12.07	23.90	0.56
2-a-b-g-h-i-u-q-t-y	12.18	23.68	1.23	10-a-d-g-h-i-u-q-t-x	11.59	23.14	−0.43
3-a-b-g-h-i-u-p-v-y	13.29	25.58	2.73	11-b-c-g-h-i-u-q-t-x	11.69	22.90	−0.65
4-a-b-g-h-i-u-q-t-x	11.23	22.74	−0.07	12-b-c-g-h-i-u-q-t-y	12.13	24.01	1.34
5-b-c-g-h-i-u-p-v-y	12.16	24.07	1.03	13-c-d-g-h-i-u-p-v-y	14.59	26.64	3.53
6-a-d-g-h-i-u-q-t-y	12.16	24.07	1.03	14-c-d-g-h-i-u-q-t-x	11.80	23.42	−0.17
7-a-d-g-h-i-u-p-v-y	13.34	25.25	2.24	15-c-d-g-h-i-u-q-v-x	12.63	24.74	1.52
8-b-c-g-h-i-u-q-v-x	11.91	23.41	0.38	16-c-d-g-h-i-u-q-t-y	12.90	25.09	2.29

solution suggest that the four lower-energy conformers might all be considered as potential responsibles for the antimalarial activity of SHA.

An approximate evaluation [39] of the octanol/water partition coefficient of SHA yields 6.49452, which suggests preference for non-polar solvents. The size of the molecule and the presence of multiple aromatic moieties also suggest poor solubility in water. Table 10 reports the solvent effect (free energy of solvation, ΔG_{solv}) for the calculated conformers of SHA in the three solvents considered, and table S6 reports both the values of ΔG_{solv} and the values of its electrostatic component G_{el}. ΔG_{solv} is positive in chloroform and acetonitrile for all the conformers and mostly positive in water (with only few slightly negative values). The values of ΔG_{solv} in acetonitrile are greater than the values in chloroform, and the values in water are the smallest. The two conformers of a symmetric pair may have slightly different ΔG_{solv} values. The electrostatic component of ΔG_{solv} (G_{el}) has negative values in all the three solvents, but significantly more negative in water.

The dipole moment of the conformers (Table S7) increases with the medium polarity, with few exceptions for conformers with the O–H···π interactions. The dipole moment of the two conformers of a symmetric pair is the same or nearly the same in the same solvent.

The HOMO-LUMO energy gap (Table S8) mostly increases with increasing medium polarity, but it decreases slightly for few high-energy conformers. The maximum increase (kcal/mol) with respect to *in vacuo* is 4.946/chlrf, 7.005/actn and 9.150/aq; the greatest decrease is −0.22/chlrf, −0.34/actn and −0.55/aq. The gap of the two conformers of a symmetric pair is the same or nearly the same in the same medium.

3.4 Comparison with Previously Studied Dimeric Naphthylisoquinoline Alkaloids

As mentioned previously, SHA, JZM and MCA (Fig. S1) are all dimeric NIQ alkaloids, which differ by the substituents in the moieties and by the fact that the two naphthalene moieties in MCA are not identical, while they are identical in SHA and JZM. The substituents in the moieties are as follows: two OHs, six CH_3 and six OCH_3 in SHA; four OHs, six CH_3 and two OCH_3 in JZM; six OHs, six CH_3 and one OCH_3 in MCA. The molecules differ also by their biological activities, as SHA and JZM are antimalarials, whereas MCA is anti-HIV. Of the two antimalarials, JZM has the lowest IC_{50} ($IC_{50} = 0.0014\ \mu M$) and SHA has $IC_{50} = 0.052\ \mu M$.

The calculations for the three molecules were performed with the same levels of theory and the conformational search utilised the same criteria. Input geometries were identified considering all the possible IHB patterns (because of the known stabilising effects of IHBs) and investigating preferred mutual orientations of the moieties. In this way, 16 conformers were identified for SHA, 80 for JZM [4] and 22 for MCA [5]. The conformational preferences of the three molecules are influenced by the IHB patterns (the number and type of IHBs present in a conformer) and by the mutual orientation of the moieties [4–6]. The symmetry of the SHA and JZM molecules yields a number of pairs of symmetric conformers on changing specific features; such pairs mostly have comparatively high relative energy; three such pairs were identified for SHA and 16 for JZM [4].

Three types of IHBs may be present in the three molecules: O–H···O, O–H···π and C–H···O, with the O–H···O IHBs having the greatest stabilizing effect. The number and positions of the IHBs is determined by the number and positions of the OH groups: only one OH for each naphthalene moiety in SHA, whereas JZM and MCA have OH groups also in the isoquinoline moieties. SHA and JZM may have O–H···O IHBs only within the naphthalene moieties, whereas MCA may have O–H···O IHBs both within each naphthalene moiety and between the two naphthalene moieties (with the two IHBs being consecutive to each other in lower energy conformers, where they are present simultaneously). SHA may have O–H···π IHBs only between an O–H in the naphthalene moiety of one unit and the closest π system in the naphthalene moiety of the other unit; JZM and MCA may have these same O–H···π IHBs and also O–H···π IHBs between the O–H in an isoquinoline moiety and the closest π system in the naphthalene moiety of the same unit [4, 5]. All the three molecules may have C–H···O IHBs within an isoquinoline moiety, but SHA may also have C–H···O IHBs between naphthalene moieties.

The bond lengths (Å) of the O–H···O IHBs within the naphthalene moieties are the same for SHA and JZM (close to 1.73/DFT and 1.77/HF), while they are slightly shorter in MCA (close to 1.70/DFT and 1.76/HF), likely because of the presence of a consecutive O–H···O IHB between the two naphthalene moieties in MCA.

The isoquinoline moieties prefer to be perpendicular to the naphthalene moieties in all the three molecules. For SHA and JZM, the naphthalene moieties also prefer to be perpendicular to one another, which is not the case for MCA (especially for the four lowest energy conformers). The difference is likely due to the presence of IHBs between moieties in MCA, whose stabilising effect overrides that of the perpendicularity of aromatic systems.

Fig. S3 compares the relative energies of the four lowest energy conformers of the three molecules in the media considered. The values show that the energy gaps are smallest for MCA (the four lowest energy conformers of MCA have relative energies below 1 kcal/mol in all the media), greater for JZM and greatest for SHA. The values remain below 1.9 kcal/mol for all these conformers, and the values in water solution remain below 1.0 kcal/mol.

Fig. S4, S5 and S6 respectively compare the dipole moments, the HOMO-LUMO energy gaps and the free energy of solvation (ΔG_{solv}) of the four lowest energy conformers of the three molecules. The conformers of SHA have the smallest dipole moments values, while the conformers of MCA have the greatest values, and the conformers of JZM have intermediate values. The conformers of MCA have smaller values of the HOMO-LUMO energy gap, and the conformers of SHA and JZM have comparable values-ranges. ΔG_{solv} is negative in water for MCA, JZM and only the fourth lowest energy conformer of SHA, with considerably greater magnitude for MCA; it is positive in chloroform and acetonitrile for all the conformers of the three molecules. For each solvent, ΔG_{solv} is lowest for MCA, higher for JZM and highest for SHA.

The comparison of the molecular properties considered in this section is not sufficient to understand the difference in the biological activities of these molecules (antimalarial for SHA and JZM, anti-HIV for MCA), and the investigation of possible activity mechanism will be crucial for better understanding. However, the comparison highlights some features that are worthy of further consideration, such as the smaller energy gaps between low-energy conformers in MCA (which increases the number of conformers with sufficient population to have non-negligible roles in the biological activity), the presence of cooperative IHBs in MCA (IHB cooperativity may considerably influence the properties and behaviour of a molecule [40–44]), the greater dipole moments (the dipole moment may have a role in the biological activity, as shown, e.g., by the case of anthracyclines [45]), the smaller HOMO-LUMO energy gap (which might favour certain types of reactivity) and the greater magnitude of ΔG_{solv} in water.

4 Discussion and Conclusions

A computational study of shuangancistrotectorine A- a naphthylisoquinoline alkaloid with antimalarial activity—has been carried out at the HF/6-31G(d,p) and DFT/B3LP/6-31+G(d,p) levels *in vacuo* and at the HF/6-31G(d,p) level in chloroform, acetonitrile and water. The results show that the conformational preferences and other properties are influenced by the presence of O–H···O IHBs (which can form within the naphthalene moieties) and other IHB-type interactions, such as O–H···π (which can form between an OH in one moiety and a π ring in another moiety) and C–H···O (which can form between a CH and an OH of the same isoquinoline moiety or between the two naphthalene moieties), and by the mutual orientations of the moieties.

Since the two units are identical, the presence or the absence of a given IHB in only one of the units has the same effects whether it occurs in unit S (O43–H52···O42) or in unit S' (O44–H54···O45). Therefore, the two conformers of a symmetric pair have the same relative energy and similar molecular properties, such as dipole moments and HOMO-LUMO energy gaps, and also rather close values of ΔG_{solv}.

The comparison of shuangancistrotectorine A, jozimine A_2 and michellamine A shows that the conformational preferences of the three molecules are dominantly influenced by the IHB patterns, with the O–H···O IHBs having principal stabilising role. The mutual orientations of the moieties also have significant influence on conformational preferences.

The solvent effect (ΔG_{solv}) is positive in chloroform and acetonitrile for all the conformers of these three molecules. It is also positive in water for shuangancistrotectorine A, whereas it is negative for the other two molecules. A planned separate study will consider adducts of these molecules with explicit water molecules, to better understand their behaviours in water solution.

Acknowledgements The authors would like to thank the Centre for High Performance Computing (South Africa) for providing computational resources used to conduct this work. M. K. Bilonda is grateful to the National Research Foundation (NRF) of South Africa for a bursary to support her PhD studies.

References

1. WHO 018 https://apps.who.int/iris/bitstream/handle/10665/275867/9789241565653-eng.pdf?ua=1
2. Xu M, Bruhn T, Hertlein B, Brun R, Stich A, Wu J, Bringmann G (2010) Chem Eur J 16:4206–4216
3. Mammino L, Bilonda MK (2016) Theor Chem Acc 135:101. https://doi.org/10.1007/s00214-016-1843-7
4. Bilonda MK, Mammino L (2018) In: Wang Y, Thachuk M, Krems R, Maruani J (eds) Concepts, methods and applications of quantum systems in chemistry and physics. Springer, Berlin, pp 305–328
5. Mammino L, Bilonda MK (2017) In: Tadjer A, Pavlov R, Maruani J, Brändas EJ, Delgado-Barrio G (eds) Quantum systems in physics, chemistry, and biology—advances in concepts and applications, Springer, pp 303–316
6. Bilonda MK, Mammino L (2018) Theor Chem Acc 137:139. https://doi.org/10.1007/s00214-018-2323-z
7. Bringmann G, Zhang G, Büttner T, Bauckmann G, Kupfer T, Braunschweig H, Brun R, Mudogo V (2013) Chem Eur J 19:916–923
8. Boyd MR (1994) J Med Chem 37:1740–1745
9. Becke AD (1993) J Chem Phys 98:5648–5662
10. Lee C, Yang W, Parr RG (1988) Phys Rev B 37:785–789
11. Mammino L, Kabanda MM (2009) J Mol Struct (Theochem) 901:210–219

12. Mammino L, Kabanda MM (2012) Int J Quantum Chem 112:2650–2658
13. Irikura K, Johnson RD III, Kacker RN (2005) J Phys Chem A 109:8430–8437
14. Barone V, Cossi M (1997) J Chem Phys 107:3210–3221
15. Tomasi J, Mennucci B, Cammi R (2005) Chem Rev 105:2999–3093
16. Barone V, Cossi M, Tomasi J (1998) J Comput Chem 19:404–417
17. Cossi M, Scalmani G, Rega N, Barone V (2002) J Chem Phys 117:43–54
18. Cancès E, Mennucci B, Tomasi J (1997) J Chem Phys 107:3032–3041
19. Tomasi J, Mennucci B, Cancès E, (1999) (Theochem) 464:211–226
20. Frisch MJ, Trucks GW, Schlegel HB, Scuseria GE, Robb MA, Cheeseman JR, Montgomery JA, Vreven T, Kudin KN, Burant JC, Millam JM, Iyengar SS, Tomasi J, Barone V, Mennucci B, Cossi M, Scalmani G, Rega N, Petersson GA, Nakatsuji H, Hada M, Ehara M, Toyota K, Fukuda R, Hasegawa J, Ishida M, Nakajima T, Honda Y, Kitao O, Nakai H, Klene M, Li X, Knox JE, Hratchian HP, Cross JB, Adamo C, Jaramillo J, Gomperts R, Stratmann RE, Yazyev O, Austin AJ, Cammi R, Pomelli C, Ochterski JW, Ayala PY, Morokuma K, Voth GA, Salvador P, Dannenberg JJ, Zakrzewski VG, Dapprich S, Daniels AD, Strain MC, Farkas O, Malick DK, Rabuck AD, Raghavachari K, Foresman JB, Ortiz JV, Cui Q, Baboul AG, Clifford S, Cioslowski J, Stefanov BB, Liu G, Liashenko A, Piskorz P, Komaromi I, Martin RL, Fox DJ, Keith T, Al-Laham MA, Peng CY, Nanayakkara A, Challacombe M, Gill PMW, Johnson B, Chen W, Wong MW, Gonzalez C, Pople JA (2003) Gaussian 03. Gaussian Inc, Pittsburgh
21. Pascual-Ahuir JL, Silla E (1990) J Comput Chem 11:1047–1047
22. Silla E, Villar F, Nilsson O, Pascual-Ahuir JL, Tapia O (1990) J Mol Graph 8:168–172
23. Silla E, Tunon I, Pascual-Ahuir JL (1991) J Comput Chem 12:1077–1088
24. Frisch MJ, Trucks GW, Schlegel HB, Scuseria GE, Robb MA, Cheeseman JR, Scalmani G, Barone V, Mennucci B, Petersson GA, Nakatsuji H, Caricato M, Li X, Hratchian HP, Izmaylov AF, Bloino J, Zheng G, Sonnenberg JL, Hada M, Ehara M, Toyota K, Fukuda R, Hasegawa J, Ishida M, Nakajima T, Honda Y, Kitao O, Nakai H, Vreven T, Montgomery JA, Jr, Peralta JE, Ogliaro F, Bearpark M, Heyd JJ, Brothers E, Kudin KN, Staroverov VN, Keith T, Kobayashi R, Normand J, Raghavachari K, Rendell A, Burant JC, Iyengar SS, Tomasi J, Cossi M, Rega N, Millam JM, Klene M, Knox JE, Cross JB, Bakken V, Adamo C, Jaramillo J, Gomperts R, Stratmann RE, Yazyev O, Austin AJ, Cammi R, Pomelli C, Ochterski JW, Martin RL, Morokuma K, Zakrzewski VG, Voth GA, Salvador P, Dannenberg JJ, Dapprich S, Daniels AD, Farkas O, Foresman JB, Ortiz JV, Cioslowski J, Fox DJ, Gaussian (2013), Gaussian 09, Revision E.01, Inc., Wallingford CT
25. Buemi G, Zuccarello F (2002) J Mol Struct (Theochem) 581:71–85
26. Simperler A, Lampert H, Mikenda W (1998) J Mol Struct 448:191–199
27. Gilli G, Bellucci F, Ferretti V, Bertolasi V (1989) J Am Chem Soc 111:1023–1028
28. Bertolasi V, Gilli P, Ferretti V, Gilli G (1991) J Am Chem Soc 113:4017–4925
29. Gilli P, Bertolasi V, Ferretti V, Gilli G (1994) J Am Chem Soc 116:909–915
30. Nolasco MM, Ribeiro-Claro PJA (2005) Chem Phys Chem 6:496–502
31. Buemi G (2002) Chem Phys 282:181–195
32. Posokhov Y, Gorski A, Spanget-Larsen J, Duus F, Hansen PE, Waluk (2004) J Chem Phys Chem 5:495–502
33. Sobczyk L, Grabowski SJ, Krygowski TM (2005) Chem Rev 105:3513–3560
34. Jablonski M, Kaczmarek A, Sadlej AJ (2006) J Phys Chem A 110:10890–10898
35. Schalley CA (2009) A. Springer, Mass spectrometry and gas-phase chemistry of non-covalent complexes. Wiley, Hoboken (NJ), p 17
36. Mammino L (2017) Molecules 22:1294. https://doi.org/10.3390/molecules22081294
37. Gu Q, Trindle C, Knee JL (2012) J Chem Phys 137:091101
38. Dasa M, Ghoshb SK (2017) J Chem Sci 129(7):975–981
39. Chem3D Ultra Version 8.0.3., ChemOffice, Cambridge Software (2003)
40. Nishiyama Y, Langan P, Chanzy H (2002) J Am Chem Soc 124:9074–9082
41. López de la Paz M, Ellis G, Pérez M, Perkins J, Jiménez-Barbero J, Vicent C (2002) Eur J Org Chem 5:840–855
42. Deshmukh MM, Bartolotti LJ, Gadre SR (2008) J Chem Phys. A 112:312–321

43. Parra RD, Gong B, Zeng XC (2001) J Chem Phys 115(13):6036–6041
44. Xing B, Yu CW, Chow KH, Ho PL, Fu D, Xu B (2002) J Am Chem Soc 124:14846–14847
45. Bushelyev SN, Stepanov NF (1989) Elektronnaya Struktura y Biologhicheskaya Aktivnost Molecul. Khimiya, Snaye, Moscow

Ab Initio and DFT Computational Study of Myristinin A and a Structurally Related Molecule

Neani Tshilande and Liliana Mammino

Abstract Anti-inflammatory drugs are drugs used for the relief of pain, fever and inflammations. Acylphloroglucinols are phloroglucinol derivatives characterised by the presence of a CRO group. The current work presents the results of a computational study of two acylphloroglucinols of plant origin having potent anti-inflammatory activity: myristinin A and a structurally-related molecule here concisely denoted as DBPO. The molecules have the same 2-(4-hydroxyphenyl)-3,4-dihydro-2H-chromen-7-ol substituent in meta to CRO and differ only by the R chain. DBPO can exist as *cis* or *trans* isomers because of the presence of a C=C double bond in R; the two isomers are studied individually. The sp^2 O atom in CRO can form an intramolecular hydrogen bond (IHB) with either phenol OH *ortho* to CRO. The phenol OHs neighbouring the substituent can form O–H···π IHBs with one of its aromatic rings, if suitably oriented. The conformational preferences and other computable molecular properties have been investigated in vacuo and in three suitably selected solvents. The results show that the O–H···O IHB is the dominant stabilising factor, followed by the O–H···π IHB. Other factors having significant influence on conformational preferences are the orientation of the rings in the substituent, the mutual orientation of the OHs in the phloroglucinol moiety, the orientation of the OHs in the substituent, and also the geometry of the R chain. Comparisons of the results of the three molecules highlight similarity of trends and frequent closeness of individual values of the properties considered. The results in solution highlight narrowing of the energy gaps between conformers and increase of the dipole moment as the medium polarity increases.

Keywords Acylphloroglucinols · Anti-inflammatories · *Cis-trans* isomers · Intramolecular hydrogen bonding · Myristinin A · Solvent effect · Solute-solvent interactions

---

Electronic supplementary material The online version of this chapter (https://doi.org/10.1007/978-3-030-34941-7_11) contains supplementary material, which is available to authorized users.

N. Tshilande · L. Mammino (✉)
School of Mathematical and Natural Sciences, University of Venda, Thohoyandou, South Africa

L. Mammino et al. (eds.), *Advances in Quantum Systems in Chemistry, Physics, and Biology*, Progress in Theoretical Chemistry and Physics 32,
https://doi.org/10.1007/978-3-030-34941-7_11

1 Introduction

Anti-inflammatory drugs are drugs which are used for the relief of pain, fever and inflammations [1]. Many of them are based on molecules isolated from plants, in extracts used in traditional medicine in various continents. However, the anti-inflammatory compounds that are currently in use have some undesirable side effects, such as an increase in the risk of gastrointestinal disorders or an increase in the risk of cardiovascular complications [2, 3]. Therefore, the search for new drugs with fewer or milder side effects is important.

Since the biological activity of a compound depends on its molecular properties [4–8], it is important to obtain detailed information about these properties to better understand how it acts and for the design of modified molecules with more potent activity or less undesirable side effects [7]. The current work considers two anti-inflammatory compounds pertaining to the class of acylphloroglucinols (ACPLs, [9], Fig. 1), namely, myristinin A (1-{2,4,6rihydroxy-3-[(2S,4R)-7-hydroxy-2-(4-hydroxyphenyl)-3,4-dihydro-2Hhromen-4-1yl]phenyl}dodecan-1-one) and a structurally related molecule (2-(4-hydroxyphenyl)-4-[2,4,6rihydroxy-3-(9etradecenoyl)phenyl]-3,4-dihydro-2H-benzopyran-7-ol); they are here concisely denoted as MYRA and DBPO respectively. They were firstly isolated from *Horsfieldia amygdaline* and exhibit a variety of biological activities, including potent anti-inflammatory activity, potent DNA-damaging activity and DNA-polymerase ß inhibitions [9]. They are non-steroidal anti-inflammatories and their structures (Fig. 2) are different from those of the anti-inflammatories currently in use or whose use has been discontinued.

This work pertains to a broad computational study of biologically active ACPLs [10–21]). ACPLs ([9], Fig. 1) are derivatives of phloroglucinol (1,3,5-trihydroxybenzene) characterised by the presence of a CRO group (acyl group). The sp^2 O atom of CRO can form an intramolecular hydrogen bond (IHB) with one to the phenol OHs *ortho* to CRO; it is here termed 'first IHB', similarly to previous works on ACPLs [10–21]). The MYRA and DBPO molecules have the same 2-(4-hydroxyphenyl)-3,4-dihydro-2H-chromen-7-ol substituent in *meta* to CRO (the

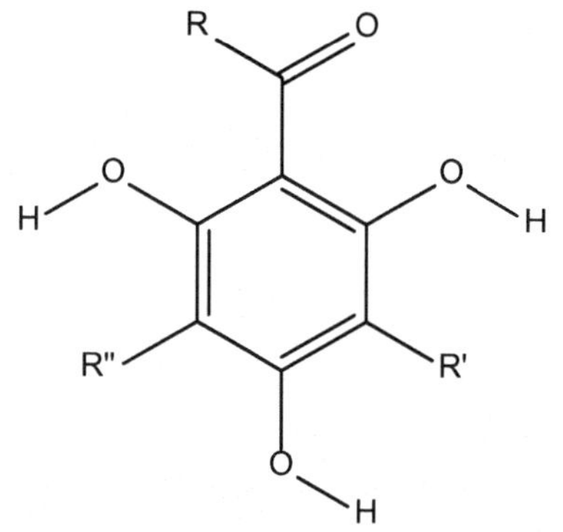

Fig. 1 General structure of acylphloroglucinols

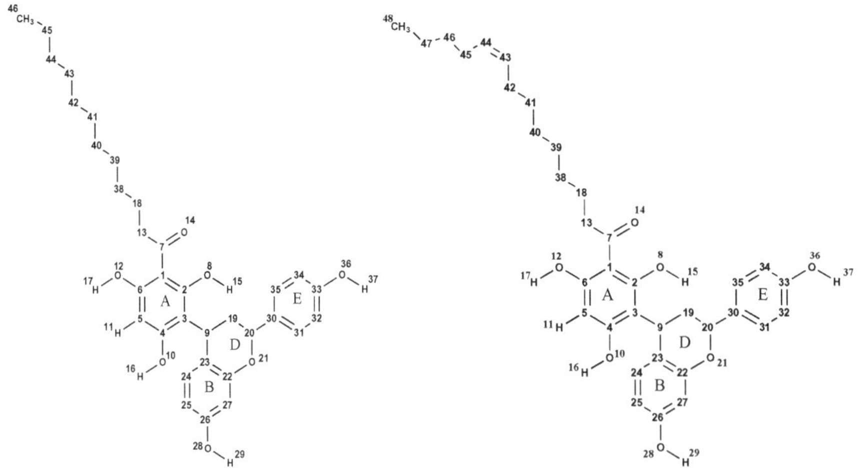

myristinin cis-DBPO

Fig. 2 Structures of myristinin A and cis-DBPO, and atom numbering utilised in this work. The C atoms in the rings and in the acyl chain are represented by the numbers denoting their positions (except the last C atom of the acyl chain) for better view of the structures. Only the H atoms attached to O atoms and to C5 are numbered individually, while the other H atoms are given the same number as the C atom to which they are attached and are not shown in the structure. The rings are denoted by uppercase letters (A, B, D and E)

substituent denoted as R′ in Fig. 1) and differ only by the long R chains in CRO. The R chain in MYRA is an alkane chain. The R chain in DBPO contains a C=C double bond, which gives rise to *cis* and *trans* geometric isomers; hereafter, they may be termed comprehensively as DBPO, or concisely distinguished as *c*-DBPO and *t*-DBPO respectively, according to the context. The expression "the three molecules considered" (or simply "the three molecules") refers to MYRA, *c*-DBPO and *t*-DBPO, in this sequence.

The *c*-DBPO and *t*-DBPO isomers are studied individually because of the growing interest in the similarities and differences in the molecular properties of geometric isomers, in relation to the possible effects on their biological activities. The activities may be substantially different, or cumulative and synergistic (for instance, the simultaneous presence of the two isomers enhances the anti-HIV properties of cinnamic acid derivatives [22, 23] and of chicoric acids [24]).

Calculations have been performed in vacuo and in three solvents with different polarities and different hydrogen bonding abilities—chloroform, acetonitrile and water. The results show that the first IHB is the dominant stabilising factor, followed by the O–H···π interaction between a phenol OH *ortho* to the substituent and one of its aromatic rings, when suitably oriented. Conformational preferences are also influenced by the orientation of the rings in the substituent, the mutual orientation of the OHs in the phloroglucinol moiety, the orientation of the OHs in the substituent, and also the geometry of the R chain. Comparisons of the results obtained for the three molecules highlight similarity of trends for most of the molecular properties considered, and frequent closeness of their individual values. The results in solution show narrowing of the energy gaps between conformers and increase of the conformers' dipole moment as the medium polarity increases.

2 Computational Details

The overall study involved three sets of calculation: preliminary calculation of a model structure, focused on the conformational preferences of the system comprising all the rings and performed at the Hartree-Fock (HF) and density Functional Theory (DFT) levels; calculation of conformers of the three molecules in which a linear geometry of the R chain is associated in turn with each identified stable geometry of the rings-system, performed at the HF level; and calculation of conformers in which different geometries of R are associated with the lowest-energy identified geometry of the rings-system, performed at the HF, DFT and Møller-Plesset Perturbation Theory (MP2) levels. All the calculations were performed with fully relaxed geometry. Each method utilised the same basis set as in previous studies of ACPLs [10–21] to enable meaningful comparisons; thus, HF and MP2 calculations utilised the 6-31G(d,p) basis set, and DFT calculations were performed with the B3LYP functional [25–27] and the 6-31+G(d,p) basis set.

Previous studies [10–21] had shown that HF/6-31G(d,p) calculations can provide a good overview of conformational preferences, and of the factors influencing them, for ACPLs, at a significantly lower computational cost. Therefore, they were performed as initial calculations for all the conformers investigated.

MP2 takes into account both the electron correlation and dispersion interactions [28–30] and, therefore, can provide good descriptions of IHBs. On the other hand, MP2 calculations are computationally highly demanding for molecules of this size. Therefore, they were added for the case when the influence of the factor under investigation (the geometry of the R chain) is expected to be small, to enable better detection of its effects; these conformers turned out to comprise most of the lower energy conformers.

Previous studies [10–21] had also shown that the presence of diffuse functions on the heavy atoms is important for the quality of DFT/B3LYP results for ACPLs. The use of diffuse and polarization functions is also important for a better description of IHBs [29]. Since DFT/B3LYP/6-31+G(d,p) calculations are considerably more expensive (computationally) than HF/6-31G(d,p), they were performed only for the 50 lower energy conformers of the model structure and for the set of calculations meant to check the influence of the geometry of R.

Frequency calculations (harmonic approximation) were performed only at the HF/6–31G(d,p) level and only for the lower energy conformers of each of the three molecules; the results were scaled by 0.8992, as recommended for this [31] level (DFT frequency calculations proved computationally highly expensive for molecules of this size).

Calculations in solution used the Conductor-like Polarizable Continuum Model (CPCM) [32]. They were performed with fully relaxed geometry, because re-optimization is important to identify conformational changes induced by the solvent [33]. They were performed only at the HF/6-31G(d,p) level and only for the lower energy conformers of each of the three molecules, because of the high computational demands for molecules of this size (DFT calculations in solution proved computationally too expensive). Three solvents with different polarities and different hydrogen bonding abilities were selected: chloroform, acetonitrile and water; their dielectric constants are 4.81, 36.64 and 78.54 respectively.

Representative adducts with explicit water molecules were calculated for the lower-energy conformers of MYRA, because they are important for a better understanding of the solute-solvent interactions when the solute can form H-bonds with the solvent molecules (as continuum solvation models do not take into explicit account directional interactions such as H-bonds). The three molecules contain five OH groups, each of which can be H-bond donor or H-bond acceptor to a water molecule, and the sp^2 O of CRO can also be H-bond acceptor; this gives rise to a high variety of possible adducts. The selection of the adducts to calculate utilised the information in [14] and pursued objectives such as the comparison of the H-bonding ability of the different sites of MYRA and the investigation of possible arrangements of water molecules around the MYRA molecule. The general equation for the calculation of the energy of the interaction between the MYRA molecule and the water molecules attached to it ($\Delta E_{interaction}$) is [34]:

$$\Delta E_{interaction} = E_{adduct} - \left(E_{MYRA} + n\,E_{aq}\right) - E_{aq\text{-}aq}$$

where E_{adduct} is the energy of the adduct, E_{MYRA} is the energy of the isolated conformer of the MYRA molecule, E_{aq} is the energy of an isolated water molecule, n is the number of water molecules in the adduct, and $E_{aq\text{-}aq}$ is the energy of the interactions among the water molecules in the adduct. $E_{aq\text{-}aq}$ needs to be considered when the water molecules are also interacting with each other, and is evaluated through a single-point calculation on the water molecules arranged in the same way as in the adduct, but without the solute molecule [34]; if the energy of this arrangement of water molecules is denoted as $E_{aq\text{-}cluster}$, then $E_{aq\text{-}aq} = E_{aq\text{-}cluster} - n\,E_{aq}$, and the equation for the calculation of $\Delta E_{interaction}$ becomes

$$\Delta E_{interaction} = E_{adduct} - E_{MYRA} - E_{aq\text{-}cluster}$$

Both E_{adduct} and $E_{aq\text{-}cluster}$ were corrected for Basis Set Superposition Error (BSSE) using the counterpoise method [35]. The adducts were calculated at the HF/6-31G(d,p) level because of the costly computational demands of the supermolecular structures.

All the calculations were performed with Gaussian 09, revision E.0.1 [36]. The visualisation of results utilised GaussView 4.1 [37] and Chem3D [38]. All the energy values reported in this work are in kcal/mol and all the distances are in angstroms (Å). For the sake of conciseness, the calculation methods and the media will be denoted by acronyms on reporting and analysing results: HF for HF/6-31G(d,p), DFT for DFT/B3LYP/6-31+G(d,p) and MP2 for MP2/6-31G(d,p) for the calculation methods; 'vac', 'chlrf', 'actn' and 'aq' respectively for vacuum, chloroform, acetonitrile and water.

Besides the tables and figures included in the text, expanded tables and figures (showing the relevant images and the values of relevant properties for all the calculated conformers), as well as comparative tables, and figures with comparative graphs for the data for which trend-comparisons can be visualised, are provided in the Electronic Supplementary Information (ESI). They may be cited in this text, and their numbers are preceded by an S. The ESI figures are organised in such a way that all the figures showing images (conformers or molecular orbitals, Figs. S1–S16) precede the set of figures showing trends' comparisons (Figs. S17–S132). The scales of the comparative diagrams are selected for each diagram, to better fit the values considered.

3 Calculations and Results

3.1 Naming of Conformers

Conformers are denoted with acronyms providing information on their characteristics. The same letters that had been used to keep track of relevant geometry features of ACPLs [10–21] are here used for the conformers of MYRA and DBPO, with some additional letters for features specific to these molecules. All the letters and their meanings are listed in Table 1. On presenting the analysis of results, it is also important to be able to mention each ring individually; thus, the rings are denoted by uppercase letters (A, B, D, E) as shown in Fig. 2. Although the substituent may collectively be called R′ (Fig. 1), it is termed 'BDE ring system', or simply 'BDE', when it is relevant to recall that it consists of three rings. The two fused rings are comprehensively termed 'BD ring system' or simply BD. The system comprising all the rings (including the phloroglucinol moiety) is termed 'ABDE ring system' or simply 'ABDE'.

The following features are taken into consideration when naming the conformers: the first IHB (whether H15···O14 or H17···O14); the O–H···π IHB (whether H15···π or H16···π); the orientation of O10–H16; the orientation of the phenol OH *ortho* to CRO and not engaged in the first IHB; the orientation of the BDE ring system with respect to the A moiety; the orientation of O28–H29; the orientation of O36–H37; the position of C19 with respect to the plane identified by C9, C23, C22 and O21 (to be considered because ring D is not planar); the angle by which the specified single bond in R is rotated with respect to the linear geometry, at input level, to generate new conformers having different geometries of R (Fig. 3 shows the numbers by which the single bonds of R are concisely denoted in the acronyms). For instance, MYRA-d-r-η-p-a-e-j is a conformer of MYRA in which the first IHB is on the same

Table 1 Symbols utilised to denote the main geometry features in the acronyms denoting the conformers of myristinin A, *cis*-DBPO and *trans*-DBPO. The meaning of the symbols is also illustrated in Fig. 4

Symbol	Geometrical feature
d	The first IHB is on the same side as R′ (H15···O14)
s	The first IHB is on the other side with respect to R′ (H17···O14)
r	The C3–C4–O10–H16 torsion angle is close to 0°
w	The C3–C4–O10–H16 torsion angle is close to 180°
u	The OH *ortho* to CRO and not engaged in the first IHB is oriented toward the acyl group
non-u	The OH *ortho* to CRO and not engaged in the first IHB is oriented away from the acyl group
η	Presence of O–H···π interaction between H16 and ring B
ε	Presence of O–H···π interaction between H15 and ring B
a	C19 in ring D is above the plane identified by C9, C23, C22 and O21
b	C19 in ring D is below the plane identified by C9, C23, C22 and O21
p	The B,D ring system is oriented 'towards us' and ring E 'towards the back' with reference to Fig. 2
q	The B,D ring system is oriented 'towards the back' and ring E 'towards us' with reference to Fig. 2
e	O28–H29 is oriented to the side of C27
f	O28–H29 is oriented to the side of the phloroglucinol moiety
j	O36–H37 is oriented to the side of C34
k	O36–H37 is oriented to the side of the phloroglucinol moiety
y	At the input level, the bond specified by the number preceding the y has been rotated by -90° with respect to the linear geometry of R shown in Fig. 3
z	at the input level, the bond specified by the number preceding the z has been rotated by +90° with respect to the linear geometry of R shown in Fig. 3
v	At the input level, the bond specified by the number preceding the v has been rotated by 180° (toward the other side with respect to R′) with respect to the linear geometry of R shown in Fig. 3
x	At the input level, the bond specified by the number preceding the x has been rotated by 180° (toward the side of R′) with respect to the linear geometry of R shown in Fig. 3
1–10	Numbers denoting the single bonds of R, as shown in Fig. 3

side as R′ (d), the C3–C4–O10–H16 torsion angle is close to 0° (r), the O10–H16···π IHB with ring B is present (η), the B,D ring system is oriented 'towards us' and ring E 'towards the back' with reference to Fig. 2 (p), C19 in ring D is above the plane identified by C9, C23, C22 and O21 (a), O28–H29 is oriented to the side of C27 (e), O36–H37 is oriented to the side of C34 (j), and the R chain is on the same plane as ring A (absence of no specifications about bond rotations); MYRA-d-r-η-p-a-e-j-2y is a conformer of MYRA with the same other features as MYRA-d-r-η-p-a-e-j, but

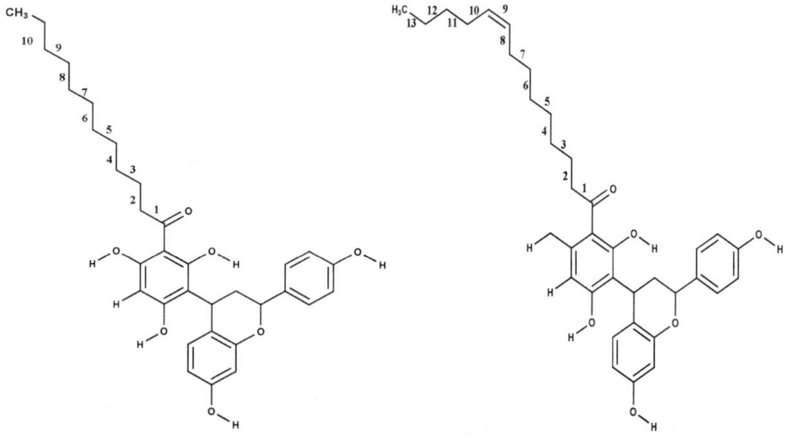

Number	Corresponding bond	Number	Corresponding bond	Number	Corresponding bond	Number	Corresponding bond
1	C7–C13	4	C38–C39	7	C41–C42	10	C44–C45
2	C13–C18	5	C39–C40	8	C42–C43		
3	C18–C38	6	C40–C41	9	C43–C44		

Fig. 3 Numbers concisely denoting the individual bonds in the R chain of MYRA (left) and DBPO (right). The numbers are shown near each bond in the R chain and their meaning is further stressed in the table. The *cis*-DBPO isomer is shown in the figure; the same numbering is used also for *trans*-DBPO

in which—at input level—the bond of R denoted as 2 has been rotated by −90° (y) with respect to the linear geometry shown in Figs. 3 and 4.

3.2 Preliminary Study of a Model Structure

A preliminary study of a model structure (MODL, Fig. 5) was performed *in vacuo*, to identify the conformational preferences of the part common to MYRA and DBPO, i.e., the ABDE ring system; the long R chain is replaced by an ethyl group meant to mimic the presence of R and the major features of its influence on the geometry features of the phloroglucinol moiety. The separate study of MODL is motivated by the fact that it comprises the major energy-influencing factors of the three molecules, because it contains all the rings, all the OH groups, and O14; therefore, it enables a preliminary investigation of the effects of the possible IHBs (including the O–H···π IHBs) and of the possible mutual orientations of the rings and of the OH groups.

The conformational study of MODL considered all the possible combinations of different geometry features, excluding only the conformers without the first IHB, since it has already been proven [10–21] that such conformers have high energy, which excludes them from the possibility of being involved in the biological activities of ACPLs (the removal of the first IHB in ACPLs with a bulky substituent at C3 causes

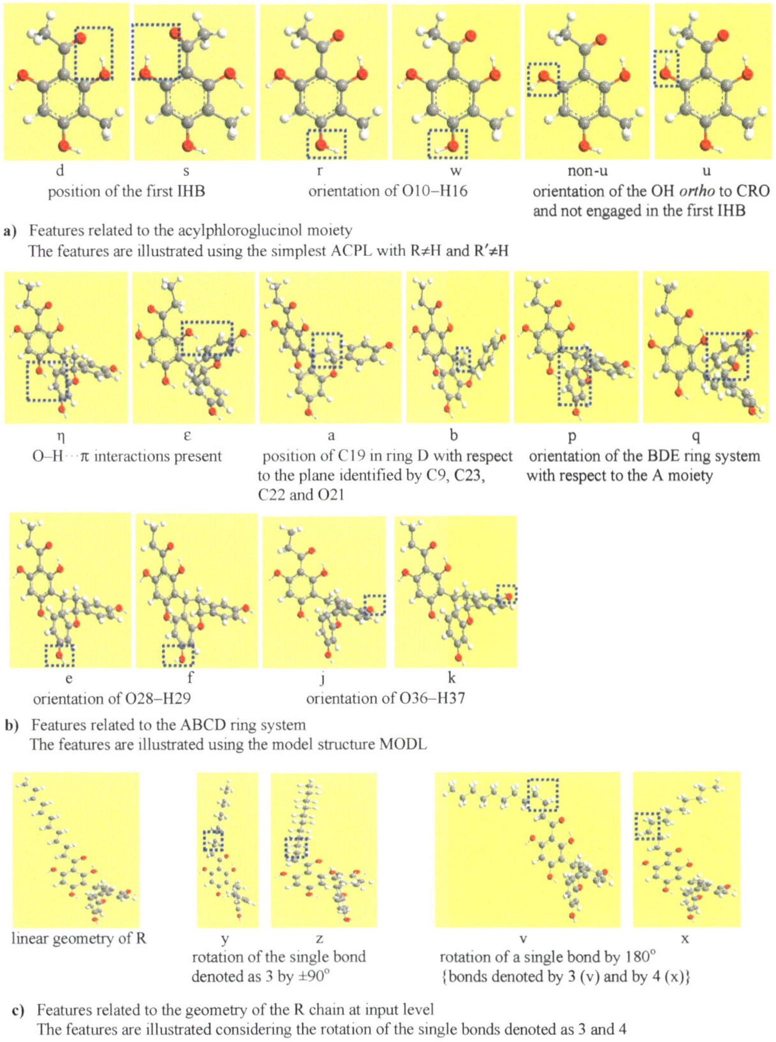

Fig. 4 Illustration of the meanings of the symbols listed in Table 1

an energy increase of at least 14 kcal/mol). Different conformers of MODL were obtained by changing, in turn, the features listed in the previous section and pertaining to ABDE. The conformers were first calculated at the HF level, and 95 conformers were thus identified. The 50 lowest energy ones were then calculated also at the DFT level. Table 2 reports the relative energies of the 50 lower energy conformers (both HF and DFT results) and Table S1 reports the HF relative energies of all the calculated conformers. Figure 6 shows the geometries of representative low energy conformers and Fig. S1 shows the geometries of all the calculated conformers; these images are relevant because they show the details of the geometry of the ABDE ring system for

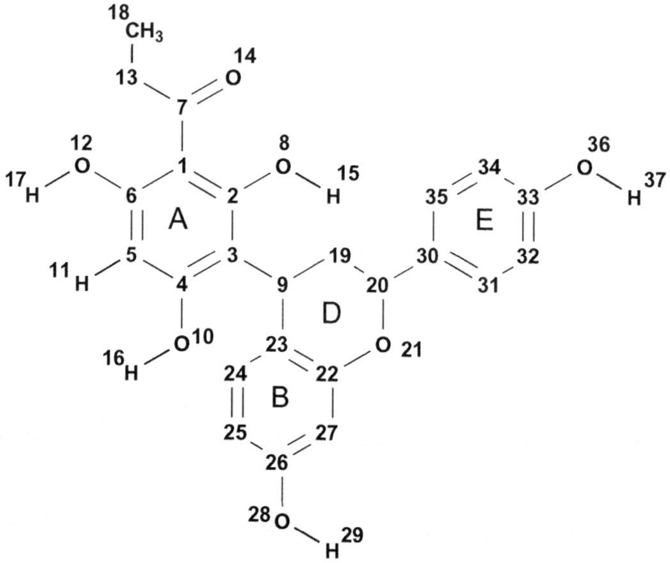

Fig. 5 Model structure utilised for a preliminary investigation of the conformational preferences of the ring systems of MYRA and DBPO. The C atoms in the rings and in the acyl chain are represented by the numbers denoting their position (except the last C atom of the acyl chain) for better view of the structures. Only the H atoms attached to O atoms and to C5 are numbered individually, while the other H atoms are given the same number as the C atom to which they are attached and are not shown in the structure. The rings are denoted by uppercase letters (A, B, D and E)

each conformer in a more visible way than the images of the conformers of MYRA and DBPO, where the size of ABDE is much smaller because of the simultaneous presence of the long R chain.

The relative energies indicate that—like in all other ACPLs studied so far—the first IHB is the dominant stabilising factor influencing conformational preferences and energetics. Other factors having significant influence are the presence of the O–H···π interaction between a phenol OH and ring B, the orientation of the BDE ring system with respect to the A moiety, the orientation of O10–H16 and the orientation of the OH *ortho* to CRO and not engaged in the first IHB. The orientations of O28–H29 and O36–H37, and the position of C19 with respect to the plane identified by C9, C23, C22 and O21 in ring D, have minor or negligible influence, and conformers differing only by one of these factors have very close or identical relative energies.

The results highlight the following trends: d conformers have lower energy than the corresponding s conformers (similarly to other findings on ACPLs having a substituent at C3 [10–21]); r conformers have lower energy than the corresponding w conformers when the O10–H16···π IHB between H16 and ring B is present, whereas they have higher energy than the w conformers when this IHB is absent; u conformers always have higher energy than the corresponding non-u ones (similarly to other findings on ACPLs [10–21]); q conformers have lower energy than the corresponding

Table 2 Relative energies of the lower energy conformers of the model structure. HF/6-31G(d,p) and DFT/B3LYP/6-31+G(d,p) results in vacuo, respectively denoted as HF and DFT in the column headings. The absolute energies of the lowest-energy conformer is −1445.7499595 and −1454.4781847 hartree in the HF and DFT results respectively

Conformers	Relative energy (kcal/mol)		Conformers	Relative energy (kcal/mol)	
	HF	DFT		HF	DFT
d-r-η-p-a-e-j	0.000	0.000	s-r-ε-q-a-e-k	5.016	4.502
d-r-η-p-a-e-k	0.013	0.015	s-r-ε-q-a-e-j	5.141	4.621
d-r-η-p-a-f-j	0.423	0.209	d-w-p-b-e-k	5.149	5.668
d-r-η-p-a-f-k	0.494	0.265	d-w-q-b-f-k	5.210	5.413
d-r-η-p-b-e-k	1.003	1.424	s-r-η-p-a-e-k	5.214	4.647
d-r-η-p-b-e-j	1.007	1.457	d-w-q-b-e-j	5.216	6.119
s-w-ε-q-a-e-k	1.140	1.345	d-w-q-a-e-j	5.216	5.453
s-w-ε-q-a-e-j	1.188	1.397	d-r-η-u-p-a-f-j	5.266	3.586
d-r-η-p-b-f-j	1.209	1.507	d-w-q-a-e-k	5.293	5.514
d-r-η-p-b-f-k	1.243	1.480	d-r-η-u-p-a-f-k	5.334	3.630
s-w-ε-q-b-e-j	1.486	2.295	s-r-η-p-a-e-j	5.349	4.787
s-w-ε-q-a-f-j	1.619	1.612	s-r-η-p-b-f-k	5.372	4.919
s-w-ε-q-a-f-k	1.633	1.601	d-w-q-b-f-j	5.469	5.956
s-w-ε-q-b-f-k	1.962	2.428	s-r-ε-q-a-f-k	5.484	4.734
d-w-p-a-e-k	4.463	4.786	d-w-q-b-e-k	5.543	6.119
d-w-p-a-f-j	4.663	4.804	s-r-ε-q-a-f-j	5.545	4.812
s-r-ε-q-b-e-j	4.693	4.656	s-w-p-b-f-j	5.566	5.634
s-r-ε-q-b-f-j	4.815	4.530	s-r-η-p-a-f-k	5.672	4.881
d-r-η-u-p-a-e-j	4.848	3.376	d-r-q-b-e-k	5.722	5.720
d-r-q-a-e-j	4.848	4.751	d-r-q-b-e-j	5.741	5.750
d-w-p-a-f-k	4.857	4.995	d-r-q-a-e-k	5.745	5.470
d-r-η-u-p-a-e-k	4.857	3.379	s-r-η-p-a-f-j	5.745	4.979
d-w-p-b-e-j	4.928	5.476	d-w-p-b-f-k	5.876	5.754
d-r-q-b-f-k	4.981	4.950	d-w-q-a-f-k	5.827	5.869
d-r-q-b-f-j	4.981	4.950	d-w-q-a-f-j	5.689	5.764

p conformers when the O8–H15···π interaction between H15 and ring B is present, whereas they have higher energy than the p conformers when the interaction is absent. No consistent conformational preference patterns are identified for the a/b, e/f and j/k pairs of conformers.

Since the properties of MODL may serve as comparison references for the other molecules, they are reported in the ESI: parameters of the first IHB (Table S2), distance between the H atom and the closest C atom in the acceptor B aromatic ring for the O–H···π IHBs (Table S3), dipole moment (Table S4), and HOMO-LUMO

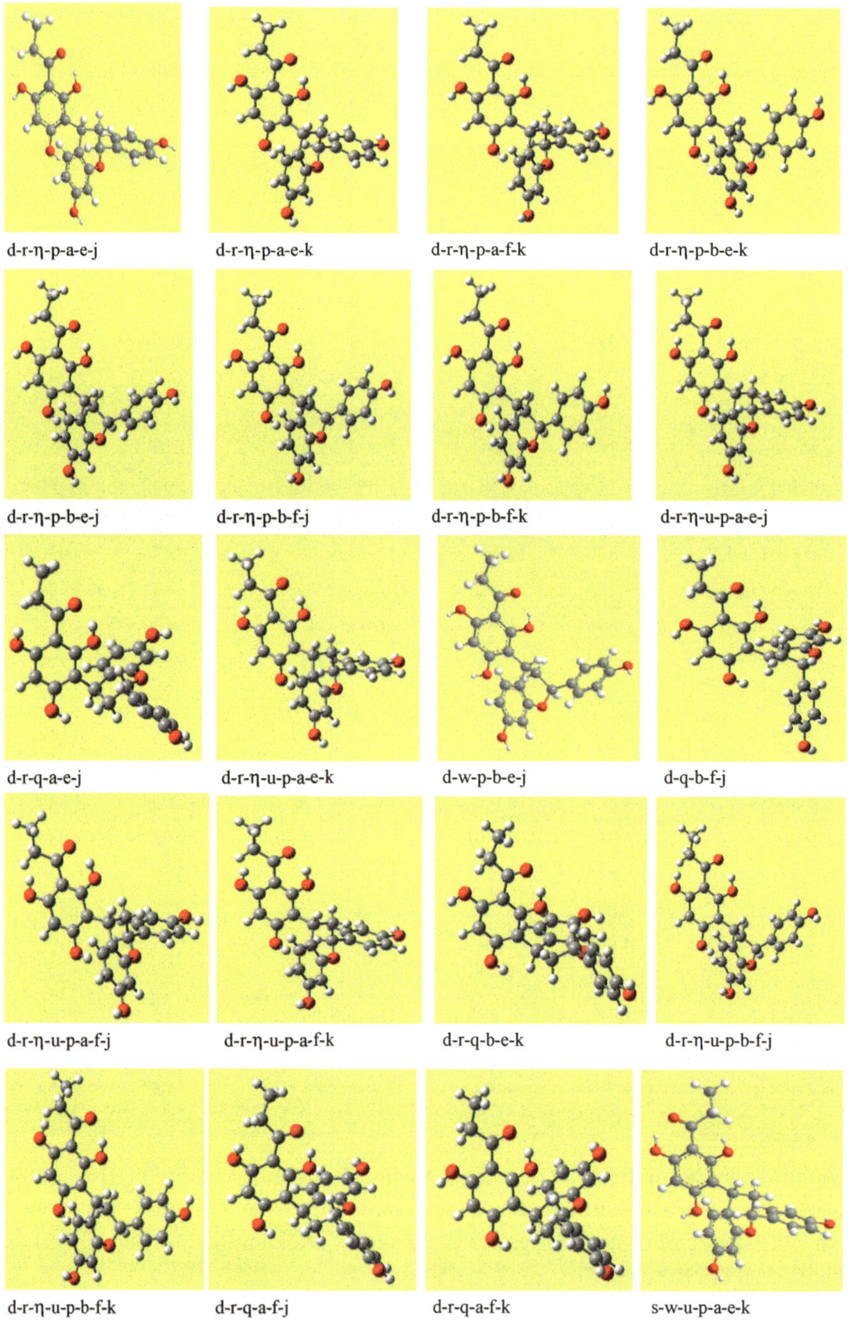

Fig. 6 Representative conformers of the model structure. HF/6-31G(d,p) results in vacuo

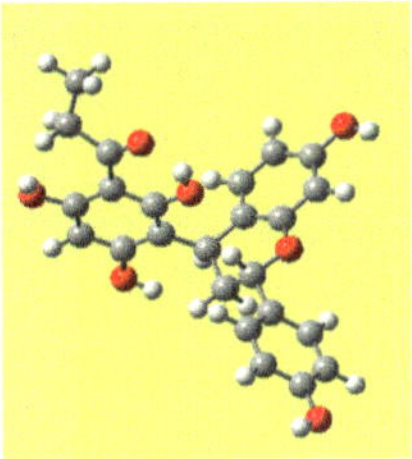

d-r-u-q-a-e-k

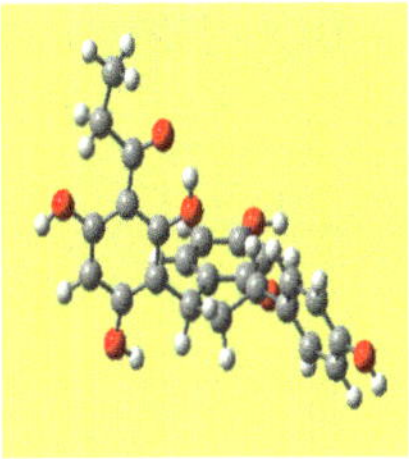

d-r-q-b-e-j

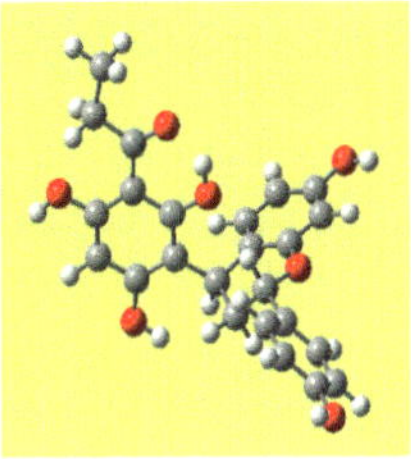

d-r-q-a-e-k

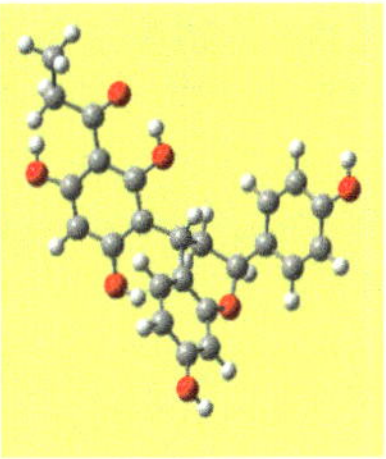

d-r-η-p-b-e-k

Fig. 6 (continued)

energy gap (Table S5). The trends of these quantities in the HF and DFT results are compared in Figs. S17–S21; the graphs highlight similarities of trends for the two methods, as well as known method-related phenomena, such as shorter IHB lengths and considerably smaller HOMO-LUMO energy gaps for the DFT results. The shapes of the HOMO and LUMO orbitals of MODL are shown in Fig. S2. Detailed comparisons of the properties of MODL and of the three molecules are presented in Sect. 3.5.

3.3 Results for the MYRA and DBPO Molecules in Vacuo

3.3.1 Types of Study Performed on the MYRA and DBPO Molecules

The results obtained from the study of MODL were utilised to investigate the MYRA and DBPO molecules, using two approaches: selecting the best geometry of the R chain and combining it in turn with the geometries of ABDE identified as the conformers of MODL; and selecting the geometry of ABDE corresponding to the best conformer of MODL and combining it with different geometries of R. The former option investigates possible influences of the size of R on the geometry of ABDE and of the geometry of ABDE on the geometry of R; the latter option investigates possible influences of the geometry of R on the geometry of ABDE. Both options investigate the influence of the various geometry features on the energetics.

The geometry of R was varied, at input level, by selecting the d-r-η-p-a-e-j conformer (a conformer in which R has linear on-plane geometry) and rotating each single bond by +90° or −90° (in which case the part of R following the rotated bond becomes perpendicular to the plane identified by the benzene ring A) and by 180° (in which case the part of R following the rotated bond remains on the plane, but with a bent geometry). In the former case, each bond can rotate both by +90° and by −90° (Fig. 4). Previous studies had shown that, for ACPLs, conformers having symmetrical geometries with respect to the plane of the benzene ring have the same energy [10]. In this perspective, only the +90° or −90° rotation of each bond would have been necessary. However, it was opted to consider both rotations, to check whether these symmetrical inputs yield symmetrical outputs on optimization, above all in view of possible influences by the bulky BDE ring system, which is not symmetric with respect to the plane of ring A. In the case of the rotation by 180°, only one rotation is possible for each bond; thus, the bonds denoted by 2, 4, 6, 8 and 10 (Fig. 4) can rotate by 180° to the side of R′ and those denoted by 3, 5, 7, and 9 can rotate by 180° to the other side with respect to R′.

The next subsections consider relevant molecular properties individually. Two sets of ESI tables are provided for each property: a set of three tables showing the values for all the calculated conformers of each of the three molecules (HF results) and a set of three tables showing the values for the calculated conformers with different geometries of R and the same geometry of ABDE (HF, DFT and MP2 results). Within each set, the tables correspond to the MYRA, *c*-DBPO and *t*-DBPO sequence; since this sequence remains the same for all cases, it is not repeated in the presentation of each set.

3.3.2 Conformational Preferences in Vacuo

Overall, 123 conformers were calculated for MYRA and 115 conformers for each of the two DBPO isomers. Tables S6–S8 report the relative energies of the calculated conformers of the three molecules; Figs. S3, S5 and S7 show the geometries of the calculated conformers having the same geometry of R and different geometries of ABDE for MYRA, *c*-DBPO and *t*-DBPO respectively, and Figs. S4, S6 and S8 show the geometries of the conformers having different geometries of R and the same geometry of ABDE. The absolute energies values show that the energy of the lowest energy conformer of *t*-DBPO is 1.843 kcal/mol lower than the energy of the lowest energy conformer of *c*-DBPO, suggesting greater stability of the *trans* isomer.

For the information about the energetics to be complete, Tables S9–S11 report the relative energies corrected for ZPE (sum of electronic and zero-point energies), the relative Gibbs free energies (sum of electronic and thermal free energy), and the corresponding corrections, for representative lower-energy conformers of the three molecules, and Figs. S22–S24 compare their trends. The ZPE corrections (kcal/mol) are close for all the conformers of the same molecule: 453.64–454.51 for MYRA, 476.32–477.20 for *c*-DBPO and 476.07–476.92 for *t*-DBPO. The thermal corrections to the Gibbs free energy have the following ranges: 405.03–408.96 for MYRA, 425.05–431.51 for *c*-DBPO and 425.16–426.50 for *t*-DBPO. For MYRA, the values of the uncorrected relative energies, ZPE-corrected relative energies and relative free energies are close for the first nine lowest energy conformers, and for conformers # 27–39 (with these numbers referring to increasing uncorrected relative energies, as used to denote conformers in Fig. S22), whereas they show significant differences for most of the # 10–26 conformers. The latter are the conformers whose uncorrected relative energies are so close as to give a nearly horizontal line in the graph, whereas the corrected ones differ. In the case of *c*-DBPO and *t*-DBPO, the trends of uncorrected and corrected energy values are similar and the values for the same conformer are close.

For all the three molecules, conformers with the same geometry of R and different geometries of ABDE show conformational preferences corresponding to those of MODL, and the preferences are influenced by the same factors; thus, the descriptions provided for MODL (the d conformers have lower energy than the s conformers, etc.) remain valid for MYRA, *c*-DBPO and *t*-DBPO. It can be inferred that the presence of a considerably longer R than the ethyl mimicking it in MODL, or the different nature of R in MYRA and DBPO, do not affect the influence of the major stabilising factors such as the IHBs, the orientation of the rings and the orientation of the OH groups.

Table 3 shows the ranges of relevant properties for the calculated conformers with different geometries of R and the same geometry of ABDE. Tables S12–S14 and Figs. S25–S27 compare the HF, DFT and MP2 relative energies of these conformers. The values show that, when R is bent, the energy of the conformers where a certain bond has been rotated by +90° or −90° is lower than the energy of the conformers

Table 3 Ranges of relevant quantities for conformers having the same geometry of the rings systems and differing by the geometry of the R chain

Quantity considered	Method	Ranges of values		
		MYRA	*c*-DBO	*t*-DBO
Energy changes (kcal/mol) when the geometry of the R chain changes from linear to bent	HF	0.323–1.067	0.489–1.054	0.342–1.076
	DFT	0.551–0.956	0.489–1.008	0.507–1.013
	MP2	−0.303 to 0.558	−0.898 to 0.516	−0.398 to 0.498
H15···O14 bond length (Å) for the first IHB	HF	1.644–1.654	1.638–1.652	1.644–1.583
	DFT	1.516–1.530	1.456–1.531	1.518–1.530
	MP2	1.574–1.583	1.574–1.585	1.573–1.585
O···O distance (Å) for the first IHB	HF	2.503–2.510	2.502–2.509	2.503–2.509
	DFT	2.458–2.468	2.445–2.467	2.459–2.468
	MP2	2.498–2.505	2.498–2.506	2.498–2.505
O$\hat{H}$O bond angle (°) for the first IHB	HF	146.3–146.6	146.4–147.4	146.3–146.6
	DFT	151.7–152.3	151.7–152.1	151.5–152.3
	MP2	151.3–151.8	151.3–151.8	151.3–152.3
H16···C23 distance (Å) for the O–H···π IHB	HF	2.128–2.129	2.122–2.129	2.127–2.129
	DFT	2.038–2.057	1.998–2.057	1.039–2.059
	MP2	2.008–2.012	2.007–2.011	2.007–2.012
Dipole moment (debye)	HF	2.273–2.439	2.097–2.558	2.247–2.435
	DFT	2.144–2.511	1.946–2.663	2.181–2.497
	MP2	2.282–2.528	2.120–2.606	2.264–2.520
HOMO-LUMO energy gap (kcal/mol)	HF	255.45–257.48	255.36–257.51	255.40–257.56
	DFT	103.32–105.27	103.26–105.25	103.33–105.85
	MP2	247.68–249.14	247.57–249.08	247.62–249.18

where the same bond has been rotated by 180° (to one or the other side, depending on the rotated bond), suggesting that geometries in which R is bent out of the plane identified by the benzene ring A are preferred to geometries in which R is bent but remains on the plane. The values also show that the energy changes related to changes in the geometry of R are not dramatic. The values from different calculation methods show a discrepancy in the identification of the lowest energy conformer: while HF and DFT identify the lowest energy conformer as the one in which R is linear (outstretched) and on the plane of ring A, MP2 identifies it as one of the lowest energy conformers in which R has an out-of-plane bent geometry (although its energy difference with the linear geometry is so small that it can be considered marginal).

3.3.3 Characteristics of the Intramolecular Hydrogen Bonds

The parameters of the first IHB (Tables S15–S17) do not appear to be influenced by the nature of R: the O14···H15 and H17···O14 lengths are the same for the same type of conformers in the three molecules. Consistently with the findings for other ACPLs with a substituent at C3 and no substituent at C5, the H15···O14 length is slightly shorter than the H17···O14 length, the O14···O8 distance is slightly shorter than the O14···O12 distance and the $\widehat{OHO}$ bond angle is greater for the H15···O14 IHB, suggesting that the H15···O14 IHB is somewhat stronger than the H17···O14 IHB. The values of the parameters suggest that the first IHB is moderate-to-strong.

Tables S18–S20 compare the results of the three calculation methods for the parameters of the first IHB in the calculated conformers with different geometries of R and the same geometry of ABDE. The narrow ranges (Table 3) of their values within each calculation method indicate only marginal influence by the geometry of R on the IHB parameters. Figures S28–S30 compare the trends of the IHB length for the different methods. The IHB length is longest in the HF results, shortest in the DFT results and intermediate in the MP2 results. This is consistent with the tendency of HF to underestimate H-bond strength and of DFT to overestimate it; the MP2 results are thus expected to be closer to the actual values. The trends across conformers are very similar for the three methods .

Tables S21–S23 report the distance between the H atom of the donor OH and the closest C atom (C23) in the acceptor aromatic B ring, for the O–H···π IHBs. The H15···C23 distances are shorter than the H16···C23 distances, suggesting that the

Table 4 Letters used to indicate the site to which water molecules are attached, in the acronyms denoting adducts of MYRA with explicit water molecules

Letter	Site	Letter	Site	Letter	Site	Letter	Site
J	O8	M	O14	P	O21	U	H17
L	O10	N	O28	F	H15	T	H29
K	O12	Q	O36	G	H16	Z	H37

interaction is somewhat stronger when O8–H15 is the donor. Tables S24–S26 and Figs. S31–S33 compare the H16···C23 distance in the HF, DFT and MP2 results, for the conformers with different geometries of R and the same geometry of ABDE. For a given conformer, the distance is longest in the HF results, intermediate in the DFT results and shortest in the MP2 results; this is consistent with the fact that MP2 takes into account also dispersion contributions, which have a relevant role in O–H···π interactions.

The red shift (lowering of the vibrational frequency of the donor) caused by an H-bond provides indication on the relative strengths of H-bonds, as stronger H-bonds cause greater red shifts. Tables S27–S29 report the vibrational frequencies of the OH groups of the three molecules, and Tables S30–S32 report the red shifts of the OHs engaged in IHBs. A red shift is calculated as difference between the frequency of an OH engaged in an IHB and the frequency of the same OH when it is free (not engaged in an IHB); in this work, the reference value for a free OH is taken as the average of the vibrational frequencies of that OH in all the conformers in which it is free. It has also to be taken into account that the red shifts calculated here are surely smaller than the actual ones, because of the tendency of HF to underestimate H-bonds strengths; on the other hand, they enable realistic comparisons of trends (as proven in studies of ACPLs, where also DFT frequencies and red shifts had been calculated). The red shift (cm^{-1}) of O8–H15 is 352.6-427.0/MYRA, 278.6-318.6/*c*-DPBO and 237.1-319.5/*t*-DPBO when it forms the O8–H15···O14 first IHB, and 94.0-112.9/MYRA, 94.2-113.1/*c*-DPBO and 92.9-113.1/*t*-DPBO when it forms the O8–H15···π IHB with the B ring; the red shift of O12–H17 is 273.1-306.6/MYRA, 277.2-306.1/*c*-DPBO and 277.3-306.2/*t*-DPBO when it forms the O12–H17···O14 first IHB; and the red shift of O10–H16 is 49.2-76.1/MYRA, 68.2-76.2./*c*-DPBO and 69.8-76.4./*t*-DPBO when it forms the O10–H16···π IHB with the B ring. These values are consistent with the inferences (from IHB lengths) that the H15···O14 IHB is slightly stronger than the H17···O14 IHB, and that the H15···π IHB is slightly stronger than the H16···π IHB.

3.3.4 Dipole Moments of the Conformers

The dipole moment of the calculated conformers of the three molecules (Tables S33–S35) is mostly influenced by the orientation of the OHs, being greater when more OHs are oriented in the same direction. The nature of R does not appear to influence the dipole moment significantly. Tables S36–S38 and Figs. S34–S36 compare the results from different calculation methods for the calculated conformers with different geometries of R and the same geometry of ABDE. The values from the same method are very close, confirming that the geometry of R influences the dipole moment only marginally. The HF and MP2 values are mostly very close, whereas DFT values differ more noticeably and are often somewhat smaller than the HF and MP2 ones.

3.3.5 HOMO-LUMO Energy Gap of the Conformers

Tables S39–S41 report the values of the HOMO-LUMO energy difference for all the calculated conformers. The difference is mostly influenced by the presence of O–H···π IHBs. The orientation of the BDE ring system and the orientation of the OH *ortho* to CRO and not engaged in the first IHB also have significant influence. The position of the first IHB (whether H15···O14 or H17···O14) has minor influence and the other factors, such as the orientation of the OHs in rings B and E, have no significant influence. Figures S9, S11 and S13 show the shapes of the HOMO and LUMO of the calculated conformers of the three molecules having the same geometry of R and different geometries of ABDE, and Figs. S10, S12 and S14 show the shapes for the conformers having different geometries of R and the same geometry of ABDE. The HOMO shapes show greater electron density concentration in the phloroglucinol moiety and the BDE ring system for all conformers, with few exceptions for d-w, d-r-q and s-w-u conformers, where the electron density is concentrated in the BDE ring system only. The LUMO shapes show greater electron density concentration in the phloroglucinol moiety.

Tables S42–S44 and Figs. S37–S39 compare the results from different calculation methods for the HOMO-LUMO energy gap of the calculated conformers having different geometries of R and the same geometry of ABDE. The HF and MP2 values are very close. The DFT values are much smaller than the values obtained with the other two methods, consistently with the known tendency of DFT to greatly underestimate the HOMO-LUMO energy gap. The trends are similar with the three methods.

3.4 Results in Solution

3.4.1 Results from the CPCM Calculations

CPCM calculations in solution were performed at the HF level on conformers with *in-vacuo* relative energy $\leq$1.984 kcal/mol. Sets of tables (S45–S74) and corresponding figures (S40–S69) compare the properties in vacuo and in the three solvents considered. Although acetonitrile has greater dipole moment than water, its effects on the properties of the solute molecules are usually intermediate between those of chloroform and those of water; therefore, on analysing results, the wording "as the medium polarity increases" is meant to refer to the vacuum-chloroform-acetonitrile-water sequence and the wording "as the solvent polarity increases" is meant to refer to the chloroform-acetonitrile-water sequence.

Tables S45–S47 and Figs. S40–S42 compare the relative energies of the conformers having the same geometry of R and different geometries of ABDE in the different media; the values show that the relative energy decreases as the medium polarity increases. The identification of the lowest energy conformer appears to differ in vacuo and in the three solvents; however, the difference with the second lowest-energy conformer is so small as to be below experimental errors. The parameters of the first IHB (Tables S48–S50; Figs. S43–S45) show some changes in solution with respect to *in vacuo*. Both the H15···O14 and the H17···O14 lengths are slightly smaller in chloroform than *in vacuo*, slightly smaller in acetonitrile than in chloroform, and the same in acetonitrile and water, for nearly all conformers. A similar trend appears for the O···O distance. The O$\hat{H}$O bond angle increases slightly as the medium polarity increases. The distance between the H atom of the donor OH and the closest C atom in the acceptor aromatic B ring for the O–H···π IHBs (H15···C23 or H16···C23, Tables S51–S53; Figs. S46–S48) decreases slightly from vacuum to chloroform to acetonitrile for the first four lowest energy conformers and increases slightly for higher energy conformers, with the only exception of *cis*-DBPO-s-w-ε-q-a; the values in acetonitrile and in water are the same. The dipole moment of the conformers (Tables S54–S56; Figs. S49–S51) increases as the medium polarity increases, consistently with commonly observed trends; the difference between the values in acetonitrile and in water is small. The HOMO-LUMO energy gap (Tables S57–S59; Figs. S52–S54) slightly decreases as the medium polarity increases; the minimum and maximum HOMO-LUMO energy gaps in solution occur for the same conformers despite the different permittivities of the solvents.

Tables S60–S74 and Figs. S55–S69 compare the same properties in the four media for the conformers with different geometries of R and the same geometry of ABDE: conformers' relative energies (Tables S60–S62; Figs. S55–S57); parameters of the H15···O14 first IHB (Tables S63–S65; Fig. S58–S60); H16···C23 distance for the H16···π interaction (Tables S66–S68; Figs. S61–S63); dipole moment of the conformers (Tables S69–S71; Figs. S64–S66); and HOMO-LUMO energy gap (Tables S72–S74; Figs. S67–S69). The relative energies decrease slightly as the medium polarity increases. The H16···C23 distance for the H16···π interaction decreases slightly from vacuum to chloroform to acetonitrile and has the same value in acetonitrile and in water. The trends of the other quantities (parameters of the H15···O14 first IHB, dipole moment, HOMO-LUMO energy gap) are similar to those of the conformers with the same geometry of R and different geometries of ABDE. The values of these quantities in acetonitrile and in water are nearly always so close that their curves overlap in the graphs highlighting trends, for both the conformers with the same geometry of R and different geometries of ABDE and the conformers with different geometries of R and the same geometry of ABDE.

The solvent effect (free energy of solvation, ΔG_{solv}, Tables S75–S80; Figs. S70–S75) is positive in acetonitrile and negative in chloroform and water for all the three molecules, with the magnitude of ΔG_{solv} being considerably greater in water than in chloroform. This is consistent with the other findings for ACPLs [14, 15]. A quick evaluation of the octanol/water partition coefficients [38] yields 8.30663 for MYRA and 8.88063 for both *c*-DBPO and *t*-DBPO, which suggests very poor solubility in water; this can be related to the bulk and to the high molecular mass of the molecules. The negative values of ΔG_{solv} in water are likely related to the presence of five OH groups, which favour interactions with water molecules; they also suggest the possibility of some presence of these molecules in the aqueous medium in a living organism. A similar quick evaluation of the octanol/water partition coefficient for MODL yields 3.54563, confirming the greater affinity for water of the OH-rich ABDE ring system, and the solubility-decreasing effect of the bulky hydrocarbon R in MYRA, *c*-DBPO and *t*-DBPO.

3.4.2 Adducts with Explicit Water Molecules

Adducts with explicit water molecules were calculated for low energy conformers of MYRA. Four types of adducts were considered: adducts with only one water molecule binding to one of the donor or acceptor sites of MYRA; adducts with two water molecules binding to the two atoms of an OH; adducts with two water molecules binding to the two atoms of an OH and a third water molecule bridging them; and adducts with several explicit water molecules. The first type of adducts is meant to provide indications about the H-bonding strength of individual sites (or binding-site preferences for the water molecules, [14, 39]); the second type indicates whether the two water molecules binding to a given OH bind also to each other; the third type provides indications about the preference for the presence of a bridging water molecule and about the preferred arrangements of three water molecules in the vicinity of an OH; and the last type is meant to approximate possible arrangements in the first solvation layer (which comprises the water molecules attached to MYRA and those bridging them). Figure 7 shows the geometries of representative adducts and fig. S16 shows the geometries of all the calculated adducts.

The adducts are denoted with acronyms providing information on the conformer of MYRA, on the number of water molecules and on their binding sites. Each binding site is denoted by an uppercase letter, as listed in Table 4. The acronym contains the symbols denoting the conformer of MYRA, the list of the binding sites to which the water molecules are H-bonded and the total number of water molecules (which may be greater than the number of binding sites, when water molecules bridging those attached to sites of MYRA are present). For example, d-r-η-p-a-e-j-L-1aq denotes an adduct of conformer d-r-η-p-a-e-j with one water molecule (1aq) attached to O10 (L); d-r-η-p-a-e-j-LG-2aq denotes an adduct of conformer d-r-η-p-a-e-j with two water molecules (2aq), attached to O10 (L) and H16 (G) respectively; d-r-η-p-a-e-j-LG-3aq denotes an adduct of conformer d-r-η-p-a-e-j with three water molecules (3aq), two of which are attached to O10 (L) and H16 (G) respectively, while the third one bridges them.

Table S81 reports the relative energies of the calculated adducts and the MYRA-water interaction energy ($\Delta E_{interaction}$), calculated as explained in Sect. 2. Since relative energies have a meaning only for adducts with the same number of water molecules, the adducts are grouped according to the number of water molecules. Table S82 reports the distances between the water molecules and the atoms of MYRA to which they are attached (bond lengths of the intermolecular H-bonds).

For the adducts with one water molecule H-bonded to different donor or acceptor sites of the MYRA molecule, the MYRA-water interaction energy corresponds to the energy of the MYRA-water intermolecular H-bond. Similarly to other findings

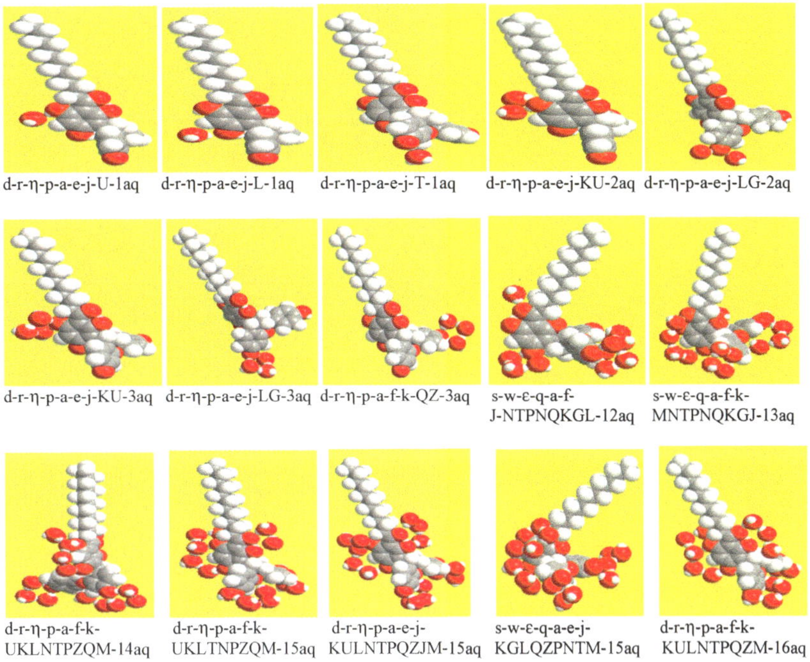

Fig. 7 Representative calculated adducts of conformers of myristinin A with explicit water molecules. HF/6-31G(d,p) results in vacuo

for ACPLs [14, 39], this H-bond is stronger when a phenol OH acts as donor (and the water molecule as acceptor) than when the phenol OH acts as acceptor to a water molecule; correspondingly, the H-bond length is shorter when water acts as acceptor. Among the adducts in which water acts as acceptor, the shortest H-bond length corresponds to H17 being the donor (d-r-η-p-a-e-j-U-1aq), and this adduct has the best $\Delta E_{interaction}$. When water acts as donor, the shortest H-bond length and the best $\Delta E_{interaction}$ correspond to O21 being the acceptor (d-r-η-p-a-e-j-P-1aq).

The geometry features of the conformers may influence geometry and energy aspects of the adducts. For instance, comparison of adducts with the water molecule attached to H37 and differing by the orientation of O36–H37 shows that the adduct with O36–H37 oriented to the side of C34 (d-r-η-p-a-e-j-M-1aq) has greater relative energy than the adduct with O36–H37 oriented to the opposite side (d-r-η-p-a-e-k-M-1aq); however, the H-bond length is the same.

Comparison of corresponding adducts with two water molecules binding to neighbouring sites of MYRA (e.g., the two atoms of an OH) and adducts with three water molecules, two of which binding to the same sites while the third one bridges them, shows better $\Delta E_{interaction}$ for the latter, consistently with the known stabilising role of the bridging water molecule. The last two columns of Table S81 lists the number of MYRA-water molecules H–bonds and the number of bridging water molecules

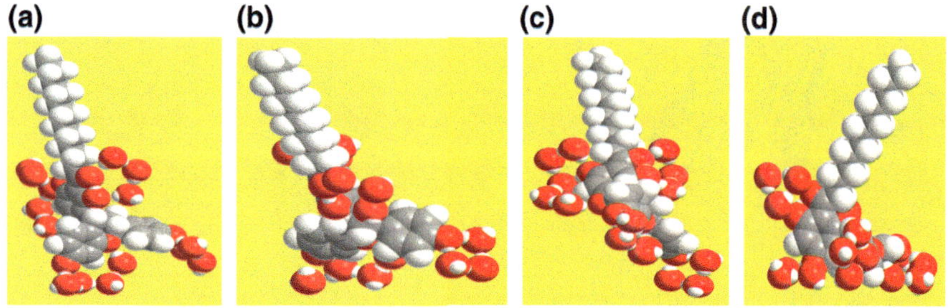

Fig. 8 Relevant arrangements of water molecules in the adducts of Myristinin A with explicit water molecules. The first three images show typical arrangements of water molecules in the vicinity of certain sites of the MYRA molecule: pentagon of O atoms in the vicinity of the first IHB (**a**); square of O atoms in the vicinity of O36-H37 (**b**) and in the vicinity of O12-H17 (**c**), in the d-r-η-p-a-e-j-KULNTPQZJM-15aq adduct. Image (**d**) shows a water molecule not attached to the MYRA molecule and not bridging water molecules attached to it, i.e., the water molecule external to the water molecules attached to O12 and O14

for the adducts with 3 or more water molecules, to facilitate quick comparisons of the major adducts-stabilising factors.

The first three images of Fig. 8. highlight preferred arrangements around specific sites, frequently encountered in adducts of ACPLs with three or more water molecules [14, 19]. A square of O atoms when two water molecules bind to a certain OH and a third water molecule bridges them (the O of the OH being one of the four O atoms of the square); and a pentagonal arrangement of O atoms in the region of the first IHB (the two O atoms involved in the IHB and the O atoms of three water molecules).

The optimization of adducts with several water molecules may be intriguing because of the tendency of water molecules to cluster together. This may result in some water molecules moving to external positions, thus not being part of the first solvation layer and not contributing significantly to the interaction energy between the solute molecule and the surrounding water molecule. An example is shown in Fig. 8d. While a water molecule bridging two water molecules bonded to the central molecule has a stabilising role and can be considered part of the first solvation layer, a water molecule that is totally external (attached to water molecules that are attached to the central molecule, but not bridging them) contributes to $E_{aq\text{-}cluster}$ but not to $\Delta E_{interaction}$. Therefore, when such water molecules appear on optimisation of a certain adduct, a new adduct is calculated, in which the external water molecule is removed (e.g., d-r-η-p-a-f-k-UKLTNPZQM-15aq from d-r-η-p-a-f-k-UKLTNPZQM-16aq, s-w-ε-q-a-f-k-MNTPNQKGJ-13aq from s-w-ε-q-a-f-k-MNTPZQJKG-14aq and s-w-ε-q-a-f-J-NTPNQKGL-12aq from s-w-ε-q-a-f-J-MNTPZQKGL-13aq).

On the average, $\Delta E_{interaction}$ increases as the number of water molecules in the adduct increases. Comparison of adducts with the same number of water molecules, the same number of MYRA-water H-bonds and the same number of bridging water molecules, highlights the dependence of $\Delta E_{interaction}$ on the type of conformer of

MYRA. It has also to be considered that the possible MYRA-water H-bonds depend on the characteristics of the conformer.

The calculated adducts with 14, 15 or 16 water molecules are too few to enable trend identification, above all with regard to conformer-type (given the high number of possible adducts with several water molecules for each conformer, considering an adequate number would become a separate study). All the same, the calculated adducts highlight preferred geometries around specific areas of MYRA (e.g., the area of the first IHB), as well as the roles of bridging water molecules. They also suggest that the MYRA-water H-bond lengths decrease as the number of water molecule attached to the MYRA molecule, and bridged by other water molecules, increases.

It is interesting to note that none of the adducts shows breaking of the first IHB. This is consistent with the findings for other ACPLs, indicating that the first IHB is maintained in water solution [14]. The adducts with explicit water molecules are suitable to signify the outcome of the competition [40] between intramolecular H-bonding and intermolecular H-bonding (formation of solute-solvent H-bonds), and studies of this type have indicated breaking of weaker IHBs in ACPLs [11].

3.5 Comparison of the Results Obtained for the Considered Molecular Structures

The role of a model structure is to provide preliminary information on the part of a molecule (or a set of molecules) to which it corresponds. It is interesting to evaluate how good the model is, i.e., how the information it provides is close to the results for the molecules it models. Since high energy conformers do not influence the biological activity of a molecule, comparisons are carried out for conformers with relative energy lower than 5.7 kcal/mol (which includes enough high energy conformers to provide sufficiently comprehensive information, but excludes the highest energy conformers).

The comparison with MODL has a meaning for the conformers of MYRA, *c*-DBPO and *t*-DBPO having the same geometry of R and different geometries of ABDE. The computed properties of MODL are compared with the properties of the corresponding conformers of the three molecules (Tables S83–S87; Figs. S76–S80). Comparisons of the properties of the three molecules are also carried out; for the conformers having different geometries of R and the same geometry of ABDE, comparisons can only concern corresponding conformers of the three molecules, because the geometry of ABDE is that of the lowest energy conformer of MODL.

The relative energies of corresponding conformers of MODL and the three molecules (Table S83; Fig. S76) are very close for conformers with relative energy lower than 4.86 kcal/mol, and some discrepancies appear only for conformers with higher energy. Similarly, the values of the H···Obond length (Table S84; Fig. S77) and of the O···O distance (Table S84) of the first IHB are very close for conformers with relative energy lower than 4.86 kcal/mol, with differences smaller than 0.002 Å;

greater differences appear only for some conformers with higher energy. The values of the O$\hat{H}$O bond angles of the first IHB are the same for MODL and the three molecules. The distance between the donor H atom and the closest C atom in the acceptor aromatic B ring for the O–H···π IHBs (Table S85; Fig. S78) is the same for corresponding conformers of MODL and the three molecules.

The values of the dipole moment are always very close for corresponding conformers of MODL, *c*-DBPO and *t*-DBPO (Table S86; Fig. S79), whereas the values for MYRA are slightly (0.04–0.172 D) smaller than those of MODL for the conformers with relative energy lower than 4.86 kcal/mol, and can be more significantly smaller or greater (by up to 4.86 D difference) for some higher energy conformers. This confirms that the dipole moments of the conformers are majorly influenced by the orientation of the OHs and other features of the ABDE ring system, and only marginally by the R chain.

The HOMO-LUMO energy gaps (Table S86, Fig. S80) are very close for the conformers with relative energy lower than 4.86 kcal/mol; minor discrepancies appear for conformers with somewhat greater relative energies and more considerable discrepancies for conformers with considerably higher relative energy. Figure S15 compares the shapes of the HOMO and LUMO for corresponding conformers of MODL and the three molecules, highlighting extensive correspondence in the orbital shape and electron density distribution.

Overall, these comparisons show that MODL is a good model for the three molecules, above all for the conformers with relative energies lower than 5 kcal/mol, which includes all the conformers that might be involved in the biological activity.

As already mentioned, comparisons of the properties of corresponding conformers having different geometries of R and the same geometry of ABDE can be carried out only for the three molecules, because no such conformers exist for MODL. Furthermore, comparisons can concern only rotations around the bonds numbered as 1–8, because no rotation is possible for *c*-DBPO and *t*-DBPO around the bond numbered as 9, as it is a double bond. Since HF, DFT and MP2 results are available for these conformers, the values of each property are compared individually for each calculation method. The relative energies (Tables S88, S89; Figs. S81, S82) are mostly very close for corresponding conformers in the HF and DFT results. For HF results, some discrepancies appear for d-r-η-p-a-e-j-2y and d-r-η-p-a-e-j-2z (where the relative energy is higher for *c*-DBPO and *t*-DBPO than for MYRA) and for d-r-η-p-a-e-j-3y, d-r-η-p-a-e-j-8y and d-r-η-p-a-e-j-8z (where the relative energy is smaller for *c*-DBPO and *t*-DBPO than for MYRA). For DFT results, some discrepancies appear only for d-r-η-p-a-e-j-3y, d-r-η-p-a-e-j-8y and d-r-η-p-a-e-j-8z. The MP2 results (Table S90, Fig. S83) show greater differences between the energies of corresponding conformers, including corresponding conformers of *c*-DBPO and *t*-DBPO.

The length of the H15···O14 first IHB (Tables S91–S93; Figs. S84–S86) has mostly close values for corresponding conformers. The greatest discrepancy concerns d-r-η-p-a-e-j-2y and d-r-η-p-a-e-j-2z, where the length for *c*-DBPO and *t*-DBPO is shorter than for MYRA, in the results of all the three calculation methods. The lengths for corresponding conformers of *c*-DBPO and *t*-DBPO are somewhat different within the HF results or the DFT results, whereas they are very close in the MP2 results.

The distance between the donor H atom and the closest C atom in the acceptor aromatic B ring for the O–H···π interaction (Tables S94–S96; Figs. S87–S89) has mostly very close values for corresponding conformers of the three molecules, with the greatest discrepancy being ≈0.007 Å in the HF results. While the values for corresponding conformers of *c*-DBPO and *t*-DBPOdiffer sufficiently in the HF and DFT results to give clearly different curves in the diagrams, they are so close in the MP2 results as to yield two mostly indistinguishable curves.

The values of the dipole moment (Tables S97–S99; Figs. S90–S92) are mostly very close for corresponding conformers of the three molecules, in the results of each method. Discrepancies involve differences smaller than 0.3 D in the HF and MP2 results, and smaller than 0.4 D in the DFT results.

The HOMO-LUMO energy gaps (Tables S100–S102; Figs. S93–S95) are also mostly close for corresponding conformers of the three molecules. Some discrepancies appear for d-r-η-p-a-e-j-2y and d-r-η-p-a-e-j-2z, where the gap is smaller for *c*-DBPO and *t*-DBPO than for MYRA. The curves for *c*-DBPO and *t*-DBPO are much closer in the HF and MP2 results, and differ more in the DFT results.

Comparisons of the properties of corresponding conformers in the three solvents considered may only concern MYRA, *c*-DBPO and *t*-DBPO, because calculations in solution were not made for MODL, as they are more meaningful for actual molecules. Comparison tables and figures consider individual properties for each solvent, resulting in three tables for each property (in chloroform, in acetonitrile and in water).

Tables S103–S121 and Figs. S96–S113 consider conformers with the same geometry of R and different geometries of ABDE. The relative energies (Tables S103–S105; Figs. S96–S98) of corresponding conformers of the three molecules are very close in the same solvent, and the corresponding curves overlap for most conformers, or are close where they do not overlap. A discrepancy appears only for s-w-ε-q-b-f-k, whose relative energy is higher for *t*-DBPO than for MYRA and *c*-DBPO.

Both the length of the first IHB (Tables S106–S108; Fig. S99–S101) and the distance between the donor H atom and the closest C atom in the acceptor aromatic B ring for the O–H···π interaction (Tables S109–S111; Figs. 102–S104) are very close for corresponding conformers in each of the solvents. The curves highlighting their trends largely overlap, and remain very close where they do not overlap. Similar considerations hold for the values of the dipole moment of corresponding conformers (Tables S112–S114; Figs. S105–S107) and for the values of their HOMO-LUMO energy gaps (Tables S115–S117; Fig. S108–S110).

The free energy of solvation (Tables S118–S120, with synopsis in Table 121, and Fig. S111–S113) is negative in chloroform, with greater magnitude for *t*-DBPO and closer values for MYRA and *c*-DBPO. It is positive in acetonitrile, with the values for *c*-DBPO being slightly higher than those of corresponding conformers of *t*-DBPO and MYRA, and the values for *t*-DBPO and MYRA being closer to each other. The values for water solution are considerably more negative than for chloroform, and the values of MYRA remain slightly higher than those of *c*-DBPO and *t*-DBPO; the values for *c*-DBPO and *t*-DBPO are closer to each other.

Tables S122–S140 and Figs. S114–S131 consider the properties of conformers of the three molecules having different geometries of R and the same geometry of ABDE. The relative energies (Tables S122–S124; Fig. S114–S116) have close values in each solvent, with only few discrepancies. The length of the first IHB (Tables S125–S127; Fig. S117–S119) shows similar trends in the three solvents, with up to 0.052 Å differences between MYRA and the other two molecules, whose values are closer. The values of the distance between the donor H atom and the closest C atom in the acceptor aromatic B ring for the O–H···π IHBs (Tables S128–S130; Figs. S120–S122) are close for all corresponding conformers of the three molecules. The values of the dipole moment (Tables S131–S133; Figs. S123–S125) are close for corresponding conformers of MYRA and *t*-DBPO, while those of *c*-DBPO differ more significantly. The HOMO-LUMO energy gaps (Tables S134–S136; Figs. S126–S128) are very close in acetonitrile and water, while *c*-DBPO-d-r-η-p-a-e-j-6y has ≈1 kcal/mol greater value that the other two molecules in chloroform. The free energy of solvation (Tables S137–S139, with synopsis in Table 140; Figs. S129–S131) shows rather similar trends within the same solvent. In chloroform, the values for *t*-DBPO are smaller than for corresponding conformers of the other two molecules; in acetonitrile, the values of MYRA are the lowest and the values for *c*-DBPO are slightly higher than for corresponding conformers of the other two molecules; in water, MYRA has the highest values and *t*-DBPO the lowest.

Since ΔG_{solv} has considerably different ranges of values in the three solvents, it is suitable for synopsis diagrams showing the trends of the three molecules in the three solvents in the same synopsis diagrams. Figure S132 shows such diagrams for the two series of conformers.

4 Discussion and Conclusions

The computational study of biologically active molecules is particularly important for drug development because it provides crucial information about the properties of a molecule, which determine its biological activities. The information needs to be as exhaustive as possible, because the activity may be influenced by the “finest details” of the molecular structure and properties [41].

The current work has provided detailed information about the conformational preferences and computable molecular properties of the molecules considered. The use of three computational methods for most of the lowest energy conformers offers validation of trends-identification. The use of a model structure provides information that can be useful for the study of modified molecules in the search for compounds with enhanced activity. The study in three solvents with different polarities and characteristics provides information useful for the consideration of the biological activities of these molecules, because the activity is exerted in a medium within living organisms. The thorough attention to intramolecular hydrogen bonding is relevant to further studies of the biological activity because of the roles often played by H-bonds.

The structures of the molecules considered here are substantially different from those of the ACPLs studied so far, above all because of the different nature of the BDE substituent. The fact that the patterns identified for these molecules are consistent with those that had been identified for ACPLs as a class of compounds expands the validity range of those findings to molecules with more complex substituents. This, in turn, enhances the overall predictive potentialities of the results on ACPLs, increasing their suitability for use in the design of new ACPLs, whether meant for pharmaceutical purposes or for other uses of interest.

Acknowledgements The authors express their gratitude to the Centre for High Performance Computing (CHPC, South Africa) for providing the computational resources used to conduct this work and to Dr. Caterina Ghio, of the Institute for Physico-chemical Processes (Pisa, Italy) for fruitful interactions. N. Tshilande is grateful to the National Research Foundation (NRF, South Africa) for a scholarship for 2018 (DST-NRF Innovation Master's Scholarship, grant number 113676) and to the Sasol Inzalo Foundation for additional support.

References

1. Simon LS (1996) Curr Opin Rheumatol 15:169–175
2. Massó González EL, Patrignani P, Tacconelli S, García Rodríguez LA (2010) Arthritis Rheum 15:1592–601
3. Hernández-Díaz S, Rodríguez LA (2000) Arch Int Med 160:2093–9
4. Cappelli C, Mennucci B, Monti S (2005) J Phys Chem A 109:1933–1943
5. Gomes CA, Girão da Cruz T, Andrade JL, Milhazes N, Borges F, Marques MPM (2003) J Med Chem 46:5395–5401
6. Sergediene E, Jonsson K, Szymusiak H, Tyrakowska B, Rietjens ICM, Cenas N (1999) FEBS Lett 462:392–396
7. Alagona G, Ghio C (2009) Phys Chem Phys 11:776–790
8. Verotta L (2002) Phytochem Rev 1:389–407
9. Singh IP, Bharate SB (2006) Nat Prod Rep 23:558–91
10. Mammino L, Kabanda MM (2007) J Mol Struct (Theochem) 805:39–52
11. Mammino L, Kabanda MM (2008) Int J Quantum Chem 108:1772–1791
12. Mammino L, Kabanda MM (2009) J Mol Struct (Theochem) 901:210–219
13. Mammino L, Kabanda MM (2009) J Phys Chem A 113:15064–77
14. Mammino L, Kabanda MM (2010) Int J Quant Chem 110:2378–2390
15. Kabanda MM, Mammino L (2012) Int J Quant Chem 112:3691–3702
16. Mammino L, Kabanda MM (2012) Int J Quant Chem 112:2650–2658
17. Mammino L, Kabanda MM (2013) Mol Sim 39:1–13
18. Mammino L (2013) J Mol Model 19:2127–2142
19. Mammino L (2014) Curr Bio Comp 10:163–180
20. Mammino L (2015) Curr Phys Chem 5:274–293
21. Mammino L (2017) Molecules 22:1294. https://doi.org/10.3390/molecules22081294
22. Peña Á, Yosa J, Cuesta-Astroz Y, Acevedo O, Lareo L, García-Vallejo F (2012) Universitas Scientiarum 17:5–15
23. Makola MM, Dubery IA, Koorsen G, Steenkamp PA, Kabanda MM, du Preez LL, Madala NE (2016) Evidence-based complementary and alternative medicine

24. Masike K, Tugizimana F, Ndlovu N, Smit E, du Preez L, Dubery I, Madala E (2017) J Chromatogr B1052:73–81
25. Lee C, Yang W, Parr RG (1998) Phys Rev B 37:785–789
26. Becke AD (1993) J Chem Phys 98:1372–1377
27. Becke AD (1993) J Chem Phys 98:5648–5652
28. Moller C, Plesset MS (1934) Phys Rev 46:618
29. Rozas I (2007) Phys Chem Chem Phys 9:2782–2790
30. Gromak VV (2005) J Mol Struc (Theochem) 726:213–224
31. Merrick JP, Moran D, Radom L (2007) J Phys Chem A 111:11683–11700
32. Cossi M (2004) Chem Phys Lett 384:179–184
33. Saracino GAA, Improta R, Barone V (2003) Chem Phys Lett 373:411–415
34. Alagona G, Ghio C (2006) J Phys Chem A 110:647–659
35. Boys SF, Bernardi F (1970) Mol Phys 19:553–566
36. Frisch MJ, Trucks GW, Schlegel HB, Scuseria GE, Robb MA, Cheeseman JR, Scalmani G, Barone V, Mennucci B, Petersson GA, Nakatsuji H, Caricato M, Li X, Hratchian HP, Izmaylov AF, Bloino J, Zheng G, Sonnenberg JL, Hada M, Ehara M, Toyota K, Fukuda R, Hasegawa J, Ishida M, Nakajima T, Honda Y, Kitao O, Nakai H, Vreven T, Montgomery JA, Jr, Peralta JE, Ogliaro F, Bearpark M, Heyd JJ, Brothers E, Kudin KN, Staroverov VN, Keith T, Kobayashi R, Normand J, Raghavachari K, Rendell A, Burant JC, Iyengar SS, Tomasi J, Cossi M, Rega N, Millam JM, Klene M, Knox JE, Cross JB, Bakken V, Adamo C, Jaramillo J, Gomperts R, Stratmann RE, Yazyev O, Austin AJ, Cammi R, Pomelli C, Ochterski JW, Martin RL, Morokuma K, Zakrzewski VG, Voth GA, Salvador P, Dannenberg JJ, Dapprich S, Daniels AD, Farkas O, Foresman JB, Ortiz JV, Cioslowski J, Fox DJ (2013) Gaussian 09, Revision E.0.1. Gaussian, Inc., Wallingford CT
37. Frisch E, Dennington II RD, Keith TA, Millam L (2007) GaussView 4.1. Gaussian Inc., Pittsburgh
38. Chem3D, version 8.0.3 (2003) ChemOffice, Cambridge Software
39. Mammino L (2018) In: Wang YA, Thachuk M, Krems R, Maruani J (eds) Concepts, methods and applications of quantum systems in chemistry and physics, book series progress in theoretical chemistry and physics, vol 31. Springer, pp 281–304
40. Bushelyev SN, Stepanov NF (1989) Elektronnaya Struktura y Biologhicheskaya Aktivnost Molecul. Khimiya, Snanye, Moscow
41. Alagona G, Ghio C (2007) Theochem 811:240–223

Current Problems in Computer Simulation of Variability of Three-Dimensional Structure of DNA

V. Poltev, V. M. Anisimov, V. Dominguez, A. Deriabina, E. Gonzalez, D. Garcia, V. Vázquez-Báez and F. Rivas

Abstract Being the main molecule of life, DNA has a simple chemical structure that is surprisingly fit to the important biological functions that it performs. In spite of its simplicity, the molecular organization of DNA as a linear copolymer of four nucleotides enables the formation of an infinite variety of three-dimensional structures depending on the monomer sequence and environmental conditions. Molecular mechanisms, general regularities, and biological importance of the diversity of DNA duplexes have been extensively studied by Quantum Mechanics (QM) and Molecular Mechanics (MM) methods at the level of simple fragments. Our previous DFT studies revealed that the important conformational characteristics of BI, BII, AI, and AII families of Watson-Crick duplexes (WCD), including sequence dependence of their three-dimensional structure, preexist in the local energy minima of the elemental single chain fragments, deoxydinucleoside monophosphates (dDMPs), and in the elemental duplex fragments, complementary dDMPs (cdDMPs). Those computations uncovered the important regularity in sequence dependence of neighbour base superposition, namely substantial superposition for purine-purine and purine-pyrimidine sequences, and negligible base overlap in pyrimidine-pyrimidine and pyrimidine-purine sequences. This regularity matches the experimental data; it can be reproduced by using various computational methods; and distinguishes WCD from other polynucleotide duplexes. We concluded that this pattern is a consequence

V. Poltev (✉) · V. Dominguez · A. Deriabina · E. Gonzalez · D. Garcia · V. Vázquez-Báez · F. Rivas
Autonomous University of Puebla, Puebla 72570, Mexico
e-mail: poltev@fcfm.buap.mx

E. Gonzalez
e-mail: gonzalez@fcfm.buap.mx

D. Garcia
e-mail: dolores@ifuap.buap.mx

F. Rivas
e-mail: rivas@ifuap.buap.mx

V. M. Anisimov
National Center for Supercomputing Applications, University of Illinois at Urbana-Champaign, Urbana, IL 61801, USA

L. Mammino et al. (eds.), *Advances in Quantum Systems in Chemistry, Physics, and Biology*, Progress in Theoretical Chemistry and Physics 32,
https://doi.org/10.1007/978-3-030-34941-7_12

of directionality and presence of preferable regions in the sugar-phosphate torsions combined with the difference of purines from pyrimidines in ring shape. It appears that along with other three-dimensional DNA structures, experimentally studied fragments of DNA contain variety of cdDMPs with Watson-Crick nucleoside pairs but with one or both dDMPs belonging to conformational families different from A and B. Comprehensive analysis of conformational characteristics of such cdDMP belonging to select families by using various computational methods, and the comparison of subunit contributions to their three-dimensional structure formation suggested that the regularities for the most of such cdDMPs differ from those of 'canonical' WCD. The geometry of the optimized conformations of those cdDMPs depends on the used computational method, and sometimes departs from the double helix structure. Comparison of the results including those from optimization of cdDMPs, dDMPs, SPB, and base-base pairwise interactions against the mutual position of DNA subunits in crystals shows that none of the computational methods is able to quantitatively describe all experimental structures of DNA, and to correctly reproduce the interaction between its subunits. Analysis of these data produces valuable information about the strengths and limitations of the employed computational methods, and suggests the pathway for their improvement. The work concludes that the complex interplay that exists between the DNA subunits requires additional investigations by using multiple computational methods each targeting particular aspects of the system.

Keywords Computer simulation · MM · QM · DFT · 3D-structure · Regularity · Multiplicity · Variability · Flexibility · Canonical · DNA · Nucleotide · Base pair interaction · Deoxyribose · Sugar-phosphate backbone · Hydrophilic · Hydrophobic · WCD · SPB · dApdG · dTpdC · cdDMP · dDMP

1 Introduction. Complexity and Simplicity of the Most Important Molecule of Life

DNA is the most important macromolecule of life. The sequence of monomer units composing the long polynucleotide chain of DNA contains all the information necessary for functioning of live organisms. Despite the enormous biological role of DNA, its chemical composition is limited to a linear arrangement of four nucleotides in the chain consisting of many thousands of monomers. Such seemingly simple construction encodes the infinite diversity of life. Figure 1 presents a fragment of DNA chain consisting of four possible nucleotides.

The fragment shown in Fig. 1 is usually designated as dApdTpdGpdC. Each nucleotide in it contains three subunits, namely, a nitrogenous base, deoxyribose, and phosphate group. Among those units, DNA base is the only variable part in the nucleotide, while deoxyribose and phosphate form chemically homogeneous sugar-phosphate backbone, SPB. We will designate bases and nucleosides, which compose bases connected to sugar moiety via N-glycoside bond, by a single letter. Two of the bases are purines: adenine, (A), and guanine, (G); and other two bases are

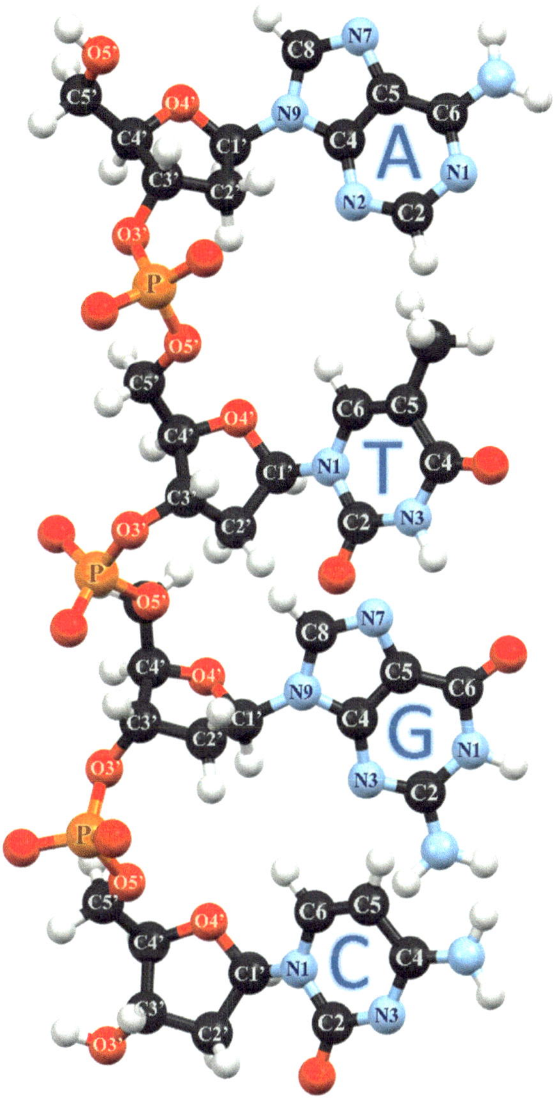

Fig. 1 Atomic structure and atom numbering of a DNA fragment dApdTdGpdC

pyrimidines: thymine, (T), and cytosine, (C). Difference in geometry between purines (Pur) and pyrimidines (Pyr) is important for the formation of sequence-dependent three-dimensional structure of DNA.

Nucleic acid strands have directional asymmetry due to non-identical connection of the phosphate group to two deoxyribose moieties. The important feature of the Watson-Crick duplex (WCD) [1] is the formation of hydrogen bonds (H-bonds) between the purine from one chain and pyrimidine from the complementary strand;

bases A and T form two H-bonds, while G and C form three H-bonds. Two complementary antiparallel polynucleotide chains of WCD assemble in right-handed double helix.

The molecular structure of DNA accommodates various intermolecular and intramolecular interactions that determine the wealth of its three-dimensional structure, and lead to the formation of complexes with other important biomolecules, including proteins. Polynucleotide chain and each of its nucleotide units contain nearly planar and rigid nitrogenous bases shown in the right-hand side of Fig. 1, as well as flexible SPB shown in the left-hand side of Fig. 1. The nucleotide structure includes both hydrophilic and hydrophobic atom groups. The hydrophilic N-H and amino groups of the bases are H-bond donors; whereas oxygen atoms of all three nucleotide subunits are acceptors. The CH, CH_2, and CH_3 groups play the role of hydrophobic units. The phosphate is an electrophilic group that carries a negative charge at biologically relevant pH conditions.

Since the discovery of DNA double helix by Watson and Crick in 1953 [1] it became clear that its molecular and three-dimensional organization is surprizingly fit to the biological functions attributed to the main molecule of life. The 3D structure of DNA supports storage, protection, replication, expression and evolutionary adjustment of genetic material among other important biological functions. Some of those features have been noticed right in the beginning and were studied during the first years after Watson and Crick's discovery. Conformational transitions between the double-stranded and single-stranded states, the dependence of helix stability on nucleotide content (due to different number of H-bonds in the two complementary pairs), the possibility of both specific (via H-bonding with the base pairs) and nonspecific (via negatively charged phosphate groups) interactions with other biologically important molecules including proteins, are examples of such features. Other important aspects of the WCD, such as several levels of organization of its three-dimensional structure, the dependence of its structure on nucleotide sequence, the regularities of this dependence, the existence of variety of conformation families with specific regions of the torsion angles of SPB for each family, were understood a few decades later. In spite of the extensive and thorough investigations of DNA structure including the mechanism in which its fragments interact one with another, not everything is sufficiently clear about the contribution of DNA subunits to its three-dimensional organization and the role that individual atoms and atomic groups play in the DNA structure and function.

For the last few decades our group has been working on deciphering general regularities of DNA three-dimensional structure organization via the application of quantum mechanics (QM) and molecular mechanics (MM) methods to the relatively simple fragments of DNA. By using density functional theory (DFT) computations, we demonstrated that the main conformational characteristics of B- and A- Watson-Crick right-handed duplexes preexist in the local energy minima of deoxynucleoside monophospates (dDMPs), which represent the elemental fragments of DNA single chain [2–8]. The discovered characteristics of dDMPs depend on nucleoside sequence and reproduce the similar dependence noticed in the crystals of WCD fragments. Specifically, base superposition patterns of Pur–Pyr and Pur–Pur sequences differ from those of Pyr–Pur and Pyr–Pyr structures corresponding to AI, AII, BI, and BII conformation families of WCD. The base rings substantially overlap in the first pair of sequences, whereas only a minor, if any, overlap exists in the second pair. These regularities also occur in complementary dDMPs (cdDMPs), which are minimal fragments of duplex, but only when both chains belong to one of the above mentioned DNA conformation families [7, 8]. Other polynucleotide duplexes, such as left-handed Z-forms, Hoogsteen and parallel-stranded duplexes, duplexes with mispairs [4–8] do not demonstrate this regularity, which is important for biological function of DNA.

Besides the duplexes mentioned above, the Nucleic acid Database, NDB [9, 10], contains DNA fragments in the variety of other conformations, such as triplexes, quadruplexes, Holliday junctions, etc. It appears that some fragments of DNA contain cdDMPs with Watson-Crick nucleoside pairs having one or both chains of SPB pertaining to conformation families different from A- and B- ones. Svozil et al. have conducted a systematic classification and analysis of various conformation families of minimal fragments of DNA single chain [11, 12].

This work is a continuation of our ongoing study of simple fragments of DNA [2–8]. We discuss the variability of three-dimensional structure of DNA duplex with Watson-Crick nucleoside pairs, the sources of this variability, and the problems of application of various QM and MM methods to the understanding and predicting conformational characteristics of such DNA fragments constructed of various nucleotide sequences.

2 Model Systems and Computational Methods

The main systems, which are considered in this work, are the minimal fragments of DNA duplex (cdDMP), and those of the single chain (dDMP) neutralized by sodium ions. Figure 2 presents the molecular structure and torsion angle designations of a sample minimal fragment of WCD, which is a complex of complementary dinucleoside monophosphates dApdG and dTpdC. Na^+ ions are omitted in this and next figures for clarity. For each fragment, an initial configuration was constructed using atomic coordinates of the corresponding fragments of a structure obtained from NDB [8, 9]. Hydrogen atoms and sodium ions were added in the position most likely for a given atomic group. For evaluation of subunit contributions to the formation of three-dimensional structure of cdDMPs (and that of longer DNA fragments) computations were performed on the minimal fragments of SPB having the bases in dDMPs substituted by hydrogen atoms. Comparison of optimized and initial cdDMP conformations conducted on select structures provide an estimate of contribution of base-base interactions. Computations of selected pairwise interactions of separate bases (1-methylpyrimidines and 9-methylpurines) locate the local energy minima that provide reference points for the comparison.

We use MM and QM methods to compute energy and the local minima. The MM method has traditionally been applied to molecular systems of substantially more

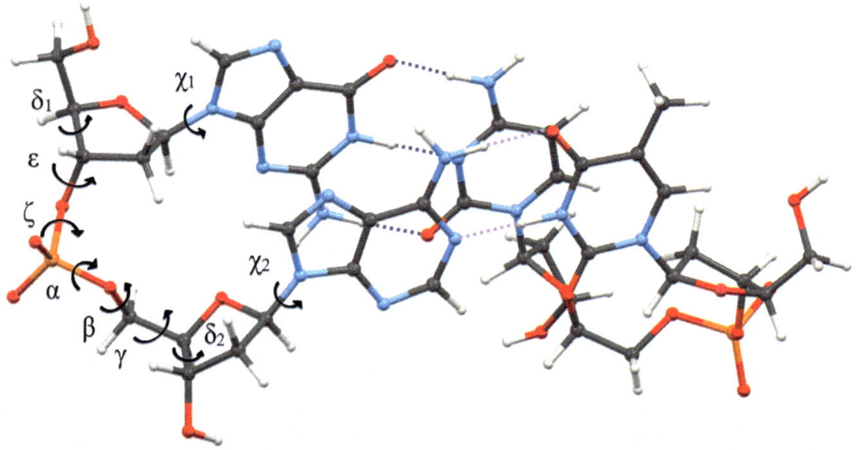

Fig. 2 Ball-and-stick model of complementary dinucleoside monophosphate dGpdA:dTpdC and the designation of torsion angles for dGpdA chain. Dotted lines indicate base-base hydrogen bonds. The A:T pair resides closer to the observer

complex composition than those considered here. This empirical method of classical physics enables to consider sufficiently large biomolecular complexes in realistic environment, e.g. fragments of biopolymers of hundreds and thousands of atoms in water solutions. However, the quality of its prediction strongly depends on the force field (FF) in use. One of the present authors has reviewed this method recently [13]. A rather simplified FF for nucleic acid computations was proposed by some of the authors of this work many years ago when it provided the only possible methodology for quantitative evaluation of interaction of simple DNA subunits (last modification of the parameters was published in 2002 [14]). One of the most popular FFs (and the corresponding software package) for biopolymer computations is AMBER [15, 16]. In this work, we use three AMBER FFs, namely ff99 [17], OL15 [18], and BSC1 [19] for computation of DNA fragments. These force fields differ only in the description of torsion angle. Even with that, the results produced by those FFs on the minimum energy structures may differ significantly. Selected structures have also been optimized by additive CHARMM force field [20] as well. We use these empirical methods to see how they compare to experiment and QM computations, to preliminarily optimize some cdDMPs, and to evaluate the possible pathways for the improvement of FF.

The replacement of classical mechanics by QM methods offers only a partial reprieve. The application of QM methods to DNA study faces several typical issues. The use of accurate coupled cluster level of theory to dDMPs and cdDMPs is prohibitively computationally expensive. The more affordable correlated ab initio methods such as MP2 happen to be insufficiently accurate for these systems even when using large basis set. Our previous studies [6, 7] using the MP2/6-311++G** level of theory for calculation of few dDMPs and cdDMPs provide an example of what can be expected from MP2 computation of the minimal fragments of DNA. The method correctly reproduces the conformational regularities mentioned in the section of Introduction, but predicts shortened contacts between the stacked bases, and that renders this level of theory unacceptable. This conclusion about the limitations of the MP2 level of theory in application to DNA fragments had further been confirmed by systematic study of pairwise interactions of DNA bases [21] at MP2/6-31G** level of theory. Those computations conclude that MP2 systematically overestimates the base stacking energy for various base positions and shortens the stacking C–C and C–N distances up to 3.0 Å for the most of pairwise combinations.

The main QM method that we use for dDMP and cdDMP optimization is based on the Density Functional Theory (DFT). This method provides reasonable compromise between the accuracy and computational cost. We use Gaussian09 [22] and ADF (Amsterdam Density Functional) [23, 24] programs for geometry optimization of the studied systems. Employed functionals are PW91 [25] and PBE [26], as well as M05-2X [27] and M06-2X [28], which are designed for computation of stacking configurations of organic molecules. We also use DFT-D method, which mixes the quantum mechanical level of theory with classic corrections, on selected dDMPs and cdDMPs. In addition to popular functionals B97-D3 [29] and B3LYP-D3 [30] implemented in GAUSSIAN09 program, and PBE-D [31] functional implemented in ADF package [24], we also use PW91-ulg functional [32] recently added to the ADF program. Computation of conformational characteristics of DNA fragments including torsion angles, deoxyribose ring puckering, and mutual base arrangements employs 3DNA software [33].

3 Sources of Variability in the DNA 3D-Structure

Analysis of the molecular structure of DNA suggests the presence of two main sources of variability in its 3D structure. These are (1) flexibility of SPB due to rotation around single bonds (including variability of sugar puckering), and (2) ability of each of the bases to participate in various base-base complexes. Molecular structure of bases enables the formation of various pairwise complexes corresponding to the energy minima of intermolecular H-bonding and stacking interactions. In this work, we limit our analysis of mutual position of subunits corresponding to the formation of double helices with Watson-Crick nucleoside pairs, i.e. with A:T and G:C complementary base pairs and anti-orientation of base and deoxyribose in both nucleosides. Some of other experimentally observed polynucleotide structures, such as left-handed Z-conformation families with purine nucleosides in syn-conformation, Hoogsteen or parallel duplexes with base pairing different from Watson-Crick one, and duplexes with mispairs arising as a result of errors of biosynthesis have been considered in our earlier studies [4–8].

3.1 Flexibility of the Sugar-Phosphate Backbone

Torsion angles of single polynucleotide chains continuously vary in relatively wide regions. Out of those bands, only selected sets of torsion ranges are compatible with double stranded structure of Watson-Crick A:T and G:C nucleoside pairs. Rather small consistent variations of the torsions can result in large changes of macroscopic parameters of the helices belonging to the same conformation family. Because of that each family can contain infinite number of conformers. We will consider here some B-like conformation families of DNA. Majority of cdDMPs of such DNA fragments deposited in NDB consist of both strands belonging to the "classical" BI family. Nevertheless many fragments contain cdDMP with one or two strands being different from BI. Here, we consider cdDMPs with SPB of BI, BII conformation families and with those classified in paper [12] as 32, 86, 110, 116, and 117 conformational classes. These classes (families) distinguish one from another due to significantly different values of two or three torsions. For instance, two of the most studied families, BI and BII, differ mainly in ε (C4′–C3′–O3′–P) and ζ (C3′–O3′–P–O5′) torsions. Mean values of ζ torsion changes from 259° for BI family to 170° for BII family, while mean ε torsion changes from 183° to 245° [12]. The conformations of SPB presented in Fig. 3 demonstrate the differences in mutual position of atoms in these two families.

Other B-like conformation families differ from the canonical one by the change in two or three torsion angles while retaining the major characteristic features of B-structure. The crank of α and γ torsions to g^+ and g^- regions respectively produce conformations of family 116, while conformation family 117 emerges as a consequence of changes in torsion angles α, β, and γ to mean values 249°, 73°, and 172° respectively. Figure 4 illustrates molecular models of two conformations of dGpdT:dApdC with SPB belonging to different families.

Figure 5 provides graphic illustration of differences in mean torsion angles for selected cdDMPs. Table 1 presents numerical values of their torsions. The BI and BII conformation families carry the SPB fragment in the conformation that matches the energy minimum of the free SPB unit. Both DFT and MM computations of SPB predict the torsion angles to be in the same region as that observed in the variety of conformations of cdDMPs in these families. The conformation family 117 is the only one among the selected systems that has all the torsions differ by less than 30° from the energy minimum conformation of the free SPB. Three other families have at least two torsions exceeding that threshold.

3.2 Multiplicity of Feasible Base-Base Complexes

Intermolecular interactions between the DNA bases have been studied since 1961 using various computational methods. Molecular structure of the bases enables the formation of variety of their pairwise complexes as well as complexes of three and four bases corresponding to energy minima. Experimentally observed or theoretically

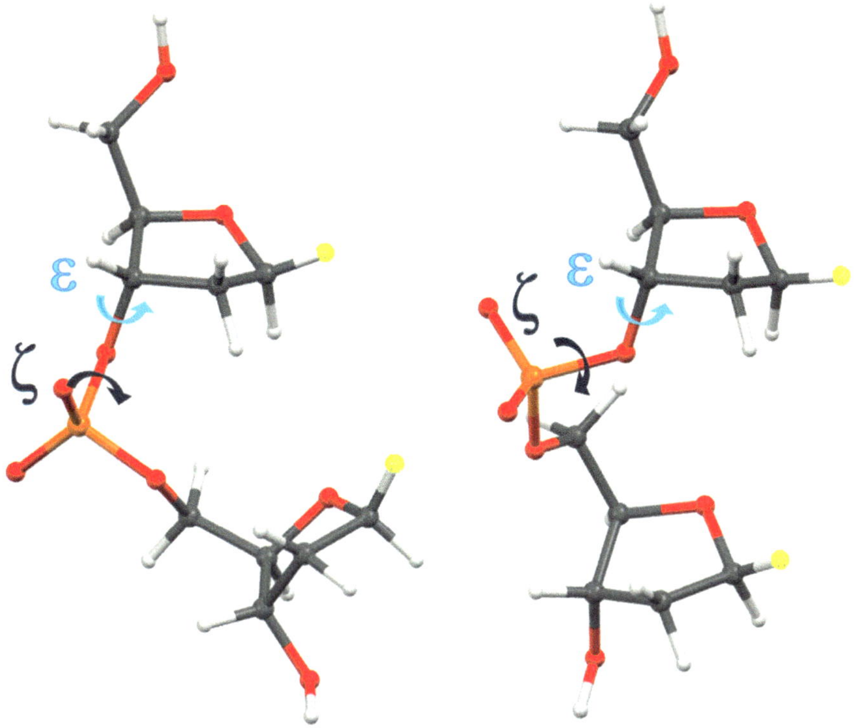

Fig. 3 Ball-and-stick model of BI (left) and BII (right) conformations of SPB for dDMPs of selected DNA fragments (NDB id PD0192 and 5SET9, respectively). The geometry of 5′-deoxyribose is the same for two families. Two Hydrogen atoms substituting the bases are shown in yellow

predicted pairwise complexes can be classified into various types. Two types of energy minima directly related to the problem of variability of DNA structure are (1) minima corresponding to the formation of two or three H-bonds, and (2) minima with nearly parallel arrangement of base ring planes (base stacking). Two other minima correspond to the formation of single H-bond, and to nearly perpendicular or in-plane position of base rings. These four types of minima are illustrated in Fig. 6 on the example of interactions of 9-methylguanine with 1-methylcytosne.

There are other energy minima on the potential energy profile of pairwise interaction of bases; but they are not as deep as the first four, and are unlikely to contribute to the conformation characteristics of cdDMPs considered here. Variations in the minimum of the first type contribute to the formation of duplexes with base pairing different from Watson-Crick A:T and G:C pairs. Examples of those are parallel or Hoogsteen duplexes, triplexes, and quadruplexes. Computational methods reproduce well the mutual H-bonded base position in duplexes with Watson-Crick and non-Watson-Crick nucleoside pairs. These complexes possess sufficient flexibility for consistent displacement and rotation of the bases without causing substantial

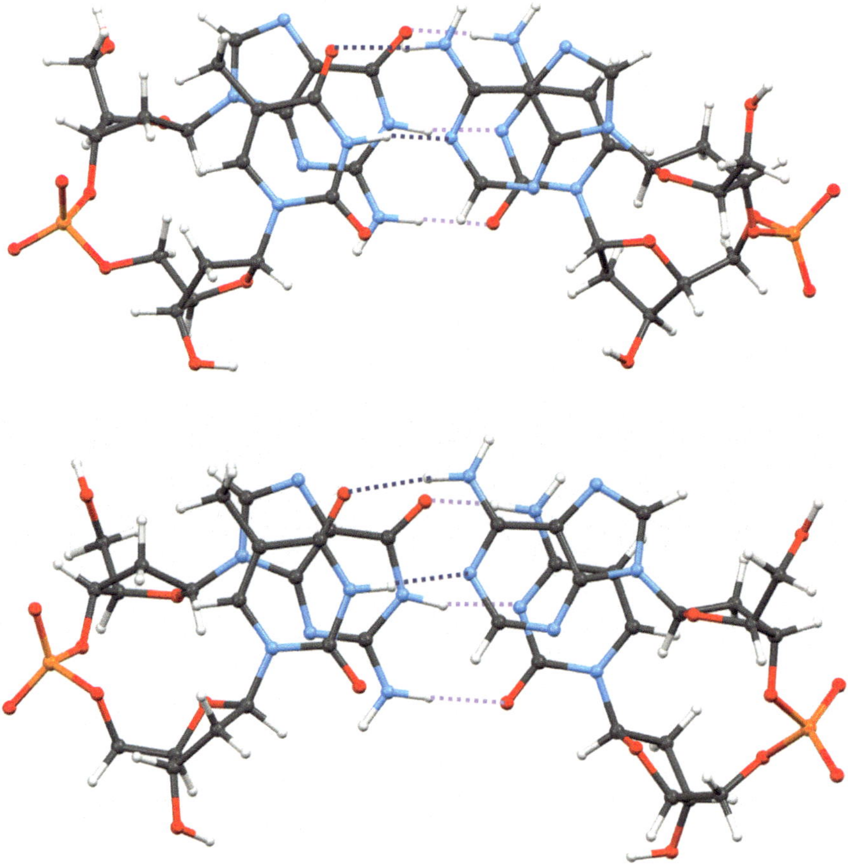

Fig. 4 Two cdDMP conformations of dGpdT:dApdC corresponding to different B-like families; A:T pair is positioned closer to the observer; upper row: dGpd CT (left chain) of 116 conformation family, and dApdC (right chain) of 117 conformation family (from the NDB id 5GUN); lower row: both chains are in BI conformations of dGpdT:dApdC (from the NDB id 5NT5)

rise in the energy. The existence of flat-bottom energy minima produces the variability regions of conformational parameters in the conformation family. Although the mutual position of stacked bases in duplexes substantially differs from that in the free stacked bases, their energy minimum has a wider flat-bottom than that in the complexes involving H-bonds; and that makes stacking interactions contributing considerably to the conformation stability of dDMPs and cdDMPs.

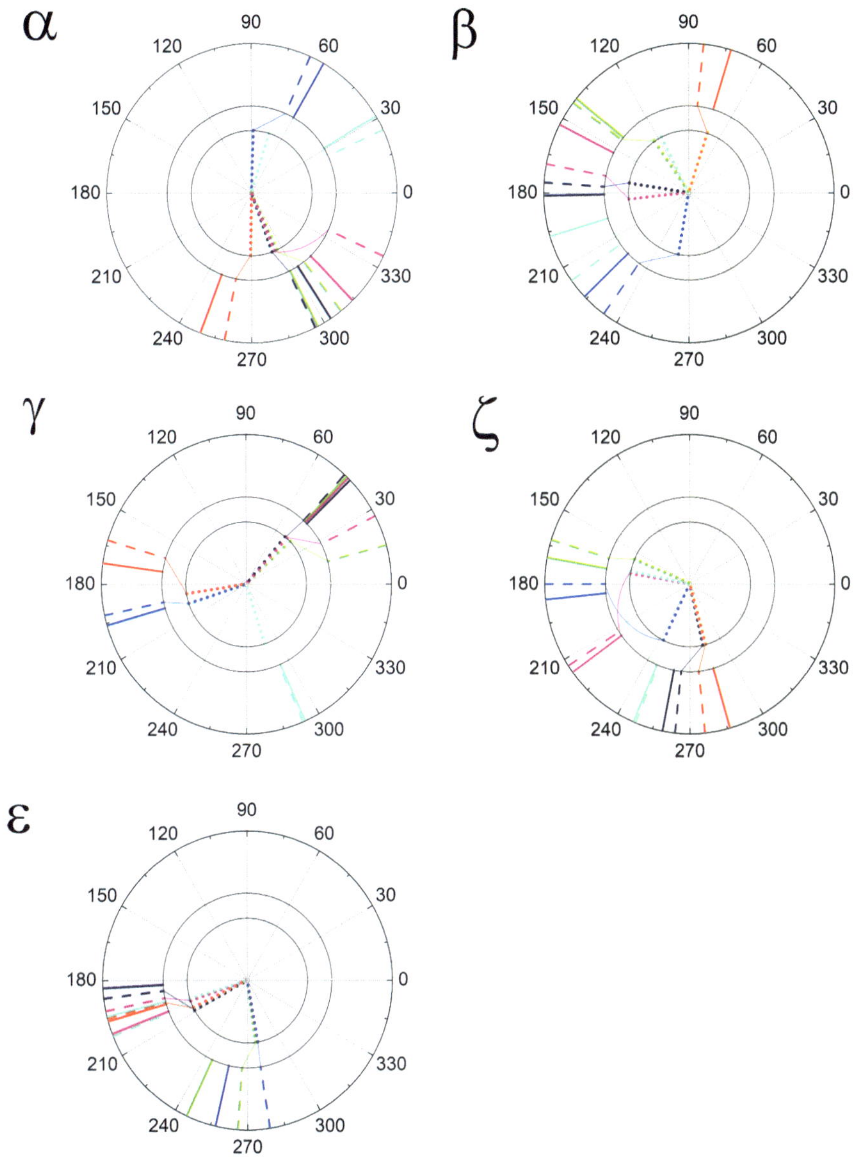

Fig. 5 Comparison of five torsion angles of the SPB. Solid lines (between the outer and intermediate circles) stand for mean values of torsions of various families as designated in [12]; dashed lines indicate the value for selected structure of NDB; dotted lines (going from the inner circle to the centre) show the PBE optimized data. Lines between the inner and the intermediate circle demonstrate the change in the torsions resulted from optimization of the SPB structure of cdDMPs extracted from the NDB. Black lines correspond to BI family, dark green ones to BII, blue to 110, orange to 117, violet to 86, bright green to 116 conformation family

Table 1 Torsion angles of SPB for selected DNA families obtained from the NDB and from geometry optimization of SPB by using various computational methods

dDMP	Method	δ1	ε	ζ	α	β	γ	δ2
dGpdG BI (116) PD0230	NDB	139	187	264	296	176	47	138
	PBE	137	208	283	290	171	48	136
	M05-2X	140	191	276	293	163	48	140
	AMBER	132	185	279	291	162	50	135
dGpdC BII (32) 5ET9	NDB	141	**266**	**163**	308	143	15	144
	PBE	129	278	157	292	125	43	138
	M05-2X	128	278	150	291	116	43	144
	AMBER	128	283	168	290	*97*	49	133
dApdC 117 (116) 5GUN	NDB	131	196	276	**259**	**84**	**163**	144
	PBE	138	207	285	269	73	189	73
	M05-2X	141	203	279	271	79	178	165
	AMBER	132	195	276	279	71	183	140
dTpdT 116 (116) PD0190	NDB	136	201	248	**25**	216	**293**	153
	PBE	134	198	*169*	*73*	*117*	288	161
	M05-2X	140	198	*174*	*67*	*119*	293	165
	AMBER	131	192	238	*64*	*178*	302	144
dCpdG 86 (BI) 5ET9	NDB	144	192	**213**	**335**	169	27	156
	PBE	136	198	*169*	*291*	186	48	137
	M05-2X	141	196	*166*	*295*	189	48	142
	AMBER	123	185	*278*	*292*	163	49	126
dApdT 110 (BI) 2MAV	NDB	143	**279**	**180**	**66**	**234**	**192**	104
	PBE	113	297	*215*	83	*277*	205	74
	M05-2X	113	*302*	*212*	68	*182*	191	75
	AMBER	122	274	167	69	*182*	182	78

Note: The dDMP column describes the sequence, the conformation family of the analyzed and complementary (in brackets) dDMP, and the NDB id of the structure. The Method column provides the data source. Bold font in the experimental value of torsion angle indicates that the deviation from the corresponding angle for BI family is greater than 30°. PBE and M05-2X functionals denote the use of GAUSSIAN program. AMBER program uses BSC1 FF. Italicized numbers indicate computed values, which differ from those of the NDB structure by more than 30°

4 Regularities in the 3D Structures of BI and BII Conformational Families. Capabilities and Limitations of Computational Methods

Our previous theoretical study of elementary units of DNA revealed important regularities in 3D structure of BI and BII conformations of WCD [2, 3]. That work included DFT computations using PW91 functional and 6-31G(d) basis set applied

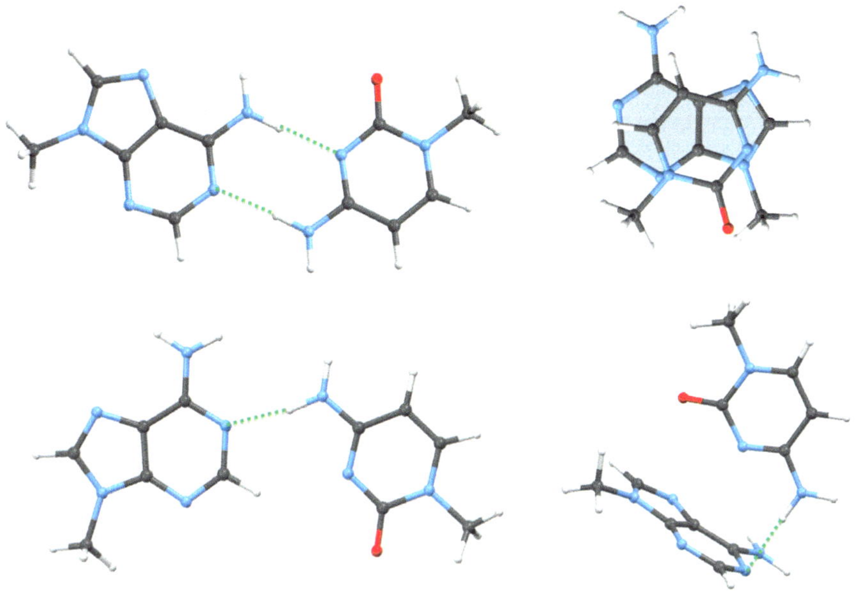

Fig. 6 Mutual base position in the four energy minima corresponding to the interaction between 9-methyl-guanine and 1-methyl-cytosine. Upper left structure corresponds to nearly in-plane base pairing with two H-bonds; upper right—base stacking; lower left—nearly in-plane base arrangement with single H-bond; lower right—nearly perpendicular (T-shaped) base arrangement with single H-bond

to all 16 possible dDMPs. The geometry optimization started from the structure extracted from DNA fragments deposited in the NDB. Since we were unable to find BII conformations of dDMPs for some nucleosides sequences we constructed them by replacing the bases in other dDMPs. It appears that all these dDMPs with any nucleoside sequence retain the conformational characteristics of the family in the optimized local energy minima. This includes the regions of SPB torsion angles, deoxyribose puckering, and nearly parallel base arrangements [3]. These results reproduce the sequence dependence of base superposition patterns on nucleoside sequence observed for BI conformation duplex fragments in crystals, namely, the substantial base ring superposition for sequences with purine nucleosides in 5′-end of dDMP and the minor superposition for sequences with pyrimidine nucleosides in 5′-end. Confirmed based on additional computations using PW91 and PBE functionals, this regularity subsequently extends to AI and AII conformation families of dDMPs [4–6] as well as to cdDMPs with both chains belonging either to BI or AI conformation families [6, 7]. This leads to conclusion that many biologically important conformational characteristics of 'canonical' WCD preexist in the local energy minima of its elementary units, dDMPs and cdDMPs. These conformational regularities and their presence in DFT computations of the elemental units distinguish WCD from other polynucleotide duplexes having non-Watson-Crick geometry of nucleoside pairs [4–8] and, as we found recently, from other families of duplexes with Watson-Crick nucleoside pairs but with changed conformation of one or both SPB chains.

The regularities mentioned above have been additionally confirmed by *ab initio* computations of selected structures using MP2 level of theory and various basis sets up to 6-311++G** [6, 7]. Nevertheless, both QM approaches failed to reproduce the fine aspects of dDMP and cdDMP structures. DFT computations using PW91 and PBE functional systematically overestimate the distance between bases of dDMPs and between the stacked base pairs of cdDMPs [6, 7]. This error may be explained by underestimation of dispersion attraction between the aromatic rings in the DFT level of theory. This limitation of the DFT method however additionally strengthened our conclusion about the critical contribution of SPB to the formation of dDMP structure reproducing the conformational characteristics of WCD and to the nucleoside sequence dependence of 3D structure. This conclusion has been validated by exploring the local energy minima of SPB corresponding to BI, BII, AI, and AII families of DNA. This analysis demonstrated that the conformations of SPB in these DNA structures correspond to those of the local energy minima of the free SPB.

The MP2 computations of dDMPs and cdDMPs produce structures with shortened distances (up to 2.9 Å) between the stacked bases in comparison to the NDB data for DNA fragments due to the known effect of overestimation of dispersion interactions at this level of theory. This limitation of the MP2 method shows up in various combinations of bases [21]. The use of counterpoise correction to the basis set superposition error, BSSE, suggested by Boys and Bernardi [34] reduces the absolute value of base-base stacking energies and improves the interatomic distances. With that, only a few moderately shortened contacts (up to 3.2 Å) remain in disagreement with the experimental data [21]. Unfortunately, the application of this method to cdDMPs remains computationally prohibitive.

The results obtained with the use of MP2 method demonstrate that the application of higher level of theory does not automatically improve the agreement with experimental data for WCD. Likewise, physically unrealistic structures having shortened atom-atom contacts arise when using DFT-D method [4] and the functionals specifically designed for computation of stacking complexes of organic molecules (such as M05-2X and M06-2X) [7]. The overall conclusion from the application of QM methods to minimal fragments of DNA is that they can reproduce and predict important conformational regularities of WCD, but neither of the methods reproduces the complete set of important characteristics of these systems.

Another topic of interest is the capability of MM method to reproduce not only the overall structure of WCD but also the regularities in sequence dependence of 3D structure revealed in DFT computations, especially on base superposition. Our calculations of selected cdDMPs with both dDMPs belonging to the BI family by using AMBER force field demonstrated that many but not all AMBER optimized structures reproduce these regularities [7]. Some of the optimized structures incorrectly show considerable base ring superposition in Pyr-Pyr or Pyr-Pur sequences. We suppose that it is because AMBER force field overestimates base stacking [7]. Systematic computations and the search for local energy minima for pairwise base stacking interactions by using AMBER and CHARMM additive force fields support that conclusion [21]. For some base combinations computation predicts stacking minima to be deeper than those for complexes with two H-bonds [21]. This observation suggests the necessity to refine the current FFs and to tune their predictions to the experimental and the most reliable QM data.

5 Computational Problems with Elementary Units of 'Non-canonical' Families of DNA

Extending the computational study of regularities of 3D structure of cdDMPs to DNA fragments with two chains belonging to different conformation families encounter additional challenges. Searching for various B-like conformations in NDB we found four fragments of dGpdC:dGpdC sequence (NDB id BDL078, BDL084, 5EZF, and 5ET9) that have two complementary chains in the duplex being attributed to different conformation families. Specifically, one chain of cdDMP belongs to BII family while the complementary chain resembles BI conformation. In that structure, puckering of 3′-end sugar is intermediate between B and A families (conformation 32 or BI-to-A according to Svozil classification [12]). These cdDMPs caught our attention because their BII chains do not demonstrate base ring superposition typical for dDMPs in BI, BII, AI and AII conformations, as well as for cdDMPs of the same sequence when both their chains are in BI or AI conformation. DFT optimization of these non-canonical cdDMPs using PW91 and PBE functionals leads to an inclination of one of the base pairs followed by formation of H-bonds between the bases belonging to different pairs that leads to significant deviation from the experimental structure. Conducting the geometry optimization by using M05-2X and M06-2X functionals reproduces the experimental conformation of the non-canonical cdDMP including the base superposition patterns of two chains. However, the optimized structures carry too short atom-atom contacts between the stacked bases.

Similar problems exist with the use of general-purpose DFT functionals, which successfully reproduce the regularities in 3D structure formation of "classical" BI and AI duplexes, but fail when applied to cdDMPs that have different B-like conformations in their two dDMPs. In those cases, SPB torsions remain in the regions of initial family while the optimized cdDMPs configurations drastically differ from the geometry of those fragments observed in the continuous DNA duplex. The use of functionals specifically adjusted for stacking interactions as well as DFT-D functionals produces structures having shortened atom-atom contacts between the neighbor bases in the chain, and in many cases leads to the increase in base superposition. In this work, we deal with two examples of cdDMPs that have one chain being in BI conformation and the complementary chain in 117 or 110 conformation. The conformations of 117 family correspond to the local energy minimum of the elementary SPB, while the SPB of 110 conformation family does not correspond to the energy minimum of the free SPB unit. The initial conformation of both cdDMPs with Pur-Pyr sequence has greater superposition of base rings in the chain of BI conformation as compared to the other chain (Tables 2, 3 and Figs. 7, 8). DFT optimization using M05-2X functional leads to the increase of base ring superposition in the dDMP, which is in BI conformation, for both cdDMPs. The other chain experiences a slight increase in the superposition for dApdT:dApdT (conformation family 110) and a decrease for dApdC:dGpdT (conformation family 117). Both optimized structures

Table 2 Torsion angles and areas of superposition of base rings (S) for the initial and DFT optimized conformations of dApdC:dGpdT with dApdC in BI conformation and dGpdC in conformation 117

cdDMPs	δ1	ε	ζ	α	β	γ	δ2	S(Å^2)
(dApdC	132	195	254	344	166	3	148	2.42
dGpdT)[1]	158	217	251	**247**	**72**	**164**	148	1.72
(dApdC	145	188	276	295	163	57	*85*	3.68
dGpdT)[2]	144	191	269	264	70	168	157	0.94
(dApdC	143	188	279	296	166	56	*89*	3.51
dGpdT)[3]	140	191	273	266	71	170	152	1.10
(dApdC	143	189	278	295	167	57	*87*	3.51
dGpdT)[4]	139	194	271	265	73	169	150	1.00
(dApdC	139	176	274	298	163	57	123	3.14
dGpdT)[5]	128	186	265	272	61	171	155	2.43

1) Initial structure extracted from UDJ060 duplex
2) GAUSSIAN-optimized structure using the M05-2X functional
3) ADF-optimized structure using dispersion correction (PW91-ulg functional)
4) ADF-optimized structure (PW91 functional)
5) AMBER-optimized structure (OL15 force field)

Table 3 Torsion angles and areas of superposition of base rings (S) for the initial and DFT optimized conformations of dApdT:dTpdT with one chain being in BI conformation (upper line in the dual row) and other chain being in 110 conformation (lower line in the dual row)

cdDMPs	δ1	ε	ζ	α	β	γ	δ2	S(Å^2)
(dApdT	146	191	278	287	171	44	143	3.63
dApdT)[1]	143	279	180	**66**	**234**	**192**	104	0.70
(dApdT	144	190	276	295	164	56	97	4.02
dApdT)[2]	153	265	180	61	236	196	73	0.92
(dApdT	142	191	279	296	166	54	101	3.98
dApdT)[3]	151	256	187	65	234	199	75	1.06
(dApdT	141	198	281	291	173	53	114	3.47
dApdT)[4]	146	272	181	76	236	193	90	1.03

1) Initial structure extracted from 2MAV duplex
2) GAUSSIAN-optimized structure using the M05-2X functional
3) ADF-optimized structure using dispersion correction (PW91-ulg functional)
4) ADF-optimized structure (PW91 functional)

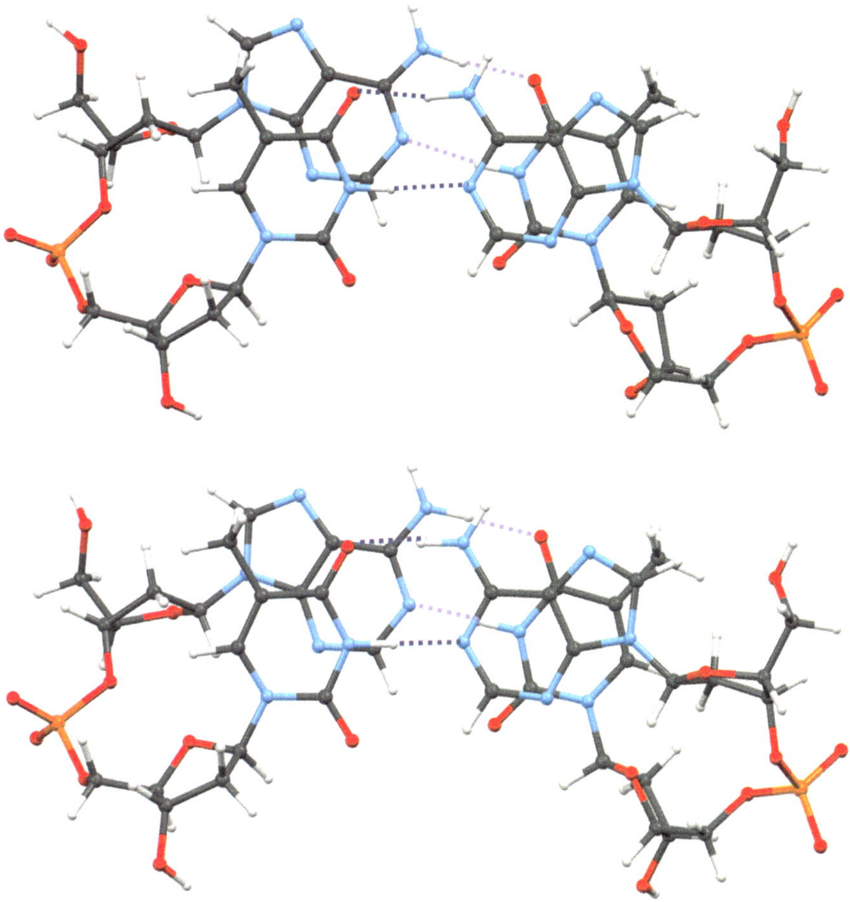

Fig. 7 Initial structure extracted from a DNA fragment with NDB id 2MAV (upper row) and DFT/PW91-ulg optimized structure (lower row) of dApdT:dApdT. Base pair closer to the observer has H-bonds marked with black dotted line; left dDMP is in 110 conformation family, right dDMP is in BI conformation family

have shortened atom-atom contacts that diminishes the quality of these models. Similar results have been obtained for a few other cdDMPs containing one or two dDMPs belonging to non-canonical B-like conformation families.

Experimentation with different DFT functionals on computation of these non-canonical cdDMPs led us to dispersion corrected functional PW91-ulg, which does not produce short atom-atom contacts between the stacked bases. Application of this functional to optimization of several cdDMPs consisting of two dDMPs belonging to different B-like conformation families produces the structures that have conformational features resembling those obtained with the use of M05-2X functional but without shortened contacts. Tables 2 and 3 demonstrate the similarity in torsion

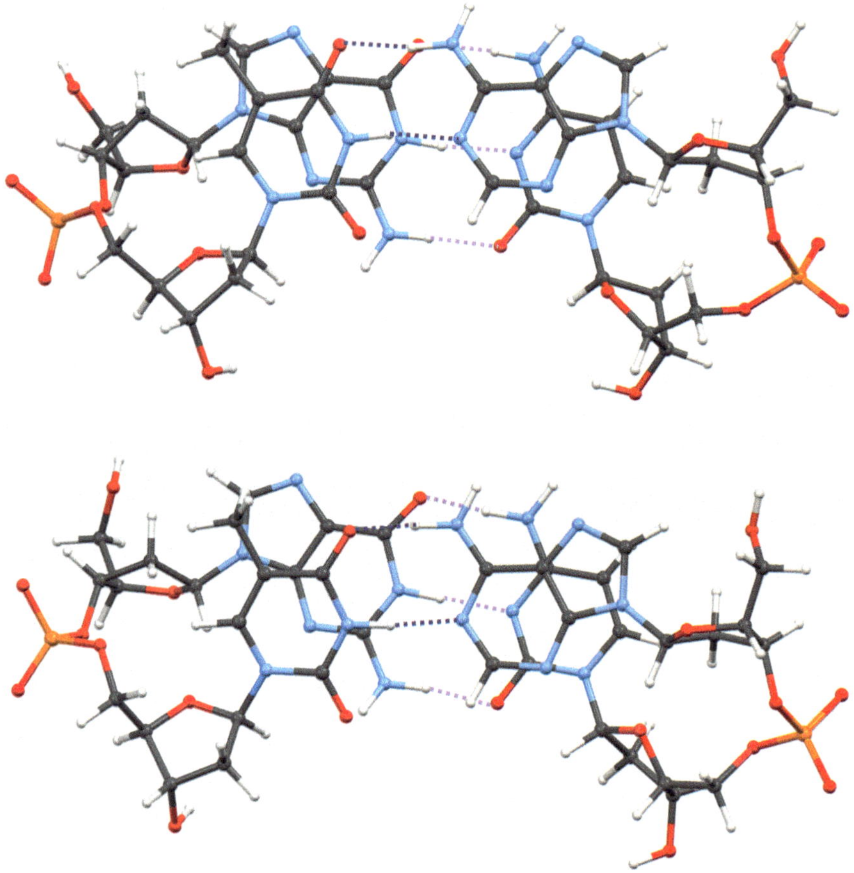

Fig. 8 Initial structure extracted from a DNA fragment with NDB id UDJ060 (upper row) and DFT/PW91-ulg optimized structure (lower row) of dApdC:dGpdT. A:T pair with H-bonds marked by black dotted line resides closer to observer; dGpdC (left chain) is in 117 conformation family; and dApdC (right chain) is in BI family

angles and the resemblance in base ring superposition for cdDMP structures optimized by using these two functionals. Based on the conducted computational study of various simple DNA fragments including various configurations of pairwise base complexes this functional could be recommended for the study of regularities in 3D structure formation of various DNA conformations. Unfortunately, we can use this functional only with the ADF program since Gaussian software does not support it.

Final observation relates to the unexpected difference in GAUSSIAN and ADF packages that has been revealed in DFT computation of cdDMPs. For some of the latter systems (shown in Tables 2 and 3) we have been able to obtain optimized structures by using the PW91 functional as implemented in the ADF program. The conformational characteristics of these structures closely resemble those obtained with the use of dispersion corrected PW91-ulg functional except the distance between stacked bases where PW91 predicts considerably larger values than those predicted by the PW91-ulg functional and observed in the experiment. Examples of the latter are cdDMP structures of experimental fragments of DNA duplexes deposited in NDB (in cdDMPs with two chains being in BI conformation). We failed to reproduce these results with GAUSSIAN software despite trying various basis sets.

6 Conclusions

Due to DNA being the principal molecule of life, deciphering the molecular mechanisms of its 3D structure organization represents one of the important tasks of natural science. In spite of extensive study of the structure of DNA including that of its fragments of various complexity quantitative evaluation of the contribution of the subunits into the macromolecule structure remains an open challenge. The presented work summarizes the results of previous investigations of simple fragments of DNA duplex and its separate chains conducted by using quantum mechanics and molecular mechanics methods. The exact and reliable approaches of quantum mechanics remain too computationally demanding to be applicable to the simplest DNA fragments. The less reliable but more affordable MP2 level of theory may produce physically inadequate structures that have shortened atom-atom contacts. Application of the Density Functional Theory methods using simple functionals has revealed interesting regularities in 3D structure formation of Watson-Crick duplexes, but failed to predict regularities for duplexes that have two complementary chains belonging to different conformation families. Existing classical molecular mechanics force fields cannot reproduce many experimental data on simple DNA fragments and require adjustment of the parameters. The study of variability of three-dimensional structure of DNA simple fragments provides the pathway to improvement of approximate computational methods for elucidation of nucleic acid biomolecular organization and biological functions.

Acknowledgements The authors thankfully acknowledge the computer resources and technical expertise and support provided by the Laboratorio Nacional de Supercómputo del Sureste de México (grant# 201901015C) and Autonomous University of Puebla (grant# 100261355-V1EP2019).

References

1. Watson JD, Crick FHC (1953) Nat 171:964–967
2. Poltev VI, Anisimov VM, Danilov VI, Deriabina A, Gonzalez E, Jurkiewicz A, Leś A, Polteva N (2008) J Biomol Struct Dyn 25:563–571
3. Poltev VI, Anisimov VM, Danilov VI, Deriabina A, Gonzalez E, Garcia D, Rivas F, Jurkiewicz A, Leś A, Polteva NA (2009) J Mol Struct: THEOCHEM 912:53–59

4. Poltev VI, Anisimov VM, Danilov VI, van Mourik T, Deriabina A, Gonzalez E, Padua M, Garcia D, Rivas F, Polteva NA (2010) Int J Quant Chem 110:2548–2559
5. Poltev VI, Anisimov VM, Danilov VI, Garcia D, Deriabina A, Gonzalez E, Salazar R, Rivas F, Polteva N (2011) Comput Theor Chem 975:69–75
6. Poltev VI, Anisimov VM, Danilov VI, Garcia D, Sanchez C, Deriabina A, Gonzalez E, Rivas F, Polteva NA (2014) Biopolymers 101:640–650
7. Poltev VI, Anisimov VM, Sanchez C, Deriabina A, Gonzalez E, Garcia D, Rivas F, Polteva NA (2016) Biophysics 61:259–269
8. Poltev V, Anisimov VM, Dominguez V, Gonzalez E, Deriabina A, Garcia D, Rivas F, Polteva NA (2018) J Mol Model 24:46
9. Berman HM, Olson WK, Beveridge DL, Westbrook J, Gelbin A, Demeny T, Hsieh SH, Srinivasan AR, Schneider B (1992) Biophys J 63:751–759
10. Narayanan BC, Westbrook J, Ghosh S, Petrov AI, Sweeney B, Zirbel CL, Leontis NB, Berman HM (2014) Nucleic Acids Res 42(Database issue):D114–D122
11. Svozil D, Kalina J, Omelka M, Schneider B (2008) Nucleic Acids Res 36:3690–3706
12. Cech P, Kukal J, Cerný J, Schneider D, Svozil D (2013) BMC Bioinformatics 14:205
13. Poltev V (2017) In: Leszczynski J, Kaczmarek-Kedziera A, Puzyn T, Papadopoulos MG, Reis H. Shukla MK (eds) Handbook of computational chemistry. Springer, International Publishing Switzerland, pp. 22–67
14. Poltev VI, Deryabina AS, Gonzalez E, Grokhlina TI (2002) Biophysics 47:920
15. Cornell WD, Cieplak P, Bayly CI, Gould IR, Merz KM, Ferguson DM, Spellmeyer DC, Fox T, Caldwell JW, Kollman PA (1995) J Am Chem Soc 117:5179–519710
16. Case DA, M. Betz RM, Cerutti DS, Cheatham TE, Darden TA, Duke RE, Giese TJ, Gohlke H, Götz AW, Homeyer N, Izadi S, Janowski P, Kaus J, Kovalenko A, Lee T-S, LeGrand S, Li P, Lin C, Luchko T, Luo R, Madej B, Mermelstein D, Merz KM, Monard G, Nguyen H, Nguyen HT, Omelyan I, Onufriev A, Roe DR, Roitberg A, Sagui C, Simmerling CL, Botello-Smith WM, Swails J, Walker RC, Wang J, Wolf RM, Wu X, Xiao L, Kollman PA (2016) Amber 16, University of California, San Francisco
17. Wang J, Cieplak P, Kollman P (2000) J Comput Chem 21:1049–1074
18. Zgarbová M, Šponer J, Otyepka M, Cheatham TE III, Galindo-Murillo R, Jurečka P (2015) J Chem Theory Comput 11:5723–5736
19. Ivani I, Dans P, Noy A, Pérez A, Faustino I, Hopsital A, Walther J, Andrió P, Goni R, Balaceanu A, Portella G, Battistini F, GelpÃ J, González C, Vendruscolo M, Laughton C, Harris S, Case D, Orozco M (2016) Nat Methods 13:55–58
20. Vanommeslaeghe K, MacKerell AD (2015) Biochim Biophys Acta. Gen Subj 1850:861–871
21. Poltev V, Deriabina A, Dominguez V, Sanchez C, Gonzalez E, Polteva N (2019) Biophysics 64:170–181
22. Frisch MJ, Trucks GW, Schlegel HB, Scuseria GE, Robb MA, Cheeseman JR, Scalmani G, Barone V, Mennucci B, Petersson GA, Nakatsuji H, Caricato M, Li X, Hratchian HP, Izmaylov AF, Bloino J, Zheng G, Sonnenberg JL, Hada M, Ehara M, Toyota K, Fukuda R, Hasegawa J, Ishida M, Nakajima T, Honda Y, Kitao O, Nakai H, Vreven T, Montgomery JA, Peralta JE, Ogliaro F, Bearpark M, Heyd JJ, Brothers E, Kudin KN, Staroverov VN, Keith T, Kobayashi R, Normand J, Raghavachari K, Rendell A, Burant JC, Iyengar SS, Tomasi J, Cossi M, Rega N, Millam JM, Klene M, Knox JE, Cross JB, Bakken V, Adamo C, Jaramillo J, Gomperts R, Stratmann RE, Yazyev O, Austin AJ, Cammi R, Pomelli C, Ochterski JW, Martin RL, Morokuma K, Zakrzewski VG, Voth GA, Salvador P, Dannenberg JJ, Dapprich S, Daniels AD, Farkas O, Foresman JB, Ortiz JV, Cioslowski J, Fox DJ (2013) Gaussian 09, Revision D.01, Gaussian Inc, Wallingford CT
23. te Velde G, Bickelhaupt FM, Baerends EJ, Fonseca Guerra C, van Gisbergen SJA, Snijders JG, Ziegler T (2001) J Comput Chem 22:931

24. Baerends EJ, Ziegler T, Atkins AJ, Autschbach J, Baseggio O, Bashford D, Bérces A, Bickelhaupt FM, Bo C, Boerrigter PM, Cavallo L, Daul C, Chong DP, Chulhai DV, Deng L, Dickson RM, Dieterich JM, Ellis DE, van Faassen M, Fan L, Fischer TH, Fonseca Guerra C, Franchini M, Ghysels A, Giammona A, van Gisbergen SJA, Goez A, Götz AW, Groeneveld JA, Gritsenko OV, Grüning M, Gusarov S, Harris FE, van den Hoek P, Hu Z, Jacob CR, Jacobsen H, Jensen L, Joubert L, Kaminski JW, van Kessel G, König C, Kootstra F, Kovalenko A, Krykunov MV, van Lenthe E, McCormack DA, Michalak A, Mitoraj M, Morton SM, Neugebauer J, Nicu VP, Noodleman L, Osinga VP, Patchkovskii S, Pavanello M, Peeples CA, Philipsen PHT, Post D, Pye CC, Ramanantoanina H, Ramos P, Ravenek W, Rodríguez JI, Ros P, Rüger R, Schipper PRT, Schlüns D, van Schoot H, Schreckenbach G, Seldenthuis JS, Seth M, Snijders JG, Solà M, Stener M, Swart M, Swerhone D, Tognetti V, te Velde G, Vernooijs P, Versluis L, Visscher L, Visser O, Wang F, Wesolowski TA, van Wezenbeek EM, Wiesenekker G, Wolff SK, Woo TK, Yakovlev AL (2018) ADF2018, SCM. Theoretical Chemistry, Vrije Universiteit, Amsterdam, The Netherlands
25. Perdew JP, Chevary JA, Vosko SH, Jackson KA, Pederson MR, Singh DJ, Fiolhais C (1992) Phys Rev B 46:6671–6687
26. Perdew JP, Burke K, Ernzerhof M (1996) Phys Rev Lett 77:3865–3868
27. Zhao Y, Truhlar DG (2007) J Chem Theory Comput 3:289–300
28. Zhao Y, Truhlar DG (2008) Theor Chem Account 120:215
29. Grimme S, Ehrlich S, Goerigk L (2011) J Comput Chem 32:1456–1465
30. Grimme S, Antony J, Ehrlich S, Krieg H (2010) J Chem Phys 132:154104
31. Grimme S (2006) J Comput Chem 27:1787–1799
32. Kim H, Choi J-M, Goddard WA (2012) J Phys Chem Lett 3:360–363
33. Zheng G, Lu X-J, Olson WK (2009) Nucleic Acids Res 37:W240–W246
34. Boys SF, Bernardi F (1970) Mol Phys 19:553–566

Fundamental Theory

Efficient "Middle" Thermostat Scheme for the Quantum/Classical Canonical Ensemble via Molecular Dynamics

Xinzijian Liu, Kangyu Yan and Jian Liu

Abstract We have recently developed a unified "middle" thermostat scheme for rationally designing molecular dynamics (MD)/path integral molecular dynamics (PIMD) algorithms for efficiently sampling the configuration space for the classical/quantum canonical ensemble with or without constraints. It significantly improves the time interval by a factor of 4–10 to achieve the same accuracy for structural and configuration-dependent thermodynamic properties for MD (as well as for any thermodynamic properties for PIMD). It has been implemented in AMBER (2018/2019 version), which is available for MD/PIMD simulations with force fields, QM/MM, or ab initio methods.

Keywords Canonical ensemble · Molecular dynamics · Path integral molecular dynamics · Thermostat algorithms · "Middle" thermostat scheme · Quantum statistics · Classical statistics · Holonomic constraints · Multi-time-step techniques · Isokinetic constraints · Sampling efficiency

1 Introduction

Molecular dynamics (MD) has been widely used in chemistry, biology, materials, environmental science, and other scientific fields [1, 2]. When nuclear quantum effects are important, MD can be implemented to perform imaginary time path integral (PIMD) by virtue of the ring-polymer isomorphism [3, 4], which in principle offers a numerically exact tool for practically studying quantum statistical properties in molecular systems [4–8] where quantum exchange effects are negligible. Many cases of interest involve the canonical ensemble, of which the number of particles (N), volume (V) and temperature (T) are constant.

Xinzijian Liu and Kangyu Yan: Both authors contributed equally to the work.

X. Liu · K. Yan · J. Liu (✉)
Beijing National Laboratory for Molecular Sciences, Institute of Theoretical and Computational Chemistry, College of Chemistry and Molecular Engineering, Peking University, Beijing 100871, China
e-mail: jianliupku@pku.edu.cn

L. Mammino et al. (eds.), *Advances in Quantum Systems in Chemistry, Physics, and Biology*, Progress in Theoretical Chemistry and Physics 32,
https://doi.org/10.1007/978-3-030-34941-7_13

In this review we focus on recent progress on the efficient "middle" thermostat [9–12] that has been established in a series of papers [7, 9, 11–16] and has been implemented in AMBER [17] (the 2018 and 2019 versions). In 2016 we first suggested the Langevin thermostat in the "middle" scheme to construct efficient sampling algorithms for PIMD [7, 13, 14]. In 2017 and 2018 we developed a unified framework [9, 11] based on either velocity-Verlet or leap-frog algorithms to apply various thermostat algorithms for MD [1, 15, 18–32] and those for PIMD [7, 13, 14, 33–39] proposed for the canonical (NVT) ensemble. In our unified theoretical framework, we can describe most conventional algorithms [1, 15, 18–27, 33–36, 38, 39] in the "side" or "end" scheme. The unified "middle" scheme we proposed in Refs. [9, 11] provides an efficient tool for configurational sampling with either stochastic or deterministic thermostats. We have further extended the "middle" thermostat scheme as an efficient configurational sampling tool for systems with holonomic or isokinetic constraints [12]. Even when multiple-time-step (MTS) techniques are employed, the "middle" scheme leads to new algorithms that outperform original ones [12]. We have recently extended PIMD to offer an exact tool for quantum statistical properties in coupled multi-electronic-state systems [40, 41] where the Born-Oppenheimer approximation breaks down, in which the "middle" thermostat scheme still offers an efficient sampling tool.

2 "Middle" scheme

The "middle" scheme for MD or PIMD allows for larger time intervals to obtain the same convergence value, which significantly reduces the computational cost. The "middle" scheme is a powerful tool for configurational sampling, thus is helpful for evaluating thermodynamic properties that depend on coordinates in classical MD simulation For PIMD, all the structural and thermodynamic properties in quantum mechanics are functions of only the configurations of the path integral beads, the "middle" scheme is particularly useful because the time interval can be increased by 4 to 10 times without loss of accuracy.

We briefly review three typical thermostats, including stochastic thermostats (the Andersen thermostat, Langevin dynamics) as well as deterministic ones (the Nosé-Hoover thermostat and Nosé-Hoover chain). The integration step with thermostats can be described as three parts, the step for updating coordinates, that for updating momenta and that for thermostatting, denoted "x", "p" and "T", respectively. Then the equations of motion are

$$\begin{pmatrix} d\mathbf{x} \\ d\mathbf{p} \end{pmatrix} = \underbrace{\begin{pmatrix} \mathbf{M}^{-1}\mathbf{p} \\ 0 \end{pmatrix} dt}_{\mathrm{x}} + \underbrace{\begin{pmatrix} 0 \\ -\frac{\partial U(\mathbf{x})}{\partial \mathbf{x}} \end{pmatrix} dt}_{\mathrm{p}} + \underbrace{\mathrm{Thermostat}}_{T} \tag{1}$$

Here, **x** and **p** are coordinates and momentum in vector forms, $U(\mathbf{x})$ is the physical potential energy, **M** the diagonal mass matrix.

The evolution of density distribution for the phase space can be described by the forward Kolmogorov equation. Define the translational propagator of coordinates as $\mathcal{L}_{\mathbf{x}}$ and that of momenta as $\mathcal{L}_{\mathbf{p}}$.

$$\mathcal{L}_{\mathbf{x}}\rho = -\mathbf{p}^T\mathbf{M}^{-1}\frac{\partial \rho}{\partial \mathbf{x}} \tag{2}$$

$$\mathcal{L}_{\mathbf{p}}\rho = \left(\frac{\partial U}{\partial \mathbf{x}}\right)^T\frac{\partial \rho}{\partial \mathbf{p}} \tag{3}$$

The definition of $\mathcal{L}_T$ depends on the thermostat employed. For a time step Δt, the phase space propagators are $e^{\mathcal{L}_{\mathbf{x}}\Delta t}$, $e^{\mathcal{L}_{\mathbf{p}}\Delta t}$ and $e^{\mathcal{L}_T\Delta t}$. The phase space propagation of velocity-Verlet "middle" scheme we propose is written as

$$e^{\mathcal{L}\Delta t} \approx e^{\mathcal{L}_{VV}^{\text{Middle}}\Delta t} = e^{\mathcal{L}_{\mathbf{p}}\Delta t/2}e^{\mathcal{L}_{\mathbf{x}}\Delta t/2}e^{\mathcal{L}_T\Delta t}e^{\mathcal{L}_{\mathbf{x}}\Delta t/2}e^{\mathcal{L}_{\mathbf{p}}\Delta t/2} \tag{4}$$

We denote it as "p-x-T-x-p", where the operations are performed from right to left. Similarly, the phase space propagation of the leap frog "middle" scheme reads

$$e^{\mathcal{L}\Delta t} \approx e^{\mathcal{L}_{\text{LF}}^{\text{Middle}}\Delta t} = e^{\mathcal{L}_{\mathbf{x}}\Delta t/2}e^{\mathcal{L}_T\Delta t}e^{\mathcal{L}_{\mathbf{x}}\Delta t/2}e^{\mathcal{L}_{\mathbf{p}}\Delta t}, \tag{5}$$

which is donated as "p-x-T-x-p", with the operations in a right-to-left sequence.

In the "middle" scheme, the thermostat process for a full time step is performed immediately after the coordinate translation for half a time step, then the coordinate translation for another half time step is followed. In contrast, traditional thermostat algorithms can be included in the "side" scheme

$$e^{\mathcal{L}\Delta t} \approx e^{\mathcal{L}_{VV}^{\text{Side}}\Delta t} = e^{\mathcal{L}_T\Delta t/2}e^{\mathcal{L}_{\mathbf{p}}\Delta t/2}e^{\mathcal{L}_{\mathbf{x}}\Delta t}e^{\mathcal{L}_{\mathbf{p}}\Delta t/2}e^{\mathcal{L}_T\Delta t/2} \tag{6}$$

$$e^{\mathcal{L}\Delta t} \approx e^{\mathcal{L}_{\text{LF}}^{\text{Side}}\Delta t} = e^{\mathcal{L}_{\mathbf{x}}\Delta t}e^{\mathcal{L}_T\Delta t/2}e^{\mathcal{L}_{\mathbf{p}}\Delta t}e^{\mathcal{L}_T\Delta t/2} \tag{7}$$

or the "end" scheme

$$e^{\mathcal{L}\Delta t} \approx e^{\mathcal{L}_{VV}^{\text{End}}\Delta t} = e^{\mathcal{L}_T\Delta t}e^{\mathcal{L}_{\mathbf{p}}\Delta t/2}e^{\mathcal{L}_{\mathbf{x}}\Delta t}e^{\mathcal{L}_{\mathbf{p}}\Delta t/2}. \tag{8}$$

2.1 *Typical thermostats*

Various thermostat methods have been developed for NVT simulations, e.g. Andersen thermostat, Nosé-Hoover chain (NHC), Langevin dynamics, etc. The "middle" scheme naturally includes the two efficient Langevin thermostat algorithms developed for MD in Refs. [29, 30, 32] and in Ref. [28] as shown in Ref. [15]. Below we briefly review the "middle" scheme with the Andersen thermostat and Nosé-Hoover chain.

2.1.1 "Middle" scheme with the Andersen thermostat

The Andersen thermostat was proposed by Andersen in 1980 for the canonical ensemble [18]. The temperature of the system is controlled by stochastic collisions with a heat bath. The time between the collisions is exponentially distributed. When a particle j undergoes a collision, its momentum is reassigned from a Maxwell momentum distribution of the target temperature T, while momenta of other particles remain unchanged. The evolution of the Andersen thermostat step is

$$\mathbf{p}^{(j)} \leftarrow \sqrt{\frac{1}{\beta}}\mathbf{M}_j^{1/2}\boldsymbol{\theta}_j, \quad \text{if } \mu_j < \nu\Delta t \left(\text{or more precisely } \mu_j < 1 - e^{-\nu\Delta t}\right) \quad \left(j = \overline{1, N}\right) \tag{9}$$

Here, ν is the collision frequency, $\boldsymbol{\theta}_j$ is a 3-dimensional Gaussian distributed random vector with zero mean and unit variance, which is independent for each particle as well as each invocation.

The phase space density propagator for the Andersen thermostat in a time interval Δt is

$$e^{\mathcal{L}_T \Delta t}\rho = \left(1 - e^{-\nu\Delta t}\right)\rho_{\text{MB}}(\mathbf{p}) \int_{-\infty}^{\infty} \rho(\mathbf{x}, \mathbf{p}) d\mathbf{p} + e^{-\nu\Delta t}\rho(\mathbf{x}, \mathbf{p}). \tag{10}$$

Consider a harmonic system $V(\mathbf{x}) = \frac{1}{2}\left(\mathbf{x} - \mathbf{x}_{eq}\right)^T \mathbf{A}\left(\mathbf{x} - \mathbf{x}_{eq}\right)$. It is easy to prove the stationary state distribution for the "middle" scheme with the Andersen thermostat is

$$\rho^{\text{Middle}} = \frac{1}{Z_N} \exp\left[-\beta\left(\frac{1}{2}\mathbf{p}^T (\mathbf{M} - \mathbf{A}\frac{\Delta t^2}{4})^{-1}\mathbf{p} + \frac{1}{2}\left(\mathbf{x} - \mathbf{x}_{eq}\right)^T \mathbf{A}\left(\mathbf{x} - \mathbf{x}_{eq}\right)\right)\right], \tag{11}$$

where Z_N is the normalization constant, while that for the traditional schemes (either "side" or "end") reads

$$\rho^{\text{Side}} = \rho^{\text{End}} = \frac{1}{Z_N} \exp\left[-\beta\left(\frac{1}{2}\mathbf{p}^T \mathbf{M}^{-1}\mathbf{p} + \frac{1}{2}\left(\mathbf{x} - \mathbf{x}_{eq}\right)^T (\mathbf{1} - \mathbf{A}\mathbf{M}^{-1}\frac{\Delta t^2}{4})\mathbf{A}\left(\mathbf{x} - \mathbf{x}_{eq}\right)\right)\right]. \tag{12}$$

That is, in the harmonic limit, either the "side" or "end" schemes with a finite time interval leads to the exact momentum distribution but an approximate configurational distribution, and in contrast, the "middle" scheme produces the exact configurational distribution but an approximate momentum distribution. Since evaluation of most properties in MD and that of all properties in PIMD depend on only the configurational distribution, the "middle" scheme is a better choice for MD/PIMD simulations.

2.1.2 "Middle" scheme with the Nosé-Hoover chain thermostat

Nosé-Hoover chain (NHC) [21] performs deterministic MD by adding extended degrees of freedom to control the temperature of the system. The equations of motion for NHC read

$$\left.\begin{aligned} \dot{x}_i &= \frac{p_i}{m_i} \\ \dot{p}_i &= -\frac{\partial U}{\partial x_i} - \frac{p_{\eta_1^{(i)}}}{Q_1} p_i \\ \dot{\eta}_j^{(i)} &= \frac{p_{\eta_j^{(i)}}}{Q_j} \\ \dot{p}_{\eta_1^{(i)}} &= \frac{p_i^2}{m_i} - k_B T - \frac{p_{\eta_2^{(i)}}}{Q_2} p_{\eta_1^{(i)}} \\ \dot{p}_{\eta_j^{(i)}} &= \frac{p^2_{\eta_{j-1}^{(i)}}}{Q_{j-1}} - k_B T - \frac{p_{\eta_{j+1}^{(i)}}}{Q_{j+1}} p_{\eta_j^{(i)}} \left(j = \overline{2, M_{\text{NHC}} - 1}\right) \\ \dot{p}_{\eta_{M_{\text{NHC}}}^{(i)}} &= \frac{p^2_{\eta_{M_{\text{NHC}}-1}^{(i)}}}{Q_{M_{\text{NHC}}-1}} - k_B T \end{aligned}\right\} \left(i = \overline{1, 3N}\right) \quad (13)$$

where M_{NHC} means the number of pairs for the additional variables $\left\{\eta_j^{(i)}, p_{\eta_j^{(i)}}\right\}$ $\left(j = \overline{1, M_{\text{NHC}}}\right)$ coupled to each degree of freedom $\left(i = \overline{1, 3N}\right)$, the parameters $Q_1, \ldots, Q_{M_{\text{NHC}}}$ are the NHC thermal masses. For achieving the accuracy, one often uses the multiple time-scale techniques including the reference system propagator algorithm (RESPA) [23] and higher-order factorizations such as the Suzuki-Yoshida decomposition framework [42–44]. When the phase space density propagator for the thermostat part $e^{\mathcal{L}_T \Delta t}$ is effectively accurate, the thermostat propagation keeps the Maxwell-Boltzmann distribution for the momentum effectively unchanged, i.e.,

$$e^{\mathcal{L}_T \Delta t} \exp\left\{-\beta\left[\frac{1}{2}\mathbf{p}^T \mathbf{M}^{-1}\mathbf{p} + \sum_{i=1}^{3N}\sum_{j=1}^{M_{\text{NHC}}} \frac{p^2_{\eta_j^{(i)}}}{2Q_j}\right]\right\} = \exp\left\{-\beta\left[\frac{1}{2}\mathbf{p}^T \mathbf{M}^{-1}\mathbf{p} + \sum_{i=1}^{3N}\sum_{j=1}^{M_{\text{NHC}}} \frac{p^2_{\eta_j^{(i)}}}{2Q_j}\right]\right\}. \quad (14)$$

The stationary state distribution of the variables $\left(\mathbf{x}, \mathbf{p}, \mathbf{p}_{\boldsymbol{\eta}}\right)$ for the harmonic system propagated with "middle-NHC" scheme is

$$\rho^{\text{Middle - NHC}} = \frac{1}{Z'_N} \exp\left[-\beta \left(\frac{1}{2} \mathbf{p}^T (\mathbf{M} - \mathbf{A}\frac{\Delta t^2}{4})^{-1} \mathbf{p} + \frac{1}{2} (\mathbf{x} - \mathbf{x}_{eq})^T \mathbf{A} (\mathbf{x} - \mathbf{x}_{eq}) + \sum_{i=1}^{3N} \sum_{j=1}^{M_{\text{NHC}}} \frac{p^2_{\eta_j^{(i)}}}{2Q_j} \right) \right], \quad (15)$$

where Z'_N is the normalization constant. The stationary marginal distribution is obtained by integration over p_η in Eq. (15) for the phase space variables $(\mathbf{x}, \mathbf{p})$, which is the same as Eq. (11). Similarly, the stationary state marginal distribution of the physical phase space variables $(\mathbf{x}, \mathbf{p})$ for the harmonic system obtained by either "side-NHC" or "end-NHC" is the same as Eq. (12).

2.2 Simulation results

2.2.1 Quartic potential

We test the quartic potential $U(x) = x^4/4$ as a typical example in the anharmonic region, where the mass is 1 au and the temperature parameter is $\beta \equiv \frac{1}{k_B T} = 8$ au.

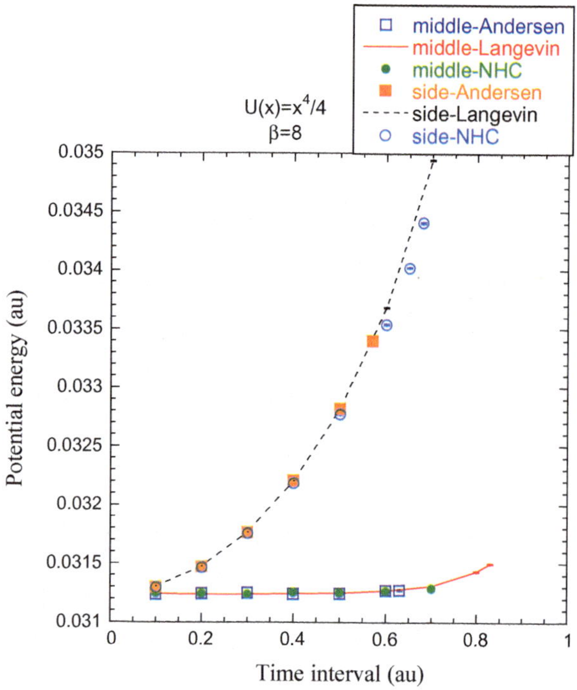

Fig. 1 MD results for the average potential energy (of the quartic potential system) with various time intervals (Reproduced with permission from Ref. [9])

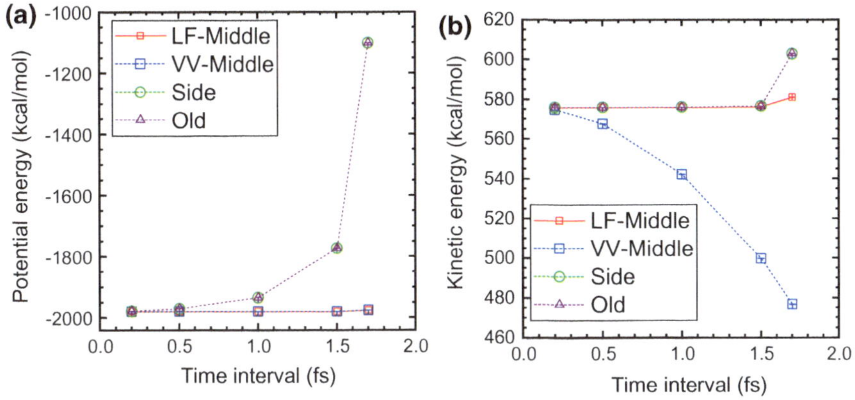

Fig. 2 MD results for liquid water at $T = 298.15$ K using different schemes, **a** average potential energy per atom $\langle U(\mathbf{x})\rangle/(N_{\text{atom}}k_B)$ (unit: Kelvin) **b** average kinetic energy per atom $\langle \mathbf{p}^T\mathbf{M}^{-1}\mathbf{p}\rangle/(2N_{\text{atom}}k_B)$ (unit: Kelvin) (Reproduced with permission from Ref. [11])

Figure 1 shows that the "middle" scheme is more efficient as well as more stable than the conventional "side" scheme, irrespective of the type of thermostat employed. To obtain the same accuracy, the "middle" scheme can employ a (much) larger time interval.

2.2.2 Liquid water

We have implemented the "middle" scheme in the AMBER2018 package [17]. A liquid water system containing a box of 216 water molecules using the TIP3P water model is used as a test case.

Figure 2 demonstrates that all algorithms lead to the same converged results as the time interval decreases to zero. As the time interval increases, both "VV-Middle" and "LF-Middle" schemes perform better than the "side" scheme or the conventional algorithm of AMBER (denoted "Old" in Fig. 2) for estimating the average potential energy. In addition, "LF-Middle" is more accurate than "VV-Middle" for evaluating the kinetic energy when the time interval becomes large.

3 Path integral molecular dynamics

Imaginary time path integral maps a quantum system into a classical ring polymer. Each bead of the ring polymer is a replica of the system, connected with adjacent beads by harmonic springs [3, 45, 46]. By assigning fictitious momenta and masses to the beads, one can employ MD to perform imaginary time path integral for the quantum system [4]. This approach is denoted path integral molecular dynamics

(PIMD), which has become a powerful tool for estimating quantum thermodynamic properties of realistic molecular systems [5, 7, 9, 38–40, 47].

3.1 Thermodynamic properties

One can write any thermodynamic property of the canonical ensemble in quantum mechanics in the general form

$$\left\langle \hat{B} \right\rangle = \frac{1}{Z} \mathrm{Tr}\left(e^{-\beta \hat{H}} \hat{B}\right), \tag{16}$$

where $Z = \mathrm{Tr}\left[e^{-\beta \hat{H}}\right]$ is the partition function and $\hat{B}$ represents a relevant operator of our interest. Eq. (16) can be expressed in the configuration space $\mathbf{x}$, i.e.,

$$\left\langle \hat{B} \right\rangle = \frac{\int d\mathbf{x} \langle \mathbf{x} | e^{-\beta \hat{H}} \hat{B} | \mathbf{x} \rangle}{\int d\mathbf{x} \langle \mathbf{x} | e^{-\beta \hat{H}} | \mathbf{x} \rangle}. \tag{17}$$

Substituting the expression of the identity operator in the configuration space $\hat{\mathbf{1}} = \int d\mathbf{x}_i |\mathbf{x}_i\rangle\langle\mathbf{x}_i|$, one can express the partition function as

$$\begin{aligned} Z &= \int_{\mathbf{x}_1 \equiv \mathbf{x}} d\mathbf{x} \langle \mathbf{x} | e^{-\beta \hat{H}} | \mathbf{x} \rangle \\ &= \lim_{P\to\infty} \int d\mathbf{x}_1 \int d\mathbf{x}_2 \ldots \int d\mathbf{x}_P \left(\frac{P}{2\pi\beta\hbar^2}\right)^{3NP/2} |\mathbf{M}|^{P/2} \\ &\quad \times \exp\left\{ -\frac{P}{2\beta\hbar^2} \sum_{i=1}^{P} \left[(\mathbf{x}_{i+1} - \mathbf{x}_i)^T \mathbf{M} (\mathbf{x}_{i+1} - \mathbf{x}_i) \right] - \frac{\beta}{P} \sum_{i=1}^{P} U(\mathbf{x}_i) \right\} \end{aligned}, \tag{18}$$

where $\mathbf{x}_{P+1} \equiv \mathbf{x}_1$ and P is the total number of path integral beads. Similarly, the numerator of Eq. (17) becomes

$$\begin{aligned} \int d\mathbf{x} \langle \mathbf{x} | e^{-\beta \hat{H}} \hat{B} | \mathbf{x} \rangle &\overset{\mathbf{x}_1 \equiv \mathbf{x}}{=} \lim_{P\to\infty} \int d\mathbf{x}_1 \int d\mathbf{x}_2 \ldots \int d\mathbf{x}_P \left(\frac{P}{2\pi\beta\hbar^2}\right)^{3NP/2} |\mathbf{M}|^{P/2} \\ &\quad \times \exp\left\{ -\frac{P}{2\beta\hbar^2} \sum_{i=1}^{P} \left[(\mathbf{x}_{i+1} - \mathbf{x}_i)^T \mathbf{M} (\mathbf{x}_{i+1} - \mathbf{x}_i) \right] - \frac{\beta}{P} \sum_{i=1}^{P} U(\mathbf{x}_i) \right\} \\ &\quad \times \tilde{B}(\mathbf{x}_1, \ldots, \mathbf{x}_P) \end{aligned} \tag{19}$$

The estimator $\tilde{B}(\mathbf{x}_1, \ldots, \mathbf{x}_P)$ for the operator $\hat{B}(\hat{\mathbf{x}})$ is

$$\tilde{B}(\mathbf{x}_1, \ldots, \mathbf{x}_P) = \frac{1}{P} \sum_{j=1}^{P} B(\mathbf{x}_j). \tag{20}$$

When operator $\hat{B}$ involves the momentum, the estimator $\tilde{B}(\mathbf{x}_1, \ldots, \mathbf{x}_P)$ can also be expressed. For instance, the primitive version of the estimator for the kinetic energy operator $\hat{B} = \frac{1}{2}\hat{\mathbf{p}}^T\mathbf{M}^{-1}\hat{\mathbf{p}}$ is

$$\tilde{B}(\mathbf{x}_1, \ldots, \mathbf{x}_P) = \frac{3NP}{2\beta} - \sum_{j=1}^{P} \frac{P}{2\beta^2\hbar^2}\left[\left(\mathbf{x}_{j+1} - \mathbf{x}_j\right)^T \mathbf{M}\left(\mathbf{x}_{j+1} - \mathbf{x}_j\right)\right], \tag{21}$$

and the virial version is

$$\tilde{B}(\mathbf{x}_1, \ldots, \mathbf{x}_P) = \frac{3N}{2\beta} + \frac{1}{2P}\sum_{j=1}^{P}\left[\left(\mathbf{x}_j - \mathbf{x}^*\right)^T \frac{\partial U\left(\mathbf{x}_j\right)}{\partial \mathbf{x}_j}\right], \tag{22}$$

where $\mathbf{x}^*$ can be selected as the centroid of the path integral beads [48]

$$\mathbf{x}^* = \mathbf{x}_c \equiv \frac{1}{P}\sum_{j=1}^{P} \mathbf{x}_j \tag{23}$$

or any specific bead.

3.2 Staging Path Integral Molecular Dynamics

Consider the staging transformation [6, 38, 47, 49]

$$\begin{aligned} \boldsymbol{\xi}_1 &= \mathbf{x}_1 \\ \boldsymbol{\xi}_j &= \mathbf{x}_j - \frac{(j-1)\mathbf{x}_{j+1} + \mathbf{x}_1}{j} \quad \left(j = \overline{2, P}\right). \end{aligned} \tag{24}$$

Define

$$\omega_P = \frac{1}{\beta\hbar}. \tag{25}$$

Eq. (18) becomes

$$\begin{aligned} Z \overset{\boldsymbol{\xi}_1 \equiv \mathbf{x}_1}{=} & \lim_{P\to\infty}\left(\frac{P}{2\pi\beta\hbar^2}\right)^{3NP/2} |\mathbf{M}|^{P/2} \int d\boldsymbol{\xi}_1 \int d\boldsymbol{\xi}_2 \ldots \int d\boldsymbol{\xi}_P \\ & \times \exp\left\{-\beta\sum_{j=1}^{P}\left[\frac{1}{2}\omega_P^2\boldsymbol{\xi}_j^T\overline{\mathbf{M}}_j\boldsymbol{\xi}_j + \frac{1}{P}U\left(\mathbf{x}_j\left(\boldsymbol{\xi}_1, \ldots, \boldsymbol{\xi}_P\right)\right)\right]\right\}, \end{aligned} \tag{26}$$

The diagonal mass matrices are given by

$$\begin{aligned}\overline{\mathbf{M}}_1 &= 0\\ \overline{\mathbf{M}}_j &= \frac{j}{j-1}P\mathbf{M}\quad \left(j=\overline{2,P}\right).\end{aligned} \tag{27}$$

Define

$$\phi(\boldsymbol{\xi}_1,\ldots,\boldsymbol{\xi}_P) = \frac{1}{P}\sum_{j=1}^{P} U\left(\mathbf{x}_j(\boldsymbol{\xi}_1,\ldots,\boldsymbol{\xi}_P)\right). \tag{28}$$

Its derivatives satisfy the chain rule

$$\begin{array}{l}\frac{\partial\phi}{\partial\boldsymbol{\xi}_1} = \sum\limits_{i=1}^{P}\frac{\partial\phi}{\partial\mathbf{x}_i} = \frac{1}{P}\sum\limits_{i=1}^{P}U'(\mathbf{x}_i)\\ \frac{\partial\phi}{\partial\boldsymbol{\xi}_j} = \frac{\partial\phi}{\partial\mathbf{x}_j} + \frac{j-2}{j-1}\frac{\partial\phi}{\partial\boldsymbol{\xi}_{j-1}}\qquad \left(j=\overline{2,P}\right)\end{array}. \tag{29}$$

Adding fictitious momenta $(\mathbf{p}_1,\ldots,\mathbf{p}_P)$ into Eq. (26) leads to

$$\begin{aligned}Z &\overset{\boldsymbol{\xi}_1\equiv\mathbf{x}_1}{=} \lim_{P\to\infty}\left(\frac{P}{4\pi^2\hbar^2}\right)^{3NP/2}|\mathbf{M}|^{P/2}\left(\prod_{j=1}^{P}\left|\tilde{\mathbf{M}}_j\right|\right)^{-1/2}\int\left(\prod_{j=1}^{P}d\boldsymbol{\xi}_j d\mathbf{p}_j\right)\\ &\times\exp\left[-\beta H_{\mathrm{eff}}(\boldsymbol{\xi}_1,\ldots,\boldsymbol{\xi}_P;\mathbf{p}_1,\ldots,\mathbf{p}_P)\right]\end{aligned} \tag{30}$$

with the effective Hamiltonian of the form

$$H_{\mathrm{eff}}(\boldsymbol{\xi}_1,\ldots,\boldsymbol{\xi}_P;\mathbf{p}_1,\ldots,\mathbf{p}_P) = \sum_{j=1}^{P}\frac{1}{2}\mathbf{p}_j^T\tilde{\mathbf{M}}_j^{-1}\mathbf{p}_j + U_{\mathrm{eff}}(\boldsymbol{\xi}_1,\ldots,\boldsymbol{\xi}_P), \tag{31}$$

where

$$U_{\mathrm{eff}}(\boldsymbol{\xi}_1,\ldots,\boldsymbol{\xi}_P) = \sum_{j=1}^{P}\frac{1}{2}\omega_{\mathrm{P}}^2\boldsymbol{\xi}_j^T\overline{\mathbf{M}}_j\boldsymbol{\xi}_j + \phi(\boldsymbol{\xi}_1,\ldots,\boldsymbol{\xi}_P). \tag{32}$$

The fictitious masses are defined as

$$\begin{aligned}\tilde{\mathbf{M}}_1 &= \mathbf{M}\\ \tilde{\mathbf{M}}_j &= \overline{\mathbf{M}}_j = \frac{j}{j-1}P\mathbf{M}\qquad \left(j=\overline{2,P}\right)\end{aligned} \tag{33}$$

such that all staging modes $(\boldsymbol{\xi}_2,\ldots,\boldsymbol{\xi}_P)$ move on the same frequency. The estimator of the thermodynamic property in Eq. (17) is

$$\left\langle \hat{B} \right\rangle = \lim_{P \to \infty} \frac{\int \left(\prod_{j=1}^{P} d\boldsymbol{\xi}_j d\mathbf{p}_j \right) \exp\left\{ -\beta H_{\text{eff}}(\boldsymbol{\xi}_1, \ldots, \boldsymbol{\xi}_P; \mathbf{p}_1, \ldots, \mathbf{p}_P) \right\} \tilde{B}(\mathbf{x}_1, \ldots, \mathbf{x}_P)}{\int \left(\prod_{j=1}^{P} d\boldsymbol{\xi}_j d\mathbf{p}_j \right) \exp\left\{ -\beta H_{\text{eff}}(\boldsymbol{\xi}_1, \ldots, \boldsymbol{\xi}_P; \mathbf{p}_1, \ldots, \mathbf{p}_P) \right\}}. \tag{34}$$

Equations (31) and (34) suggest the Hamilton equations of motion

$$\begin{aligned} \dot{\boldsymbol{\xi}}_j &= \tilde{\mathbf{M}}_j^{-1} \mathbf{p}_j \\ \dot{\mathbf{p}}_j &= -\omega_P^2 \overline{\mathbf{M}}_j \boldsymbol{\xi}_j - \frac{\partial \phi}{\partial \boldsymbol{\xi}_j} \quad \left(j = \overline{1, P} \right) \end{aligned}. \tag{35}$$

One should couple it with a thermostat to ensure the canonical distribution for $(\boldsymbol{\xi}_1, \ldots, \boldsymbol{\xi}_P, \mathbf{p}_1, \ldots, \mathbf{p}_P)$. Note that the estimator of any thermodynamic properties in Eq. (34) depends on only the configurational distribution of the beads in PIMD.

The choice of the fictitious masses in Eqs. (25), (27), and (33) is different from that in Refs. [7, 9]. The procedure in Eqs. (24)–(35) for PIMD makes it possible to use the same time interval to obtain converged results, regardless of the value of P, the total number of path integral beads. This has been suggested earlier in Ref. [6].

The conventional wisdom often employs the decomposition of the equations of motion in PIMD algorithms

$$\begin{pmatrix} \dot{\boldsymbol{\xi}}_j \\ \dot{\mathbf{p}}_j \end{pmatrix} = \underbrace{\begin{pmatrix} \tilde{\mathbf{M}}_j^{-1} \mathbf{p}_j \\ -\omega_P^2 \overline{\mathbf{M}}_j \boldsymbol{\xi}_j \end{pmatrix}}_{\mathrm{x}} + \underbrace{\begin{pmatrix} 0 \\ -\frac{\partial \phi}{\partial \boldsymbol{\xi}_j} \end{pmatrix}}_{\mathrm{p}} + \underbrace{(\text{Thermostat})}_{T} \left(j = \overline{1, P} \right) \tag{36}$$

because the fictitious ring polymer force term $-\omega_P^2 \overline{\mathbf{M}}_j \boldsymbol{\xi}_j$ varies more frequently than the physical force term $-\frac{\partial \phi}{\partial \boldsymbol{\xi}_j}$, and the exact solution to the first term of Eq. (36) is available [39]. E. g., Eq. (36) leads to exact results in the free particle limit. Our recent work [7], however, shows that

$$\begin{pmatrix} \dot{\boldsymbol{\xi}}_j \\ \dot{\mathbf{p}}_j \end{pmatrix} = \underbrace{\begin{pmatrix} \tilde{\mathbf{M}}_j^{-1} \mathbf{p}_j \\ 0 \end{pmatrix}}_{\mathrm{x}} + \underbrace{\begin{pmatrix} 0 \\ -\frac{\partial U_{\text{eff}}}{\partial \boldsymbol{\xi}_j} \end{pmatrix}}_{\mathrm{p}} + \underbrace{(\text{Thermostat})}_{T} \quad \left(j = \overline{1, P} \right) \tag{37}$$

is a more accurate and efficient decomposition scheme for developing PIMD algorithms in the "middle" thermostat scheme.

E.g., when one apply the Langevin dynamics as thermostat, it has been clarified in Appendix C of Ref. [7] (and its Supplementary Material [50]) that Eq. (37) guarantees the exact marginal configuration distribution of the path integral beads in the harmonic limit, irrespective to the time interval, while Eq. (36) does not. The conclusion holds for any thermostat as long as the thermostat faithfully maintains

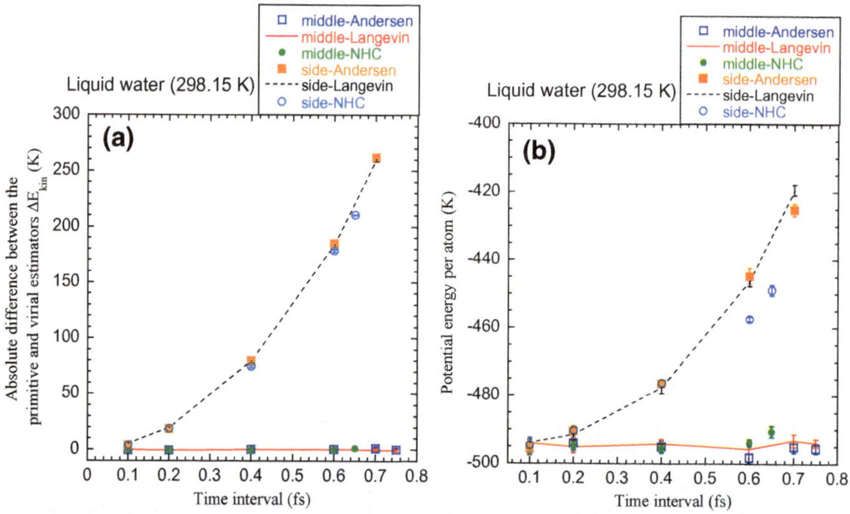

Fig. 3 PIMD results using different time intervals for liquid water at $T = 298.15$ K. **a** Absolute difference between the primitive and virial estimators for averaged kinetic energy per atom (unit: Kelvin). **b** The averaged potential energy per atom $\langle U(\mathbf{x})\rangle/(N_{\text{atom}}k_{\text{B}})$ (unit: Kelvin). Statistical error bars are included. (Reproduced with permission from Ref. [9])

the Maxwell momentum distribution. This is verified by the simulation results for PIMD in Ref. [9] when the Andersen thermostat and NHC are used.

We apply the "middle" and conventional "side" schemes with PIMD algorithms for liquid water system to study the canonical ensemble at constant temperature $T = 298.15$ K and $\rho_l = 0.997\ \text{g}\cdot\text{cm}^{-3}$. As presented in Fig. 3, in comparison to the "side" scheme, the "middle" scheme reduces the error by about an order of magnitude with the same time interval.

3.3 Normal-mode Path Integral Molecular Dynamics

Consider the normal mode transformation of path integral beads [48, 51, 52]

$$\mathbf{x} = \mathbf{C}^{norm}\mathbf{q} \tag{38}$$

with $\mathbf{q} = \left(\mathbf{q}_0 \ldots \mathbf{q}_{P-1}\right)^T$ and $\mathbf{x} = \left(\mathbf{x}_1 \ldots \mathbf{x}_P\right)^T$, where the element of the orthogonal transformation matrix $\mathbf{C}^{norm}$ is

$$C_{jk}^{norm} = \begin{cases} \sqrt{1/P}, & k = 0 \\ \sqrt{2/P}\cos(2\pi jk/P), & 1 \le k \le P/2 - 1 \\ \sqrt{1/P}(-1)^j, & k = P/2 \\ \sqrt{2/P}\sin(2\pi jk/P), & P/2 + 1 \le k \le P - 1 \end{cases} \quad \left(j = \overline{1, P}\right), \tag{39}$$

for even P and

$$C_{jk}^{norm} = \begin{cases} \sqrt{1/P}, & k = 0 \\ \sqrt{2/P}\cos(2\pi jk/P), & 1 \le k \le (P - 1)/2 \\ \sqrt{2/P}\sin(2\pi jk/P), & (P + 1)/2 \le k \le P - 1 \end{cases} \quad \left(j = \overline{1, P}\right), \tag{40}$$

for odd P. When one employs the normal mode transformation, the partition function becomes

$$Z = \lim_{P\to\infty}\left(\frac{P}{2\pi\beta\hbar^2}\right)^{NP/2} |\mathbf{M}|^{P/2} \int d\mathbf{q}_0 \int d\mathbf{q}_1 \ldots \int d\mathbf{q}_{P-1} \times \exp\left\{-\beta\left[\sum_{k=0}^{P-1}\frac{1}{2}\omega_k^2\mathbf{q}_k^T\overline{\mathbf{M}}_k^{norm}\mathbf{q}_k + \frac{1}{P}\sum_{j=1}^{P}V\left(\mathbf{x}_j(\mathbf{q}_0, \ldots, \mathbf{q}_{P-1})\right)\right]\right\}. \tag{41}$$

The mass matrices are defined as $\overline{\mathbf{M}}_0^{norm} = \mathbf{0}$ and $\overline{\mathbf{M}}_k^{norm} = P\mathbf{M}$ $\left(k = \overline{1, P - 1}\right)$, the frequency for each mode is given by

$$\omega_k = 2\omega_P \sin(k\pi/P) \quad \left(k = \overline{0, P - 1}\right), \tag{42}$$

respectively. Eq. (28) then becomes

$$\phi(\mathbf{q}_0, \ldots, \mathbf{q}_{P-1}) = \frac{1}{P}\sum_{j=1}^{P}V\left(\mathbf{x}_j(\mathbf{q}_0, \ldots, \mathbf{q}_{P-1})\right), \tag{43}$$

and the derivatives $\partial\phi/\partial\mathbf{q}_k$ is obtained from

$$\frac{\partial\phi}{\partial\mathbf{q}} = \left(\frac{\partial\mathbf{x}}{\partial\mathbf{q}}\right)^T\frac{\partial\phi}{\partial\mathbf{x}} = \left(\mathbf{C}^{norm}\right)^T\frac{\partial\phi}{\partial\mathbf{x}}. \tag{44}$$

Employing fictitious momenta $(\mathbf{p}_0, \ldots, \mathbf{p}_{P-1})$ into Eq. (41) produces

$$Z = \lim_{P\to\infty}\left(\frac{P}{4\pi^2\hbar^2}\right)^{NP/2} |\mathbf{M}|^{P/2}\left(\prod_{k=0}^{P-1}\left|\tilde{\mathbf{M}}_k^{norm}\right|\right)^{-1/2} \int\left(\prod_{k=0}^{P-1}d\mathbf{q}_k d\mathbf{p}_k\right) \times \exp\left[-\beta H_{\text{eff}}^{norm}(\mathbf{q}_0, \ldots, \mathbf{q}_{P-1}; \mathbf{p}_0, \ldots, \mathbf{p}_{P-1})\right], \tag{45}$$

where the effective Hamiltonian can be expressed by the normal mode variables

$$H_{\text{eff}}^{norm}(\mathbf{q};\mathbf{p}) = \sum_{k=0}^{P-1} \frac{1}{2}\mathbf{p}_k^T \tilde{\mathbf{M}}_{norm,k}^{-1}\mathbf{p}_k + U_{\text{eff}}^{norm}(\mathbf{q}). \quad (46)$$

In Eq. (46) the effective potential is

$$U_{\text{eff}}^{norm}(\mathbf{q}) = \sum_{k=0}^{P-1} \frac{1}{2}\omega_k^2 \mathbf{q}_k^T \overline{\mathbf{M}}_k^{norm} \mathbf{q}_k + \phi(\mathbf{q}), \quad (47)$$

and the fictitious masses $\left\{\tilde{\mathbf{M}}_{norm,k}\right\}$ can be arbitrary, which can be chosen as $\tilde{\mathbf{M}}_{norm,k=0} = \mathbf{M}$ and $\tilde{\mathbf{M}}_{norm,k} = P\mathbf{M}\left(k = \overline{1, P-1}\right)$ such that all non-zeroth normal modes $\left(k = \overline{1, P-1}\right)$ move on the same frequency.

The thermodynamic property Eq. (17) can be evaluated by

$$\left\langle \hat{B} \right\rangle = \lim_{P\to\infty} \frac{\int \left(\prod_{k=0}^{P-1} d\mathbf{q}_k d\mathbf{p}_k\right) \exp\left\{-\beta H_{\text{eff}}^{norm}(\mathbf{q}_0,\ldots,\mathbf{q}_{P-1};\mathbf{p}_0,\ldots,\mathbf{p}_{P-1})\right\} \tilde{B}(\mathbf{x}_1,\ldots,\mathbf{x}_P)}{\int \left(\prod_{k=0}^{P-1} d\mathbf{q}_k d\mathbf{p}_k\right) \exp\left\{-\beta H_{\text{eff}}^{norm}(\mathbf{q}_0,\ldots,\mathbf{q}_{P-1};\mathbf{p}_0,\ldots,\mathbf{p}_{P-1})\right\}}. \quad (48)$$

It then suggests a MD scheme to sample $(\mathbf{q}_0,\ldots,\mathbf{q}_{P-1},\mathbf{p}_0,\ldots,\mathbf{p}_{P-1})$. The equations of motion derived from the effective Hamiltonian (Eq. 46) are

$$\begin{aligned} \dot{\mathbf{q}}_k &= \tilde{\mathbf{M}}_{norm,k}^{-1}\mathbf{p}_k \\ \dot{\mathbf{p}}_k &= -\omega_k^2 \overline{\mathbf{M}}_k^{norm} \mathbf{q}_k - \frac{\partial \phi}{\partial \mathbf{q}_k} \quad \left(k = \overline{0, P-1}\right), \end{aligned} \quad (49)$$

which must be coupled with a thermostat to ensure the correct canonical distribution for $(\mathbf{q}_0,\ldots,\mathbf{q}_{P-1},\mathbf{p}_0,\ldots,\mathbf{p}_{P-1})$. This is denoted normal mode PIMD (NM-PIMD).

Similar to Eq. (37), the equations of motion of NM-PIMD should be decomposed into three parts

$$\begin{pmatrix} \dot{\mathbf{q}}_k \\ \dot{\mathbf{p}}_k \end{pmatrix} = \underbrace{\begin{pmatrix} \tilde{\mathbf{M}}_{norm,k}^{-1}\mathbf{p}_k \\ 0 \end{pmatrix}}_{\text{x}} + \underbrace{\begin{pmatrix} 0 \\ -\frac{\partial U_{\text{eff}}}{\partial \mathbf{q}_k} \end{pmatrix}}_{\text{p}} + \underbrace{(\text{thermostat})}_{\text{T}} \quad \left(k = \overline{0, P-1}\right) \quad (50)$$

in the “middle” thermostat scheme for designing efficient NM-PIMD algorithms.

3.4 Multi-electronic-state PIMD

Consider a Hamiltonian with N_s electronic states of the form $\hat{\mathbf{H}} = \hat{\mathbf{T}} + \hat{\mathbf{V}}$, where $\hat{\mathbf{V}} = \mathbf{V}(\hat{\mathbf{R}})$ is a $N_s \times N_s$ symmetric diabatic potential energy matrix and $\hat{\mathbf{T}}$ is the kinetic energy matrix. The canonical partition function is defined as

$$Z = \mathrm{Tr}_{ne}[e^{-\beta\hat{\mathbf{H}}}] \tag{51}$$

and a specific physical property of interest is evaluated by

$$\left\langle \hat{\mathbf{B}} \right\rangle = \frac{1}{Z}\mathrm{Tr}_{ne}[e^{-\beta\hat{\mathbf{H}}}\hat{\mathbf{B}}] \tag{52}$$

In Eqs. (51) and (52) the trace is integrated over both the nuclear and electronic degrees of freedom, i.e.

$$\mathrm{Tr}_e[\hat{\mathbf{O}}] = \sum_{n=1}^{N_s} \langle n|\hat{\mathbf{O}}|n\rangle \tag{53}$$

and

$$\mathrm{Tr}_n[\hat{\mathbf{O}}] = \int d\mathbf{R}\langle \mathbf{R}|\hat{\mathbf{O}}|\mathbf{R}\rangle. \tag{54}$$

Substituting the resolution of the identity into Eq. (51) yields

$$Z = \lim_{P\to\infty}\left|\frac{P\mathbf{M}}{2\pi\beta\hbar^2}\right|^{P/2} \int d\mathbf{R}_1 \dots d\mathbf{R}_P \exp\left[-\frac{\beta}{2}\omega_P^2 \sum_{i=1}^{P} (\mathbf{R}_i - \mathbf{R}_{i+1})^T \mathbf{M}(\mathbf{R}_i - \mathbf{R}_{i+1})\right]$$
$$\times \mathrm{Tr}_e\left[\prod_{i=1}^{P} \mathbf{O}^T(\mathbf{R}_i)\mathbf{O}(\mathbf{R}_i)\right] \tag{55}$$

where $\mathbf{O}(\mathbf{R}_i)$ is related to the splitting scheme. We have studied three typical decomposition schemes, namely, the "diagonalization", "first-order expansion", and "hyperbolic function" methods in Ref. [40].

Because $\mathrm{Tr}_e\left[\prod_{i=1}^{P} \mathbf{O}^T(\mathbf{R}_i)\mathbf{O}(\mathbf{R}_i)\right]$ is not always positive-definite for general multi-electronic-state (MES) systems, regardless of which decomposition scheme is employed, it is not recommended to use either $\mathrm{Tr}_e\left[\prod_{i=1}^{P} \mathbf{O}^T(\mathbf{R}_i)\mathbf{O}(\mathbf{R}_i)\right]$ or its absolute value to define an effective potential function $\phi(\mathbf{R}_1, \dots, \mathbf{R}_P)$ for performing PIMD. A reasonable effective potential is defined by

$$e^{-\beta\phi^{(\mathrm{dia})}(\mathbf{R}_1,\ldots,\mathbf{R}_P)} = \mathrm{Tr}_e\left[\prod_{i=1}^{P} e^{-\beta\mathbf{V}_{\mathrm{diag}}(\mathbf{R}_i)/P}\right]. \tag{56}$$

Note that the right-hand side of Eq. (56) is always positive-definite. Here $\mathbf{V}_{\mathrm{diag}}(\mathbf{R}_i)$ is a diagonal matrix, whose elements are the diagonal elements of $\mathbf{V}(\mathbf{R})$. The partition function [Eq.(55)] may then be expressed as

$$Z = \lim_{P\to\infty}\left|\frac{P\mathbf{M}}{2\pi\beta\hbar^2}\right|^{P/2}\int d\mathbf{R}_1\ldots d\mathbf{R}_P \exp\left[-\beta U_{\mathrm{eff}}^{(\mathrm{dia})}(\mathbf{R}_1,\ldots,\mathbf{R}_P)\right]\tilde{Z}^{(\mathrm{dia})}(\mathbf{R}_1,\ldots,\mathbf{R}_P), \tag{57}$$

of which the estimator is

$$\tilde{Z}^{(\mathrm{dia})}(\mathbf{R}_1,\ldots,\mathbf{R}_P) = \frac{\mathrm{Tr}_e\left[\prod_{i=1}^{P}\mathbf{O}^T(\mathbf{R}_i)\mathbf{O}(\mathbf{R}_i)\right]}{\mathrm{Tr}_e\left[\prod_{i=1}^{P} e^{-\beta\mathbf{V}_{\mathrm{diag}}(\mathbf{R}_i)/P}\right]} \tag{58}$$

and

$$U_{\mathrm{eff}}^{(\mathrm{dia})}(\mathbf{R}_1,\ldots,\mathbf{R}_P) = \frac{1}{2}\omega_P^2\sum_{i=1}^{P}(\mathbf{R}_i-\mathbf{R}_{i+1})^T P\mathbf{M}(\mathbf{R}_i-\mathbf{R}_{i+1}) + \phi^{(\mathrm{dia})}(\mathbf{R}_1,\ldots,\mathbf{R}_P). \tag{59}$$

Then one can obtain any specific physical property of interest in Eq. (52) from

$$\left\langle\hat{\mathbf{B}}\right\rangle = \frac{\int d\mathbf{R}_1\ldots d\mathbf{R}_P \exp\left[-\beta U_{\mathrm{eff}}^{(\mathrm{dia})}(\mathbf{R}_1,\ldots,\mathbf{R}_P)\right]\tilde{B}^{(\mathrm{dia})}(\mathbf{R}_1,\ldots,\mathbf{R}_P)}{\int d\mathbf{R}_1\ldots d\mathbf{R}_P \exp\left[-\beta U_{\mathrm{eff}}^{(\mathrm{dia})}(\mathbf{R}_1,\ldots,\mathbf{R}_P)\right]\tilde{Z}^{(\mathrm{dia})}(\mathbf{R}_1,\ldots,\mathbf{R}_P)} \tag{60}$$

The estimators $\tilde{B}^{(\mathrm{dia})}(\mathbf{R}_1,\ldots,\mathbf{R}_P)$ in the diabatic representation for some typical operators are described in Ref. [40].

Define an effective Hamiltonian

$$H_{\mathrm{eff}}^{(\mathrm{dia})}(\mathbf{R}_1,\ldots,\mathbf{R}_P;\mathbf{p}_1,\ldots,\mathbf{p}_P) = \sum_{i=1}^{P}\frac{1}{2}\mathbf{p}_i^T\tilde{\mathbf{M}}_i^{-1}\mathbf{p}_i + U_{\mathrm{eff}}^{(\mathrm{dia})}(\mathbf{R}_1,\ldots,\mathbf{R}_P) \tag{61}$$

with the fictitious masses $\tilde{\mathbf{M}}_i$ and momenta $\mathbf{p}_i$. It leads to the MES-PIMD equations of motion

$$\begin{matrix}\dot{\mathbf{R}}_i = \tilde{\mathbf{M}}_i^{-1}\mathbf{p}_i \\ \dot{\mathbf{p}}_i = -\frac{\partial}{\partial\mathbf{R}_i}U_{\mathrm{eff}}^{(\mathrm{dia})}(\mathbf{R}_1,\ldots,\mathbf{R}_P)\end{matrix}\quad\left(i=\overline{1,P}\right). \tag{62}$$

Their coupling to a thermostat produces a proper canonical distribution for $(\mathbf{R}_1, \ldots, \mathbf{R}_P; \mathbf{p}_1, \ldots, \mathbf{p}_P)$, which changes Eq. (60) into

$$\langle \hat{\mathbf{B}} \rangle = \lim_{P\to\infty} \frac{\int \left(\prod_{i=1}^{P} d\mathbf{R}_i d\mathbf{p}_i \right) \exp\left\{ -\beta H_{\text{eff}}^{(\text{dia})}(\mathbf{R}_1, \ldots, \mathbf{R}_P; \mathbf{p}_1, \ldots, \mathbf{p}_P) \right\} \tilde{B}^{(\text{dia})}(\mathbf{R}_1, \ldots, \mathbf{R}_P)}{\int \left(\prod_{i=1}^{P} d\mathbf{R}_i d\mathbf{p}_i \right) \exp\left\{ -\beta H_{\text{eff}}^{(\text{dia})}(\mathbf{R}_1, \ldots, \mathbf{R}_P; \mathbf{p}_1, \ldots, \mathbf{p}_P) \right\} \tilde{Z}^{(\text{dia})}(\mathbf{R}_1, \ldots, \mathbf{R}_P)}. \tag{63}$$

Similar to Eq. (37), the MES-PIMD equations of motion are decomposed into three parts

$$\begin{pmatrix} \dot{\mathbf{R}}_i \\ \dot{\mathbf{p}}_i \end{pmatrix} = \underbrace{\begin{pmatrix} \tilde{\mathbf{M}}_i^{-1}\mathbf{p}_i \\ 0 \end{pmatrix}}_{\mathbf{x}} + \underbrace{\begin{pmatrix} 0 \\ -\frac{\partial U_{\text{eff}}^{(\text{dia})}}{\partial \mathbf{R}_i} \end{pmatrix}}_{\mathbf{p}} + \underbrace{(\text{Thermostat})}_{T} \quad \left(i = \overline{1, P}\right). \tag{64}$$

The "middle" scheme then yields efficient MES-PIMD algorithms. As discussed in Sects. 3.2 and 3.3, the staging or normal-mode transformation of path integral beads can be used in Eq. (64).

We have investigated a seven-state system in Ref. [40]. It is shown in Fig. 4 that the "middle" thermostat scheme greatly improves the efficiency over the conventional "side" scheme.

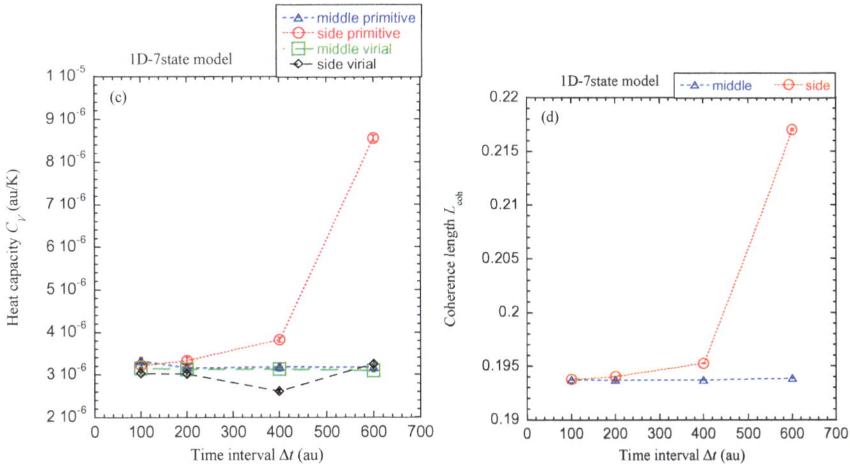

Fig. 4 Results for the "middle" scheme in comparison to those for conventional thermostat schemes for MES-PIMD for a 1-D seven-state system. Results for the heat capacity and coherence length are plotted as functions of the time interval Δt. Atomic units (au) are used (Reproduced with permission from Ref. [40])

4 "Middle" scheme with constraints

4.1 Holonomic constraint

Define the holonomic constraint

$$\boldsymbol{\sigma}(\mathbf{x}) = \mathbf{0}, \tag{65}$$

where $\boldsymbol{\sigma}(\mathbf{x})$ is n_c-dimensional vector function of the configuration $\mathbf{x}$. Its derivative yields the constraint for the momentum $\mathbf{p}$

$$\frac{d}{dt}\boldsymbol{\sigma}(\mathbf{x}) = \left(\frac{\partial \boldsymbol{\sigma}}{\partial \mathbf{x}}\right)^T \mathbf{M}^{-1}\mathbf{p} = \mathbf{0}. \tag{66}$$

SHAKE [53] and RATTLE [54] are two typical algorithms for applying constraints to molecular systems. While SHAKE guarantees only the position constraint, RATTLE satisfies both the position and momentum constraints. It is straightforward to employ the "middle" scheme with the SHAKE or RATTLE algorithm for sampling the canonical (NVT) ensemble. Various versions can be constructed to guarantee that the position and momentum satisfy Eqs. (65)–(66) at the end of a time step. Our recommended "VV-Middle" scheme with holonomic constraints is

$$\tilde{\mathbf{p}}\left(\frac{\Delta t}{2}\right) \leftarrow \mathbf{p}(0) - \frac{\partial U}{\partial \mathbf{x}(0)}\frac{\Delta t}{2}$$

$$\mathrm{C}_2 : \begin{cases} \text{Solve } \boldsymbol{\mu} : \left(\frac{\partial \boldsymbol{\sigma}}{\partial \mathbf{x}(0)}\right)^T \mathbf{M}^{-1}\left(\tilde{\mathbf{p}}\left(\frac{\Delta t}{2}\right) + \frac{\partial \boldsymbol{\sigma}}{\partial \mathbf{x}(0)}\boldsymbol{\mu}\right) = \mathbf{0} \\ \tilde{\tilde{\mathbf{p}}}\left(\frac{\Delta t}{2}\right) \leftarrow \tilde{\mathbf{p}}\left(\frac{\Delta t}{2}\right) + \frac{\partial \boldsymbol{\sigma}}{\partial \mathbf{x}(0)}\boldsymbol{\mu} \end{cases}$$

$$\tilde{\mathbf{x}}\left(\frac{\Delta t}{2}\right) \leftarrow \mathbf{x}(0) + \mathbf{M}^{-1}\tilde{\tilde{\mathbf{p}}}\left(\frac{\Delta t}{2}\right)\frac{\Delta t}{2}$$

$$\textit{Thermostat for a full time step } \Delta t \textit{ (in which } \tilde{\tilde{\mathbf{p}}}\left(\frac{\Delta t}{2}\right) \textit{ is updated)}$$

$$\tilde{\mathbf{x}}(\Delta t) \leftarrow \tilde{\mathbf{x}}\left(\frac{\Delta t}{2}\right) + \mathbf{M}^{-1}\tilde{\tilde{\mathbf{p}}}\left(\frac{\Delta t}{2}\right)\frac{\Delta t}{2}$$

$$\mathrm{C}_1 : \begin{cases} \text{Solve } \boldsymbol{\lambda} : \boldsymbol{\sigma}\left(\tilde{\mathbf{x}}(\Delta t) + \mathbf{M}^{-1}\frac{\partial \boldsymbol{\sigma}}{\partial \mathbf{x}(0)}\boldsymbol{\lambda}\right) = \mathbf{0} \\ \mathbf{x}(\Delta t) \leftarrow \tilde{\mathbf{x}}(\Delta t) + \mathbf{M}^{-1}\frac{\partial \boldsymbol{\sigma}}{\partial \mathbf{x}(0)}\boldsymbol{\lambda} \\ \mathbf{p}\left(\frac{\Delta t}{2}\right) \leftarrow \tilde{\tilde{\mathbf{p}}}\left(\frac{\Delta t}{2}\right) + \frac{1}{\Delta t}\frac{\partial \boldsymbol{\sigma}}{\partial \mathbf{x}(0)}\boldsymbol{\lambda} \end{cases}$$

$$\tilde{\mathbf{p}}(\Delta t) \leftarrow \mathbf{p}\left(\frac{\Delta t}{2}\right) - \frac{\partial U}{\partial \mathbf{x}(\Delta t)}\frac{\Delta t}{2}$$

$$\mathrm{C}_2 : \begin{cases} \text{Solve } \boldsymbol{\mu} : \left(\frac{\partial \boldsymbol{\sigma}}{\partial \mathbf{x}(\Delta t)}\right)^T \mathbf{M}^{-1}\left(\tilde{\mathbf{p}}(\Delta t) + \frac{\partial \boldsymbol{\sigma}}{\partial \mathbf{x}(\Delta t)}\boldsymbol{\mu}\right) = \mathbf{0} \\ \mathbf{p}(\Delta t) \leftarrow \tilde{\mathbf{p}}(\Delta t) + \frac{\partial \boldsymbol{\sigma}}{\partial \mathbf{x}(\Delta t)}\boldsymbol{\mu} \end{cases}, \tag{67}$$

which is denoted "C2-p-C1-x-T-x-C2-p", where the operations are performed from right to left. When "LF-Middle" is used, our recommended version is

$$
\begin{aligned}
&\tilde{\mathbf{p}}\left(\frac{\Delta t}{2}\right) \leftarrow \mathbf{p}\left(-\frac{\Delta t}{2}\right) - \frac{\partial U}{\partial \mathbf{x}(0)}\Delta t \\
&C_2 : \begin{cases} \text{Solve } \boldsymbol{\mu} : \left(\frac{\partial \sigma}{\partial \mathbf{x}(0)}\right)^T \mathbf{M}^{-1}\left(\tilde{\mathbf{p}}\left(\frac{\Delta t}{2}\right) + \frac{\partial \sigma}{\partial \mathbf{x}(0)}\boldsymbol{\mu}\right) = \mathbf{0} \\ \tilde{\tilde{\mathbf{p}}}\left(\frac{\Delta t}{2}\right) \leftarrow \tilde{\mathbf{p}}\left(\frac{\Delta t}{2}\right) + \frac{\partial \sigma}{\partial \mathbf{x}(0)}\boldsymbol{\mu} \end{cases} \\
&\tilde{\mathbf{x}}\left(\frac{\Delta t}{2}\right) \leftarrow \mathbf{x}(0) + \mathbf{M}^{-1}\tilde{\tilde{\mathbf{p}}}\left(\frac{\Delta t}{2}\right)\frac{\Delta t}{2} \\
&\textit{Thermostat for a full time step } \Delta t \textit{ (in which } \tilde{\tilde{\mathbf{p}}}\left(\frac{\Delta t}{2}\right) \textit{ is updated)} \\
&\tilde{\mathbf{x}}(\Delta t) \leftarrow \tilde{\mathbf{x}}\left(\frac{\Delta t}{2}\right) + \mathbf{M}^{-1}\tilde{\tilde{\mathbf{p}}}\left(\frac{\Delta t}{2}\right)\frac{\Delta t}{2} \\
&C_1 : \begin{cases} \text{Solve } \boldsymbol{\lambda} : \boldsymbol{\sigma}\left(\tilde{\mathbf{x}}(\Delta t) + \mathbf{M}^{-1}\frac{\partial \sigma}{\partial \mathbf{x}(0)}\boldsymbol{\lambda}\right) = \mathbf{0} \\ \mathbf{x}(\Delta t) \leftarrow \tilde{\mathbf{x}}(\Delta t) + \mathbf{M}^{-1}\frac{\partial \sigma}{\partial \mathbf{x}(0)}\boldsymbol{\lambda} \\ \tilde{\tilde{\tilde{\mathbf{p}}}}\left(\frac{\Delta t}{2}\right) \leftarrow \tilde{\tilde{\mathbf{p}}}\left(\frac{\Delta t}{2}\right) + \frac{1}{\Delta t}\frac{\partial \sigma}{\partial \mathbf{x}(0)}\boldsymbol{\lambda} \end{cases} \\
&C_2 : \begin{cases} \text{Solve } \boldsymbol{\mu} : \left(\frac{\partial \sigma}{\partial \mathbf{x}(\Delta t)}\right)^T \mathbf{M}^{-1}\left(\tilde{\tilde{\tilde{\mathbf{p}}}}\left(\frac{\Delta t}{2}\right) + \frac{\partial \sigma}{\partial \mathbf{x}(\Delta t)}\boldsymbol{\mu}\right) = \mathbf{0} \\ \mathbf{p}\left(\frac{\Delta t}{2}\right) \leftarrow \tilde{\tilde{\tilde{\mathbf{p}}}}\left(\frac{\Delta t}{2}\right) + \frac{\partial \sigma}{\partial \mathbf{x}(\Delta t)}\boldsymbol{\mu} \end{cases}
\end{aligned}
\tag{68}
$$

We denote it "C2-C1-x-T-x-C2-p".

We use AMBER2018 [55] to simulate a liquid water system subject to intramolecular bond length and angle constraints. The system contains a box of 216 water molecules (with periodic boundary conditions). The TIP3P model are employed as the force field. Fig. 5a implies that the two "middle" schemes (Eqs. 67 and 68) perform better in configurational sampling than the conventional "side" scheme as well as the BBK algorithm [25] for systems subject to constraints. Fig. 5b shows that "LF-Middle" leads to a more accurate marginal momentum distribution than "VV-Middle".

4.2 Isokinetic constraints in the "middle" scheme

Nonholonomic constraints (that involve the constraint of momenta) are also widely used in MD simulation in the canonical ensemble. In Ref. [12] we have recently extended the application of the "middle" thermostat scheme to the isokinetic constraint as well as to the MTS technique, which leads to a more efficient version for the Stochastic-Iso-NH-RESPA [SIN(R)] algorithm [56].

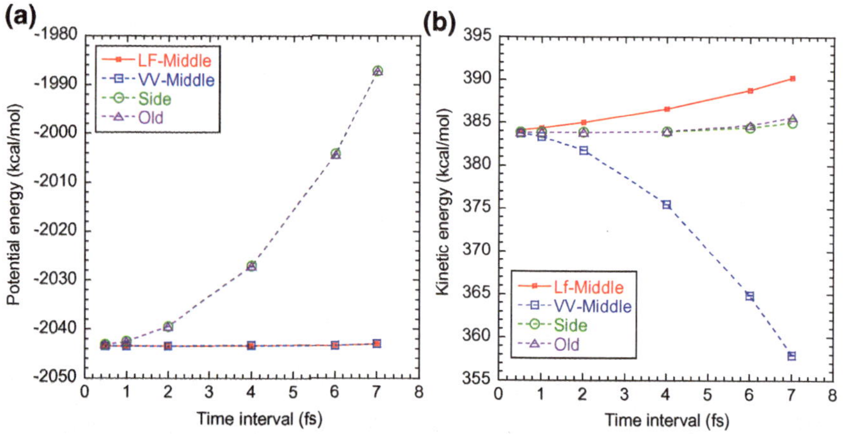

Fig. 5 MD results of liquid water with intramolecular O–H bond length and H–O–H angle constraints at $T = 298.15$ K using different schemes, **a** average potential energy per atom $\langle U(\mathbf{x})\rangle/(N_{\text{atom}}k_B)$ (unit: Kelvin) **b** average kinetic energy per atom $\langle \mathbf{p}^T\mathbf{M}^{-1}\mathbf{p}\rangle/(2N_{\text{atom}}k_B)$ (unit: Kelvin). "Old" stands for the BBK algorithm [25] used in AMBER (Reproduced with permission from Ref. [11])

Given $\Delta t = n\delta t$, with Δt as the outer time interval and δt the inner time interval, respectively. The propagating order of original SIN(R) [56] is expressed as

$$
\begin{aligned}
e^{\mathcal{L}\Delta t} \approx\ & e^{\mathcal{L}_N\delta t/2}e^{\mathcal{L}_{\mathbf{p}}^{(f)}\delta t/2+\mathcal{L}_{\mathbf{p}}^{(s)}\Delta t/2}e^{\mathcal{L}_{\mathbf{x}}\delta t/2}e^{\mathcal{L}_O\delta t}e^{\mathcal{L}_{\mathbf{x}}\delta t/2}e^{\mathcal{L}_{\mathbf{p}}^{(f)}\delta t/2}e^{\mathcal{L}_N\delta t/2} \\
& \times \left(e^{\mathcal{L}_N\delta t/2}e^{\mathcal{L}_{\mathbf{p}}^{(f)}\delta t/2}e^{\mathcal{L}_{\mathbf{x}}\delta t/2}e^{\mathcal{L}_O\delta t}e^{\mathcal{L}_{\mathbf{x}}\delta t/2}e^{\mathcal{L}_{\mathbf{p}}^{(f)}\delta t/2}e^{\mathcal{L}_N\delta t/2}\right)^{n-2} \\
& \times e^{\mathcal{L}_N\delta t/2}e^{\mathcal{L}_{\mathbf{p}}^{(f)}\delta t/2}e^{\mathcal{L}_{\mathbf{x}}\delta t/2}e^{\mathcal{L}_O\delta t}e^{\mathcal{L}_{\mathbf{x}}\delta t/2}e^{\mathcal{L}_{\mathbf{p}}^{(f)}\delta t/2+\mathcal{L}_{\mathbf{p}}^{(s)}\Delta t/2}e^{\mathcal{L}_N\delta t/2},
\end{aligned}
\tag{69}
$$

where the Kolmogorov operators in Eq. (69) are explicitly defined in Ref. [12]. Refs. [9, 11] have already suggested that applying thermostat in the middle of the propagation step always leads to much better performance in sampling the configuration space. When the middle thermostat scheme is used to design new SIN(R) algorithms, the propagating order of the "VV-Middle" version is

$$
\begin{aligned}
e^{\mathcal{L}\Delta t} \approx\ & e^{\mathcal{L}_{\mathbf{p}}^{(f)}\delta t/2+\mathcal{L}_{\mathbf{p}}^{(s)}\Delta t/2}e^{\mathcal{L}_{\mathbf{x}}\delta t/2}e^{\mathcal{L}_N\delta t/2}e^{\mathcal{L}_O\delta t}e^{\mathcal{L}_N\delta t/2}e^{\mathcal{L}_{\mathbf{x}}\delta t/2}e^{\mathcal{L}_{\mathbf{p}}^{(f)}\delta t/2} \\
& \times \left(e^{\mathcal{L}_{\mathbf{p}}^{(f)}\delta t/2}e^{\mathcal{L}_{\mathbf{x}}\delta t/2}e^{\mathcal{L}_N\delta t/2}e^{\mathcal{L}_O\delta t}e^{\mathcal{L}_N\delta t/2}e^{\mathcal{L}_{\mathbf{x}}\delta t/2}e^{\mathcal{L}_{\mathbf{p}}^{(f)}\delta t/2}\right)^{n-2} \\
& \times e^{\mathcal{L}_{\mathbf{p}}^{(f)}\delta t/2}e^{\mathcal{L}_{\mathbf{x}}\delta t/2}e^{\mathcal{L}_N\delta t/2}e^{\mathcal{L}_O\delta t}e^{\mathcal{L}_N\delta t/2}e^{\mathcal{L}_{\mathbf{x}}\delta t/2}e^{\mathcal{L}_{\mathbf{p}}^{(f)}\delta t/2+\mathcal{L}_{\mathbf{p}}^{(s)}\Delta t/2},
\end{aligned}
\tag{70}
$$

[denoted "VV-Middle-SIN(R)"] and that of the "LF-Middle" version becomes

$$
e^{\mathcal{L}\Delta t} \approx e^{\mathcal{L}_{\mathbf{x}}\delta t/2}e^{\mathcal{L}_N\delta t/2}e^{\mathcal{L}_O\delta t}e^{\mathcal{L}_N\delta t/2}e^{\mathcal{L}_{\mathbf{x}}\delta t/2}e^{\mathcal{L}_{\mathbf{p}}^{(f)}\delta t}
$$

$$\times \left(e^{\mathcal{L}_{\mathbf{x}}\delta t/2} e^{\mathcal{L}_{N}\delta t/2} e^{\mathcal{L}_{O}\delta t} e^{\mathcal{L}_{N}\delta t/2} e^{\mathcal{L}_{\mathbf{x}}\delta t/2} e^{\mathcal{L}_{\mathbf{p}}^{(f)}\delta t} \right)^{n-2}$$
$$\times\, e^{\mathcal{L}_{\mathbf{x}}\delta t/2} e^{\mathcal{L}_{N}\delta t/2} e^{\mathcal{L}_{O}\delta t} e^{\mathcal{L}_{N}\delta t/2} e^{\mathcal{L}_{\mathbf{x}}\delta t/2} e^{\mathcal{L}_{\mathbf{p}}^{(f)}\delta t + \mathcal{L}_{\mathbf{p}}^{(s)}\Delta t} \quad (71)$$

[denoted "LF-Middle-SIN(R)"]. Eqs. (70) and (71) yield the same efficiency and accuracy for sampling the configuration space. These versions in the "middle" scheme are all easily extended with more than two time steps for SIN(R).

We have simulated a system of liquid water [216 H2O molecules in a box with periodic boundary conditions at the state point $T = 298.15$ K(temperature) and 0.997 g·cm^{-3} (density)] to compare the original version of SIN(R) [Eq. (69)] and its new version with the "middle" scheme ("VV-Middle-SIN(R)") [Eq. (70)]. POLI2VS-a polarizable and flexible force field [57] for liquid water is used. In MTS, one may treat the force contributed by the intramolecular interactions as the fast part and consider the force derived from the noncovalent interactions as the slow part. Fig. 6 demonstrates the results with the inner time interval δt varied from 0.1 to 2fs while the outer time interval Δt are fixed at 2 or 4fs. As the inner time step δt increases, "VV-Middle-SIN(R)" performs more efficiently than conventional SIN(R).

All versions in the "middle" scheme for MD in Sect. 4.1 (with holonomic constraints) and Sect. 4.2 (with nonholonomic constraints as well as resonance-free multiple time-step techniques) can easily be extended to PIMD. We have already applied some of them in PIMD simulations for liquid water [10].

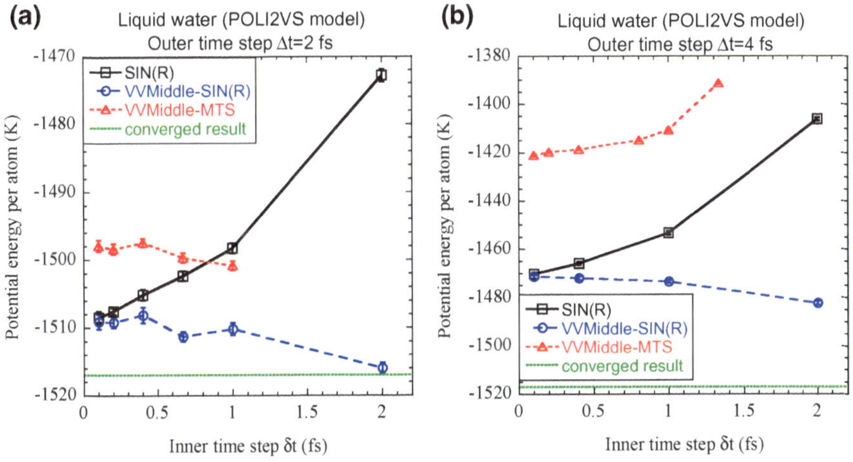

Fig. 6 Average potential energy per atom $\langle U(\mathbf{x})\rangle/(N_{\mathrm{atom}}k_B)$ (unit: Kelvin) of liquid water at $T = 298.15$ K as a function of the inner time step δt with a fixed outer time step. **a** $\Delta t = 2$ fs **b** $\Delta t = 4$ fs (Reproduced with permission from Ref. [12])

5 Conclusions

In many remarks, the "middle" thermostat scheme [7, 9, 11, 12, 15, 16] (either "VV-Middle" or "LF-Middle") provides a promising approach to design efficient MD/PIMD algorithms for sampling the configuration space of the canonical ensemble with or without constraints. Combination of the "middle" scheme with resonance-free MTS techniques leads to more efficient and robust algorithms for sampling the configuration space for multi-time-scale systems [12].

It is straightforward to employ the "middle" thermostat scheme with various advanced enhanced sampling techniques [55, 58–68] to accelerate configurational sampling for molecular systems where rare events become important [12]. Since the "middle" thermostat scheme is useful for any types of potential energy surfaces or force fields (for realistic molecular systems), it will be helpful to implement the "middle" scheme for ab initio MD or ab initio PIMD[21, 69] to reduce the computation cost. It will be interesting to develop more efficient MD and PIMD algorithms in the unified theoretical framework for these purposes.

Acknowledgements This work was supported by the Ministry of Science and Technology of China (MOST) Grants No. 2016YFC0202803 and No. 2017YFA0204901, by the National Natural Science Foundation of China (NSFC) Grants No. 21373018 and No. 21573007, by the Recruitment Program of Global Experts, by Specialized Research Fund for the Doctoral Program of Higher Education No. 20130001110009, and by Special Program for Applied Research on Super Computation of the NSFC-Guangdong Joint Fund (the second phase) under Grant No. U1501501.

References

1. Allen MP, Tildesley DJ (1989) Computer simulation of liquids. Clarendon Press
2. Frenkel D, Smit B (2002) Understanding molecular simulation, 2nd edn. Academic Press, San Diego
3. Chandler D, Wolynes PG (1981) Exploiting the isomorphism between quantum theory and the classical statistical mechanics of polyatomic fluids. J. Chem. Phys. 74(7):4078–4095
4. Parrinello M, Rahman A (1984) Study of an f center in molten kcl. J. Chem. Phys. 80(2):860–867
5. Berne BJ, Thirumalai D (1986) On the simulation of quantum systems: path integral methods. Annu Rev Phys Chem 37:401–424
6. Tuckerman ME (2010) Statistical mechanics: theory and molecular simulation. Oxford University Press, New York
7. Liu J, Li D, Liu X (2016) A simple and accurate algorithm for path integral molecular dynamics with the langevin thermostat. J Chem Phys 145(2):024103
8. Markland TE, Ceriotti M (2018) Nuclear quantum effects enter the mainstream. Nat Rev Chem 2(3):14
9. Zhang Z, Liu X, Chen Z, Zheng H, Yan K, Liu J (2017) A unified thermostat scheme for efficient configurational sampling for classical/quantum canonical ensembles via molecular dynamics. J Chem Phys 147(3):034109
10. Liu X, Liu J (2018) Critical role of quantum dynamical effects in the raman spectroscopy of liquid water. Mol Phys 116(7–8):755–779
11. Zhang Z, Yan K, Liu X, Liu J (2018) A leap-frog algorithm-based efficient unified thermostat scheme for molecular dynamics. Chin Sci Bull 63(0023–074X):3467

12. Zhang Z, Liu X, Yan K, Tuckerman ME, Liu J (2019) Unified efficient thermostat scheme for the canonical ensemble with holonomic or isokinetic constraints via molecular dynamics. J Phys Chem A 123(28):6056–6079
13. Liu J, Li D, Liu X (2016) Further study of path integral liouville dynamics
14. Liu J, Zhang Z (2016) Path integral liouville dynamics: Applications to infrared spectra of oh, water, ammonia, and methane. J Chem Phys 144(3):034307
15. Li D, Han X, Chai Y, Wang C, Zhang Z, Chen Z, Liu J, Shao J (2017) Stationary state distribution and efficiency analysis of the langevin equation via real or virtual dynamics. J Chem Phys 147(18):184104
16. Li D-z, Chen Z-f, Zhang Z-j, Liu J (2017) Understanding molecular dynamics with stochastic processes via real or virtual dynamics. Chin J Chem Phys 30(6):735–760
17. Case DA, Ben-Shalom IY, Brozell SR, Cerutti DS, Cheatham TE III, Cruzeiro VWD, Darden TA, Duke RE, Ghoreishi D, Gilson MK, Gohlke H, Goetz AW, Greene D, Harris R, Homeyer N, Izadi S, Kovalenko A, Kurtzman T, Lee TS, LeGrand S, Li P, Lin C, Liu J, Luchko T, Luo R, Mermelstein DJ, Merz KM, Miao Y, Monard G, Nguyen C, Nguyen H, Omelyan I, Onufriev A, Pan F, Qi R, Roe DR, Roitberg A, Sagui C, Schott-Verdugo S, Shen J, Simmerling CL, Smith J, Salomon-Ferrer R, Swails J, Walker RC, Wang J, Wei H, Wolf RM, Wu X, Xiao L, York DM (2018) Kollman PA Amber 2018. University of California, San Francisco
18. Andersen HC (1980) Molecular dynamics simulations at constant pressure and/or temperature. Journal of Chemical Physics 72(4):2384–2393
19. Nosé S (1984) A molecular dynamics method for simulations in the canonical ensemble. Mol Phys 52(2):255–268
20. Hoover WG (1985) Canonical dynamics: equilibrium phase-space distributions. Phys Rev A 31(3):1695–1697
21. Martyna GJ, Klein ML, Tuckerman M (1992) Nosé–hoover chains: The canonical ensemble via continuous dynamics. J Chem Phys 97(4):2635–2643
22. Martyna GJ, Tuckerman ME, Tobias DJ, Klein ML (1996) Explicit reversible integrators for extended systems dynamics. Mol Phys 87(5):1117–1157
23. Tuckerman M, Berne BJ, Martyna GJ (1992) Reversible multiple time scale molecular dynamics. J Chem Phys 97(3):1990–2001
24. Andrea TA, Swope WC, Andersen HC (1983) The role of long ranged forces in determining the structure and properties of liquid water. J Chem Phys 79(9):4576–4584
25. Brünger A, Brooks Iii CL, Karplus M (1984) Stochastic boundary conditions for molecular dynamics simulations of st2 water. Chem Phys Lett 105(5):495–500
26. Goga N, Rzepiela AJ, de Vries AH, Marrink SJ, Berendsen HJC (2012) Efficient algorithms for langevin and dpd dynamics. J Chem Theory Comput 8(10):3637–3649
27. Bussi G, Parrinello M (2007) Accurate sampling using langevin dynamics. Phys Rev E 75(5):056707
28. Grønbech-Jensen N, Farago O (2013) A simple and effective verlet-type algorithm for simulating langevin dynamics. Mol Phys 111(8):983–991
29. Leimkuhler B, Matthews C (2012) Rational construction of stochastic numerical methods for molecular sampling. Appl Math Res Express 2013(1):34–56
30. Leimkuhler B, Matthews C (2013) Robust and efficient configurational molecular sampling via langevin dynamics. J Chem Phys 138(17):174102
31. Leimkuhler B, Matthews C (2016) Efficient molecular dynamics using geodesic integration and solvent–solute splitting. Proc R Soc A: Math, Phys Eng Sci 472(2189)
32. Leimkuhler B, Matthews C (2015) Molecular dynamics with deterministic and stochastic numerical methods. Springer
33. Hall R, Berne BJ (1984) Nonergodicity in path integral molecular dynamics. J. Chem. Phys. 81(8):3641–3643
34. Gillan MJ (1987) Quantum simulation of hydrogen in metals. Phys Rev Lett 58(6):563–566
35. Singer K, Smith W (1988) Path integral simulations of condensed phase lennard-jones systems. Mol Phys 64(6):1215–1231

36. Müser MH (2002) On new efficient algorithms for pimc and pimd. Comput Phys Commun 147(1–2):83–86
37. Drozdov AN, Talkner P (1998) Path integrals for fokker-planck dynamics with singular diffusion: accurate factorization for the time evolution operator. J Chem Phys 109(6):2080–2091
38. Tuckerman ME, Marx D, Klein ML, Parrinello M (1996) Efficient and general algorithms for path integral car-parrinello molecular dynamics. J Chem Phys 104(14):5579–5588
39. Ceriotti M, Parrinello M, Markland TE, Manolopoulos DE (2010) Efficient stochastic thermostatting of path integral molecular dynamics. J Chem Phys 133(12):124104
40. Liu X, Liu J (2018) Path integral molecular dynamics for exact quantum statistics of multi-electronic-state systems. J Chem Phys 148(10):102319
41. Wang H, Liu X, Liu J (2018) Accurate calculation of equilibrium reduced density matrix for the system-bath model: A multilayer multiconfiguration time-dependent hartree approach and its comparison to a multi-electronic-state path integral molecular dynamics approach. Chin J Chem Phys 31(4):446–456
42. Suzuki M (1985) Decomposition formulas of exponential operators and lie exponentials with some applications to quantum mechanics and statistical physics. J Math Phys 26(4):601–612
43. Yoshida H (1990) Construction of higher order symplectic integrators. Phys Lett A 150(5):262–268
44. Suzuki M (1991) General theory of fractal path integrals with applications to many-body theories and statistical physics. J Math Phys 32(2):400–407
45. Ceperley DM (1995) Path integrals in the theory of condensed helium. Rev Mod Phys 67(2):279–355
46. Feynman RP (1953) Atomic theory of the lambda-transition in helium. Phys Rev 91(6):1291–1301
47. Tuckerman ME, Berne BJ, Martyna GJ, Klein ML (1993) Efficient molecular-dynamics and hybrid monte-carlo algorithms for path-integrals. J Chem Phys 99(4):2796–2808
48. Herman MF, Bruskin EJ, Berne BJ (1982) On path integral monte-carlo simulations. J Chem Phys 76(10):5150–5155
49. Pollock EL, Ceperley DM (1984) Simulation of quantum many-body systems by path-integral methods. Phys. Rev. B 30(5):2555–2568
50. Liu J, Li D, Liu X (2016) Supplementary material for the paper 'a simple and accurate algorithm for path integral molecular dynamics'. J Chem Phys 145:024103. ftp://ftp.aip.org/epaps/journ_chem_phys/E-JCPSA6-145-007626
51. Feynman RP, Hibbs AR (1965) Quantum mechanics and path integrals. McGraw-Hill, New York
52. Cao J, Berne BJ (1993) A Born-Oppenheimer approximation for path-integrals with an application to electron solvation in polarizable fluids. J Chem Phys 99(4):2902–2916
53. Ryckaert J-P, Ciccotti G, Berendsen HJC (1977) Numerical integration of the cartesian equations of motion of a system with constraints: Molecular dynamics of n-alkanes. J Comput Phys 23(3):327–341
54. Andersen HC (1983) Rattle: A "velocity" version of the shake algorithm for molecular dynamics calculations. J Comput Phys 52(1):24–34
55. Hamelberg D, Mongan J, McCammon JA (2004) Accelerated molecular dynamics: A promising and efficient simulation method for biomolecules. J Chem Phys 120(24):11919–11929
56. Leimkuhler B, Margul DT, Tuckerman ME (2013) Stochastic, resonance-free multiple time-step algorithm for molecular dynamics with very large time steps. Mol Phys 111(22–23):3579–3594
57. Hasegawa T, Tanimura Y (2011) A polarizable water model for intramolecular and intermolecular vibrational spectroscopies. J Phys Chem B 115(18):5545–5553
58. Torrie GM, Valleau JP (1977) Nonphysical sampling distributions in monte carlo free-energy estimation: Umbrella sampling. J Comput Phys 23(2):187–199
59. Laio A, Parrinello M (2002) Escaping free-energy minima. Proc Natl Acad Sci 99(20):12562
60. Gao YQ (2008) An integrate-over-temperature approach for enhanced sampling. J Chem Phys 128(6):064105

61. Valsson O, Parrinello M (2014) Variational approach to enhanced sampling and free energy calculations. Phys Rev Lett 113(9):090601
62. Valsson O, Tiwary P, Parrinello M (2016) Enhancing important fluctuations: Rare events and metadynamics from a conceptual viewpoint. Annu Rev Phys Chem 67(1):159–184
63. Sugita Y, Okamoto Y (1999) Replica-exchange molecular dynamics method for protein folding. Chem Phys Lett 314(1):141–151
64. Peters B (2017) Reaction rate theory and rare events simulations. Elsevier, Amsterdam, Netherlands
65. Carter EA, Ciccotti G, Hynes JT, Kapral R (1989) Constrained reaction coordinate dynamics for the simulation of rare events. Chem Phys Lett 156(5):472–477
66. Sprik M, Ciccotti G (1998) Free energy from constrained molecular dynamics. J Chem Phys 109(18):7737–7744
67. Sergi A, Ciccotti G, Falconi M, Desideri A, Ferrario M (2002) Effective binding force calculation in a dimeric protein by molecular dynamics simulation. J Chem Phys 116(14):6329–6338
68. Bello-Rivas JM, Elber R (2015) Exact milestoning. J Chem Phys 142(9):094102
69. Anandakrishnan R, Drozdetski A, Walker Ross C, Onufriev Alexey V (2015) Speed of conformational change: Comparing explicit and implicit solvent molecular dynamics simulations. Biophys J 108(5):1153–1164

Megascopic Quantum Phenomena

A Critical Study of Physical Interpretations

Michal Svrček

Abstract A historical study of metaphysical interpretations is presented. A megascopic revalidation is offered providing responses and resolutions of current inconsistencies and existing contradictions in present-day quantum theory. As the core of this study we present an independent proof of the Goldstone theorem for a quantum field formulation of molecules and solids. Along with phonons two types of new quasiparticles appear: rotons and translons. In full analogy with Lorentz covariance, combining space and time coordinates, a new covariance is necessary, binding together the internal and external degrees of freedom, without explicitly separating the centre-of-mass, which normally applies in both classical and quantum mechanical formulations. The generally accepted view regarding the lack of a simple correspondence between the Goldstone modes and broken symmetries, has significant consequences: an ambiguous BCS theory as well as a subsequent Higgs mechanism. The application of the archetype of the classical spontaneous symmetry breaking, i.e. the Mexican hat, as compared to standard quantum relations, i.e. the Jahn-Teller effect, superconductivity or the Higgs mechanism, becomes a disparity. In short, symmetry broken states have a microscopic causal origin, but transitions between them have a teleological component. The different treatments of the problem of the centre of gravity in quantum mechanics and in field theories imply a second type of Bohr complementarity on the many-body level opening the door for megascopic representations of all basic microscopic quantum axioms with further readings for teleonomic megascopic quantum phenomena, which have no microscopic rationale: isomeric transitions, Jahn-Teller effect, chemical reactions, Einstein-de Haas effect, superconductivity-superfluidity, and brittle fracture. We demonstrate how the megascopic extension of the microscopic theory deals with the various paradoxes, such as the arrow of time, the Jordan-von Weizsäcker problem, to explain classicality, Santilli's "no reduction theorem", the Schrödinger cat paradox and Wigner's friend, and the Kauzmann paradox of residual entropy. Only

M. Svrček (✉)
CMOA Czech Branch, Carlsbad, Czech Republic
e-mail: m.sv@o2active.cz

L. Mammino et al. (eds.), *Advances in Quantum Systems in Chemistry, Physics, and Biology*, Progress in Theoretical Chemistry and Physics 32,
https://doi.org/10.1007/978-3-030-34941-7_14

the Copenhagen interpretation seems to be able to incorporate telicity via megascopic mirroring of each of its basic axioms, and in that way does confront the above mentioned paradoxes.

Keywords Goldstone bosons · Bohr complementarity · Centre-of-mass · Spontaneous symmetry breaking · Superconductivity · Jahn-Teller effect · Higgs mechanism · Matter-mind dualism

1 Introduction

Quantum theory was at first formulated and construed within the framework of the Copenhagen School. This interpretation was a result of long years of intensive exploration of the microworld answering one of the greatest scientific paradoxes of all times: on the one hand maintaining classical concepts in the description of experimental results, and on the other hand to build a novel perception of the microworld, the latter not the world of reality as we know from Newton's equations of the macroworld, but rather the abstract world of Aristotelian potentiality. The primary consequences of this conception were indeterminism and complementarity, which were introduced for the first time in science. Since classical physics has influenced philosophy for centuries, leading to such extreme views as mechanical materialism, it was indeed difficult for many scientists at first to accept the Copenhagen interpretation in its entirety. There were numerous unsuccessful attempts to explain microprocesses in a deterministic way by means of hidden parameters. The sharpest discussions proceeded between Bohr and Einstein, which culminated in the most ingenious objection of Einstein, known as the EPR paradox [1]. Later Bell discussed this paradox and proved what is known today as Bell's theorem [2]. Based on this insight, subsequent experimental tests fully confirmed the validity of all quantum mechanical predictions also in concert with the celebrated Copenhagen interpretation.

The initial work of Bohr and later realized as the Copenhagen interpretation basically concerned the hydrogen atom, but how to continue? What about molecules, crystals, liquids, and other forms of macroscopic matter? Do we understand them as well, too? Detailed many body treatments were developed during the last half century, recommending how to deal with complex systems composed of elementary particles. These many body treatments have one common feature: they respect one classical rule, i.e. knowing the physical law of motion of the smallest parts, we can unambiguously predict the motion of the whole. Although the majority of scientists believe this, one may still ask: is this really true? We have of course no reason to doubt this belief, yet what is in full accordance with our deep-rooted experience, we might still find such examples or paradoxes that convince us otherwise. Surprisingly such examples and paradoxes do exist, although they usually are considered only to be of peripheral interest. Nevertheless this will be the topic of the presented article.

One may ask whether the Copenhagen interpretation is bad, as Einstein and all supporters of hidden parameter schemes tried to prove, or whether the elementary particle equations are at fault? Of course, no! On the contrary, the Copenhagen interpretation was formed in a very accurate, tangible and original way. However the simple characteristics of the hydrogen atom, did not reveal the whole truth about a many-body quantum system alongside yielding an inappropriate relationship between the micro-, the macro- and the mega-world. As will be demonstrated in this paper, there is also the controversial problem regarding the occurrence of the alleged quantum jumps, which have no origin in the original Copenhagen microscopic formulation. Hence this article will promote a so-called ‘second quantum floor’ describing megascopic quantum phenomena, in contrast to the ‘first floor’, defined by the standard description of microscopic phenomena within the Copenhagen interpretation. We will in addition expound, which heretofore being unexplained or even misinterpreted, upon the phenomena that belong to this ‘second floor’.

2 The Clamped-Nuclei Paradox

Let us start with the simplest formulation of the many-body problem based on the Schrödinger equation. Solving it exactly is in practice impossible, and therefore its simplified solution known as the Born-Oppenheimer (B-O) approximation [3] is widely used. Quite recently Sutcliffe and Woolley [4] offered an extensive historical justification of the development of quantum theory, in particular the quantum chemical solution of Schrödinger’s equation, discussing the applicability as well as the limitations of Potential Energy Surfaces (PES) in quantum chemistry. As conclusion they wrote: “This qualitative modification of the internal Hamiltonian, the extra choice of fixed nuclear positions in the ‘electronic’ Hamiltonian, is ad hoc in the same sense that Bohr’s quantum theory of the atom is an ad hoc modification of classical mechanics. An essential feature of the answer is put in by hand. We know that both modifications have been tremendously useful and our point is not that something else must be done in practical calculations on molecules. The point is how the successful description of molecules involving the clamped-nuclei modification at some stage can best be understood in terms of quantum mechanics. In the case of the Bohr atom the resolution of the inconsistency in mechanics applied to the microscopic realm was achieved quite quickly with the formulation of quantum mechanics; in the molecular case, no such resolution is at present known.”

In other words, we have enough numerical verifications available, but what we need, is the elucidation why the clamped-nuclei modification works so well. Their statement is an imperative challenge to everybody. In passing we should not forget that two scientific sub-disciplines, quantum chemistry and solid state physics, are mostly dependent on the B-O approximation, making them two sand-castles. This begs the question: What can we expect: the confirmation of the B-O approximation that satisfies our desire for perfection—or a disclosure of something new that goes beyond our contemporary knowledge of physics and chemistry? We will contend in this paper that the latter is closer to the truth.

What about the objection that we can avoid all the above mentioned problems by solving the Schrödinger equation exactly. I quote from Monkhorst's article [5] an analogous critique of the above mentioned PES concept: "The BO approximation is strictly valid only near the multidimensional PES minimum. On the other hand, the BH (Born-Huang) treatment inspired the thinking that the entire PES is usually valid to describe the dynamics of a molecule, including dissociation into various fragments, i.e., chemical reactions. However, this view is only acceptable if the PES-associated wave functions interact only weakly for all nuclear configurations. It is usually quite impossible to verify this, and most consumers of the PES concept assume it. Their only "defense" is the adiabatic nature of a molecular wave function, where the electronic wave functions adjust "instantaneously" to the evolving nuclear configuration. In fact, this view is central to the teachings of quantum chemistry."

Monkhorst also describes his coupled-cluster method for the solution of the system of electrons and nuclei on the same footing, completely evading the B-O approximation: "Therefore, when the need arises to address the limitations of the PES, it seems valuable to remove it entirely from the toolbox of the molecular scientist. One option is the molecular coupled-cluster (MCC) method I formulated 10 years ago [6]. This method takes a very atomic view of a molecule: instead of fixing the nuclei as in the BO approximation, the electrons and nuclei are both described quantum-dynamically within a centrosymmetric shell structure. The coupled-cluster method, duly generalized, is used to describe the correlation among all particles."

Being aware of the practical computational limitations of this exact method, he adds: "Even though the MCC method seems attractive, and (I hope) computationally tractable once implemented, it can be only practical for small molecules. I cannot see how the PES concept, and its attendant molecular structure ideas, will be superseded with this method for large molecules. Its qualitative appeal, its semiquantitative success, and its deep roots in the chemists' minds will keep the PES as real as it is now."

The reasons given above are most likely why most of chemists prefer calculations based on the B-O approximation. As far as the error caused by this calculation, on the adiabatic level the Born-Huang (B-H) ansatz [7] is applied, and as Kutzelnigg proved [8], the B-H ansatz fully compensates for this error. Note however that Kutzelnigg's proof indicates only the numerical equivalence, and solely on the adiabatic level. There are no objective equivalences between the exact solution of the Schrödinger equation and solutions based on the whole of the B-O approximation and B-H ansatz at all, even on the adiabatic level.

Fortunately, a small number of scientists were inspired by Monkhorst's idea, and performed exact calculations of some of the simplest molecules. Cafiero and Adamowitz wrote [9]: "The model of the molecule… is quite similar to an atom, as has been noted by Monkhorst [5]. We have the analogue of the nucleus with the heavy particle at the center of the internal coordinate system, and we have the analogues of electrons in the internal particles. The main difference between this model and an atom is that the internal particles in an atom are all electrons and in the molecular atom or atomic molecule the internal particles may be both electrons and nuclei (or, as we should more correctly say, pseudo-particles resembling the electron and the nuclei). Formally this difference manifests itself in the effective masses of the pseudo-particles and in the way the permutational symmetry is implemented in the wave function."

And further looking at the conclusion from their calculations: "Since in the nonadiabatic treatment we have included the nuclei in the wave function, we must determine the molecular structure by calculating expectation values of the distances between the nuclei. Since the operators representing the internuclear distances... do not commute with the Hamiltonian..., we cannot measure these distances exactly for stationary states of the system. Furthermore, since the ground state is spherically symmetric, we cannot gain additional information from the wave function such as the expectation values of the x, y, z coordinates of any of the particles, since these will average out to zero in all cases. Lastly, the expectation values for the distances between any two particles in each subset of identical particles (including the distance to the particle at the origin, if that particle belongs to the subset) will be equal. This is a widely known fact for electrons, i.e., due to antisymmetry, or indistinguishability, all of the electrons are on average at the same distance from each other, as well as from each nucleus in the molecule. That this same fact applies to nuclei is, perhaps, less known."

These first calculations of their kind gave us a remarkable insight into the actual differences between the exact and the clamped-nuclei approaches. The exact solution communicates the concept of the "molecular atom or atomic molecule", where we can only see one undivided isolated composite object, while we are unable to say anything more about the individual molecules involved in such a system. The hierarchy elementary particles → atoms → molecules doesn't exist here as we know it from the B-O approximation; instead it is replaced with the reduced hierarchy elementary particles → molecular atoms or atomic molecules. The exact solution leads always to a full symmetry picture of the molecule, and the indistinguishability principle, valid for nuclei as well, doesn't allow any "symmetry breaking", such as e.g. an isomerism. As Sutcliffe and Woolley state: "The Isolated Molecule model doesn't capture isomerism, nor optical activity." It is indeed a pity that exact solutions are so complex that till now they have only been tested on very small systems, i.e. up to ten particles, electrons and nuclei included. It would be very interesting to test them on something bigger at least on the simplest Jahn and Teller [10] systems. It seems that the indistinguishability of the nuclei and their shell nature, no J-T symmetry breakings are possible at all, cf. the case of isomerism, and that we will obtain fully symmetrical solutions for the ground states. Although standard theories of the J-T effect [11, 12] go beyond the B-O approximation and are nonadiabatic, they are nevertheless all of clamped-nuclei type.

Thus we can see that even exact solutions of Schrödinger's equation do not yield the full picture of all quantum chemical phenomena. Yet we cannot avoid the clamped-nuclei concept in the B-O approximation, leading to a serious paradox: how can an approximation produce results that by no means follow from the exact solution? What is the nature of the B-O model when it is not viewed as an approxima tion? Where is the true origin of the clamped-nuclei concept, allowing the description of individual molecules? This is indeed a quantum paradox, since we don't know any classical analogies. How do we reach the B-O approximation? Perhaps by simply transgressing the law of nature due to the hierarchical way of quantization—first the electrons are quantized and the nuclei remain classical, and then they are quantized a posteriori. But quantum mechanics demands simultaneous quantization, regardless how heavy the nuclei are compared to the electrons. Sutcliffe and Woolley ask: "The interesting question is how to get from the quantum theory of an Isolated Molecule to a quantum theory of an individual molecule by rational mathematics." The problem is thus, on "what" quantum theory the mentioned rational mathematics do apply. If we understand quantum theory as quantum mechanics based on the original Copenhagen interpretation, employing the Schrödinger equation for particles in the many-body systems, we find no answer. The input of electrons and nuclei in the mechanical Schrödinger equation exhibits a full symmetry regarding both types of particles. As we will see in this paper, many-body systems are running in dual mode, and we need to search the answer not under the mechanical pattern, but under the field concept where simultaneous quantization can be performed.

The concepts "isolated system" and "individual system" are well defined in the classical world, but only the first one has a similar meaning in the microworld, whereas the second one needs a radical redefinition on the quantum level. However, in speaking about elementary particles as "individual systems", there is no problem. The problems start when we try to apply this concept to composite systems like molecules or crystals.

In regard to the phenomena mentioned above, the explanation calls for the concept of clamped-nuclei, but, as just said, this concept, from the viewpoint of an exact quantum mechanical many-body formulation, does not work. The Copenhagen interpretation must be incomplete in this respect.

3 The Paradox of Time Irreversibility

The fundamental problem of both classical and quantum physics is still the question of the arrow of time. Why does time have a direction?

A half century ago, Santilli, then a nuclear physicist, stated, what he called a no-reduction theorem [13]: "A macroscopic irreversible system cannot be consistently reduced to a finite number of elementary constituents all in reversible conditions and, vice versa, a finite number of elementary particles all in reversible conditions cannot consistently characterize a macroscopic irreversible system." He further said, claiming that he has discussed this problem with Heisenberg and Dirac and that none of the quantum theory founders knew the answer: "The above theorem establishes that irreversibility originates at the most ultimate structure of nature." Till now the scientific community does not seem to take this paradox seriously, nevertheless Santilli continues: "During the 20th century it was generally believed that the irreversibility over time of our macroscopic environment was "illusory" (sic) because, when macroscopic events are reduced to their elementary particle constituents, irreversibility "disappears" (sic) and one recovers nice elementary particles in the reversible conditions necessary for the applicability of special relativity, quantum mechanics and quantum chemistry." Although he was correct with the formulation of the "no reduction theorem", he is certainly wrong with his later interpretation. His approach is to introduce the so-called "Lie-admissible algebras" for the construction of irreversible equations of the microworld, seemingly giving up the concept of reversible quantum equations. As a consequence the macroscopic limit of his irreversible equations will lead also to classical irreversible expressions, e.g. replacing the standard Newtonian laws, unravelling puzzles arising from his earlier theorem once and for all. In doing so he does not object to microscopic irreversibility even when no irreversible equations are needed. This appears justified since any intervention during measurement is an irreversible act, causing the reduction of the wave function of the system, ultimately in agreement with microscopic irreversibility. Yet he destroys the philosophical logic of a consistent quantum philosophical microscopic view simply for the limited purpose to explain macroscopic irreversibility. A better alternative would be to accept the incompleteness of quantum theory as stated by the Copenhagen interpretation, rather than rewriting physics in its entirety.

Looking into the history of quantum mechanics, Bohr initially described the Heisenberg principle of uncertainty as a purely epistemic problem. In this context there is no problem to accept the "objective reality" in the macroscopic limit. But today we know that the principle of uncertainty is also ontological, and there is unfortunately no classical limit relating to some "objective reality", even if, on the macroscopic scale, "objective potentiality" still persists. This raises the question: how to relate the various forms of "objective potentiality" at the macroscopic level into one reality. The problem is that classical physics describes events at the scale of continuous time. On the contrary, quantum equations cannot describe any event, since they arise only after outside intervention, i.e. after a measurement which can lead to irreversible processes. However, there are also many irreversible processes, e.g. chemical reactions that occur independently on any observer, i.e. without any specific measurement process. Consequently, if there is no macroscopic ontological limit of quantum laws that applies to the microworld, then there must exist yet another type of events, that occurs on the macroscopic level, which are responsible for the time irreversibility, but have no roots in the original Copenhagen School. This is of fundamental importance, since if and when we find this, until now unknown type of events, we will recover the concept of unitary transformations as pertaining to reversible equations on both the microscopic as well as the macroscopic levels.

Classical physics recognizes reversible processes as the primary ones along with the full reversibility on the time axis, whereas irreversible processes violate this symmetry with the arrow of time oriented only in the "forward" direction. This stance also carries over in microscopic quantum theory. Santilli recognized the ascending order of irreversibility, but nevertheless made an inaccurate inference in the assumption of irreversibility conceiving reversibility as a constricted limit of his irreversible equations. There is essentially no irreversibility in the primary equations of physics, which leads us to explore deep seated formulations that might reveal precise relations between reversibility/irreversibility allowing solutions that provide the proper characteristics depending on the case at hand. Therefore the classical concept of the "arrow of time" becomes inadequate in the quantum world and without such readings the Copenhagen interpretation remains incomplete.

4 Bohm's Five Prophetic Statements

Bohm is largely known for his attempts to invigorate the theory of hidden parameters, originally formulated by de Broglie as the contentious pilot-wave theory, adding the well-known conception of the famed quantum potential [14]. However, this appears to lead nowhere. The theory of hidden parameters cannot be helpful for a solution of the incompleteness problem associated with the Copenhagen interpretation. Nevertheless, as is less known, thirty years after he published his new version of non-local hidden parameters, Bohm arrived at five important critical conclusions regarding the aforementioned lack of completeness. He himself never found the resolution, since his own Bohmian Mechanics indeed suffers from the same shortcomings as the Copenhagen interpretation, believing that it will be answered in the future by some new modifications of the hidden variable idea. As can be seen, it is easier to be critical rather than finding the correct solution.

These five critical points of Bohm are of extraordinary relevance, because neither the Copenhagen interpretation nor any other rendering is able to cope with them, to take their ideas into consideration fulfilling their requirements. I will cite them all and give my reflections and detailed remarks below. The first four are taken from his book Wholeness and the Implicate Order [15], the last fifth from his dialogue with the biologist Sheldrake [16].

The first quote runs:

1. "Nevertheless, in most of the work that is being done in physics today the notions of formative and final cause are not regarded as having primary significance. Rather, law is still generally conceived as a self-determined system of efficient causes, operating in an ultimate set of material constituents of the universe (e.g. elementary particles subject to forces of interaction between them). These constituents are not regarded as formed in an overall process, and thus they are not considered to be anything like organs adapted to their place and function in the whole (i.e. to the ends which they would serve in this whole). Rather, they tend to be conceived as separately existent mechanical elements of a fixed nature."

Here Bohm discerns two categories of causes—efficient and final, according to the ancient classification introduced by Aristotle. Nature is full of phenomena based on final causes, but unfortunately the science of physics (classical, relativistic or quantum) is unable to incorporate them and represent them in some form of meaningful equations. What is even worse, physicists often do not recognize final causes at all treating the relevant phenomena as they would be induced by efficient causes. A nice example is the spontaneous symmetry breaking. Since it is believed that nothing in nature is spontaneous, physicists not seeing any explicit efficient cause, they call such symmetry breakings "spontaneous" imparting adverse formulations of such phenomena that are based on this type of symmetry violation. We will discuss this problem in more detail below.

The second statement reads:

2. "When it comes to the informal language and mode of thought in physics, which infuses the imagination and provokes the sense of what is real and substantial, most physicists still speak and think, with an utter conviction of truth, in terms of the traditional atomistic notion that the universe is constituted of elementary particles which are 'basic building blocks' out of which everything is made. In other sciences, such as biology, the strength of this conviction is even greater, because among workers in these fields there is little awareness of the revolutionary character of development in modern physics. For example, modern molecular biologists generally believe that the whole of life and mind can ultimately be understood in more or less mechanical terms, through some kind of extension of the work that has been done on the structure and function of DNA molecules. A similar trend has already begun to dominate in psychology."

The quantum physics community, as understood today by a vast majority of physicists, constitutes an unluckily fulfilment of the materialistic ideas of Democritus. As concerns ancient Democritus' atomism, the concepts of quantum physics differs little or not at all from classical mechanics. In fact the great men of classical physics (Newton, Leibniz, Laplace and others) sincerely believed that if someone gives them the laws of motion for all existing constituents of matter, they could predict the motion of the whole universe. No matter how much the laws of motion of elementary particles in quantum physics are profoundly dissimilar, both mathematically and philosophically, to the laws of classical physics, yet by introducing complementarity and uncertainty no one dares to contest any unambiguousness in the formulation of many-body quantum equations. As a result many physicists, particularly cosmologists, believe in the existence of a wave function describing the whole of the universe, one of the first and main anticipation in the Everett relative state interpretation [17]. Bohm's conclusion is absolutely correct, and we will prove here that the conventional many-body formulation in quantum physics is ambiguous as well, and that microscopic quantum laws are not the ultimate and unique platform for understanding all natural phenomena.

The third citation becomes:

3. "So fragmentation is in essence a confusion around the question of difference and sameness (or one-ness), but the clear perception of these categories is necessary in every phase of life. To be confused about what is different and what is not, is to be confused about everything... The question of fragmentation and wholeness is a subtle and difficult one, more subtle and difficult than those which lead to fundamentally new discoveries in science. To ask how to end fragmentation and to expect an answer in a few minutes makes even less sense than to ask how to develop a theory as new as Einstein's was when he was working on it, and to

expect to be told what to do in terms of some programme, expressed in terms of formulae or recipes."

This is indeed a very important objection. We will demonstrate a relation between Bohm's dilemma of wholeness versus fragmentation and the problem of isolated versus individual as formulated by Sutcliffe and Woolley. The microscopic quantum world simply does not match our concepts of individual or fragmented, albeit these words have clear and well-defined meanings on the classical level. And further it is most astonishing to compare how the authors mentioned characterize this problem and evaluate its complexity. Bohm emphasized that the challenge is more difficult than the problem of relativity confronting Einstein, while Sutcliffe and Woolley emphasized that it is harder to solve than the problem of quantum theory for Bohr.

The fourth excerpt reads:

4. "Here, we may hope to get some clues by considering problems in a domain where current theories do not yield generally satisfactory results, i.e. one connected with very high energies and very short distances. With regard to such problems, we first note that the present relativistic quantum field theory meets severe difficulties which raise serious doubts as to its internal self-consistency. There are the difficulties arising in connection with the divergences (infinite results) obtained in calculations of the effects of interactions of various kinds of particles and fields. It is true that for the special case of electromagnetic interactions such divergences can be avoided to a certain extent by means of the so-called 'renormalization' techniques. It is by no means clear, however, that these techniques can be placed on a secure logical mathematical basis. Moreover, for the problem of mesonic and other interactions, the renormalization method does not work well even when considered as a purely technical manipulation of mathematical symbols, apart from the question of its logical justification. While it has not been proved conclusively, as yet, that the infinities described above are essential characteristics of the theory, there is already a considerable amount of evidence in favour of such a conclusion."

This is entirely a quantum field problem. Quantum mechanical patterns, derived from the classical Hamilton-Jacobi formalism, don't seem to lead to any similar difficulties. The question is rather, whether an analogical quantization of the classical electro-magnetic field Langrangian will provide us with the full quantum field pattern, especially when it is used in closed systems such as nuclei in quantum chromodynamics. Note that original field patterns of quantum electrodynamics represent only the scattering process of a few free particles in vacuum. We will here analyze these inconsistencies in applying such a limited pattern to bound-state massive systems like molecules and crystals.

The final Bohm quotation contains the exchange with Sheldrake:

5. Bohm: "'How is time to be understood?' Now, in terms of the totality beyond time, the totality in which all is implicate, what unfolds or comes into being in any present moment is simply a projection of the whole. That is, some aspect of the

whole is unfolded into that moment and that moment is just that aspect. Likewise, the next moment is simply another aspect of the whole. And the interesting point is that each moment resembles its predecessors but also differs from them. I explain this using the technical terms 'injection' and 'projection'. Each moment is a projection of the whole, as we said. But that moment is then injected or introjected back into the whole. The next moment would then involve, in part, a re-projection of that injection, and so on in-definitely."

Sheldrake: "But don't you get time in physics when you have a collapse of the wave function?"
Bohm: "Yes, but that's outside the framework of quantum physics today. That collapse is not treated by any law at all, which means that the past is, as it were, wiped out altogether... You see, the present quantum mechanics does not have any concept of movement or process or continuity in time; it really deals with one moment only, one observation, and the probability that one observation will be followed by another one..."

In short the question is whether those "injections", promoted by Bohm, would be able to solve the problem of time in quantum mechanics? And further what are those "injections" and what do they represent, and how could we formulate them by rigorous mathematics? They should first of all represent quantum jumps in a similar way as "projections". We also know that "projections" are the results of our conscious observations of the quantum systems, but if "injections" are quantum jumps as well, how do we observe and measure them? Is it possible, that some complementation or extension of the Copenhagen interpretation could finally lead to an ontological justification of those "injections", yet without having any epistemic access to them? In other words, is there some, until now unknown, kind of quantum jumps in nature, which can neither be explained on the basis of present-day quantum theory, nor directly measured, and perhaps with no microscopic origin at all? As already said, this will be the permeating topic of our paper, where in particular we will show how Bohm's "injections" will be closely related to the time reversibility/irreversibility problem of Santilli.

5 The Paradox of Schrödinger's Cat

This is the well-known paradox from the time when Einstein and Schrödinger exchanged arguments regarding the indeterministic features of the newly conceived quantum theory. Nevertheless today, against the background of the mentioned incompleteness of the Copenhagen interpretation, many alternatives are being developed with Schrödinger's cat playing an important testing role. J. D. Norton, the well-known historian of physics, highlights the significance in the following terms [18]: "This paradox of the Schrödinger's cat is the most vivid expression of a lingering problem in the foundations of quantum theory. In the last two decades especially, there has been a huge amount of work devoted to finding variations to standard quantum theory or just new ways to think about the same theory that avoid this problem. There is no consensus on which approach is the correct one or even if some sort of repair is needed."

Norton describes the history of the Schrödinger's cat problem, which followed shortly after the appearance of the famous EPR paradox [1] in March 1935: "In the aftermath of this paper, Einstein and Schrödinger exchanged letters in which they aired their common concerns about quantum theory. In that correspondence, Einstein put to Schrödinger what we now see is an early version of the cat paradox. He outlined a "crude macroscopic example" in a letter to Schrödinger of August 8, 1935: "The system is a substance in chemically unstable equilibrium, perhaps a pile of gunpowder that, by means of intrinsic forces, can spontaneously combust, and where the average life span of the whole setup is a year. In principle this can quite easily be represented quantum-mechanically. In the beginning the ψ-function characterizes a reasonably well-defined macroscopic state. But, according to your equation, after the course of a year this is no longer the case at all. Rather, the ψ-function then describes a sort of blend of not-yet and of already-exploded systems. Through no art of interpretation can this ψ-function be turned into an adequate description of a real state of affairs; [for] in reality there is just no intermediary between exploded and non-exploded" [19]".

As can readily be seen, Santilli's paradox previously mentioned, is a remake of the old gunpowder pile account due to Einstein, as both paradoxes deal with the factual irreversibility of chemical reactions. As we know that Einstein was not successful with his famed EPR objection, this time he did formulate a problem still unresolved.

Norton continues [18]: "Erwin Schrödinger published his "cat" thought experiment in a lengthy paper in the November 29, 1935, issue of the journal Die Naturwissenschaften. Here's the entirety of his original account: "One can even make quite ludicrous examples. A cat is enclosed in a steel chamber, together with the following infernal machine (which one must secure against the cat's direct reach): in the tube of a Geiger counter there is a tiny amount of a radioactive material, so small that although one of its atoms might decay in the course of an hour, it is just as probable that none will. If the decay occurs, the counter tube fires and, by means of a relay, sets a little hammer into motion that shatters a small bottle of prussic acid. When the entire system has been left alone for an hour, one would say that the cat is still alive provided no atom has decayed in the meantime. The first atomic decay would have poisoned it. The ψ-function of the total system would yield an expression for all this in which, in equal measure, the living and the dead cat (sit venia verbo ["pardon the expression"]) blended or smeared out. The characteristic of these examples that an indefiniteness originally limited to atomic dimensions gets transformed into gross macroscopic indefiniteness, which can then be reduced by direct observation. This prevents us from continuing naively to give credence to a "fuzzy model" as a picture of reality. In itself this contains nothing unclear or contradictory. There is a difference between a blurred or unsharply taken photograph and a shot of clouds and mist.""

Due to the Schrödinger's cat narrative, the scientific community was divided: some believed that living and dead cats are really in the state of a quantum superposition until the moment of a conscious measurement, others claimed that the cat had to be either alive or dead yet before the box was opened. Carpenter and Anderson further analysed the laboratory thought experiment with the following conclusion [20]: "The implications arising from the "Schrödinger's cat" thought experiment have led some authors to argue that observation of a measurement by a conscious observer is required to collapse quantum wave-functions. Here we combine Schrödinger's experimental paradigm with a system for splitting the information about the quantum state between two observers, thereby allowing distinct outcomes to be recorded without either observer knowing the state of the measured quantum event. Our results imply that to collapse a quantum wave-function, measurement alone, rather than conscious observation of a measurement, is sufficient."

The last objection made by Einstein against the Copenhagen interpretation was published in his Reply to Criticisms in 1949 [19], which runs in analogy with Schrödinger's cat argument: "As far as I know, it was E. Schrödinger who first called attention to a modification of this consideration, which shows an interpretation of this type to be impracticable. Rather than considering a system which comprises only a radioactive atom (and its process of transformation), one considers a system which includes also the means for ascertaining the radioactive transformation—for example, a Geiger-counter with automatic registration-mechanism. Let this latter include a registration-strip, moved by a clockwork, upon which a mark is made by tripping the counter. True, from the point of view of quantum mechanics this total system is very complex and its configuration space is of very high dimension. But there is in principle no objection to treating this entire system from the standpoint of quantum mechanics. Here too the theory determines the probability of each configuration of all its co-ordinates for every time instant. If one considers all configurations of the coordinates, for a time large compared with the average decay time of the radioactive atom, there will be (at most) one such registration-mark on the paper strip. To each coordinate configuration corresponds a definite position of the mark on the paper strip. But, inasmuch as the theory yields only the relative probability of the thinkable co-ordinate-configurations, it also offers only relative probabilities for the positions of the mark on the paper strip, but no definite location for this mark."

Yet again, cf. the case of gunpowder pile, Einstein's critique was justifiable, and hence there exists until now no resolution to this paradox. Nevertheless, for Einstein the only way out was the acceptance of the statistical interpretation of quantum mechanics, which today is not the accepted majority standpoint. "One arrives at very implausible theoretical conceptions, if one attempts to maintain the thesis that the statistical quantum theory is in principle capable of producing a complete description of an individual physical system. On the other hand, those difficulties of theoretical interpretation disappear, if one views the quantum-mechanical description as the description of ensembles of systems" [19].

As the Schrödinger's cat has inspired many new interpretations of quantum theory, Schreiber in his Master of Science thesis, made an extensive comparison of them and wrote a very nice parody "The Nine Lives of Schrödinger's Cat" [21] where he describes the fate of the poor cat in the nine most popular alternative interpretations. In his conclusion, he says: "Discussion: The orthodox interpretation works well in practice but is ambiguous and therefore unacceptable. Bohr's epistemological views are also somewhat vague and require classical physics to be assumed. For those who are satisfied with epistemology, the decoherent histories approach admirably ties down any ambiguity or vagueness. It may be used to recover classical physics. Those who would like to see an ontological model of the world must look elsewhere. Neither many-world nor many-minds provide such models. The idea that the mind causes collapse does but the model is problematic. Bohm's interpretation does seem to work although it is rather awkward, especially in the context of quantum field theory. The best hope at the moment for an ontological theory is that the world's state collapses according to some fixed set of consistent histories. However, no clear criterion has emerged for the appropriate set of histories. Finally, there are approaches which involve not interpretation of quantum mechanics but modification. These approaches have not been considered here. The most successful of these is the stochastic scheme of 'GRW' (this scheme was put in a very nice form by Bell) although it is not without its problems. Several other attempts have involved introducing non-linear terms in the Schrödinger equation but none of these seems to work."

Schreiber then ends with his final anticipation—what should be done, but it never was: "Final word: Quantum mechanics is a theory lacking an ontological picture of the world. The search for such an ontology has been long, hard and appallingly haphazard. It is time that the entire programme was defined and analysed in a systematic and uniform mathematical way. Strange as this idea may seem, I am convinced that it is possible. When this programme is completed, theoretical physicists will finally be able to put the cat out and take a well-earned rest."

For Norton as well as for Schreiber, none of the present known interpretations is fully acceptable [18]: "My own feeling is that none of these responses is satisfactory… However, if there are new physical laws that would resolve the measurement problem, we can be pretty sure that they are quite exotic and not produced by a small adjustment in our existing theories. For, if these small adjustments are there to be found, eight decades of work by many of the brightest minds in quantum physics has failed to find them. Nonetheless, this "new, as yet unknown, physics" response is the one that all the other responses have to beat. Ask each of them, is this response more plausible than the "new physics" response? In each case, I answer no."

The two aforementioned conclusions suggest that we have to take on an apparent challenge searching for a radical new ontology and a new physics, based on deductive reasoning and rigorous mathematics. As we will show in this paper, this "new physics" imparts neither some new artificial topological spaces beyond the Hilbert space, nor some speculative multidimensional superstring proposals. All we need, should be to find nature's hidden laws, which have not yet been discovered and not yet revealed.

6 What Is the Copenhagen Interpretation?

The Copenhagen interpretation reflects a standard reference of our philosophical and physical views of quantum theory, yet there are remarkable differences of perspectives regarding the perception and understandings between the actual founders of quantum physics. Recognizing that the Copenhagen interpretation may not be complete, it will be valuable to remind the reader of the most important stances before starting to make any attempts to construct a more general theory.

Schlosshauer and Camilleri [22] takes up the unresolved disagreements between Bohr and Heisenberg: "In order to meaningfully speak of observation in quantum mechanics, Bohr concluded, "one must therefore cut out a partial system somewhere from the world, and one must make 'statements' or 'observations' just about this partial system" [23] … Bohr's epistemological demand for a "cut between the observed system on the one hand and the observer and his apparatus on the other hand" also became a key theme of Heisenberg's thinking. However, for Heisenberg, the object–instrument divide was coincident with the quantum–classical divide… It is important to realize that Heisenberg's views on the cut were somewhat at odds with Bohr's own view of the problem. Indeed, in an exchange of correspondence in 1935 Heisenberg and Bohr argued the point without resolution. Some twenty years later Heisenberg would report that "Bohr has emphasized that it is more realistic to state that the division into the object and rest of the world is not arbitrary" and that the object is determined by the very nature of the experiment [24]. In a letter to Heelan in 1975, Heisenberg also explained that he and Bohr had never really resolved their disagreement. Heisenberg remained convinced "that a cut could be moved around to some extent while Bohr preferred to think that the position is uniquely defined in every experiment" [25]."

Heisenberg's argument for the arbitrary position of the cut reads as follows: "The dividing line between the system to be observed and the measuring apparatus is immediately defined by the nature of the problem but it obviously signifies no discontinuity of the physical process. For this reason there must, within certain limits, exist complete freedom in choosing the position of the dividing line [26]."

Heisenberg's quantum-classical divide is also supported by von Weizsäcker, who tries to relate irreversibility and classicality: "Having thus accepted the falsity of classical physics, taken literally, we must ask how it can be explained as an essentially good approximation [when describing objects at the macrolevel]… This amounts to asking what physical condition must be imposed on a quantum-theoretical system in order that it should show the features which we describe as "classical."… I am unable to prove mathematically that the condition of irreversibility would suffice to define a classical approximation, but I feel confident it is a necessary condition [27]."

The position of an arbitrary cut was also shared by von Neumann [28]. He suggested that the entangled state of the object and the instrument collapses to a determinate state whenever a measurement takes place. So what finally causes such a collapse seems to be the mind of the observer. Other physicists also argued the latter boundary, i.e. the observer's consciousness, such as London and Bauer [29], Peierls [30], Wigner [31], etc.

The philosopher Jan Faye discusses, in his Stanford Encyclopedia article [32], the ambiguities, significances and implications of the Copenhagen interpretation, in particular Howard's and Henderson's careful analyses: "Don Howard [33] goes as far as concluding that "until Heisenberg coined the term in 1955, there was no unitary Copenhagen interpretation of quantum mechanics." The term apparently occurs for the first time in Heisenberg [34]. In addition, Howard also argues that it was Heisenberg's exposition of complementarity, and not Bohr's, with its emphasis on a privileged role for the observer and observer-induced wave packet collapse that became identical with that interpretation. Says he: "Whatever Heisenberg's motivation, his invention of a unitary Copenhagen view on interpretation, at the center of which was his own, distinctively subjectivist view of the role of the observer, quickly found an audience". This audience included people like Bohm, Feyerabend, Hanson, and Popper who used Heisenberg's presentation of complementarity as the target for their criticism of the orthodox view."

Furthermore he says: "Recently, Henderson [35] has come to a similar conclusion. He makes a distinction between different versions of Copenhagen interpretations based on statements from some of the main characters. On one side of the spectrum there is Bohr who did not think of quantum measurement in terms of a collapse of the wave function; in the middle we find Heisenberg talking about the collapse as an objective physical process but thinking that this couldn't be analyzed any further because of its indeterministic nature, and at the opposite side Johann von Neumann and Eugene Wigner argued that the human mind has a direct influence on the reduction of the wave packet. Unfortunately, von Neumann's dualistic view has become part of the Copenhagen mythodology by people opposing this interpretation."

Wigner seems to be the last contributor of the philosophical conception of quantum theory, known today as the orthodox Copenhagen interpretation. Let us take a closer look on a more detailed study of Wigner's writings, such as the preprint of Primas and Esfeld [36], where Wigner's original definition of the measurement process reads:

"[The reduction of the wave packet] takes place whenever the result of an observation enters the consciousness of the observer—or, to be even more painfully precise, my consciousness, since I am the only observer, all other people being only subjects of my observations... The measurement is not completed until its result enters our consciousness. This last step occurs when a correlation is established between the state of the last measuring apparatus and something which directly affects our consciousness. This last step is, at the present state of our knowledge, shrouded in mystery and no explanation has been given for it so far in terms of quantum mechanics, or in terms of any other theory" [31].

Accordingly Primas and Esfeld exhibit Wigner's attempt to avoid the position of excessive subjectivity allowing the person's mind to be in a state of superposition, figuratively known under the name of Wigner's friend.

Moreover they continue: "Wigner's main argument for his view is that it is "difficult to accept the possibility that a person's mind is in a superposition of two states... We ourselves never have felt we were in such superpositions." [37] In this context, Wigner discusses the famous "paradox of Wigner's friend": If one observer describes another observer who observes something like a von Neumann chain and if the first observer describes this whole process in the terminology of quantum mechanics, he will end up with ascribing a superposition of different states of consciousness to the second observer. Wigner avoids this paradox by maintaining that quantum theory does not apply to consciousness so that there are no superpositions of different states of consciousness. A state reduction occurs at the level of the friend's consciousness

[38] ... Wigner's proposal is a move away from the Copenhagen idea that the quantum state represents knowledge available to a community of communicating observers, who have a common knowledge that is useful for making predictions about their combined future experiences. Wigner suggests that each conscious being is able to collapse one single objective quantum state, regardless of whether the information is actually physically shared. It is a move away from an essentially subjective pragmatic interpretation toward a more objective absolute one." From his later years Wigner's view matures in that he expresses a hope that a more general theory will be developed in the future, while being unsatisfied with the present incompleteness of his interpretation. In [39] it is cited: "In his later papers, Wigner expressed his "desire for a less solipsistic theory" but only in last papers did he consider the consequence of solipsism as a sufficient reason to repudiate his earlier views... At the end of a paper published in 1977, Wigner expresses the hope "that quantum mechanics will also turn out to be a limiting case, limiting in more than one regard, and that the philosophy which an even deeper theory of physics will support, will give a more concrete meaning to the word 'reality', will not embrace solipsism, much truth as this may contain, and will let us admit that the world really exists" [39]".

But what does this hope really mean in a more "realistic" description of the material world? Is it identical with Einstein's belief, quote—unquote: „Es gibt so etwas wie den 'realen Zustand' eines physikalischen Systems, was unabhängig von jeder Beobachtung oder Messung objektiv existiert und mit den Ausdrucksmitteln der Physik im Prinzip beschrieben werden kann." [40]? Primas and Esfeld [36] are convinced that the majority of contemporary scientists share this view with Einstein. Even if they are right, this is clearly not a good sign, since it means the comeback of 19th century materialism when classical physics fulfilled all requirements of "objective reality" as inferred above by Einstein.

The authors further polemize with Wigner, quoting some of their most interesting disagreements with him: "Wigner never discussed spontaneously broken symmetries, which are nowadays on all levels of physics, from elementary particle physics to molecular and solid state physics, of utmost importance... On the other hand, Wigner mostly analysed fictitious isolated systems with a finite number of degrees of freedom whose infinite environment he assumed to have a negligible effect. Yet, the interactions of an object system with its environment can have dramatic qualitative effects such as dressing, symmetry breakdown, and the emergence of qualitatively new properties. It is well known that the coupling of small molecular systems with the environment can lead to symmetry breakings which are of great biological significance such as molecular chirality."

Moreover: "Curiously for a scientist with an excellent background in chemistry, Wigner denies the objective reality of molecular chirality: "It should be admitted, first that the concept of the positions of atoms continues to be a useful concept but it has turned out to be an approximate one. It is useful under certain condition but it is not difficult to produce situations in which it is meaningless. A particular striking example is an optically active organic molecule in its normal state. This is neither right handed, nor left handed—in fact, the atoms have no clearly defined positions with respect to each other, not even approximately." [41]"

- and: “Wigner showed much courage in relating the then unresolved questions of the measurement problem to the much deeper problem of consciousness. In view of this very unorthodox proposal it is astonishing that Wigner was very reactionary with respect of the dogmas of orthodox quantum mechanics. In contrast to von Neumann himself, he took the old von-Neumann codification of quantum mechanics as authoritative and not to be questioned. Much of the efforts to interpret the meaning of this codification and to prove no-go theorems, such as the insolubility of the measurement problem or the impossibility of a quantum theory of individual objects, are physically irrelevant since they are based on a codification of quantum mechanics that is valid only for strictly closed systems with finitely many degrees of freedom.”
- and: “It is hard to believe that Wigner did not recognize the importance of the many possible physically inequivalent representations of the canonical commutation relations and the associated symmetry breakings for our understanding of molecular, mesoscopic, and macroscopic phenomena.”

So where is the problem? Wigner knew of course what he was talking about. However, his final challenge for a more “realistic” theory does not meet the above mentioned counterarguments at all. Yet the incompleteness of the orthodox Copenhagen interpretation does not mean that it should be wrong, its incompleteness rather imparts a possible incompleteness of the concepts it is dealing with. It will be shown in this article that we must tackle the very incompleteness of our knowledge of (i) time irreversibility, (ii) the symmetry breaking, and (iii) the role of the environment and (iv) the quantum definition of individual objects. This means, that we cannot confront the incomplete orthodox interpretation by means of a reasoning based on incompletely defined concepts. In other words: these concepts must first be carefully re-examined before we can even start to construct any proposal regarding how to deal with an improved Copenhagen interpretation.

7 The Paradox of Quantum Decoherence

In 1980 a novel decoherence program started to emerge, based on Everett’s Ph.D. thesis from 1957. Everett first summarised the Copenhagen interpretation in its final von Neumann–Wigner form [17]: “We take the conventional or “external observation” formulation of quantum mechanics to be essentially the following: A physical system is completely described by a state function ψ, which is an element of a Hilbert space, and which furthermore gives information only to the extent of specifying the probabilities of the results of various observations which can be made on the system by external observers. There are two fundamentally different ways in which the state function can change:

Process 1: The discontinuous change brought about by the observation of a quantity with eigenstates $\varphi_1, \varphi_2, \ldots$, in which the state ψ will be changed to the state φ_j with probability $|(\psi, \varphi_j)|^2$.

Process 2: The continuous, deterministic change of state of an isolated system with time according to a wave equation $\partial\psi/\partial t = A\psi$, where A is a linear operator. This formulation describes a wealth of experience. No experimental evidence is known which contradicts it."

Everett questioned the whole process of measurement—the problem already well-known at the time—namely the fact that the von Neumann view did only fit a certain type of measurement but is not able the explain all of them: "Von Neumann showed how to treat a special class of approximate measurements by the method of projection operators. However, a general treatment of all approximate measurements by the method of projections operators can be shown to be impossible... von Neumann's example is only a special case of a more general situation. Consider any measuring apparatus interacting with any object system. As a result of the interaction the state of the measuring apparatus is no longer capable of independent definition. It can be defined only relative to the state of the object system. In other words, there exists only a correlation between the two states of the two systems. It seems as if nothing can ever be settled by such a measurement" [17].

Furthermore Everett concentrated only on a set of measurement problems that differs from the von Neumann class, i.e. those excluding observers with a conscious mind and taking only into the consideration devices in the role of "observers". Quoting: "Not all conceivable situations fit the framework of this mathematical formulation. Consider for example an isolated system consisting of an observer or measuring apparatus, plus an object system. Can the change with time of the state of the total system be described by Process 2? If so, then it would appear that no discontinuous probabilistic process like Process 1 can take place. If not, we are forced to admit that systems which contain observers are not subject to the same kind of quantum-mechanical description as we admit for all other physical systems. The question cannot be ruled out as lying in the domain of psychology. Much of the discussion of "observers" in quantum mechanics has to do with photoelectric cells, photographic plates, and similar devices where a mechanistic attitude can hardly be contested. For the following one can limit himself to this class of problems, if he is unwilling to consider observers in the more familiar sense on the same mechanistic level of analysis" [17].

Up to this point the Everett's conclusion is certainly correct. But what about his proposal for a generalization of von Neumann's method based on projection operators? One should perhaps anticipate some generalizations where observers with a conscious mind together with the devices in the role of "observers" should be put side by side in some wider context. But in contrast he simply attempts to remove "Process l" from the Copenhagen postulates. In consequence it means to get rid of two axioms, i.e. the Born rule and the von Neumann–Wigner rule, downgrading the Born rule, not as an axiom but as an emergent rule from "Process 2". Everett finally arrives at his own eccentric and surreal interpretation of quantum mechanics, known now as the MWI (many-worlds interpretation): "We thus arrive at the following picture: Throughout all of a sequence of observation processes there is only one physical system representing the observer, yet there is no single unique state of the observer (which follows from the representations of interacting systems). Nevertheless, there is a representation in terms of a superposition, each element of which contains a definite observer state and a corresponding system state. Thus with each succeeding observation (or interaction), the observer state "branches" into a number of different states. Each branch represents a different outcome of the measurement and the corresponding eigenstate for the object-system state. All branches exist simultaneously in the superposition after any given sequence of observations" [17].

One can assume that Everett's understanding was inspired by Mach's principle of psychophysical paralysis [42] which otherwise also might have inspired Einstein in his theory of relativity. Everett assigns a similar role and position to the observer as ocurring in Einstein's theory of relativity. But there is one significant difference: Whereas in the theory of relativity the observer can be a "body without soul", quantum theory does not allow this in general. For instance the subject (the observer with conscious mind) and the object (the physical word) are not completely interchangeable, since there always must remain "something" from the object as well as from the subject. Goethe was not only a great poet but a great mystic as well. Once he said: "Alles, was im Subject ist, ist im Object und noch etwas mehr. Alles, was im Object ist, ist im Subject und noch etwas mehr." [43] And just this is the reason why the external reality hypothesis (ERH) should not be valid any more in quantum physics, which incorporates the roles of both—object and subject, in contrast with classical physics that concentrates fully only on the external material world. ERH was reformulated by Tegmark [44] as follows: "There exists an external physical reality completely independent of us humans." Hence Tegmark still believes in its validity and claims, and, if it is true, it is a sufficient condition for the justification of the MWI interpretation.

The MWI interpretation has recently inspired many physicists in their efforts to explain classicality as an emergent property of open quantum systems induced by their environments. This is best known as the quantum decoherence program. Zurek [45] introduced a new vernacular to describe the process of wavefunction collapse as well as elucidate the emergence of classical descriptions of reality appearing from quantum descriptions viz. einselection, the latter standing for environment—induced superselection. Although an excellent idea, it was inaptly set into the framework of the Everettian MWI. Such a concept will not work since it leads to a circular argument and therefore was rejected by a large part of the scientific community. Kastner [46] did perform a detailed analysis of this circularity quote, unquote: "In attempting to derive irreversible macroscopic thermodynamics from reversible microscopic dynamics, Boltzmann inadvertently smuggled in a premise that assumed the very irreversibility he was trying to prove: 'molecular chaos.' The program of 'Einselection' (environmentally induced superselection) within Everettian approaches faces a similar 'Loschmidt's Paradox': the universe, according to the Everettian picture, is a closed system obeying only unitary dynamics, and it therefore contains no distinguishable environmental subsystems with the necessary 'phase randomness' to effect einselection of a pointer observable. The theoretically unjustified assumption of distinguishable environmental subsystems is the hidden premise that makes the derivation of einselection circular. In effect, it presupposes the 'emergent' structures from the beginning. Thus the problem of basis ambiguity remains unsolved in Everettian interpretations."

It's surprising how many physicists, seeing the difficulties with the wave function collapse and the von Neumann–Wigner interpretation, yet jump happily into the above-mentioned decoherence program, not realizing the circularity of all these problems. For instance, before we can start to speak about some environmental subsystems, we must first attend to the third of Bohm's objections—wholeness versus fragmentation—discussed in Sect. 4. Since quantum mechanics deals with isolated subsystems, hence we must decipher the puzzle formulated by Sutcliffe and Woolley, i.e. the isolated versus the individual, as considered in Sect. 2. This is crucial to understanding the origin of 'emergent' structures, which if not will be falsely presupposed in the MWI interpretation, as Kastner correctly did notice. Only then can we begin a reasonable examination about the true meaning of Zurek's einselection. In passing one notes that Zurek later came forward with another promising idea, which he called quantum Darwinism [47] as a part of universal Darwinism. The thesis of universal Darwinism claims that Darwinian processes are not confined to biology and can be applied in other scientific branches. The Darwinian algorithm consists of three parts: (1) random variation in mutations or recombinations; (2) selection of the fittest variants; (3) heredity of these features in offsprings.

Darwin worried about randomness in the variation, since in his time classical physics determinism prevailed in philosophy and random processes descending from indeterministic principles were impermissible. This is not a problem today, when quantum physics has revealed such a principle. On the other hand, Darwin did not worry about the selective process, although he was convinced that it must be of a teleological origin, but in this case he was not in conflict with contemporary science. Exact sciences such as physics simply were and still are not able to incorporate teleology in an authentic, precise logical and mathematical manner, but at the same time it cannot be denied. In contrast to indeterminancy, which is opposite to determinancy, teleology is not opposite to causality (the opposite of "causal" is "acausal").

In a historical perspective Darwin was a teleologist. Lennox, for instance, shows in his paper [48] that Darwin uses the term 'Final Cause' consistently in his Species Notebook, the Origin of Species and also afterwards, and that evolution—as conceived by Darwin—is the result of mutations arising by chance and the selection which is teleological in nature. Here Lennox also quotes from the correspondence between Darwin and his close friend and colleague Asa Gray: "In an appreciation of Charles Darwin published in Nature in 1874, Asa Gray noted "Darwin's great service to Natural Science" lies in bringing back Teleology "so that, instead of Morphology versus Teleology, we shall have Morphology wedded to Teleology". Darwin quickly responded, "What you say about Teleology pleases me especially and I do not think anyone else has ever noticed the point."... Francis Darwin and T. H. Huxley reiterate this sentiment. The latter wrote that "...the most remarkable service to the philosophy of Biology rendered by Mr. Darwin is the reconciliation of Teleology and Morphology, and the explanation of the facts of both, which his view offers.""

Only a few biologists seem at present to be aware of the true meaning of teleology, at least to the extent that was understood and expressed by Darwin. Today the majority of them appears quite confused. We can find many absurd propositions, such as "artificial selection is teleological and natural selection is not, since in the latter there is no intentional choice." Teleology is defined as downward causation or the final cause, and is quite independent of the fact whether there are some intentions behind or not. As examples one might mention statements of two prominent biologists: (a) the famous evolutionary biologist Ernst Mayr, although introducing a more explicit terminology, e.g. teleonomic processes as those owing its goal-directedness to the influence of an evolved program, expressed [49]: "adaptedness… is an a posteriori result rather than an a priori goal-seeking", (b) Madrell [50]: "the proper but cumbersome way of describing change by evolutionary adaptation may be substituted by shorter overtly teleological statements for the sake of saving space, but this should not be taken to imply that evolution proceeds by anything other than from mutations arising by chance, with those that impart an advantage being retained by natural selection." One might wonder how much these attitudes of teleology of contemporary biologists really agree with Darwin's original view.

The opinions and attitudes of most physicists regarding teleology are both negative and ignorant. In contrast Bohm (cf. the first objection mentioned in Sect. 4) warns us that disrespecting teleological principles prevent the construction of a full quantum theory. What is most surprising indeed, is that the teleology problem appeared for the first time already in classical physics. For instance Maxwell became immediately aware of it, when he studied the work of Boussinesq, Cournot, and St Venant dealing with the solution of hydrodynamic equations. Maxwell was fascinated by the occurrence of singular points leading to complete unpredictability of future states.

On 26 February 1879, he wrote a letter to Charles Darwin's half-cousin Francis Galton [51]: "Do you take any interest in Fixed Fate, Free Will etc. If so Boussinesq [of hydrodynamic reputation] 'Conciliation du véritable déterminisme mécanique avec l'existence de la vie et de la liberté morale' (Paris 1878) does the whole business by the theory of singular solutions of the differential equations of motion… There are certain cases in which a material system, when it comes to a phase in which the particular path which it is describing coincides with the envelope of all such paths may either continue in the particular path or take to the envelope (which in these cases is also a possible path) and which course it takes is not determined by the forces of the system (which are the same for both cases) but when the bifurcation of path occurs, the system, ipso facto, invokes some determining principle which is extra physical (but not extra natural) to determine which of the two paths it is to follow. When it is on the enveloping path it may at any instant, at its own sweet will, without exerting any force or spending any energy, go off along that one of the particular paths which happens to coincide with the actual condition of the system at that instant… But I think Boussinesq's method is a very powerful one against metaphysical arguments about cause and effect and much better than the insinuation that there is something loose about the laws of nature, not of sensible magnitude but enough to bring her round in time."

One must admire Maxwell's brilliant insights observing how he mentions teleological principle: principle which is extra physical but not extra natural; principle acting without exerting any force or spending any energy! Certainly, classical physics was unable to incorporate such a principle, but it begs the question: Is it possible to accomplish this task in quantum physics? Unfortunately most physicists are reluctant to tackle the problem, perhaps being victims of "causal fundamentalism", a wording introduced by Norton [52]: "Nature is governed by cause and effect; and the burden of individual sciences is to find the particular expressions of the general notion in the realm of their specialized subject matter."

Norton continues with a pessimistic feeling of vanity and emptiness that hovers over the current debate on the subject of teleology: "Aristotle described four notions of cause: the material, efficient, final and formal; with the efficient and final conforming most closely to the sorts of things we would now count as a cause. The final cause, the goal towards which a process moves, was clearly modelled on the analogy between animate processes and the process of interest. In the seventeenth century, with the rise of the mechanical philosophy, it was deemed that final causes simply did not have the fundamental status of efficient causes and that all science was to be reconstructed using efficient causes alone. Although talk of final causes lingers on, this is a blow from which final causes have never properly recovered."

Here lies a profound challenge for all of us. Until the teleological principle is adequately implemented in the mathematical formalism of quantum theory, the whole program of quantum decoherence, based on universal Darwinism, suggests nothing more than many years of squandering time. It imparts, e.g. that the von Neumann–Wigner interpretation can never be removed and that the causal Born rule cannot be replaced by einselection, the latter being inspired by teleological Darwinian adaptedness. Instead, the Born and einselection rules should coexist side by side, but first after the true teleological meaning and quantum origin of the einselection have been revealed.

8 The Paradox of Free Will

What is meant by the notion of free will? Does free will really exist at all? Observe that in the era of classical physics it was generally accepted that all things in the universe do evolve in a strictly deterministic manner, and hence there was obviously no room for any justification of a free will. But what happened after the arrival of quantum mechanics, and the knowledge of the indeterministic laws of quantum physics? Is this agreement sufficient to give a reasonable explanation of free will? As will be seen below, the problem of a proper understanding of our free will still remains.

The information philosopher, Robert Doyle summarizes, in his 2011 book "The Scandal in Philosophy" [53], in great detail the whole history of opinions of famous philosophers and scientists on the topic of free will. He maintains that there prevails a sceptical view regarding possible satisfactory free will explanations, which he calls the "standard argument against free will" [53]: "The standard argument has two parts. (1) If determinism is the case, the will is not free. We call this the Determinism Objection. (2) If indeterminism and real chance exist, our will would not be in our control. We could not be responsible for our actions if they are random. We call this the Randomness Objection. Together, these objections can be combined as a single Responsibility Objection, namely that no Free Will model has yet provided us an intelligible account of the agent control needed for moral responsibility."

From the standpoint of a historical retrospective analysis, it seems that only the two-stage model could be a true candidate for the explanation of free will. This two-stage model consists of two processes: the process of random alternatives followed by the selection process of one choice. Doyle writes in his book that this idea appeared for the first time in 1884 and that it was formulated by William James. As a psychologist he felt no problem in denying determinism in the psychological domain of free will: "The stronghold of the determinist argument is the antipathy to the idea of chance… This notion of alternative possibility, this admission that any one of several things may come to pass is, after all, only a roundabout name for chance" [54].

It is interesting to note how physicists in general accepted the two-stage model. The first reaction came from Poincaré, at the time when quantum indeterminism had not yet arrived. Hence he does not refer to the laws of physics, concentrating mostly on the role of our unconscious mind as a source of random combinations and possibilities, but also on our conscious mind where the definite process of validation occurs [55]: "It is certain that the combinations which present themselves to the mind in a kind of sudden illumination after a somewhat prolonged period of unconscious work are generally useful and fruitful combinations… All the combinations are formed as a result of the automatic action of the subliminal ego, but those only which are interesting find their way into the field of consciousness… A few only are harmonious, and consequently at once useful and beautiful, and they will be capable of affecting the geometrician's special sensibility I have been speaking of; which, once aroused, will direct our attention upon them, and will thus give them the opportunity of becoming conscious… In the subliminal ego, on the contrary, there reigns what I would call liberty, if one could give this name to the mere absence of discipline and to disorder born of chance."

The physicist Sir Roger Penrose shares the synoptical view of the two-stage model with Poincaré, particularly as concerns the role of our unconscious and conscious minds [56]: "In relation to this, the question of what constitutes genuine originality should be raised. It seems to me that there are two factors involved, namely a 'putting-up' and a 'shooting-down' process. I imagine that the putting-up could be largely unconscious and the shooting-down largely conscious. Without an effective putting-up process, one would have no new ideas at all. But, just by itself, this procedure would have little value. One needs an effective procedure for forming judgements, so that only those ideas with a reasonable chance of success will survive. In dreams, for example, unusual ideas may easily come to mind, but only very rarely do they survive the critical judgements of the wakeful consciousness. In my opinion, it is the conscious shooting-down (judgement) process that is central to the issue of originality, rather than the unconscious putting-up process; but I am aware that many others might hold to a contrary view."

A similar description of the two-stage model was also promoted by Compton [57]: "A set of known physical conditions is not adequate to specify precisely what a forthcoming event will be. These conditions, insofar as they can be known, define instead a range of possible events from among which some particular event will occur. When one exercises freedom, by his act of choice he is himself adding a factor not supplied by the [random] physical conditions and is thus himself determining what will occur." Compton was the first physicist, who discussed openly the relationship between free will and the quantum indeterminism and the randomness on the first stage of the two-stage model, i.e. what Penrose calls the 'putting-up' process. On the other hand, he was very pessimistic about the scientific explanation of the second stage—the act of choice, or Penrose's 'shooting-down' process.

It is notable that present-day quantum physics is still incomplete and unable to create a paradigmatic background for the justification of the second stage of free will [57]: "Let me then summarize how a physicist now views man's freedom. It is not from scientific observation that we know man is free. Science is incapable of telling whether a person's acts are free or not. Freedom is not something that one can touch or measure. We know it through our own innermost feelings. The first essential of freedom is the desire to attain something that one considers good. But desire lies outside the realm of science—at least outside of physics. You can't locate desire as somewhere in space. Similarly our recognition that within limits we can do what we try to do is not a matter of measurement or of external observation. It is a matter of immediate awareness. There is nothing in such awareness of freedom that is inconsistent with science. Freedom does, however, involve the additional determining factor of choice, about which science tells us nothing."

The citation above means that quantum theory, based on standard interpretations such as the orthodox Copenhagen view, forms only a necessary, but not a sufficient foundation for the explanation of the free will. This circumstance is clearly confirmed by Flanagan [58]: "Free actions, if there are any, are not deterministically caused nor are they caused by random processes of the sort countenanced by quantum physicists or complexity theorists. Free actions need to be caused by me, in a non-determined and non-random manner… I am open to there being genuine ontological indeterminacy at both the quantum level and the level of neural processing. But the attempt to gain free will from indeterminacy at the quantum level or at the level of global brain processes is a bad idea. The last thing anyone wants is for free will to be the result of random causal processes."

It is fetching to learn that Flanagan's representation of "non-determinism and non-randomness" as the main characteristics of free actions, appears under the name "adequate determinism" in Doyle's Cogito model [53]: "Free will is a combination of microscopic randomness and macroscopic adequate determinism, in a temporal sequence—first chance, then choice. Creative thoughts come to us freely. But our actions come from our adequately determined wills. Randomness is the "free" in free will. Adequate determinism explains the "will" in free will." One can understand adequate determinism as something indeterministic in principle, but unlike the laws of microscopic physics, it must not be random. Doyle thinks that a macroscopic ensemble, which statistically restrains randomness, fulfils this requirement. But our decisions are not statistical in character, since we often have to decide in a binary style such as "either—or". Moreover, our decisions have teleological traits, while the fortitude of macroscopic physics has nothing to do with it. One might think that Doyle is not aware of the incompleteness of quantum theory and therefore tries to explain free will at any cost, by only using the traditional understanding of quantum physics. The issue of adequate determinism must be solved exclusively on the quantum level. However, up till now, presently known interpretations of quantum theory are not capable of dealing with phenomena, which are simultaneously non-deterministic and non-random.

One notes that Compton's discussion of the closeness between quantum indeterminism and randomness in free will finally did convince Popper, who originally denied any influence of microphysical laws upon the psychological domain of free will. Besides, the similarity of the two-stage models, used in Darwin theory and in the concept of free will, caused a radical change of Popper's attitude towards natural selection, which he at first deemed to be a "tautology" making it unfalsifiable. In his 1977 Darwin Lecture, Darwin College Cambridge, Popper confessed his change of mind [59] by stating: "The selection of a kind of behavior out of a randomly offered repertoire may be an act of choice, even an act of free will. I am an indeterminist; and in discussing indeterminism I have often regretfully pointed out that quantum indeterminacy does not seem to help us; for the amplification of something like, say, radioactive disintegration processes would not lead to human action or even animal action, but only to random movements. This is now the leading two-stage model of free will. I have changed my mind on this issue. A choice process may be a selection process, and the selection may be from some repertoire of random events, without being random in its turn. This seems to me to offer a promising solution to one of our most vexing problems, and one by downward causation."

This formulation anticipates the concept of unification of a single two-stage model that holds both for Darwinism and free will. Eccles and Popper [60] formulate the view as: "New ideas have a striking similarity to genetic mutations. Now, let us look for a moment at genetic mutations. Mutations are, it seems, brought about by quantum theoretical indeterminacy (including radiation effects). Accordingly, they are also probabilistic and not in themselves originally selected or adequate, but on them there subsequently operates natural selection which eliminates inappropriate mutations. Now we could conceive of a similar process with respect to new ideas and to free-will decisions, and similar things. That is to say, a range of possibilities is brought about by a probabilistic and quantum mechanically characterised set of proposals, as it were—of possibilities brought forward by the brain. On these there then operates a kind of selective procedure which eliminates those proposals and those possibilities which are not acceptable to the mind."

Although one might, in such a way, achieve a unification of nature's laws of biology and psychology, there still remains a fundamental problem: There appears a void or lack of a hitherto unravelled fundamental natural law in science in general and quantum theory in particular, which is mandatory for a proper understanding and justification of the second part of two-stage model. This is the conundrum of the selection process in biological evolution and the choice process in the psychology of free will. Contemporary quantum theory allows only an explanation of the first part of two-stage model, which is not sufficient for a free will justification. In fact this is exactly in line with Bohr's opinion [61]: "To connect free will more directly with the limitation of causality in atomic physics, as it is often suggested, is, however, entirely foreign to the tendency underlying the remarks made here about biological problems." Bohr knew that quantum physics gives no answer to the selection process (the second part of two-stage model), and he therefore only called upon the recognition of relevant psychological experiences [61]: "To illustrate the argument, we may briefly refer to the old problem of free will. From what has already been said it is evident that the word volition is indispensable to an exhaustive description of psychical phenomena, but the problem is how far we can speak about freedom to act according to our possibilities. As long as unrestricted deterministic views are taken, the idea of such freedom is of course excluded. However, the general lesson of atomic physics, and in particular of the limited scope of mechanistic description of biological phenomena, suggests that the ability of organisms to adjust themselves to environment includes the power of selecting the most appropriate way to this purpose. Because it is impossible to judge such questions on a purely physical basis, it is most important to recognize that psychological experience may offer more pertinent information on the problems."

Bohr's statement above represents a synoptical vision, cf. the one of Eccles and Popper, unifying the principles behind biological and psychical phenomena. However, the problem of the second part of the two-stage model must indeed have a solution within a satisfactory extension of the present understanding of quantum theory, originally formulated for the description of inanimate nature, but such a solution does not yet exist on the level of quantum physics as restricted by the Copenhagen interpretation.

9 The Paradox of the Meissner Effect

The strange phenomenon of superconductivity was coined by Heike Kamerlingh Onnes in his two ground breaking papers [62], describing the complete disappearance of electrical resistance of mercury observed in 1911 by his assistant Gilles Holst at the Leiden research lab later known as the Kamerlingh-Onnes laboratory. Since then superconductors have been regarded as ideal conductors with zero resistance. The first theoretical articles on the subject of superconductivity, by Becker et al. [63], reflected this ideal conductivity, deriving the relation between the current and the electrical field, replacing Ohm's law in normal conductors. Starting with the expression for current

$$\mathbf{j} = en\mathbf{v} \tag{9.1}$$

where e is the electron charge, n the density of superconducting carriers and $\mathbf{v}$ the superfluid velocity, and further using Newton law for the carriers with mass m, they obtained the final expression, Becker's law

$$\frac{m}{ne^2}\frac{\partial \mathbf{j}}{\partial t} = \mathbf{E} \tag{9.2}$$

From (9.2) the expulsion of the magnetic field from superconductors can be predicted assuming that the external magnetic field was switched on after cooling below the critical temperature. However, if the order of these events was altered, the magnetic field was expected to remain inside the superconductor without any change. The subsequent experiment by Meissner and Ochsenfeld [64] brought a huge surprise, i.e. real superconductors, unlike ideal conductors, expelled the magnetic field from their interior even in the presence of a constant field with the succeeding transition into the superconducting state. The Meissner effect is hence in strong contradiction with Becker's law, since a constant field cannot produce any electromotive force bringing superconducting carriers in motion.

Two years later the so-called London equations were derived by London and London [65] yielding phenomenological equations in compliance with the Meissner effect. First by using Faraday's law

$$\nabla \times \mathbf{E} = -\frac{1}{c}\frac{\partial \mathbf{B}}{\partial t} \tag{9.3}$$

they rewrote the Becker's Eq. (9.2) in its curl form

$$\frac{m}{ne^2}\nabla \times \frac{\partial \mathbf{j}}{\partial t} = -\frac{1}{c}\frac{\partial \mathbf{B}}{\partial t} \tag{9.4}$$

and then integrating it over time

$$\nabla \times \mathbf{j} = -\frac{ne^2}{mc}\mathbf{B} \tag{9.5}$$

putting the integration constant to zero as a specific ansatz for true superconductors, where the London Eq. (9.5) replaces Ohm's law in conductors. Applying Ampere's law

$$\nabla \times \mathbf{B} = \frac{4\pi}{c}\mathbf{j} \tag{9.6}$$

this equation can be written in the form

$$\lambda^2 \Delta \mathbf{B} = \mathbf{B} \tag{9.7}$$

where λ stands for the London penetration depth

$$\lambda = \sqrt{\frac{mc^2}{4\pi ne^2}} \tag{9.8}$$

The London Eq. (9.5) differs from the Becker's one (9.4) in one small detail: an ansatz in the form of zero integration constant, corresponding to the simple but important fact, that real superconductors, in contrast to ideal conductors, exhibits no memory effects.

A similar distinction is observed for the case of rotating objects, i.e. both ideal conductors and superconductors exhibit the same behavior, if they are accelerated below the critical temperature. Becker et al. [63] derived the relation between the angular velocity $\boldsymbol{\omega}$ and a generated magnetic field in the interior of ideal conductor. Substituting for $\mathbf{v}$ in Eq. (9.1) i.e.

$$\mathbf{v} = \boldsymbol{\omega} \times \mathbf{r} \tag{9.9}$$

one obtains from the Becker's Eq. (9.4) after integration the final relation

$$\mathbf{B} = -\frac{2mc}{e}\boldsymbol{\omega} \tag{9.10}$$

If an ideal conductor is first accelerated and subsequently cooled, instead of (9.10) after the integration, we get only the trivial solution with $\mathbf{B}$ equal to zero. Related to the Meissner effect, London [66] predicted an analogous effect also for rotating superconductors, where the Becker's relation (9.10) holds for superconductors regardless of its history, i.e. including the case when acceleration occurred first, followed by the cooling. London's derivation of Eq. (9.10) is based on a direct substitution in his Eq. (9.5), so that the generated magnetic field would always be the same, independent of the order of the acceleration and the cooling events. The generated magnetic moment is called the London moment. London's prediction for rotating superconductors was for the first time experimentally verified in 1964 by Hildebrandt [67].

It is intriguing to ask whether London's ansatz to Becker's equations is necessary? Alternatively, can we have an ansatz-free classical description of superconductors with an *ab initio* derivation of the London equations based solely on the laws of classical mechanics and classical electrodynamics? This question has in fact divided the scientific community into two camps. For instance Eq. (9.5) was elegantly derived by de Gennes [68] by minimizing the total energy. He wrote the expressions for the kinetic energy and the energy of magnetic field:

$$E_k = \frac{1}{2}\int nm\mathbf{v}^2 dV = \frac{1}{2}\int \frac{m}{ne^2}\mathbf{j}^2 dV \tag{9.11}$$

$$E_f = \frac{1}{8\pi}\int \mathbf{B}^2\, dV \tag{9.12}$$

Using Ampere's law (9.6) the total energy reads

$$E = E_k + E_f = \frac{1}{8\pi} \int \left[\mathbf{B}^2 + \lambda^2(\nabla \times \mathbf{B})^2\right] dV \tag{9.13}$$

and by minimizing this expression with respect to **B** he got

$$\mathbf{B} + \lambda^2 \nabla \times (\nabla \times \mathbf{B}) = 0 \tag{9.14}$$

which is exactly the London Eq. (9.7). This derivation was recapitulated by Essén and Fiolhais quoting [69]: "The conclusion of de Gennes is clearly stated in his 1965 book [68] (emphasis from the original): "The superconductor finds an equilibrium state where the sum of the kinetic and magnetic energies is minimum, and this state, for macroscopic samples, corresponds to the expulsion of magnetic flux." In spite of this, most textbooks continue to state that "flux expulsion has no classical explanation" as originally stated by Meissner and Ochsenfeld [64] and repeated in the influential monographs by London [66] and Nobel laureate von Laue [70]." They continue with a list of authors preferring classical derivations of the Meissner effect [69]: "Many other authors have also reached the conclusion that the equilibrium state of a superconductor is in fact the state of minimum magnetic energy of an ideal conductor. Some of these are, in chronological order, Cullwick [71], Pfleiderer [72], Karlsson [73], Badía-Majós [74], Kudinov [75], and Mahajan [76]. For example, Cullwick writes "The well-known Meissner effect in pure superconductors is shown to be an expected rather than an unexpected phenomenon…". " And finally quoting [69]: "The reader may get the impression from our investigations above that we consider superconductivity to be a classical phenomenon. Nothing could be further from the truth… Since quantum physics must lead to classical physics in some macroscopic limit, it must be possible to derive our classical result from a quantum perspective. What we want to correct is the mis-statement that the Meissner effect proves that superconductors are "not just perfect conductors." According to basic physics and a large number of independent investigators, the specific phenomenon of flux expulsion follows naturally from classical physics and the zero resistance property of the superconductors—they are just perfect conductors."

Summarizing this is the paradox of the Meissner effect. We have two camps of physicists—the first follows Meissner and London, and claims, that superconductivity in contrast to ideal conductivity has no classical explanation, and that quantum physics is necessary for its explanation. The second camp on the other hand identifies superconductors with perfect classical conductors. Who is right? A personal answer might appear shocking: no one! Perhaps a more correct resolution of this dilemma, caused by the two factions of physicists, might open the door for an understanding of the true nature of superconductivity, helping us to find accurate equations and precise descriptions and explanation.

As one can see, Becker's and London's equations are almost identical except one small but important difference: the latter ones have no history, indicating that causality is broken. Now, does some archetype of causality breaking occur in classical physics? Yes, it does and it refers back to Norton's dome, described in Chapter 3 "A-causality in Classical Physics", Ref. [52]: "The dome sits in a downward directed gravitational field, with acceleration due to gravity g. The dome has a radial coordinate r inscribed on its surface and is rotationally symmetric about the origin $r = 0$, which is also the highest point of the dome. The shape of the dome is given by specifying h, how far the dome surface lies below this highest point, as a function of the radial coordinate in the surface, r. For simplicity of the mathematics, we shall set $h = (2/3\ g)r^{3/2}$. (Many other profiles, though not all, exhibit analogous a-causality.)" Norton's dome has two kinds of solutions. The first one is trivial and describes the situation where the mass simply remains at rest at the apex for all time:

$$r(t) = 0 \tag{9.15}$$

The second corresponds to another large class of unexpected solutions. For any radial direction the following equations hold:

$$r(t) = 0; \quad t \leq T \tag{9.16a}$$

$$r(t) = (1/144)(t - T)^4; \quad t \geq T \tag{9.16b}$$

where T is an arbitrarily chosen, positive constant.

For this reason, the paradox associated with the Meissner effect appears to be of the same origin as the Norton dome paradox. From the point of view of classical physics ideal conductors and superconductors represent one common Norton-dome-like system with two solutions: one causal for ideal conductivity with the trivial solution, and one teleonomic for superconductivity, allowing the application of the principle of energy minimum reminiscent of de Gennes' derivation of London's equation. It is really a curiosity of classical physics, that both solutions—causal and telic—are still entirely consistent with the mathematics of Newton's laws of motion. Unfortunately, the teleological features of superconductivity have not yet been revealed, so a true understanding of this phenomenon has remained concealed.

The Nobel Laureates Ginzburg and Landau elaborated the description and the structure of the classical phenomenological description of superconductors by simulating the free energy F in terms of a complex order parameter Ψ [77]:

$$F_{GL} = \alpha|\Psi|^2 + \frac{\beta}{2}|\Psi|^4 + \frac{\hbar^2}{2m}\left|\left(\nabla - \frac{2ie}{\hbar c}\mathbf{A}\right)\Psi\right|^2 + \frac{1}{8\pi}|\mathbf{B}|^2 \tag{9.17}$$

which after minimizing with respect to variations in the order parameter Ψ and the vector potential $\mathbf{A}$ leads to the celebrated Ginzburg-Landau equations. Equation (9.17), where α, β are phenomenological parameters, contains both quadratic and biquadratic functions of the order parameter, forming the well-known Mexican hat, which is similar to the Norton dome, albeit being fully Lipschitz continuous. Actually, the solution of Eq. (9.17) based on the minimization process corresponds to the acausal Norton dome solution (9.16).

Returning to the controversy, related to the interpretation of the Meissner effect, it is true that both camps believe in the universal validity and consequences of Bohr's correspondence principle, i.e. that provided ideal conductors are (are not) identical with superconductors in classical physics, they are (are not) identical in quantum physics as well, and vice versa. Adding the fact, that superconductivity has a teleonomic component, Bohr's correspondence principle does not hold in this case, since quantum physics, at least in its Copenhagen version, is solely causal. This limitation of the Copenhagen interpretation was confronted by Bohm in his first critical point discussed in Sect. 4. The consequences of this ignorance are significant. As already stated, quantum physics cannot deal with teleological phenomena implying that it, in view of what has been said above, cannot explain superconductivity. It can hence only describe superconductors and distinguish them from ideal conductors. Taking the point further, it cannot realize telicity in the superconducting phenomenon, indicating that the only possibility to describe superconductors within quantum physics would be to regard them as insulators. Of course, the official attitude of physicists today is different and the majority still believes that superconductivity is essentially explained within the framework of quantum physics. Nonetheless, such calculations were not fully logical since they conflated the classical teleological Norton-dome-like potentials with microscopic quantization procedures.

So we reach a fascinating conclusion: Classical physics does not distinguish superconductors from ideal conductors (it views them as one system with two possible solutions—one teleological, one causal), whereas quantum physics on the other hand does not distinguish them from insulators (since the Bohr principle of correspondence cannot be applied to teleological phenomena). We bring forward now three other confirmations that contemporary quantum physics, based on the Copenhagen interpretation, cannot really explain superconductivity.

Quantum theory of superconductivity must necessarily explain the Meissner effect. But it was never done: either in the first officially accepted microscopic BCS theory [78], or in any of its clones or replacements. The problem is that the electromotive forces described by Faraday's law of induction are equal to zero in stationary conditions of the Meissner effect, whereas the existing theories do not suggest any other electric forces needed to accelerate the electrons until the steady state supercurrent described by the London equation is achieved. Many physicists were inspired by this problem, and in the last decade we have observed several attempts to solve it, but such solutions go beyond the boundaries of contemporary classical or quantum physics. E.g. Kozynchenko [79] proposed a way out that requires an ad hoc reformulation of the laws of classical electrodynamics for superconductors. On the other hand, Hirsch [80] did submit another solution, transcending ad hoc laws of quantum physics. As he wrote many notable and significant articles devoted to this topic, we will quote and discuss some of them below.

In his paper "The origin of the Meissner effect in new and old superconductors" Hirsch [81] summarizes the position of all contemporary theories of superconductivity particularly regarding their inability to explain the Meissner effect: "In a somewhat circular argument, a 'conventional' superconductor is defined to be a superconductor described by BCS theory. In addition there are by now at least 10 different classes of materials that are generally believed to be 'unconventional', i.e. not described by BCS theory... And certainly there is no single 'unconventional mechanism' proposed to describe all unconventional superconductors: new mechanisms are being proposed that apply specifically to one family only, e.g. the cuprates, or the iron pnictides, or the heavy fermion superconductors. However all superconductors, whether conventional or not, exhibit the Meissner effect. I argue that BCS theory cannot explain the Meissner effect, so it cannot explain any superconductor. Furthermore, none of the unconventional mechanisms proposed to explain 'unconventional' superconductivity has addressed the question of how to explain the Meissner effect. I argue that none of these mechanisms describe any superconductor because they cannot explain the Meissner effect."

The problem of the transition into the superconducting state after cooling is not the only one. For instance, what goes on when, beyond the critical temperature, a superconductor becomes conductor again. What happens with the momentum of the supercurrent? Hirsch, in his paper "The disappearing momentum of the supercurrent in the superconductor to normal phase transformation" [82] writes: "A conundrum that didn't exist before was thus created by Meissner's discovery: if there are no collision processes that dissipate Joule heat in the superconductor-to-normal transition in the presence of a magnetic field, what happens to the mechanical momentum of the disappearing current? The kinetic energy of the current is 'stored' in the normal state electronic state, but its momentum is not. Of course the only possible answer is that the momentum of the current is transmitted to the body as a whole. But what is the physical mechanism by which this transfer of momentum happens with no energy transfer and no energy dissipation? Surprisingly this basic and fundamental question has never been asked (nor answered) in the extensive literature on superconductivity since 1933 (213,616 papers according to the Web of Science)."

In the author's opinion, Hirsch is absolutely right, as far as the two statements given above. Does it mean that the Meissner effect and superconductivity have no quantum physical solutions? Hirsch thinks that they have, and revives in his work "The Bohr superconductor" [80] some old ideas from the pre-BCS era. He shows that any superconductor, described classically, is equivalent to a system of charge carriers rotating in the whole bulk in orbits of radius 2λ. He presents several ways of solution; the simplest and most transparent is the following: The total angular momentum of the Meissner current in a long cylinder of radius R and height d with applied magnetic field parallel to the cylinder axis can be written in the two equivalent forms

$$L = (2\pi R d\lambda n)(mvR) = \left(\pi R^2 dn\right)(mv(2\lambda)) \tag{9.18}$$

where the first expression describes the angular momentum of the supercurrent flowing within λ of the surface, and the second one describes the angular momentum of all the charge carriers in the bulk in their orbits of radius 2λ. The angular momentum of each electron in its circular orbit yields

$$l = mv(2\lambda) \tag{9.19}$$

From these results Hirsch infers: "It is natural to conclude that electrons reside in such orbits even in the absence of an applied magnetic field, as opposed to assuming that the 2λ orbits are somehow 'created' by the applied field… It should be noted that the hypothesis that superconducting electrons reside in large orbits was made by several researchers in the pre-BCS era [83–85]."

At this point a serious problem appears, i.e. when the classical concept of mesoscopic orbits is quantized (for details of the derivation of quantum equations see the original paper). This forces Hirsch to declare that such a solution exceeds the rules of quantum physics: "The Bohr atom starts from some simple assumptions and deduces that the angular momentum of the electron in Bohr orbits is quantized in integer units of $\hbar$… Similarly we point out here that from some simple assumptions it can be deduced that electrons in superconductors reside in mesoscopic orbits with orbital angular momentum $\hbar/2$… The fact that the orbital angular momentum in these orbits in the superconductor is found to be a half integer rather than an integer multiple of $\hbar$ is very remarkable. The correct interpretation of this finding could have profound implications. We have suggested that it indicates an intrinsic double-valuedness of the electron wavefunction, in contradiction with conventional quantum mechanics. Other less radical interpretations may be possible." Hirsch closes his paper, expressing his desire to find the wavefunction, that will obey the above mentioned concept of mesoscopic electron orbits: "What we are proposing is that the correct wavefunction of the superconductor, when it is found, will necessarily show physical properties consistent with the picture provided by quantized 2λ orbits with angular momentum $\hbar/2$ and the associated macroscopically inhomogeneous charge distribution. The BCS wavefunction does not."

No one up to now has been lucky to discover such a wavefunction, including Hirsch himself. According to Bohr's principle of correspondence, this wavefunction must exist being derived ab initio from first principles, i.e. only from the knowledge of the Schrödinger equation taking into account a certain number of nuclei and electrons. One should note that superconductors are crystals with the same discrete translational symmetry as other solids, yet from a mathematical point of view it is impossible to find some type of "rotational symmetry" leading to the solution of mesoscopic circle orbits. How is it possible to unravel such a puzzle? The answer is simple repeating the statement that Bohr's principle of correspondence for superconductors is broken and quantum physics is unable to explain them.

Continuing the focus will be set on the problem of rotating superconductors, especially on the relation between the magnetic field and the angular velocity (9.10). The obvious question befalls: what type of mass m does appear in this equation? Is it a bare or an effective electron mass? Hirsch argues in his paper "The London moment: what a rotating superconductor reveals about superconductivity" [86] and quotes a huge number of experimental work dealing with measurements of both the low-T_c as well as the high-T_c superconductors. The results confirm without exception the validity of using the bare electron mass in Eq. (9.10). Hirsch continues: "BCS theory does not describe any change in the character of the electronic states other than the pairing correlations, and the electric current in the normal state is carried by carriers with effective mass m^* rather than bare mass m, so BCS theory is in disagreement with experiment. Furthermore, BCS theory attributes the pairing of the carriers to the electron-phonon interaction, which is completely inconsistent with the evidence that superfluid carriers become free of interactions with the ionic lattice."

Hirsch further proposes the concept of mesoscopic orbits as a possible solution for the mass problem in the London moment. As already mentioned, no one has found any microscopic wave function describing such mesoscopic orbits, since it is impossible to derive from the mathematical aspects of crystal symmetry. Does there exist some microscopic explanations in favor of the bare mass? Naturally one cannot avoid the universal concept of Bloch states for the description of carriers, and it does not matter if these carriers are paired or unpaired. If they represent original electrons or are replaced by polarons, bipolarons etc., we will always arrive at the effective mass which contradicts the experimental facts of the bare mass in the London moment. For the third time we conclude that Bohr's principle of correspondence is broken for superconductors and quantum physics is unable to explain them.

Finally we turn our attention to the experimental constituent in Eq. (9.1). While the macroscopic current is measurable, the density and velocity of carriers are not. In his last paper [87] Hirsch proposes a method for the velocity measurement: "Experimentally the speed of the supercurrent in superconductors has never been measured and has been argued to be non-measurable, however we point out that it is in principle measurable by a Compton scattering experiment. We predict that such experiments will show that superfluid carriers respond to an applied magnetic field according to their bare mass, in other words, that they respond as free electrons, undressed from the electron-ion interaction, rather than as Bloch electrons."

If superconductivity is going to have a microscopic explanation, Hirsch's request is absolutely legitimate. The velocity of carriers in superconductors must be measurable in the same manner as in conductors. Does this indicate that experimental physicists are lazy, incompetent or badly equipped? Certainly not! They must have deeper reasons why such measurements are infeasible. Hence, for the fourth time we conclude that for superconductors Bohr's principle of correspondence is broken and quantum physics is unable to explain them.

Since the discovery of high-T_c superconductivity dozens, of various microscopic theories have been proposed. Every theory starts from some effective Hamiltonian, from which the macroscopic phase of the supercurrent should be derived. To the present day there is no agreement among scientists, which one is natural, correct, proper, exact or authentic. Every theory attempts to see something in place of the supercurrent carriers. We argue that a true microscopic theory must not see anything, i.e. must describe superconductors as insulators, because the explanation of teleological traits in the phenomenon of superconductivity goes beyond quantum physics as being limited by the Copenhagen interpretation.

In this section we have discussed various attempts to explain and understand the Meissner effect, requiring always some ad hoc extension beyond either the classical or the quantum physical formulation. Moreover, if looking at any author, who tries to find some causal origin of the Meissner effect, one always sees huge complicated drawings of possible mechanisms along with long causal chains. One can interminably ask what is behind the specification of the first link of the chain and what is its own cause. Such drawings remind us of medieval designs for the construction of a perpetuum mobile.

However, the problem of so-called teleological phenomena is quite different. Its characteristics are not ad hoc only for the Meissner effect and superconductivity. As we will demonstrate in the next section, it has played an enormous role in understanding the electroweak and the strong interactions, and, as we have mentioned in Sect. 5, it also concerns Universal Darwinism. From here it has inspired quantum decoherence and, as outlined in Sect. 6, contributed to the elucidation of our decisions in the problem of free will. Phrased differently, in the Meissner effect we have encountered a teleological principle, which exceeds the domains of superconductivity and solid state physics in general and, in an astonishing way, reaches three major provinces of the world; inanimate nature, animate nature, and psychology.

10 The Paradox of Higgs' Boson

As we have seen the teleological principle has not been acknowledged as a compelling argument for the understanding of the phenomenon of superconductivity, which led the scientific community to accepting causal quantum-physical explanations such as the BCS theory. This did impact the construction of elementary particle mechanisms, which, to a high degree, were inspired by the BCS theory of superconductivity. The most famous example is the celebrated Higgs mechanism, incorporated in the Standard model, with the aim to explain how particles in the electro-weak interaction gain their mass. Since teleological phenomena violate Bohr's principle of correspondence, it is not surprising, when Comay [88], after a careful analysis of Higgs equations, came to the conclusion that they violate the Bohr principle as well. Comay's analysis is based on well-known cornerstones such as the variational principle, special relativity, Maxwellian electrodynamics and the fundamental elements of quantum field theory. Additionally, Comay insists on the necessity to test every quantum-physical equation, according to a wider sense of Bohr's correspondence principle, stated by the present version of Wikipedia as: "The term is also used more generally, to represent the idea that a new theory should reproduce the results of older well-established theories (which become limiting cases) in those domains where the old theories work."

Three basic testings of the quantum-physical equations are mentioned, i.e. the equation for the action

$$S = \int L(\psi, \psi_{,\mu})\, d^4x \tag{10.1}$$

the Euler-Lagrange equation of the Lagrangian density L

$$\frac{\partial}{\partial x^{\mu}}\frac{\partial L}{\partial\frac{\partial\psi}{\partial x^{\mu}}}-\frac{\partial L}{\partial\psi}=0 \tag{10.2}$$

and the Hamiltonian density H, obtained from the Lagrangian density L by means of the Legendre transformation

$$H=\sum_{i}\dot{\psi}_{i}\frac{\partial L}{\partial\dot{\psi}_{i}}-L \tag{10.3}$$

Comay [88] made four main contentions against the quantum theoretical consistency of the Higgs mechanism. Using units, $\hbar=c=1$, the Higgs boson Lagrangian density writes in the following compact form:

$$L_{Higgs}=\phi^{+}_{,\mu}\phi_{,\nu}g^{\mu\nu}+m^{2}\phi^{+}\phi+OT \tag{10.4}$$

where ϕ is the scalar function of the Higgs boson, m denotes the Higgs mass and OT denotes other terms (cubic etc.).

The first Comay argument avers: The action S in Eq. (10.1) is a dimensionless Lorentz scalar, while the Lagrangian density L in (10.1) and (10.4) is a Lorentz scalar whose dimension is $[L^{-4}]$. Hence, the first term of (10.4) proves that the dimension of the Higgs function ϕ is $[L^{-1}]$. Since the density of a quantum particle is given in terms of its wave function

$$\rho=|\psi|^{2} \tag{10.5}$$

the Schrödinger density satisfies the continuity equation, with the dimension of the wave function is $[L^{-3/2}]$. The difference in dimensions between the Higgs and the Schrödinger functions indicates that Higgs theory does not have a non-relativistic limit. This is an inconsistency in the broader sense of Bohr's correspondence principle.

The second Comay argument reads: The Higgs function ϕ is a scalar function whose dimension is $[L^{-1}]$. Hence, the product $\phi^{+}\phi$ is a scalar with dimension $[L^{-2}]$. On the other hand, the density is the zeroth component of a 4-vector whose dimension is $[L^{-3}]$. Hence, the product $\phi^{+}\phi$ does not represent the density of the Higgs boson. This is another inconsistency of the Higgs theory with the Bohr's correspondence principle.

The third Comay argument says: Examining the Euler-Lagrange equation of the Higgs boson, one finds, applying Eq. (10.2) to Higgs Lagrangian density (10.4), the following equation for ϕ:

$$\Box\phi-m^{2}\phi+OT=0 \tag{10.6}$$

The Euler-Lagrange equation of the particle's Lagrangian density must agree with the fundamental quantum mechanical equation, the Schrödinger equation, which is a first order differential equation with respect to the time variable. But the Euler-Lagrange equation of the Higgs boson is a second order differential equation with respect to the time variable. Hence, the Bohr's correspondence principle is violated again.

The fourth Comay argument yields: Applying (10.3) to (10.4), one obtains for the Higgs Hamiltonian density

$$H_{Higgs} = \dot{\phi}^{+}\dot{\phi} + (\nabla\phi^{+}).\nabla\phi - m^2\phi^{+}\phi + OT \quad (10.7)$$

The first term in (10.7) proves that the Higgs Hamiltonian must depend on a time-derivative of the Higgs function ϕ. This is yet another inconsistency of Higgs theory in relation to the broader sense of Bohr's correspondence principle, since the Schrödinger Hamiltonian is independent of the time derivatives of the wave function.

Further arguments can be found in Comay's original paper, e.g. the analysis of the Higgs energy-momentum tensor that depends quadratically on its mass m contradicting its classical correspondence of linear dependency on the mass. Comay himself views the main problem of Higgs theory by the fact that Higgs equation belongs to the Klein-Gordon family. He cites Dirac's opinion on this issue [89]: "I found this development quite unacceptable. It meant departing from the fundamental ideas of the non-relativistic quantum mechanics, ideas which demanded a wave equation linear in $\partial/\partial t$. It meant abandoning the whole beautiful mathematical scheme for the sake of introducing certain physical ideas." In fact Comay at last appeals: "The present work provides further arguments that support Dirac's opinion... The quite large numbers of contradictions of the Higgs boson theory which are described in this work make a basis for the expectation that this theory will be abandoned."

In fact physicists usually do concentrate on the problem of renormalizability of field theories, yet it seems that the significance of Comay's results is the proof of a total failure of Bohr's principle of correspondence in relation to the Higgs mechanism. It has remained unnoticed or fully ignored by the scientific community. Comay's analysis focuses on the final stage of the long-termed development of the Higgs mechanism in its resulting equations. It is clear that one have to find the crucial mistake in the whole process proving that quantum physics must be "on guard" as regards the principle of causality, and further not allowing the incorporation of the teleological "Mexican hat" potential into the quantum equation. This will prohibit the Higgs equation to follow from "true" quantum physics. We will begin with a brief recapitulation of the history from the main papers, the ideas of which were later used in the formulation of the Higgs mechanism, continuing with the simplest example of the Abelian version of the mechanism, which is now regarded as the model for the explanation of the phenomenon of superconductivity.

As we have mentioned in the previous section, the Mexican hat in a classical setting was for the first time adopted into a macroscopic formulation of superconductivity in 1950 by Ginzburg and Landau [77], leading to a spontaneous breaking of symmetry. The first accepted microscopic theory of superconductivity [78], the BCS model from 1957 did not mention the Mexican hat at all and was derived without any reference to any symmetry violations. The ideas leading to the phenomenological

Ginzburg-Landau theory including the Mexican hat were for the first time translated into quantum field theory in 1961 by Nambu and Jona-Lasinio [90]. However, during the same year, serious problems surfaced when Goldstone formulated his theorem [91] stating: “The spontaneous breaking of a continuous symmetry can be associated with a massless and spinless particle, the so called Nambu-Goldstone boson“. The Goldstone theorem, proved one year later by Goldstone et al. [92], disqualified the Nambu–Jona-Lasinio model. Since it is difficult to find, from the general proof [92], the particles that could serve as ‘Goldstone bosons’, it was believed that the application of spontaneous symmetry violations in field theory, such as the model of Nambu and Jona-Lasinio, would be impossible.

In the same year, 1962, Schwinger [93] proposed the possibility of combining mass with gauge fields. It caused an immediate revival of the old gauge theories from 1954 of Yang and Mills [94], dealing with an extension of the concept of gauge theory considered for Abelian groups used in quantum electrodynamics, to non-Abelian groups, in an attempt to explain strong interactions. In order to preserve gauge invariance, the Yang–Mills field must only give rise to massless particles. One year later, Anderson being inspired by the Goldstone theorem, Schwinger’s idea and the Yang-Mills theory, wrote an article [95] about the possible violation of Goldstone’s theorem in superconductors: “Schwinger has pointed out that the Yang-Mills vector boson implied by associating a generalized gauge transformation with a conservation law… does not necessarily have zero mass… We show that the theory of plasma oscillations is a simple nonrelativistic example exhibiting all of the features of Schwinger’s idea… The boson, which appears as a result of the Goldstone theorem and has zero unrenormalized mass, is converted into a finite-mass plasmon by the interaction with the appropriate gauge field, which is the electromagnetic field.” In his article Anderson suggested that a similar process in particle physics could be responsible for giving mass to Yang-Mills gauge bosons, further adding: “It is likely, then, considering the superconducting analogue, that the way is now open for a degenerate-vacuum theory of the Nambu type without any difficulties involving either zero-mass Yang-Mills gauge bosons or zero-mass Goldstone bosons. These two types of bosons seem capable of “cancelling each other out” and leaving only finite mass bosons.” It is symptomatic that Anderson’s statement became the hint that soon led to the development of the Higgs mechanism.

It is straightforward to work out the derivation of the Abelian Higgs equations for superconductors, which the reader can find in most current textbooks dealing with the Higgs mechanism. The Langrangian density is formulated in a quantum version of the classical Ginzburg-Landau theory, with quadratic and quartic powers of the potential, imitating the Mexican hat.

$$L = -\frac{1}{4} F^{\mu\nu} F_{\mu\nu} + D_{\mu}^{*} \phi^{*} D^{\mu} \phi - \mu^{2} \phi^{*} \phi - \lambda (\phi^{*} \phi)^{2} \tag{10.8}$$

where F and D are defined by means of the electromagnetic potential A and the charge q:

$$F^{\mu\nu} \equiv \partial^\mu A^\nu - \partial^\nu A^\mu; \quad D_\mu \equiv \partial_\mu + iqA_\mu \tag{10.9}$$

The complex field $\phi(x)$, simulating the Mexican hat, may then be expressed by means of two real fields $\eta(x)$ and $\xi(x)$ and a real number v:

$$\phi(x) = \frac{1}{\sqrt{2}}(\eta(x) + v)e^{i\xi(x)/v} \tag{10.10}$$

The Lagrangian density (10.8) then takes the form:

$$L = \left[-\frac{1}{4}F^{\mu\nu}F_{\mu\nu} + \frac{q^2v^2}{2}A_\mu A^\mu\right] + \left[\frac{1}{2}\partial_\mu\eta\partial^\mu\eta + \mu^2\eta^2\right] + \frac{1}{2}\partial_\mu\xi\partial^\mu\xi + qvA^\mu\partial_\mu\xi + OT \tag{10.11}$$

where *OT* stands for cubic and quartic terms. The terms between the second square brackets represent a massive scalar particle with mass $m_\eta = (-2)^{1/2}\mu$. The third term is a massless scalar particle. The occurrence of this particle was inferred from the shape of the Mexican hat potential with the degree of freedom connected to the angular displacement. A motion in this direction does not face any resistance since the energy in the adjacent state is the same. This field is therefore massless and was identified with the Goldstone boson.

After the application of the gauge transformation for the electromagnetic field

$$\phi \to \phi' = e^{i\theta(x)}\phi; \quad A \to A' = A_\mu - \frac{1}{q}\partial_\mu\theta \tag{10.12}$$

with the specific substitution for $\theta(x)$

$$\theta = -\frac{\xi}{v} \tag{10.13}$$

one arrives at the final Abelian Higgs equation for the Lagrangian density

$$L = \left[-\frac{1}{4}F^{\mu\nu}F_{\mu\nu} + \frac{q^2v^2}{2}A'_\mu A'^\mu\right] + \left[\frac{1}{2}\partial_\mu\eta\partial^\mu\eta + \mu^2\eta^2\right] + OT \tag{10.14}$$

The Higgs theory contains no massless particles, since the field $\xi(x)$ has entirely disappeared. The gauge symmetry, unlike any global symmetry, does not lead to massless Goldstone bosons, but instead ends up with a massive gauge field. The final Lagrangian contains only a massive gauge boson $A'_\mu(x)$ (first part) and a massive component of the Higgs field $\eta(x)$ (second part). The first part is viewed as the quantum representation of the classical London Eq. (9.7) and is usually interpreted as the formation of massive photons inside superconductors.

Starting the examination of the Higgs mechanism with Anderson's original paper [95], we will pay special attention to the last sentence in the following quote: "I should like to close with one final remark on the Goldstone theorem. This theorem was initially conjectured, one presumes, because of the solid-state analogues, via the work of Nambu and of Anderson. The theorem states, essentially, that if the Lagrangian possesses a continuous symmetry group under which the ground or vacuum state is not invariant, that state is, therefore, degenerate with other ground states. This implies a zero-mass boson. Thus, the solid crystal violates translational and rotational invariance, and possesses phonons; ..."

One may ask whether Anderson, in view of what has been said here, really appreciated the content of the Goldstone theorem? Quasiparticles, representing collective oscillations in condensed matter were known long before the Goldstone theorem was formulated. When it was proved, it was sometimes easy to assign them to Goldstone bosons and corresponding broken symmetries, and sometimes not. For instance, magnons in ferromagnets are Goldstone bosons descending from broken rotational symmetry, or longitudinal phonons in liquids are Goldstone bosons of the broken Galilean symmetry. But what about solids? Anderson claims, that "the solid crystal violates translational and rotational invariance, and possesses phonons". However, not so, phonons are the consequence of a broken Galilean symmetry, and further, which Goldstone bosons correspond to the broken translational and rotational symmetry in crystals? Unfortunately one can find no answers to this question in Anderson's papers.

One might try to find an answer to the question: what are the Goldstone bosons in solids in Wikipedia. A recent reply reads: "In solids, the situation is more complicated; the Goldstone bosons are the longitudinal and transverse phonons and they happen to be the Goldstone bosons of spontaneously broken Galilean, translational, and rotational symmetry with no simple one-to-one correspondence between the Goldstone modes and the broken symmetries."

Although Wikipedia occasionally must be taken with a grain of salt, the sentence "No simple one-to-one correspondence between the Goldstone modes and broken symmetries" is surprising! Since the Goldstone theorem is a complement of Noether's first theorem, the claim is equivalent to saying that the conservation laws for linear and angular momentum may not be in a simple one-to-one correspondence with translational and rotational symmetry. Consequently a unique and unambiguous one-to-one correspondence between the Goldstone modes and spontaneously broken symmetries in solids must necessarily exist! The conclusion must be that phonons in solids do not represent the full set of Goldstone bosons, and some bosons are still missing. It is better if we in the future would add this issue to the list of unsolved problems of quantum physics rather than sweeping it under the carpet.

Principally we should concentrate our energies on hunting for the lost Goldstone bosons in solids. Supposing that a solid is composed of N nuclei, the whole system has $3N$ degrees of freedom. The quantum mechanical treatment eliminates 6 degrees associated with the centre of mass, and the remaining $3N - 6$ degrees after the Galilean symmetry breaking represent the vibrational modes, the phonons. At the first sight it seems that the violation of translational and rotational invariance has no counterpart in the form of the existence of the corresponding Goldstone bosons. One may ask whether this is really true. Does the quantum field share the centre of mass separation with quantum mechanics considering the same system? The answer is no. We will provide a proof in the next section and show: first, qualitatively the appearance of the 6 lost Goldstone bosons, and second, quantitatively their contributions to the quantum field equations together with the experimental evidence. It means that every solid composed of N nuclei has precisely $3N$ spontaneously broken symmetries with the exact correspondence to the emerging $3N$ Goldstone bosons. This rule holds universally, regardless of the size of N, i.e. if it is small or great, if it represents finite molecule or infinite crystal, and irrespective of the solid character, i.e. if it is insulator, conductor, semiconductor, or superconductor.

Returning to Eq. (10.10) where the field $\xi(x)$ relates to the degrees of freedom connected with the angular displacement in the Mexican hat. The field corresponding to the Lagrangian density in Eq. (10.11) results in a massless scalar particle that was identified with the Goldstone boson, or is it really so? As just said above, every solid produces exactly $3N$ Goldstone bosons. Thus the massless scalar particle descending from the field $\xi(x)$ cannot really be a Goldstone boson! It rather looks like the following. Instead of loosing six authentic Goldstone bosons in the solid suddenly one imposter Goldstone boson pops up in the superconductor. Thus we end up with the illusive Eq. (10.11) without the chance to discuss the rest of the Higgs equations which became misleading as well. This should be the end of Anderson's textual description about the gauge bosons eating the Goldstone bosons, on the basis of which the anticipated Higgs mechanism was created.

We will close this section with a philosophical consideration of the Higgs mechanism. One of the best reflections I have found was written by Earman [96]: "Higgs further showed that the gauge could be chosen so that the Goldstone bosons are suppressed and that in this "unitary gauge" the new field had acquired a mass. As the semi-popular presentations put it, "Particles get their masses by eating the Higgs field." Readers of Scientific American can be satisfied with these just-so stories. But philosophers of science should not be. For a genuine property like mass cannot be gained by eating descriptive fluff, which is just what gauge is. Philosophers of science should be asking the Nozick question: What is the objective (i.e. gauge invariant) structure of the world corresponding to the gauge theory presented in the Higgs mechanism? …consider the following three-tiered dilemma. First tier: Either the gauge invariant content of the Higgs mechanism is described by local quantum fields satisfying the standard assumptions of Poincaré invariance, local commutativity, spectrum condition, etc. or not. If not, then the implementation of the Higgs mechanism requires a major overhaul of conventional QFT. If so, go to the second tier. Second tier: Either the gauge invariant system admits a finite dimensional Lie group as an internal symmetry group or not. If so, Noether's first theorem applies again. But (since the other standard assumptions are in place) Goldstone's theorem also applies and, hence, Goldstone bosons have not been suppressed after all. If not, go to the third tier. Third tier: Either the gauge invariant system admits no non-trivial symmetries at all or else it admits only discrete symmetries. In either case Goldstone bosons are quashed. In the former case spontaneous symmetry breaking is not an issue since there is no symmetry to break. In the latter case it is possible that the discrete symmetry is spontaneously broken. But the usual argument for symmetry breaking using the conserved Noether current does not apply. And while it is possible that some completely different sort of construction will demonstrate the spontaneous breakdown of the hypothesized discrete symmetry there are no extant demonstrations that have more than a hand waving force."

If we compare our analysis of the Goldstone theorem with the Earman's multi-tier reflection, it is obvious that the whole chain ends at the second tier, i.e. the Goldstone bosons have not been suppressed after all. The rest is nothing but metaphysics. In fact Higgs named his Nobel lecture "Evading the Goldstone theorem" [97], but this was no evasion rather a total misreading of this theorem. Anderson's stimulating paper [95] was indeed an incorrect marking, misleading Higgs and his co-authors on a lost course.

11 The Paradox of the Centre of Mass of Quantum Systems

In this section we will show an independent proof of the Goldstone theorem for condensed matter, molecules and solids. Since the original Goldstone-Salam-Weinberg proof [92] is admittedly the most general, it is not, for the purpose of this study, sufficiently transparent regarding the fundamental question of the one-to-one correspondence between broken symmetries and associated massless and spinless bosons. The problem of translational and rotational symmetry violations in solids is a nice example.

Solid state physics has adopted some adequate patterns from quantum electrodynamics. The quantum field electron-photon Feynman diagrammatic technique has found many successful applications in solid state electron-phonon treatments. Many years ago, when I finished my studies in solid state physics, I did start to work in a quantum chemistry department. In this environment quantum field methods are not so often exercised in comparison to its use in solid state physics, and moreover mostly on the fermionic electron-hole level. At that time I was familiar with the works of Czech emigrants Paldus and Čížek, who were inspired by Goldstone's revision [98] of Brueckner's many-body theory [99] with the solution expressed in a manner that avoids the problem of unlinked clusters, introducing Goldstone's diagrammatic electron-hole mechanism into quantum chemistry [100].

The goal of my Ph.D. thesis [101] was to introduce, in a diagrammatic treatment, the full electron-phonon interactions into quantum chemistry, i.e. to elaborate the mathematical framework for second quantization of the full electron-vibrational Hamiltonian. Goldstone's theorem was at first beyond my interest, since it was mostly discussed in the domain of elementary particle physics, but hardly not mentioned at the time in solid state physics and quantum chemistry. As a result, I thought that electron-phonon mechanisms really constituted a complete description of solids, i.e. exactly how we learned in school as it was described in all textbooks.

Let us start with the full Hamiltonian of the system of nuclei and electrons in second quantization electron-hole formalism, as it is conventionally used in quantum chemistry.

$$H = T_N(\dot{\mathbf{R}}) + E_{NN}(\mathbf{R}) + \sum_{PQ} h_{PQ}(\mathbf{R})\, a_P^+ a_Q + \frac{1}{2} \sum_{PQRS} v_{PQRS}^0 a_P^+ a_Q^+ a_S a_R \quad (11.1)$$

Here T_N stands for kinetic energy of nuclei, E_{NN} for nuclear potential energy, h_{PQ} for one-electron matrix elements and v_{PQRS}^0 for two-electron matrix elements. The terms E_{NN} and h_{PQ} can be expanded in the Taylor series

$$E_{NN}(\mathbf{R}) = \sum_{n=0}^{\infty} E_{NN}^{(n)}(\mathbf{R}) \quad (11.2)$$

$$h_{PQ}(\mathbf{R}) = h_{PQ}^0 + \sum_{n=1}^{\infty} u_{PQ}^{(n)}(\mathbf{R}) = h_{PQ}^0 + \sum_{n=1}^{\infty} \langle P | \sum_i \frac{-Z_i e^2}{|\mathbf{r}-\mathbf{R}_i|} | Q \rangle^{(n)} \quad (11.3)$$

where E_{NN}^0 and h_{PQ}^0 are nuclear potential and one-electron terms for fixed (equilibrium) nuclear coordinates. In full analogy with the effective electron-phonon solid state Hamiltonian

$$H = \sum_{\mathbf{k},\sigma} \varepsilon_{\mathbf{k}} a_{\mathbf{k},\sigma}^+ a_{\mathbf{k},\sigma} + \sum_{\mathbf{q}} \hbar\omega_{\mathbf{q}} \left(b_{\mathbf{q}}^+ b_{\mathbf{q}} + \tfrac{1}{2}\right) + \sum_{\mathbf{k},\mathbf{q},\sigma} u^{\mathbf{q}} \left(b_{\mathbf{q}} + b_{-\mathbf{q}}^+\right) a_{\mathbf{k}+\mathbf{q},\sigma}^+ a_{\mathbf{k},\sigma} \quad (11.4)$$

we can rewrite the Hamiltonian (11.1) in a complete second quantization form for electronic and as well as vibronic modes:

$$H = T_N(\widetilde{B}) + E_{NN}(B) + \sum_{PQ} h_{PQ}(B)\, a_P^+ a_Q + \frac{1}{2} \sum_{PQRS} v_{PQRS}^0 a_P^+ a_Q^+ a_S a_R \quad (11.5)$$

where B and $\widetilde{B}$ represent the coordinate and momentum oscillator operators. If we require that these and all following equations to be applicable to both quantum chemistry and solid state physics, we need to introduce a cross-platform notation for B and $\widetilde{B}$, viz $B_r = b_r + b_{\breve{r}}^+$ and $\widetilde{B}_r = b_r - b_{\breve{r}}^+$. In quantum chemistry $r = \breve{r}$ holds, while in solid state physics one assumes that for any vibrational mode r there exists corresponding mode $\breve{r}$ that fulfills the identity $\omega_r = \omega_{\breve{r}}$. Comparing this notation with the usual solid state notation the simple transition $P \to \mathbf{k}, \sigma, r \to \mathbf{q}, \breve{r} \to -\mathbf{q}$ is supposed.

The potential energy of the motion of the nuclei is defined through the quadratic part of the internuclear potential plus some additive term representing the self-consistent influence of the electron-nuclear potential

$$E_{pot} = E_{NN}^{(2)}(B) + V_N^{(2)}(B) \tag{11.6}$$

In the adiabatic limit the kinetic energy E_{kin} is identical to the kinetic energy of the nuclei T_N. If we want to include the nonadiabatic case in our treatment, it is necessary to incorporate a new additive kinetic term originating from the kinetic energy of electrons. The resulting kinetic energy of the system has the form

$$E_{kin} = T_N(\widetilde{B}) + W_N^{(2)}(\widetilde{B}) \tag{11.7}$$

so that the vibrational part of the total Hamiltonian can be expressed as

$$H_V = E_{kin}(\widetilde{B}) + E_{pot}(B) \tag{11.8}$$

where the kinetic and potential energies are given by

$$E_{pot} = \frac{1}{4} \sum_{r \in V} \hbar\omega_r B_r^+ B_r \tag{11.9}$$

$$E_{kin} = \frac{1}{4} \sum_{r \in V} \hbar\omega_r \widetilde{B}_r^+ \widetilde{B}_r \tag{11.10}$$

Finally we get the well-known vibrational Hamiltonian

$$H_V = \frac{1}{4} \sum_{r \in V} \hbar\omega_r \left(B_r^+ B_r + \widetilde{B}_r^+ \widetilde{B}_r\right) = \sum_{r \in V} \hbar\omega_r \left(b_r^+ b_r + \tfrac{1}{2}\right) \tag{11.11}$$

Then the total Hamiltonian of the system reads:

$$H = E_{NN}(B) - E_{NN}^{(2)}(B) - V_N^{(2)}(B) - W_N^{(2)}(\widetilde{B}) + \sum_{PQ} h_{PQ}(B)\, a_P^+ a_Q$$
$$+ \frac{1}{2} \sum_{PQRS} v_{PQRS}^0 a_P^+ a_Q^+ a_S a_R + \frac{1}{4} \sum_{r \in V} \hbar\omega_r \left(B_r^+ B_r + \widetilde{B}_r^+ \widetilde{B}_r\right) \tag{11.12}$$

However there is one obstacle here. The Hamiltonian (11.12) cannot be directly used for quantum chemical calculations like the Hamiltonian (11.4) in solid state physics. Both are in the form of a "crude" representation, to be precise, the representation with fixed nuclear positions. Although this makes no problems in solids, where only valence and conducting bands are to be taken into account, in molecular calculations all electrons must be incorporated, including also the inner shell ones. Unfortunately direct application of the Hamiltonian (11.12) leads to a divergent series expansion.

Nevertheless there is a straightforward way to solve this difficulty, namely one must find a suitable renormalization where all the divergent terms vanish. Instead of the original electrons and bosons (vibrational modes or phonons) we introduce a new set of fermionic and bosonic operators, with the transformed Hamiltonian optimized on the Born-Oppenheimer (B-O) level for adiabatic molecules.

The most general quasiparticle transformation, which "simulates" the B-O approximation, can be introduced in the following form:

$$\bar{a}_P = \sum_Q c_{PQ}(B) a_Q \quad \bar{b}_r = b_r + \sum_{PQ} d_{rPQ}(B) a_P^+ a_Q \tag{11.13}$$

with the unitary conditions

$$\sum_R c_{PR}(B) c_{QR}^+(B) = \delta_{PQ} \quad d_{rPQ} = \sum_R c_{RP}^+(B)[b_r, c_{RQ}(B)] \tag{11.14}$$

For nonadiabatic systems we add to the previous Q (coordinate)-dependent transformation yet another P (momentum)-dependent one:

$$\bar{a}_P = \sum_Q \tilde{c}_{PQ}(\widetilde{B}) a_Q \quad \bar{b}_r = b_r + \sum_{PQ} \tilde{d}_{rPQ}(\widetilde{B}) a_P^+ a_Q \tag{11.15}$$

with the unitary conditions

$$\sum_R \tilde{c}_{PR}(\widetilde{B}) \tilde{c}_{QR}^+(\widetilde{B}) = \delta_{PQ} \quad \tilde{d}_{rPQ} = \sum_R \tilde{c}_{RP}^+(\widetilde{B})[b_r, \tilde{c}_{RQ}(\widetilde{B})] \tag{11.16}$$

The successive application of both transformations, adiabatic and nonadiabatic ones is equivalent to the unitary transformation of the whole Hamiltonian

$$\overline{H} = e^{-S_2(P)} e^{-S_1(Q)} H e^{S_1(Q)} e^{S_2(P)} \tag{11.17}$$

Note that I have preferred quasiparticle transformations to unitary transformation of the Hamiltonian for the sake of greater transparency. Incidentally, Fröhlich used a similar unitary transformation in his derivation of the effective two-electron interaction [102] that was later incorporated into the BCS theory:

$$\overline{H} = e^{-S(Q,P)} H e^{S(Q,P)} \tag{11.18}$$

For details of the derivation of the field equations, as based on the quasiparticle transformations (11.13)–(11.16), see the book [103]. For a long time I considered this theory to be essentially correct. To sum up, there were four well-known formulae, one in quantum mechanics, and three in quantum field theory, which could have been rederived in exact agreement:

(1) the Pople's equations for the ab initio calculation of vibrational frequencies [104]
(2) the Fröhlich's expression for the correction of the ground state energy [102, 105]

(3) the Fröhlich's effective two-electron interaction [102]
(4) the Lee-Low-Pines polarons and their self-energy [106].

During the numerical tests in 1998 the above mentioned theory totally failed in the calculations of the first correction beyond the B-O approximation, i.e. the adiabatic correction known as the Born-Huang ansatz [7]:

$$\Delta E_{0(ad)} = \langle \psi_0(\mathbf{R}) | T_N | \psi_0(\mathbf{R}) \rangle_{\mathbf{R}_0} \tag{11.19}$$

This was a great disappointment for me. In order to directly compare the field formula for the adiabatic correction with the mechanical Born-Huang ansatz, I did rewrite the latter into its field form. In the adiabatic case the N-electron function $\psi_0(\mathbf{R})$ can be expanded as a single Slater determinant though the one-electron functions $\varphi_I(\mathbf{R})$:

$$\psi_0(\mathbf{R}) = \frac{1}{\sqrt{N!}} \left\| \prod_I^N \varphi_I(\mathbf{R}) \right\| \tag{11.20}$$

The function $\varphi_P(\mathbf{R})$ can be expanded in terms of the coefficients $c_{PQ}(\mathbf{R})$, dependent on the nuclear coordinates, and the orthonormal set of one-electron wavefunctions defined in the equilibrium position $\mathbf{R}_0$:

$$\varphi_P(\mathbf{R}) = \sum_Q c_{PQ}(\mathbf{R})\, \varphi_Q(\mathbf{R}_0) \tag{11.21}$$

In the following work we will use the following notation for the spin-orbitals: I, J, K, L—occupied; A, B, C, D—virtual (unoccupied); P, Q, R, S—the arbitrary ones. One obtains then a very simple expression for the Born-Huang ansatz (11.19)

$$\Delta E_{0(ad)} = \sum_{AIi\alpha} \frac{\hbar^2}{2M_i} \left| c_{AI}^{i\alpha} \right|^2 \tag{11.22}$$

where M_i stands for the nuclear mass and α for the Cartesian coordinates.

It is now useful to proceed from the Cartesian to the normal coordinate system using the following transformation

$$c_{PQ}^r = \sum_{i\alpha} c_{PQ}^{i\alpha} \alpha_{i\alpha}^r \tag{11.23}$$

One wants to substitute for $c_{PQ}^{i\alpha}$ in (11.22), so we need to know the inverse matrix $\beta_{i\alpha}^r$. From

$$\alpha^+ \beta = I \tag{11.24}$$

Equation (11.22) can be expressed in normal coordinates as

$$\Delta E_{0(ad)} = \sum_{AIrsi\alpha} \frac{\hbar^2}{2M_i} c^r_{AI} c^{s*}_{AI} \beta^{r*}_{i\alpha} \beta^s_{ia} \tag{11.25}$$

The matrices α and β have dimension $3N \times 3N$. The first one diagonalizes the potential energy

$$\alpha^+ E_{pot} \alpha = \left\{ \frac{1}{2} \hbar\omega, 0, 0 \right\} \tag{11.26}$$

resulting in $3N - 5$ or $3N - 6$ vibrational frequencies (the first holds for diatomic molecules) and the 5 or 6 accounting for zero energy. On the other hand, the second matrix diagonalizes the kinetic energy

$$\beta^+ E_{kin} \beta = \left\{ \frac{1}{2} \hbar\omega, \rho, \tau \right\} = \sum_{i\alpha} \frac{\hbar^2}{4M_i} \beta^*_{i\alpha} \beta_{i\alpha} \tag{11.27}$$

with all $3N$ nonzero values: aside the $3N - 5$ or $3N - 6$ vibrational modes we have 2 or 3 nonzero values for rotational modes and 3 nonzero values for translational ones. So the final field expression for the Born-Huang ansatz reads:

$$\Delta E_{0(ad)} = 2 \sum_{AI} \left(\sum_{r \in V} \frac{1}{2} \hbar\omega_r + \sum_{r \in R} \rho_r + \sum_{r \in T} \tau_r \right) \left| c^r_{AI} \right|^2 \tag{11.28}$$

It is now evident that all electron-vibrational and electron-phonon field theories fail beyond the B-O approximation, because they do not pass the Born-Huang test. In Eq. (11.28) they are only able to justify the first term that consists of vibrational modes, but they are unable to arrive at the second and third terms dealing with rotational and translational quanta. The electron-phonon mechanism has its roots in the B-O approximation, where only the diagonalization procedure (11.26) of the potential energy via the α matrices is relevant. Therefore, in this specific case the quantum field can only share with quantum mechanics the centre-of-mass (COM) separation of the external and the internal degrees of freedom.

Even though I immediately discovered this serious problem of the quantum field formulation after the 1998 numerical tests, I have hesitated for many years to publish the revised version of employing quantum field theories for condensed matter systems such as molecules and solids. Naturally I did not want to make a fool of myself by claiming that there are some new particles in physics, and that there is a new type of covariance in the quantum field, binding together the internal and external degrees of freedom in a way similar to the way Lorentz covariance binds together space and time. Finally I did present this topic in Cambridge in 2010, and this lecture was published in 2012 [107].

Without any knowledge of the Goldstone theorem, we can see that the existence of $3N - 5$ or $3N - 6$ massless spinless bosons—vibrations/phonons—results from Eq. (11.26). The rest of the massless spinless bosons which I have called rotons and translons, follows from Eq. (11.27). However, the complete lack of interest from the scientific community in this topic has indeed been astonishing, but the reason may be quite banal. Whenever I speak with solid state physicists, they all admit that they have always thought that the adiabatic approximation is synonymous with the B-O approximation and that they see no difference between them, but it is in actual fact the almost negligible difference between them that generates the 5 or 6 lost Goldstone bosons! As a consequence the original concept of the electron-vibrational field theory [103] must be abandoned and replaced by a new field theory with two major upgrades:

The first upgrade: Eqs. (11.9)–(11.10) for the potential and the kinetic vibrational energies, must be rewritten using a complete set of $3N$ Goldstone bosons—phonons, rotons and translons—in agreement with the Born-Huang field transcription (11.28). We shall call these bosons hyperphonons or hypervibrations.

$$E_{pot} = \frac{1}{4} \sum_{r \in V} \hbar\omega_r B_r^+ B_r \tag{11.29}$$

$$E_{kin} = \frac{1}{2}\left(\frac{1}{2} \sum_{r \in V} \hbar\omega_r + \sum_{r \in R} \rho_r + \sum_{r \in T} \tau_r \right) \widetilde{B}_r^+ \widetilde{B}_r \tag{11.30}$$

Defining the hyper-vibrational double-vector:

$$\boldsymbol{\omega} = \begin{pmatrix} \omega_r \\ \tilde{\omega}_r \end{pmatrix} = \begin{pmatrix} \omega_r & 0 & 0 \\ \omega_r & \frac{2}{\hbar}\rho_r & \frac{2}{\hbar}\tau_r \end{pmatrix} \tag{11.31}$$

we get covariant expressions for the boson part of the hyper-vibrational Hamiltonian with respect to all $3N$ Goldstone modes:

$$H_B = \frac{1}{4} \sum_r \left(\hbar\omega_r B_r^+ B_r + \hbar\tilde{\omega}_r \widetilde{B}_r^+ \widetilde{B}_r \right) \tag{11.32}$$

Instead of the original total Hamiltonian (11.12) we obtain

$$\begin{aligned} H = {} & E_{NN}(B) - E_{NN}^{(2)}(B) - V_N^{(2)}(B) - W_N^{(2)}(\widetilde{B}) + \sum_{PQ} h_{PQ}(B)\, a_P^+ a_Q \\ & + \frac{1}{2} \sum_{PQRS} v_{PQRS}^0 a_P^+ a_Q^+ a_S a_R + \frac{1}{4} \sum_r \left(\hbar\omega_r B_r^+ B_r + \hbar\tilde{\omega}_r \widetilde{B}_r^+ \widetilde{B}_r \right) \end{aligned} \tag{11.33}$$

The second upgrade: We will apply the same quasiparticle transformations (11.13)–(11.16) as in the original electron-vibrational concept, but with entirely different interpretation. In full analogy with Lorentz covariance, binding together space and time coordinates, we introduce a new covariance, binding together the internal and external degrees of freedom, i.e. Eqs. (11.13)–(11.16) will comprise all

$3N$ degrees of freedom without the mechanical COM separation. We will denote the quality of this unitary transformation as the field COM covariance.

Here we will only sketch the derivation, since it is, however, very time-consuming. Details of the derivation have been given in previous work [107]. The final formula for the change of the ground state energy has a surprisingly simple analytical form:

$$\Delta E_0 = \sum_{AIr} \left(\hbar\tilde{\omega}_r \left|c^r_{AI}\right|^2 - \hbar\omega_r \left|\tilde{c}^r_{AI}\right|^2 \right) \tag{11.34}$$

where the summation refers to virtual spin-orbitals A, occupied spin-orbitals I, and all hyper-vibrational modes r, $r \in \{\mathrm{V, R, T}\}$. The coefficients c resp. $\tilde{c}$ are related to the adiabatic and the non-adiabatic transformation, respectively, and determined by the set of equations

$$\begin{aligned} u^r_{PQ} + (\varepsilon^0_P - \varepsilon^0_Q)c^r_{PQ} + \sum_{AI} [(v^0_{PIQA} - v^0_{PIAQ})c^r_{AI} - (v^0_{PAQI} - v^0_{PAIQ})c^r_{IA}] \\ - \hbar\omega_r \tilde{c}^r_{PQ} = \varepsilon^r_P \delta_{PQ} \end{aligned} \tag{11.35}$$

$$\begin{aligned} (\varepsilon^0_P - \varepsilon^0_Q)\tilde{c}^r_{PQ} + \sum_{AI} [(v^0_{PIQA} - v^0_{PIAQ})\tilde{c}^r_{AI} - (v^0_{PAQI} - v^0_{PAIQ})\tilde{c}^r_{IA}] \\ - \hbar\tilde{\omega}_r c^r_{PQ} = \tilde{\varepsilon}^r_P \delta_{PQ} \end{aligned} \tag{11.36}$$

where u are the coefficients of the electron-hyperphonon interaction, ε^0 are one-electron energies, and v^0 two-electron potential energies. Finally the set of equations in the second order of the Taylor expansion results in the ab initio self-consistent equations for hyper-vibrational frequencies ω and $\tilde{\omega}$, namely for the unknown potential and kinetic matrix elements in Eqs. (11.6)–(11.7) and in the total Hamiltonian (11.33):

$$V^{rs}_N = \sum_I u^{rs}_{II} + \sum_{AI} [(u^r_{IA} + \hbar\omega_r \tilde{c}^r_{IA})c^s_{AI} + (u^s_{IA} + \hbar\omega_s \tilde{c}^s_{IA})c^r_{AI}] \tag{11.37}$$

$$W^{rs}_N = 2\hbar\tilde{\omega}_r \sum_{AI} c^r_{AI} \tilde{c}^s_{IA} \tag{11.38}$$

The fermionic one-particle correction $\Delta H'_F$ is more complex and therefore we select only those terms which are decisive for excitation mechanism

$$\begin{aligned} \Delta H'_F = \sum_{PQr} \left[\hbar\tilde{\omega}_r \left(\sum_A c^r_{PA} c^{r*}_{QA} - \sum_I c^r_{PI} c^{r*}_{QI} \right) - \hbar\omega_r \left(\sum_A \tilde{c}^r_{PA} \tilde{c}^{r*}_{QA} - \sum_I \tilde{c}^r_{PI} \tilde{c}^{r*}_{QI} \right) \right] N[a^+_P a_Q] \\ + \sum_{PRr} \left[\left(\varepsilon^0_P - \varepsilon^0_R \right) \left(|c^r_{PR}|^2 + |\tilde{c}^r_{PR}|^2 \right) - 2\hbar\tilde{\omega}_r \mathrm{Re}\left(\tilde{c}^r_{PR} c^{r*}_{PR} \right) \right] N[a^+_P a_P] \end{aligned} \tag{11.39}$$

The first part (11.39) is of a pure one-fermion origin and has a non-diagonal form. The second part is a vacuum value of type $\langle 0| B_r B_s |0\rangle$ and/or $\langle 0| \widetilde{B}_r \widetilde{B}_s |0\rangle$ of the mixed fermion-boson terms, where the bosonic part is of the quadratic form of coordinate and/or momentum operators.

We can now demonstrate the simplest applications of these general equations where the field COM covariant theory reduces to the classical mechanical COM separation limit, i.e. $\omega = \tilde{\omega}$ and only the Goldstone bosons descending from the Galilean broken symmetries—vibrations/phonons—are taken into consideration. We will present the limits of the four examples leading up to Pople's equations, Fröhlich's ground state energy correction and the effective two-electron Hamiltonian, and the Lee-Low-Pines polarons. These four cases were particularly mentioned as they immediately follow from the pure electron-vibrational theory [103]. We start with the four cases and continue with three more related to the Born-Huang ansatz.

(1) Pople's equations for the ab initio calculation of vibrational frequencies [104]:

From Eqs. (11.35) and (11.37) in the adiabatic limit, where the coefficients $\tilde{c}$ equal zero we get

$$u^r_{PQ} + (\varepsilon^0_P - \varepsilon^0_Q)c^r_{PQ} + \sum_{AI} [(v^0_{PIQA} - v^0_{PIAQ})c^r_{AI} - (v^0_{PAQI} - v^0_{PAIQ})c^r_{IA}]$$
$$= \varepsilon^r_P \delta_{PQ};\ c^r_{PP} = 0 \tag{11.40}$$

$$V^{rs}_N = \sum_I u^{rs}_{II} + \sum_{AI} \left(u^r_{IA}c^s_{AI} + u^s_{IA}c^r_{AI}\right) \tag{11.41}$$

One observes that Ref. [104] contains these equations, not in the MO (molecular orbital) basis, but for programming purposes in the LCAO (linear combination of atomic orbitals) basis of "moving" atomic orbitals that follow the adiabatic "motion" of the nuclei. Both approaches are fully equivalent; the MO basis notation is simpler and shorter, and therefore more suitable for a theoretical treatment, whereas the LCAO basis is more practical in numerical calculations. One can look at Eqs. (11.35)–(11.38) as a generalization of Pople's CPHF equations [104] for the case of a general field COM covariant theory, which includes also the case of the break-down of the B-O approximation.

(2) Fröhlich's expression for the correction to the ground state energy [102, 105]:

Neglecting two-electron terms, the formula for the change of the ground state energy (11.34) takes the following explicit simple form:

$$\Delta E_0 = \sum_{AI,r\in V} \left(\hbar\omega_r |c^r_{AI}|^2 - \hbar\omega_r |\tilde{c}^r_{AI}|^2\right) = \sum_{AI,r\in V} |u^r_{AI}|^2 \frac{\hbar\omega_r}{(\varepsilon^0_A - \varepsilon^0_I)^2 - (\hbar\omega_r)^2} \tag{11.42}$$

After a changeover from a quantum chemical to a solid state physics notation, one obtains exactly the same result as originally derived by Fröhlich [105] from perturbation theory and then rederived by him again on the basis of the unitary transformation (11.18) [102].

$$\Delta E_0 = 2 \sum_{\mathbf{k},\mathbf{k}';\mathbf{k}\neq\mathbf{k}'} \left|u^{\mathbf{k}'-\mathbf{k}}\right|^2 f_{\mathbf{k}}(1-f_{\mathbf{k}'}) \frac{\hbar\omega_{\mathbf{k}'-\mathbf{k}}}{(\varepsilon^0_{\mathbf{k}'}-\varepsilon^0_{\mathbf{k}})^2-(\hbar\omega_{\mathbf{k}'-\mathbf{k}})^2} \tag{11.43}$$

In his first paper [105] Fröhlich tried to interpret the new state as a superconducting one, since the optimization of the occupation factors $f_{\mathbf{k}}$ yields a decrease of the total energy. But as soon the experimental verification of the existing gap appeared Fröhlich, in response in his second paper [102] admittedly rederived this result once more, but this time laying stress on the effective two-electron terms resulting from electron-phonon interactions as a possible source of the gap formation. He did publish it as a challenge that somehow, by means of a true many-body treatment, going beyond the Hartree-Fock approximation, the expected gap would be achieved.

(3) Fröhlich's effective two-electron interaction (Fröhlich Hamiltonian) [102]:

The electron-hyper-vibrational Hamiltonian in the electron-vibrational limit leads to the following result for the effective two-electron interaction [107]:

$$\Delta H''_F = \sum_{\substack{PQRSr\\ P\neq R,Q\neq S}} u^r_{PR} u^{r*}_{SQ} \frac{\hbar\omega_r[(\varepsilon^0_P-\varepsilon^0_R)(\varepsilon^0_S-\varepsilon^0_Q)-(\hbar\omega_r)^2]}{[(\varepsilon^0_P-\varepsilon^0_R)^2-(\hbar\omega_r)^2][(\varepsilon^0_S-\varepsilon^0_Q)^2-(\hbar\omega_r)^2]} N[a^+_P a^+_Q a_S a_R] \tag{11.44}$$

In solid state notation this sum reads ($r \to \mathbf{q}$; $P \to \mathbf{k}+\mathbf{q}, \sigma$; $Q \to \mathbf{k}', \sigma'$; $R \to \mathbf{k}, \sigma$; $S \to \mathbf{k}'+\mathbf{q}, \sigma'$):

$$\Delta H''_F = \sum_{\substack{\mathbf{k},\mathbf{k}',\mathbf{q},\sigma,\sigma'\\ \mathbf{q}\neq 0}} \frac{|u^{\mathbf{q}}|^2\hbar\omega_{\mathbf{q}}[(\varepsilon^0_{\mathbf{k}+\mathbf{q}}-\varepsilon^0_{\mathbf{k}})(\varepsilon^0_{\mathbf{k}'+\mathbf{q}}-\varepsilon^0_{\mathbf{k}'})-(\hbar\omega_{\mathbf{q}})^2]}{[(\varepsilon^0_{\mathbf{k}+\mathbf{q}}-\varepsilon^0_{\mathbf{k}})^2-(\hbar\omega_{\mathbf{q}})^2][(\varepsilon^0_{\mathbf{k}'+\mathbf{q}}-\varepsilon^0_{\mathbf{k}'})^2-(\hbar\omega_{\mathbf{q}})^2]} N[a^+_{\mathbf{k}+\mathbf{q},\sigma} a^+_{\mathbf{k}',\sigma'} a_{\mathbf{k}'+\mathbf{q},\sigma'} a_{\mathbf{k},\sigma}] \tag{11.45}$$

In comparison to the expression obtained by Fröhlich, known as the Fröhlich Hamiltonian

$$\Delta H''_{F(Fr)} = \sum_{\substack{\mathbf{k},\mathbf{k}',\mathbf{q},\sigma,\sigma'\\ \mathbf{q}\neq 0}} \left|u^{\mathbf{q}}\right|^2 \frac{\hbar\omega_{\mathbf{q}}}{(\varepsilon^0_{\mathbf{k}+\mathbf{q}}-\varepsilon^0_{\mathbf{k}})^2-(\hbar\omega_{\mathbf{q}})^2} a^+_{\mathbf{k}+\mathbf{q},\sigma} a^+_{\mathbf{k}',\sigma'} a_{\mathbf{k}'+\mathbf{q},\sigma'} a_{\mathbf{k},\sigma} \tag{11.46}$$

one can see that both expressions are equivalent, but not exactly identical. As the coordinate and momentum operators do not commute, the Hamiltonians (11.17) and (11.18) lead to a slightly different expression for the Fröhlich effective two-electron interaction. This type of ambiguity of the Fröhlich Hamiltonian was extensively discussed by Lenz and Wegner [108] by means of continuous unitary transformations.

(4) Lee-Low-Pines polarons and their self-energy [106]:

Neglecting two-electron terms in Eq. (11.39), one obtains a simple analytical expression for the fermionic one-particle excitation energies

$$\Delta\varepsilon_P = \sum_{r\in V}\left(\sum_{A\neq P}\frac{|u^r_{PA}|^2}{\varepsilon^0_P-\varepsilon^0_A-\hbar\omega_r}+\sum_{I\neq P}\frac{|u^r_{PI}|^2}{\varepsilon^0_P-\varepsilon^0_I+\hbar\omega_r}\right)$$
$$=\sum_{r\in V}\left(\sum_{R\neq P}|u^r_{PR}|^2\frac{1}{\varepsilon^0_P-\varepsilon^0_R-\hbar\omega_r}-2\sum_{I\neq P}|u^r_{PI}|^2\frac{\hbar\omega_r}{(\varepsilon^0_A-\varepsilon^0_I)^2-(\hbar\omega_r)^2}\right) \quad (11.47)$$

and in the solid state notation ($r\to\mathbf{q}$; $P\to\mathbf{k},\sigma$; $R\to\mathbf{k}-\mathbf{q},\sigma$; $I\to\mathbf{k}-\mathbf{q},\sigma$ with the occupation factor $f_{\mathbf{k}-\mathbf{q}}$)

$$\Delta\varepsilon_{\mathbf{k}} = \sum_{\mathbf{q}\neq 0}|u^{\mathbf{q}}|^2\frac{1}{\varepsilon^0_{\mathbf{k}}-\varepsilon^0_{\mathbf{k}-\mathbf{q}}-\hbar\omega_{\mathbf{q}}}-2\sum_{\mathbf{q}\neq 0}|u^{\mathbf{q}}|^2 f_{\mathbf{k}-\mathbf{q}}\frac{\hbar\omega_{\mathbf{q}}}{(\varepsilon^0_{\mathbf{k}}-\varepsilon^0_{\mathbf{k}-\mathbf{q}})^2-(\hbar\omega_{\mathbf{q}})^2} \quad (11.48)$$

The electron energies $\varepsilon^0_{\mathbf{k}}$ with the corrections (11.48) represent the quasiparticles/polarons that were originally derived via the Lee-Low-Pines transformation [106]. The first part of (11.48) refers to individual polarons, whereas the second part represents correlations originating from the effective field of other polarons.

The previous examples were based on a simplistic model of the electron-vibrational/electron-phonon Hamiltonian, taking only Galilean broken symmetries from the Goldstone theorem into account. Investigating the consequences of including all Goldstone bosons, descending from all three types of symmetry breakings in condensed matter, our attention will be focused on three most interesting and important examples: the Born-Huang ansatz, and ground state energy and excitation spectra of the B-O degenerate systems [107, 109].

(5) The Born-Huang ansatz [7]:

In the adiabatic limit, which means that all non-adiabatic coefficients $\tilde{c}$ will be equal to zero, the change of the ground state energy (11.34) yields the adiabatic correction

$$\Delta E_{0(ad)} = \sum_{AIr}\hbar\tilde{\omega}_r|c^r_{AI}|^2 = 2\sum_{AI}\left(\sum_{r\in V}\frac{1}{2}\hbar\omega_r+\sum_{r\in R}\rho_r+\sum_{r\in T}\tau_r\right)|c^r_{AI}|^2 \quad (11.49)$$

that is exactly identical with the Born-Handy ansatz (11.28). We see that only the field COM covariant theory can pass the test of the Born-Huang ansatz. Noting that the roton terms in (11.49) has nothing to do with the energies of the rotational degrees of molecular freedom as well as the translon terms with the de Broglie wave of the translational freedom of the whole system, the contribution of roton and translon quanta occurs even if the molecule or crystal is completely in rest and does not rotate or move.

(6) The ground state energy of the B-O degenerate systems:

Let us consider the complete non-adiabatic and field COM covariant case, where we only omit two-electron terms in order to obtain transparent analytical expressions:

$$\Delta E_0 = \sum_{AIr} \left|u^r_{AI}\right|^2 \frac{\hbar\tilde{\omega}_r}{(\varepsilon^0_A - \varepsilon^0_I)^2 - (\hbar\omega_r)^2} \tag{11.50}$$

which in the form of the sum of vibrational, rotational and translational parts finally reads

$$\begin{aligned}\Delta E_0 = &\sum_{AI,r\in V} \left|u^r_{AI}\right|^2 \frac{\hbar\omega_r}{(\varepsilon^0_A - \varepsilon^0_I)^2 - (\hbar\omega_r)^2} \\ &+ 2\sum_{AI,r\in R} \left|u^r_{AI}\right|^2 \frac{\rho_r}{(\varepsilon^0_A - \varepsilon^0_I)^2} + 2\sum_{AI,r\in T} \left|u^r_{AI}\right|^2 \frac{\tau_r}{(\varepsilon^0_A - \varepsilon^0_I)^2}\end{aligned} \tag{11.51}$$

After the rewriting Eq. (11.51) in solid state notation one obtains

$$\begin{aligned}\Delta E_0 = 2&\sum_{\mathbf{k},\mathbf{k}'} \left|u^{\mathbf{k}'-\mathbf{k}}\right|^2 \frac{\hbar\omega_{o,\mathbf{k}'-\mathbf{k}}}{(\varepsilon^0_{c,\mathbf{k}'} - \varepsilon^0_{v,\mathbf{k}})^2 - (\hbar\omega_{o,\mathbf{k}'-\mathbf{k}})^2} \\ &+ 4\sum_{\mathbf{k},r\in R} \left|u^r\right|^2 \frac{\rho_r}{(\varepsilon^0_{c,\mathbf{k}} - \varepsilon^0_{v,\mathbf{k}})^2} + 4\sum_{\mathbf{k},r\in T} \left|u^r\right|^2 \frac{\tau_r}{(\varepsilon^0_{c,\mathbf{k}} - \varepsilon^0_{v,\mathbf{k}})^2}\end{aligned} \tag{11.52}$$

where o denotes the optical branches and c, v the conducting and the valence bands respectively. This is a fascinating result. It demonstrates how the true quantum field, respecting fully the Goldstone theorem, copes with the B-O degenerate systems such as J-T molecules and superconductors: the superposition principle for degeneracy removal is simply bypassed, since rotons and translons are actually responsible for symmetry breaking. In a superconductor, such symmetry breakings produce several geometrically different symmetry broken states, and split the original half-occupied conducting band of a conductor into two bands—one fully occupied valence band and one empty conducting band.

(7) The excitation spectra of B-O degenerate systems:

Looking at the degeneracy removal and a gap formation in the B-O degenerate systems, we will only take the diagonal form in Eq. (11.39) into account. Further we neglect the second part of this equation, which does not depend on the electron distribution defined by the occupation/virtual states as does the first part. After omitting of two-electron terms we get:

$$\Delta\varepsilon_P = \sum_r \hbar\tilde{\omega}_r \left(\sum_{A\neq P} \frac{|u^r_{PA}|^2}{(\varepsilon^0_P-\varepsilon^0_A)^2-(\hbar\omega_r)^2} - \sum_{I\neq P} \frac{|u^r_{PI}|^2}{(\varepsilon^0_P-\varepsilon^0_I)^2-(\hbar\omega_r)^2} \right) \tag{11.53}$$

The diagonal form of the J-T one-particle excitation expression (11.53) is fully justified in solid state physics where translational symmetry is supposed. Since we have two bands, in solid state notation the one-particle Hamiltonian (11.53) reads

$$\begin{aligned}\Delta\varepsilon_{v,\mathbf{k}} &= \sum_{\mathbf{q}\neq 0} |u^{\mathbf{q}}|^2 \left(\frac{\hbar\omega_{o,\mathbf{q}}}{(\varepsilon^0_{v,\mathbf{k}} - \varepsilon^0_{c,\mathbf{k-q}})^2 - (\hbar\omega_{o,\mathbf{q}})^2} - \frac{\hbar\omega_{a,\mathbf{q}}}{(\varepsilon^0_{v,\mathbf{k}} - \varepsilon^0_{v,\mathbf{k-q}})^2 - (\hbar\omega_{a,\mathbf{q}})^2} \right)\\ &+ 2\sum_{r\in R} |u^r|^2 \frac{\rho_r}{(\varepsilon^0_{v,\mathbf{k}} - \varepsilon^0_{c,\mathbf{k}})^2} + 2\sum_{r\in T} |u^r|^2 \frac{\tau_r}{(\varepsilon^0_{v,\mathbf{k}} - \varepsilon^0_{c,\mathbf{k}})^2}\end{aligned} \tag{11.54}$$

$$\begin{aligned}\Delta\varepsilon_{c,\mathbf{k}} &= -\sum_{\mathbf{q}\neq 0} |u^{\mathbf{q}}|^2 \left(\frac{\hbar\omega_{o,\mathbf{q}}}{(\varepsilon^0_{c,\mathbf{k}} - \varepsilon^0_{v,\mathbf{k-q}})^2 - (\hbar\omega_{o,\mathbf{q}})^2} - \frac{\hbar\omega_{a,\mathbf{q}}}{(\varepsilon^0_{c,\mathbf{k}} - \varepsilon^0_{c,\mathbf{k-q}})^2 - (\hbar\omega_{a,\mathbf{q}})^2} \right)\\ &- 2\sum_{r\in R} |u^r|^2 \frac{\rho_r}{(\varepsilon^0_{c,\mathbf{k}} - \varepsilon^0_{v,\mathbf{k}})^2} - 2\sum_{r\in T} |u^r|^2 \frac{\tau_r}{(\varepsilon^0_{c,\mathbf{k}} - \varepsilon^0_{v,\mathbf{k}})^2}\end{aligned} \tag{11.55}$$

leading to two sets of one-particle corrections, one set for the valence band electronic corrections and the latter set for the conducting ones.

Taking notice of the inner-band frequencies $\omega_{a,\mathbf{q}}$ that are not involved in the ground state energy equation, but are present in one-particle correction terms, these terms are the same as those in the Fröhlich's Hamiltonian (11.46), i.e. the denominators of them can achieve both positive and negative values. On the other hand the terms with inter-band optical frequencies $\omega_{o,\mathbf{q}}$ are optimized by means of Eq. (11.52), and therefore the negative denominators will be prevailing. This will result in negative values of $\Delta\varepsilon_{v,\mathbf{k}}$ and positive values of $\Delta\varepsilon_{c,\mathbf{k}}$. Of course, from the general form of Eqs. (11.54)–(11.55) one cannot uniquely predict the existence of a gap. Not all conductors might become necessary superconductors at absolute zero. It depends on many factors but the most important factor is the bandwidth. It is clear from (11.54) to (11.55) that the narrow bands (high T_C superconductors) result in greater gaps than broad bands (low T_C superconductors).

We have presented here an independent proof of the Goldstone theorem with a one-to-one correspondence between broken symmetries and associated massless and spinless bosons in condensed matter, such as molecules and solids, and we have found the lost bosons, i.e. the rotons and the translons. We have shown that a field theory based only on phonons is insufficient for the description of B-O degenerate states such as J-T systems and superconductors where the general field COM covariant theory, incorporating all Goldstone bosons—phonons, rotons and translons is unavoidable. The Born-Huang ansatz plays the same role for the field COM covariance as the Maxwell equations do for Lorentz covariance.

One small question remains: Is the Born-Huang ansatz experimentally credible to the same extent as the Maxwell equations? If so we can authorize the formulation of a new type of field covariance. It was the merit of Handy [110] and many others, that thousands of Born-Huang ansatz calculations and their comparison with experimental data were performed, since for a long time a reasonable doubt prevailed regarding the equivalency of these results, and those of the exact quantum mechanical COM separation and the Born-Huang ansatz as the first correction to the B-O approximation. Kutzelnigg therefore renamed the Born-Huang ansatz as the Born-Handy ansatz [8]: "Handy and co-workers have never claimed to have invented the ansatz referred to here as the "Born-Handy ansatz", but they certainly convinced a large audience that this ansatz is of enormous practical value, even if it has not been completely obvious why it leads to correct results. Handy and co-workers realized that the difficulties with the traditional approach come from the separation of the COM motion (and the need to define internal coordinates after this separation has been made). They therefore decided to renounce the separation."

Once the Born-Huang (Born-Handy) ansatz was experimentally confirmed, we have subsequently experimental evidence of the lost Goldstone bosons—rotons and translons—as well. Their role in quantum systems is quite curious. Their contribution to the corrections of the ground state energy is already significant in small molecules like hydrogen molecule with only one vibrational mode and the five lost Goldstone boson modes, and is even several times greater than the contribution of the pure vibrations. The bigger the molecule, the lesser effect have these rotons and translons. In most of solids, with a huge amount of phonons, such as conductors, semiconductors or insulators, the effect of the six rotons/translons becomes negligible and the electron-phonon field theory is sufficient. However, in B-O degenerate systems suddenly a surprise appears. We have arrived at the same formula for the ground states of finite as well as infinite systems, both molecules and crystals, in the case of a B-O electronic degeneracy. However, Goldstone bosons arising from the violation of rotational and translational symmetries, rotons and translons, produce singularities in the original symmetrical positions, and the system is forced to avoid them, removing the degeneracy in an effectively one-particle manner at new asymmetric positions. This principle unifies the formation of the ground states of J-T molecules and superconductors, showing the way how quantum field formulations deal with virtual degeneracies originating from approximative B-O solutions. They are profoundly different from real degeneracies, where the principle of superposition takes place after the removal by some external perturbation (Stark or Zeeman effect). Quantum field theories simply do not share the centre of mass defined in quantum mechanics but solves the centre of mass problem "on its own".

If the lost Goldstone bosons—rotons and translons—are responsible for the mechanism of the formation of quantum states with spontaneously broken symmetries, it means, that the spontaneous symmetry breaking (SSB) was only phenomenologically described by classical physics, but it was completely misunderstood on the quantum level. We will continue with the analysis of the SSB in the next section.

12 The Paradox of Spontaneous Symmetry Breaking

Goldstone's theorem implies a quantum field reformulation of the J-T effect which is also valid for superconductors [107]: "Molecular and crystallin entities in a geometry of electronically degenerate ground states are unstable at this geometry except for the case when all matrix elements of the electron-rotational and electron-translational interactions equal zero." When I published this definition [107], I did not, at first, realize that it was a direct consequence of the Goldstone theorem, and secondly I did look upon it as an alternative to the official quantum mechanical definition. Later I saw that only the field J-T definition should be correct, since it leads to a truly broken symmetry for the ground state.

The official mechanical definition starts with the clamped-nuclei concept of the B-O approximation (see Sect. 2), whereas the exact mechanical solution, based on the Monkhorst-Cafiero-Adamowitz shell method for nuclei, leads to no symmetry breaking since the nuclei can be delocalized. On the other hand, the BCS theory of superconductivity is a quantum field theory where a Bloch-type field is used that does not respect the Goldstone theorem, and therefore the BCS theory cannot be correct. Yet the Bloch field was successfully applied to insulators, conductors, and semiconductors, but from the perspective of materials with B-O degenerate ground states and broken symmetries, such a field will be crippled and not usable. In such a case it permits linear superpositions of degenerate states, as it was originally described in the BCS paper [78]: "The normal phase is described by the Bloch individual-particle model. The ground state of a superconductor, formed from a linear combination of normal state configurations in which electrons are virtually excited in pairs of opposite spin and momentum, is lower in energy than the normal state by amount that is proportional to the average $(\hbar\omega)^2$, which is consistent with the isotope effect. A mutually orthogonal set of excited states, in one-to-one correspondence with those of the normal phase, obtains by specifying the occupation of certain Bloch states and by using the rest to form a linear combination of virtual pair configurations."

It is interesting to observe that, the BCS paper [78] does not mention any symmetry breakings. Four decades later, Weinberg wondered how Bardeen, Cooper, and Schrieffer could not see it [111]: "A superconductor is simply a material in which electromagnetic gauge invariance is spontaneously broken. This is not the way that most experts have historically thought about superconductivity. Early phenomenological theories were known to violate electromagnetic gauge invariance, but this was regarded as more annoying than enlightening. Broken symmetry is never mentioned in the seminal paper by Bardeen et al. [78] that first gave us a microscopic theory of superconductivity. Anderson [95] subsequently stressed the important role of broken symmetry in superconductors, but even today most textbooks explain superconductivity in terms of detailed dynamical models, with broken symmetry rarely mentioned."

Looking closely at the structure of the BCS theory, there is no asymmetry input/output, and hence there was absolutely no reason for Bardeen, Cooper, and Schrieffer to discuss any violations of symmetry. But we have already seen above that using the correct field, fully respecting the Goldstone theorem, the microscopic theory for superconductors will yield true asymmetric ground states, which are geometrical in accordance with the J-T effect. It means, that superconductors also share the common spontaneous symmetry breaking (SSB) with the J-T molecules. The quantum chemist and expert on the J-T effect, Isaac Bersuker came exactly to the same conclusion, writing in his book [112]: “Moreover, since the JTE has been shown to be the only source of spontaneous distortion of high-symmetry configurations, we come to the conclusion that the JTE is a unique mechanism of all the symmetry breakings in condensed matter.” Bersuker further restated the prominence and role of the J-T effect in connection with the discovery of high-Tc superconductivity [112]: “The next significant resurgence of interest in the Jahn-Teller effect is related to the late 1980s and is still continuing. It was triggered by one of the most important Nobel Prize discoveries in physics of our times inspired by the Jahn-Teller effect: the high-temperature superconductivity. As explained by the authors of this discovery, “the guiding idea in developing this concept was influenced by the Jahn-Teller polaron model” (J.G. Bednorz and K.A. Müller, in Nobel Lectures: Physics, Ed. G. Ekspong, World Scientific, Singapore, 1993, p. 424).”

Since the mechanism of SSB lies behind both the J-T effect and superconductivity, it entails that both phenomena must be described by similar mathematical equations, cf. the presentation given in the previous section. It is therefore somewhat surprising that Bersuker ignored this fact accepting the Cooper pair based explanation for superconductors. In his book [112] he suggests further efforts to find a bridge between J-T distortions and Cooper pairing citing: “However, the electron–phonon coupling is not the only factor that influences the superconductivity, and it does not determine the SC transition temperature directly. The path from local JT distortions to formation of polarons via lattice dynamics, to formation of Cooper pairs, their density, stability, and mobility, temperature dependence, to perfect diamagnetism, to structural phase transition, and finally to SC, is very long and thorny, and therefore a comprehensive JT theory of this important phenomenon has not yet been accomplished.”

In this connection it should be worth recalling Newton’s second antimetaphysical rule [113]: „Hypoth. II. Ideoque effectuum naturalium ejusdem generis eædem sunt causæ. Uti respirationis in Homine & in Bestia; descensus lapidum in Europa & in America; Lucis in Igne culinari & in Sole; reflexionis lucis in Terra & in Planetis.” It means that to the same natural effect the same causes must be assigned. Hence J-T distortions are the common causes of both phenomena, the J-T effect and superconductivity, which are observed on systems with B-O degenerate ground states and broken symmetries. Then it is not necessary to prompt the “very long and thorny path” from J-T distortions to the formation of Cooper pairs. In order to improve the understanding of the identity of these two phenomena, we will concentrate on their common cause, i.e. the J-T distortions, and especially on their underlying intrinsic concepts, the B-O approximation and the SSB, which both need to be carefully revised.

Starting first with the SSB and the problem of broken symmetries in the early period of classical physics, it led to a theological controversy about the question of its origin, between Newton and Leibniz. Hermann Weyl in his book [114], describes the conflict in detail: "The net result is that in all physics nothing has shown up indicating an intrinsic difference of left and right. Just as all points and all directions in space are equivalent, so are left and right. Position, direction, left and right are relative concepts. In language tinged with theology this issue of relativity was discussed at great length in a famous controversy between Leibniz and Clarke, the latter a clergyman acting as the spokesman for Newton [115]. Newton with his belief in absolute space and time considers motion a proof of the creation of the world out of God's arbitrary will, for otherwise it would be inexplicable why matter moves in this rather than in any other direction. Leibniz is loath to burden God with such decisions as lack "sufficient reason." Says he, "Under the assumption that space be something in itself it is impossible to give a reason why God should have put the bodies (without tampering with their mutual distances and relative positions) just at this particular place and not somewhere else; for instance, why He should not have arranged everything in the opposite order by turning East and West about. If, on the other hand, space is nothing more than the spatial order and relation of things then the two states supposed above, the actual one and its transposition, are in no way different from each other... and therefore it is a quite inadmissible question to ask why one state was preferred to the other.""

Weyl attempts to reconcile these opposite views from a new perspective of the theory of relativity after three centuries: "If nature were all lawfulness then every phenomenon would share the full symmetry of the universal laws of nature as formulated by the theory of relativity. The mere fact that this is not so proves that contingency is an essential feature of the world. Clarke in his controversy with Leibniz admitted the latter's principle of sufficient reason but added that the sufficient reason often lies in the mere will of God. I think, here Leibniz the rationalist is definitely wrong and Clarke on the right track. But it would have been more sincere to deny the principle of sufficient reason altogether instead of making God responsible for all that is unreason in the world. On the other hand Leibniz was right against Newton and Clarke with his insight into the principle of relativity. The truth as we see it today is this: The laws of nature do not determine uniquely the one world that actually exists, not even if one concedes that two worlds arising from each other by an automorphic transformation, i.e., by a transformation which preserves the universal laws of nature, are to be considered the same world."

Does Weyl's statement hold in universally physics? In classical physics including Einstein's relativity, we need not have to care about Leibniz' principle of sufficient reason, since this principle is of teleological origin. We have also seen that in the classical solution of superconductivity this principle is hidden behind a multiplicity of solutions, the first one leading to ideal conductivity and the second one to superconductivity, in full agreement with Weyl's interpretation. However, this construal is not valid in quantum physics, since quantum physics is causal and hence cannot observe teleological phenomena. After a rigorous use of Goldstone's theorem, we have obtained a superconducting ground state, which is identical to the ground states of insulators with J-T distorted nuclear positions. This is of course not enough for a quantum explanation of superconductivity. The quantum nature of Leibniz' principle must be first discovered, and only then might the quantum essence of superconductivity and all other teleological phenomena be revealed and seriously discussed.

This evokes the question: Does there exist a general formulation of the relationship between broken symmetries and the respective physical phenomena, valid both in classical and quantum physics? Pierre Curie proposed already in 1894 a formulation [116], fully based on causality, predating an analysis, which according to Bohr's principle of correspondence would apply to an extension from classical to quantum physics. The English translation of Curie's formulation with further extensive analysis can be found in the book of Brading and Castellani [117]. We quote here only the essential part: "Curie was led to reflect on the question of the relationship between physical properties and symmetry properties of a physical system by his studies on the thermal, electric and magnetic properties of crystals, these properties being directly related to the structure, and hence the symmetry, of the crystals studied... His conclusions, systematically presented in his 1894 work "Sur la symétrie dans les phénomènes physiques", can be synthesized as follows:

(a) A phenomenon can exist in a medium possessing its characteristic symmetry or that of one of its subgroups. What is needed for its occurrence (i.e. for something rather than nothing to happen) is not the presence, but rather the absence, of certain symmetries: "Asymmetry is what creates a phenomenon".
(b) The symmetry elements of the causes must be found in their effects, but the converse is not true; that is, the effects can be more symmetric than the causes.

Conclusion (a) clearly indicates that Curie recognized the important function played by the concept of symmetry breaking in physics (he was indeed one of the first to recognize it). Conclusion (b) is what is usually called "Curie's principle" in the literature, although one should notice that (a) and (b) are not independent of each other."

One might observe that Curie uses solely the pair of concepts: symmetry—asymmetry, and does not explicitly speak about SSB. Therefore Curie's formulation reveals only half of the truth about SSB, i.e. it deals only with causal phenomena arising from the multiplicity of asymmetric solutions of physical equations. It says nothing about the teleological phenomenon descending from the transitions between these asymmetric states. As an example the causal effect, related to ferromagnetism, is well explained in quantum physics. On the other hand, the associated teleological Einstein-de Haas effect has no quantum physical explanation, even if it was well described in the framework of classical physics in 1915 [118], before the concept of the electron spin was known, leaving only Ampère's hypothesis that magnetism is caused by microscopic circular motions of electric charges as a sufficient background. Superconductivity is an another example: Quantum physics can only solve the causal condensation of a conductor to the superconducting phase with multiple asymmetric ground states, but the explanation of the teleological transition between these ground states lies beyond the limits of quantum physics.

The SSB is primarily associated with two phenomena, i.e. one causal and one teleological. Curie's formulation deals with the first one. The phenomena in question correspond to the following symmetry argument that was formulated by van Fraassen

[119]: "There are two forms of argument which reach their conclusion 'on the basis of considerations of symmetry'. One, the symmetry argument proper, relies on a meta-principle: that structurally similar problems must receive correspondingly similar solutions. A solution must 'respect the symmetries' of the problem. The second form, rather less important, assumes a symmetry in its subject, or assumes that an asymmetry can only come from a preceding asymmetry. Both exert a strong and immediate appeal, that may hide substantial tacit assumptions."

The first part of van Fraassen's argument is quite clear and there is full agreement in the scientific community regarding its interpretation. Even if we get several asymmetric states as solutions of symmetric equations, i.e. if the original symmetry is broken, the asymmetric states together form a set which respects the original symmetry of the problem. However, the second form is not clear at all, and there is no unified view how to interpret it. One group of physicists attempts to explain it within the causal laws of physics. Usually imperfections and random fluctuations are taken into account. Stewart and Golubitsky in their book [120] describe the role of imperfections as follows: "We've said that mathematically the laws that apply to symmetric systems can sometimes predict not just a single effect, but a whole set of symmetrically related effects. However, Mother Nature has to choose which of those effects she wants to implement. How does she choose? The answer seems to be: imperfections. Nature is never perfectly symmetric. Nature's circles always have tiny dents and bumps. There are always tiny fluctuations, such as the thermal vibration of molecules. These tiny imperfections load Nature's dice in favour of one or other of the set of possible effects that the mathematics of perfect symmetry considers to be equally possible."

There is no doubt that imperfections play an important role in macroscopic objects. The Norton dome is a nice example of this. Nonetheless, Norton emphasises the difference between the real and ideal dome [121]: "However, we do not even have a falsified prediction. The dome is not intended to represent a real physical system. The dome is purely an idealization within Newtonian theory. On our best understanding of the world, there can be no such system. For an essential part of the setup is to locate the mass exactly at the apex of the dome and exactly at rest. Quantum mechanics assures us that cannot be done. What the dome illustrates is indeterminism within Newtonian theory in an idealized system that we do not expect to be realized in the world."

Such idealizations are of course not convincing enough for many physicists to take the Norton dome seriously, particularly when there are no applications to the real world. Norton reckons all objections against the dome and the indeterminism in classical physics [121]: "What's wrong with the dome? Many believe that the dome somehow lies outside what is proper in Newtonian theory. Three distinct bases for this judgment are described below, along with my reasons for finding them unconvincing. (1) Does the dome employ an incomplete formulation of Newtonian physics? (2) Is the dome "unphysical"? (a) Unphysical as gauge (over-description). (b) Unphysical as false. (c) Unphysical as pathological. (d) Unphysical through under-description. (3) Does the dome use inadmissible idealizations?"

Although Norton disproves very carefully the above mentioned objections, a curious situation arises: On one hand Norton is correct in his arguments against his opponents, but on the other his opponents are right about the determinism of classical physics. Malament [122] has expressed this quandary in a very elegant way: "But I am not sure that the full complexity and interest of the breakdown is adequately captured by saying, either, "Newtonian particle mechanics is an indeterministic theory" (full stop) or "Norton's example is not a well-defined Newtonian system" (full stop). Indeed, I am not convinced we have clearly posed alternatives here—because we do not have a sufficiently clear idea in the first place what should count as a "Newtonian system" (or count as falling within the "domain of application" of Newtonian theory). My inclination is to avoid labels here and direct attention, instead, to a rich set issues that the example raises."

In order to gain a deeper insight into this problem one should look for analogies such as the Norton dome. A historical review of comparable examples was written by van Strien. In her article "The Norton Dome and the Nineteenth Century Foundations of Determinism" [123] she says: "Lipschitz-indeterministic systems played an important role in the work of Boussinesq, who, in 1878, discussed several such systems, one of which was similar to the Norton dome [124]." This paper contains also an observation of an unexpected importance: "Though determinism as a general metaphysical principle was seldom doubted, it was not necessarily founded on a physical theorem about the uniqueness of solutions to certain differential equations."

It's quite astonishing to notice the difference in the way 19th century scientists were thinking! Maxwell, in his letter to Darwin's half-cousin Galton (quoted in Sect. 7), being fascinated by the works of Boussinesq and al., does not speak about the problem of indeterminism in classical physics at all. Yet he mentions "some determining principle which is extra physical (but not extra natural)" instead. It indicates that Maxwell was well aware of a missing law of nature, when he continues: "Boussinesq's method is a very powerful one against metaphysical arguments about cause and effect and much better than the insinuation that there is something loose about the laws of nature, not of sensible magnitude but enough to bring her round in time."

It appears that Norton already used the correct word "acausal" in his first paper dealing with the dome [52]. Using the word "acausal" as a proper opposite to the word "causal" is to be preferred, because it leaves the door open for teleological phenomena. Any attempt to analyse indeterminism on the classical level is meaningless since classical physics is unable to deal with it, perhaps at best can somehow "tolerate" it as in the case of Norton's dome. We may not even need to discuss the problem of Lipschitz continuity: the Mexican hat in the Ginzburg-Landau equations has no such discontinuity at the apex and, after all, we observe the same type of double solutions for the Norton dome and ideal conductors (first solution) and superconductors (second solution) the latter leading to the Meissner effect. Classical physics is able to describe the Norton dome as well as the Meissner effect, but is unable to determine what evoked the phenomenon.

If we take into account all the known examples of SSB in classical physics, i.e. the Mexican hat, Norton's dome, or the systems discussed by Boussinesq, we conclude in relation to the SSB archetype, forming the "falling down from the apex" in the classical physics, that the trigger of falling down is unknown. According to Maxwell's contemporary, mathematician and philosopher Peirce, no physics based on a corpuscular philosophy can solve the problem of the SSB trigger [125]: "Now I maintain that the original segregation of levo-molecules, or molecules with a left-handed twist, from dextro-molecules, or molecules with a right-handed twist, is absolutely incapable of mechanical explanation... The three laws of motion draw no dynamical distinction between right-handed and left-handed screws, and a mechanical explanation is an explanation founded on the three laws of motion. There then is a physical phenomenon absolutely inexplicable by mechanical action. This single instance suffices to overthrow the Corpuscular Philosophy."

Although these lines were written in the era of classical physics, Peirce's argument is applicable to quantum physics as well: It is built on the atomic ideas of Democritus, with the same corpuscular philosophy as in classical physics. It means that quantum physics in its Copenhagen interpretation is incapable of solving the problem of the SSB trigger. As we have realized this is a teleological issue, related to Leibniz' principle of sufficient reason, the second form of van Fraassen's argument, and Peirce's corpuscular philosophy argument, which can be fully ignored in classical physics. Unfortunately physicists usually ignore this difficulty also in quantum physics, and this attitude is a mistake, since it is motivated by the incorrect belief, that quantum physics employs the same SSB archetype as the classical one.

On the quantum level, physical systems can be described in a dual way: either mechanically or as a field system, in accordance with the Weinberg's statement [126]: "If it turned out that some physical system could not be described by a quantum field theory, it would be a sensation; if it turned out that the system did not obey the rules of quantum mechanics and relativity, it would be a cataclysm." Note that we demonstrated a similar picture in Sect. 11 particularly for condensed matter cases (molecules and crystals): Side by side the precise Monkhorst's quantum mechanical approach to the solution of the Schrödinger equation we present the true quantum field solution fully respecting the Goldstone theorem.

We are now in the position to ask whether this dual description of physical systems would lead to the same result. The answer is certainly not, because different results obtain for symmetry broken states. Quantum mechanics admits the superposition of multiple degenerate states, while Weinberg, on the other hand, has proven their orthogonality for quantum fields within the infinite volume limit. The Weinberg's argument is often quoted by various authors, e.g. in the article of Brading and Castellani [127]: "In quantum physics SSB actually does not occur in the case of finite systems: tunnelling takes place between the various degenerate states, and the true lowest energy state or "ground state" turns out to be a unique linear superposition of the degenerate states. In fact, SSB is applicable only to infinite systems—many-body systems (such as ferromagnets, superfluids and superconductors) and fields—the alternative degenerate ground states being all orthogonal to each other in the infinite volume limit and therefore separated by a "superselection rule" (see for example Weinberg [111], pp. 164–165)."

Though this citation reflects Weinberg's opinion, one can also find other interpretations, emphasizing the difference between the mechanical and the field descriptions, rather than the condition of the volume infinity of fields, as e.g. in the paper of van Dam [128]: "For the SSB in the rod and in the ferromagnet two comments were made that apply equally to this case: each of the ground states has an equal chance to be the ground state of the physical system, and the ground states are related to each other by the U(1) symmetry of the Lagrangian. Relating this situation to quantum mechanics might provoke suspicion: through tunnelling the real ground state could surely be a superposition of the individual ground states, which would not be degenerate. But although such a situation could occur in ordinary quantum mechanics, it does not apply to quantum field theory. The fields live in an infinite volume and thus have infinitely many degrees of freedom. Tunnelling cannot happen in this case. All vacuum states are orthogonal. This is extensively discussed by, for example, Weinberg [111, pp. 163–167]."

The results presented in the previous section, testify in favour of van Dam's interpretation: there is no SSB in quantum mechanics, it can solely be described in quantum field theory. Moreover, one might not need to rely on Weinberg's proof, which is valid only for infinite-volume fields, since the Goldstone bosons, associated with broken translational and rotational symmetries, earlier in this work called translons and rotons [107], are responsible for singularities at symmetric points. Hence the manifestation of SSB, in order to avoid these singularities, does appear for any system, regardless of the system being finite or infinite.

As a matter of fact, if we have two legitimate physical descriptions, the quantum mechanical and the quantum field one, these two formulations must necessarily yield the same energies for the system under examination. Therefore the classical SSB archetype, here denoted as "falling down from the apex", is not workable for quantum systems. Quantum physics can only calculate the symmetry broken states, while the SSB transition must somehow follow from the transitions between the mechanical and the field states of the system proceeding without any energy gain or loss, since symmetrical mechanical points and symmetry broken field points represent two descriptions with the same energy. A new law of nature is indispensable for this purpose, since this type of transitions has no support within the Copenhagen interpretation. As a consequence, from this new law, the origin of asymmetry must emerge, and evidently, according to the second form of the van Fraassen's argument, this asymmetry cannot arise *ex nihilo*.

So where should one start searching for this new law of nature? One way would be to focus on the simplest cases of SSB. We will, however, not start with complex nonadiabatic phenomena such as superconductivity or the J-T effect, nor with the adiabatic, but infinite systems like ferromagnets. The simplest example of SSB for our purpose would be small adiabatic systems such as the formation of isomers. The exact quantum mechanical solution, according to the Monkhorst-Cafiero-Adamowitz approach [6, 9], cannot lead to isomerism. The latter appears only after the introduction of the B-O approximation, or the clamped-nuclei concept. It seems like this approximation somehow mimics the quantum field properties where SSB can only appear. This is probably why Sutcliffe and Woolley, see Sect. 2, have put questions related to the transition from the isolated to the individual systems on the list of unsolved problems in quantum theory. In fact the solution of this problem, as presented here, is the key to providing a description of the SSB archetype on the quantum level.

As stated here the application of the classical SSB, i.e. the Mexican hat, to quantum equations, such as the J-T effect, superconductivity or the Higgs mechanism, is a disparity. This results in a misinterpretation of related phenomena, and it reminds of the reincarnation of the notorious medieval puzzle, namely how many angels can dance on the apex of a needle, i.e. a comparable mismatch by combining a material piece of a needle with the spiritual world of angels. Consequently quantum equations, grounded on the classic SSB, lead to a paradoxical situation. Hence the scientific community appears divided in regard to the question whether SSB is presently accounted for or not. According to Bersuker, the J-T effect is a unique mechanism of all existing SSBs in condensed matter, while, according to Weinberg, quantum mechanics cannot lead to any SSB. Bardeen, Cooper and Schrieffer did not find any SSB in their theory of superconductivity, yet Weinberg wondered how they could miss it. In standard text books, the Higgs mechanism is described as a SSB mechanism, but gradually a strong opposition against this view has emerged, usually quoting the Elitzur's theorem [129], maintaining that local gauge symmetries cannot be spontaneously broken. The tip of the iceberg is the statement by Anderson, that ferromagnetism is not a case of SSB. His opinion was condemned by Peierls and Kaplan, see their discussions in Ref. [130].

Finally, Unzicker in his book "The Higgs Fake: How Particle Physicists Fooled the Nobel Committee" [131] quotes the prophetic pronouncement of Rudolf Clausius: "Once the error is based like a foundation stone in the ground, everything is built thereupon, nevermore it returns to light." In the following section we will return to the beginning of the many-body treatment in quantum physics, revealing an overlooked secret regarding the B-O approximation. Here one may anticipate the starting point of the whole inconsistent chain that culminates in the uncertainty of many physicists, which phenomenon is of the SSB type and which one is not.

13 The Paradox of Bohr's Complementarity

We begin by looking more carefully at the qualitative nature of the B-O approximation. Even if it has been successfully proven in most molecular many-body calculations as well as its accuracy verified times and again, it is nevertheless a very non-standard approximation. In its original derivation, via the m/M (electron mass/nuclear mass) series expansion, it is unique in the sense that no competing series exists, cf. the Brillouin-Wigner or the Rayleigh-Schrödinger perturbation theories. In contrast to the latter, the B-O approximation is not only the basis of almost all molecular quantum chemical calculations, but it provides also a fundamental concept for molecular structure [132]. Since one knows that physical quantities, such as mass, charge, energy, momentum etc., need exact definitions, how is it then possible that molecular and crystalline structures are defined by an approximation. The latter must either be erroneous or alternatively not principally an approximation at all, hiding a precise and fundamental meaning. One reason for accepting the second choice, would be the argument by Sutcliffe and Woolley, see Sect. 2, regarding the simplest case of broken symmetries. For instance the Monkhorst-Cafiero-Adamowitz approach, dealing with isolated molecules, does not recognize any isomerism. They appear only on the B-O level, that deals exclusively with individual molecules. The conclusion is clear, no approximation will lead to new phenomena such as symmetry violations or even to making an ontological shift from isolated to individual order of the systems under investigation. Quoting again the challenge by Sutcliffe and Woolley, see Sect. 2: "The interesting question is how to get from the quantum theory of an Isolated Molecule to a quantum theory of an individual molecule by rational mathematics."

Pioneering quantum mechanics, describing a system of nuclei and electrons, cannot answer the above mentioned dilemma. We need a field theoretical description of the fermions as renormalized electrons and the nuclei replaced by Goldstone bosons as vibrational, rotational and translational modes. As was shown in Sect. 11, the field Hamiltonian (11.33) satisfies this request finally resulting in the clamped-nuclei concept, the first step of the B-O approximation, yielding all the equations known from this approximation, such as those of Pople for the ab initio calculation of vibrational frequencies (11.40), (11.41) and the Born-Huang ansatz (11.49). However, the clamped-nuclei concept is a *contradictio in adjecto*. For instance either one keeps the nuclei in mind, which can never be "clamped" since the nuclear positions do not commute with the total Hamiltonian, or one becomes fixed on the adjective "clamped", emphasizing instead of nuclei rather to speak about some traces or footprints, such as e.g. traces of electrons on the screen or traces of photons on a photographic plate. In a previous paper [109] I did introduce the notion of property-object dualism as follows: In quantum mechanics, nuclei and electrons represent objects to be described on an equal footing, and the vibrational modes are their common property. On the other hand, in quantum field theories the objects are represented by electrons and the Goldstone bosons, and "nuclear" positions or "clamped nuclei" are properties of the pertinent field equations.

As far as adiabatic systems are concerned, we have shown a proof of the equivalency between the field equations and the mechanical equations based on the B-O concept. What happens for nonadiabatic systems? While the practice of the B-O model is limited to adiabatic systems, there are no such limitations for the field equations. They are valid on the whole scale from adiabatic to nonadiabatic. For example we proved in Sect. 11 how the true field equations, based on Goldstone's theorem, do lead to symmetry broken ground states with the degeneracy removed on the one-particle level in both J-T systems and superconductors. In contrast the B-O approach to nonadiabatic systems leads to metaphysical solutions that are based on linear combinations of degenerate states. For the case of an infinite number of degenerate states the B-O concept gives rise to Mexican hats and the Berry-phase problem in so-called J-T molecules, see Bersuker [112]: "...one of the simplest JT $E \otimes e$ problems with linear vibronic coupling yields an APES (Adiabatic Potential Energy Surface) in the form of a "Mexican hat"... For a long time, up to the last decade, not very much attention was paid to these conical intersections; rather they were viewed largely as a "theoretical curiosity." This perception changed recently in view of the latest achievements in the treatment of such systems. One of these achievements is a generalization directly related to the JTE and now known as the topological (geometric) phase, or the Berry-phase problem... An important feature of the Berry-phase implications in JT problems is that the peculiar phase factor that changes the sign of the electronic wavefunctions and makes the ground vibronic state degenerate occurs only when one or an odd number of conical intersections are encircled, while it retains the same sign if an even number (including zero) are encircled. This was shown by direct calculation of the phase in the $E \otimes e$ problem. In fact, the phase is $\phi_0 = n\pi$, where n is the number of conical intersections encircled; for $n = 0, 2, 4, \ldots$ the sign of the wavefunction does not change."

The Berry-phase was mentioned above. The exact definition can be found in Berry's original paper [133]: "A quantal system in an eigenstate, slowly transported round a circuit C by varying parameters $\mathbf{R}$ in its Hamiltonian $H(\mathbf{R})$, will acquire a geometrical phase factor $\exp[i\gamma\ (C)]$ in addition to the familiar dynamical phase factor... If C lies near a degeneracy of H, $\gamma\ (C)$ takes a simple form which includes as a special case the sign change of eigenfunctions of real symmetric matrices round a degeneracy." However, there is a problem, since the Berry-phase, based on the definition above, cannot be applied to J-T molecules at all. For instance the B-O approximation can either be derived as a result of the m/M expansion or in an equivalent form using the hierarchical quantization process, i.e. first quantize the electronic motion at fixed classical nuclear positions and then a posteriori quantize the nuclear motion. Applying the Berry-phase to J-T molecules is just a product of this type of hierarchical quantization. A more exact simultaneous quantization method, i.e. Monkhorst's approach, see Sect. 2, or the quantum field approach using the full set of Goldstone bosons, presented here Sect. 11, can never lead to a meaningful Berry-phase in J-T systems.

In Sect. 12 we mentioned the often quoted Weinberg attitude to the occurrence of SSB: Only infinite many-body systems and fields can be spontaneously broken, whereas finite systems due to the possible tunnelling between degenerate states cannot. But from the perspective of our analysis this attitude does not reflect the whole truth. Physicists often reject considerations that have even if only a little sniff of chemistry and do not realize that the simplest examples of SSB, such as isomerism arises on the platform of quantum chemistry in finite systems and that a revealing of the true nature of the B-O approximation is the correct response to a correct understanding of SSB. In fact, fields with infinite degrees of freedom are not necessary for SSB occurance at all.

We have here argued that a field theoretical formulation is valid for molecules with finite degrees of freedom as well. Moreover, this theory is able to describe isomerism, i.e. SSB. On the other hand, many-body quantum mechanical applications, regardless of being finite or infinite, can never lead to any SSB. Based on Monkhorst's quantum mechanical approach, fully ignoring the B-O concept, one concludes that there is no isomerism, no ferromagnetism, no J-T effect, and no superconductivity. Ferromagnets, in calculations similar to those of Cafiero and Adamowitz, would look like giant molecules with full spherical symmetry and therefore without any SSB. J-T molecules would look like any other molecules with a non-degenerate electronic spectrum and superconductors looking like insulators. Only after the introduction of the B-O approximation, here a "dirty trick", the isomerism appears, and we can construct the simplest Heisenberg model of ferromagnetism. But this simplification makes only sense for adiabatic systems. However, in cases such as J-T molecules and superconductors, it yields ontologically incorrect results and here field methods based on the full set of Goldstone bosons must be applied.

In agreement with Weinberg's statement that every quantum system can be described by two profoundly different descriptions, either as a mechanical system or as a field system, and that only the second one recognizes SSB, one may ask if these two descriptions are still equivalently valid or whether only one of them is fundamentally correct. There is also the temptation to believe that the field approach supersedes the mechanical one, i.e. that the quantum field description should be seen as a generalization of quantum mechanics in the same way as general relativity becomes a generalization of special relativity. But listen to what Einstein said about this topic [19]: "Newton's theory deserves the name of a classical theory. It has nevertheless been abandoned since Maxwell and Hertz have shown that the idea of forces at a distance has to be relinquished and that one cannot manage without the idea of continuous "fields." The opinion that continuous fields are to be viewed as the only acceptable basic concepts, which must also be assumed to underlie the theory of the material particles, soon won out. Now this conception became, so to speak, "classical"; but a proper, and in principle complete, theory has not grown out of it. Maxwell's theory of the electric field remained a torso, because it was unable to set up laws for the behaviour of electric density, without which there can, of course, be no such thing as an electro-magnetic field. Analogously the general theory of relativity furnished then a field theory of gravitation, but no theory of the field-creating masses."

Einstein's quote above indicates that we must accept both the mechanical- and the field description as equally valid. However, the mechanical description pictures every system as isolated with no SSB, while the field formulation allows SSB at the same time being responsible for the individual character of the system. The question becomes whether there is any mathematical transformation between them, as inquired by Sutcliffe and Woolley. This reminds on the first stage of the development of quantum physics, when it turned out that elementary entities had both mechanical properties (particles) and field properties (waves). Even if Bohr had introduced the concept of complementarity into physics, some physicists developed a negative attitude towards it and attempted to find direct transformations between particles and waves without the need for any complementarity, such as Einstein or de Broglie. In analogy with just how a century ago the different aspects of the mechanical and the field descriptions of elementary entities led to Bohr's concept of complementarity, we are now in a similar situation with respect to the descriptions of the many-body system. Hence a second type of Bohr complementarity for the many-body level is requested.

This suggestion does not sound weird, since a second type of complementarity was already requested a long time ago by one of the co-founder of quantum mechanics Pascual Jordan. In his now almost forgotten paper [134] he wrote: "We assume here an idealised photographic plate: Each photon hitting it will be absorbed, and a single photon will with certainty activate a certain silver grain... If we assume the exposed silver grain to be indeed in a state of well-defined decision as to its developability, then we must conclude that it is not merely voluntary resignation on our part if we do not describe the silver grain in terms of wave functions of its single atoms. Doing so would entangle us in contradictions, as we have already seen above. Therefore the physical situation itself must contain guarantees that such contradictions cannot take place—and only a second type of complementarity can give this guarantee. There must exist in the silver grain a certain situation by which its description in terms of atomic wave functions is made impossible—only in this manner can the grain function as it does... It is then apparent that the situation—though it is clear to a certain extent—does not allow a complete and final analysis; there remain open certain questions. For one cannot avoid the difficulties merely by describing the silver grain (or an analogous part of any observational instrument) as a "mixture" of the form of the statistical matrix; this would not help us much, for it cannot describe an increase of entropy any better than the Schrödinger equation of a "pure case." It seems to me that entirely new conceptions are necessary."

This infers an extra quality of complementarity imparting a new type of quantum transitions on the many-body level, which cannot be deduced from any known rules of quantum physics. It entails a new axiom, and this axiom was requested by Jordan in the same paper [134]: "This leads us to acknowledge that it is both possible and necessary to formulate a physical axiom not formulated hitherto. Above we held it to be part of the definition of macrophysics, to show no complications in the manner of complementarity, but to allow a complete "objectivation" of phenomena in space and time. But usually one defines macrophysics only by stating that it deals with great numbers of microphysical individuals—and this is another and a different definition. We need therefore a special axiom to express the empirical fact that these two definitions define the same thing—that really each large accumulation of microphysical individuals always shows a well-defined state in space and time—that a stone never, unlike an electron, has indeterminate coordinates. One often vaguely believes this to be guaranteed already by Heisenberg's $\Delta p \cdot \Delta q > h$; but in fact this relation only provides a possibility and not a necessity for the validity of our axiom. Let us assume that, in our experiment involving the photon, the photographic plate be removed, but that we have an arrangement whereby a macro-physical stone will fall according to the decision of the photon. Then, if we strictly assume v. Neumann's view, the stone comes to possess a wave function which makes it undecided whether it does fall or does not, and an observer has the opportunity to compel the stone to a decision by the mental process of forgetting that interference between the two wave functions of the falling stone would be possible. Schrödinger's famous cat is another illustration of this point. I think we can summarize the situation by saying that indeed a new feature—to be formulated by a new axiom—lies in the fact that such things do not happen; all formulations of quantum mechanics hitherto given do not suffice to exclude them. We are unable to make a clock with a hand which does not always point to a definite figure on the dial. This is a well-known fact, but a fact of which present theory gives no sufficient account."

The suggested axiom above, is in our case relevant to describe quantum jumps between the isolated and individual characteristics of the many-body system, and provides a response to the request of Sutcliffe and Woolley. There is, however, one main obstacle regarding the definition of an isolated system. Every finite isolated system is subjectively defined and since transitions between isolated and individual systems are in principle immeasurable, one must necessarily exclude the subjectivity of the observer. As a consequence the isolated system must be identical with the whole Universe, with the individual systems then identical with its fragmentations. We have arrived at the problem of fragmentation and wholeness, cf. how it was framed by Bohm in the 3rd statement quoted in Sect. 4. The second complementarity, embracing the whole Universe, is the megascopic mirror of the first Bohr microscopic complementarity. We have just touched ancient knowledge, i.e. presenting the Universe as a mosaic where the smallest one resembles the Greatest One.

The new axiom presented here is necessary for the justification of quantum jumps of the Universe. From its state of total wholeness to its total fragmented states and vice versa, it certainly goes beyond the scope of the Copenhagen interpretation. Yet it neither contradicts nor denies this interpretation, but rather appears to be its natural extension. Then again, acceptance of this axiom implies the end of Everett's MWI (discussed in Sect. 7). The MWI was promoted as a deterministic theory for the physical Universe and attempted to explain why a world appears to be indeterministic for human observers. The basic stipulation of MWI was to represent the whole Universe by a deterministic wave function. Although still popular, many physicists now have doubts about the completeness of such a description. Vaidman comments on this problem, i.e. whether the wave function is or not sufficient, in the following way [135]: "As mentioned above, the gap between the mathematical formalism of the MWI, namely the wave function of the Universe, and our experience is larger than in other interpretations. This is the reason why many thought that the ontology of the wave function is not enough. Bell 1987 (p.201) felt that either the wave function is not everything, or it is not right. He was looking for a theory with local "beables" [136]."

Surely, the ontology of the wave function is not enough. It does not mean that it is not right, since no known experiment is in conflict with it. In fact the wave function is an incomplete description of the Universe. Most probably the protagonists of the MWI never gave a thought to the ontological problem of the B-O approximation, and therefore do not realize that their wave function is unable to describe real objects like molecules or solids, but only the subjectively defined isolated objects of different ontological meanings, where, according to the language of Cafiero and Adamowitz, instead of individual molecules one has only molecular atoms or atomic molecules, and instead of crystals only crystalline atoms or atomic crystals.

However, the incomplete character of the wave function of the Universe should not set us off to look for some local "beables". As we have seen, there is a complementary field description of the Universe, leading ultimately to real objects, the symmetries of which can be broken. The second complementarity has an original significance for a proper understanding of all SSB phenomena. These processes have a sign of complexity and emergence, and therefore one must ask which fundamental principle lies behind. Rowlands has adequately formulated this request [137]: "This is symmetry, which we see all about us in the laws of physics, the fundamental interactions and the fundamental particles. Finding symmetries can help us to decomplexify our explanations, and, if we can understand where symmetry comes from, lead to more profound understanding. We should also note that some symmetries are broken; that is, what is fundamentally symmetric appears, under certain conditions, to display some asymmetry. The reason for this cannot be arbitrary, and if we can discover it, along with the reason for symmetry, this will be a big step in our understanding of the foundations. One thing it can't be is fundamental because nature never acts in such an arbitrary way. It has to be, in some way, a sign of complexity or emergence."

With the up till now known laws of quantum physics one can simply only arrive at a mechanical many-body description with no symmetries violated and a field description exhibiting possible broken symmetries. But to suggest an explanation for the broken symmetry, understood as a transition from unbroken to broken states, on needs to accept the new axiom. We continue quoting Rowlands giving his opinion [137]: "Again, if we decide that simplicity is to be preferred over complexity, we should be looking for something that is staggeringly simple, yet somehow capable of generating complexity. If we think our basic idea is a complicated construction, say a 10-dimensional space-time, then we have no way of knowing how this breaks up into the simpler component parts that must exist because we have no fundamental mechanism for doing this. We should certainly take notice of what the string theorists say about the symmetries required by nature, and we should expect to find them, even those expressed in 10 dimensions, but we should expect to find them by working out how such a complex idea emerges from simpler ones, in which the structures of the components reveal them as diverse in origin, the 'brokenness' of the larger symmetry coming from its inherent complexity, not by some arbitrarily-imposed concept of 'symmetry-breaking'. Broken symmetries are a sure signature of complexity, not of simplicity."

In the above citation Rowlands has touched upon the crucial problem how symmetry breaking is comprehended today as some arbitrarily imposed conception. Yet if the second complementarity is original, then every SSB is an emergent process, i.e. the succession of events, appearing either on the microscopic or the macroscopic level, as a consequence of the perpetual sequence of megascopic quantum jumps between states that represent the wholeness and the fragmentation of the Universe.

Unfortunately megascopic quantum jumps are not measurable. This fact is reflected in emergent SSB phenomena, such as superconductivity and thc J-T effect being the best-known examples. Knowing, that the density and velocity of superconducting carriers are non-measurable in principle, see Sect. 9, Bersuker describes a similar situation in J-T systems [112]: "Among other things Van Vleck [138] wrote that "it is a great merit of the J-T effect that it disappears when not needed." This declaration reflects the situation when there was very poor understanding of what observable effects should be expected as a consequence of the J-T theorem. The point is that the simplified formulation of the consequences of the J-T theorem as "spontaneous distortion" is incomplete and therefore inaccurate, and may lead to misunderstanding. In fact, there are several (or an infinite number of) equivalent directions of distortion, and the system may resonate between them (the dynamic J-T effect)... It does not necessarily lead to observable nuclear configuration distortion, and this explains why such distortions often cannot be observed directly... Even in 1960 Low in his book [139] stated that "it is a property of the J-T effect that whenever one tries to find it, it eludes measurements.""

Bersuker's countering argument against Van Vleck and Low can only be acceded from the point of view of the microscopic origin of the J-T effect. But as we have realized, the J-T effect is a megascopic phenomenon. Any other attempt to explain this effect microscopically leads inevitably to paradoxes. From a microscopic perspective quantum tunnelling between differently distorted J-T states restores the original symmetry provided one considers a sufficiently long period of time. And this opens directly the problem discussed at the end of the previous section, i.e. is there any SSB or not. Moreover consider the simplest microscopic example of quantum tunnelling, the particle in a box, divided into two parts by a barrier. If the particle is initially located in the first part, after an extended time period it will be found with the same probability in both parts. But in the case of the J-T effect, there is no observer who selects one of the possible distorted J-T states. Therefore the microscopic concept of quantum tunnelling cannot be applied to the J-T effect, and Van Vleck and Low were absolutely right, when they challenged its direct measurability. Incidentally microscopic processes are causal, while megascopic ones are teleological. Teleology interpreted as the final cause can be then understood as downward causality, descending from the Greatest One, the whole Universe, in contrast to the upward causality, ascending from the smallest entities, the elementary particles. In this way teleology finally gains a primary significance, as it was requested by Bohm, quoted in his 1st statement in Sect. 4. Causality in physics has its well-established empirical basis, but the principal non-measurability of megascopic phenomena might perhaps be the main reason why teleology is not till now explicitly incorporated in the exact sciences.

Approaching the next questions, i.e. what identifies the probabilities of megascopic quantum jumps, and what are the megascopic mirrors of the microscopic projection and the Born rule, Bohm in his 5th cited statement, see above, speaks about projections and injections. Indeed the latter phrase agrees perfectly with a requested megascopic mirror of the former. One can observe in various international forums that some scientists do feel that there is a connection between the mechanism of SSB and Zurek's einselection, but that no one knows exactly how the latter might explicitly be incorporated. The quantum decoherence program is focused on replacing the Born rule by einselection and entirely removing the von Neumann–Wigner rule, see Sect. 7. This program, however, was inspired by Universal Darwinism, and Darwin was himself a teleologist, with a telic notion of adaptation. Therefore an environmentally induced superselection rule must be of teleological origin and co-exists with the Born law being its megascopic mirror.

Unlike the Born rule einselection is not defined by some simple formula. If the environmental perturbation is negligible, the probability of transition into each of the n distorted J-T states equals $1/n$. Alternatively, if we have an isomer with two possible configurations, e.g. a left-handed and right-handed molecule, and the environmental influence prefers the right-handed one, then the transition probability for the latter equals one and for the other equalling zero. In superconductors transition probabilities between distorted ground states are given by the magnitude of the external magnetic field.

Note that Bohm's 2nd statement, implicates a limited world view, based on microscopic quantum physics, dealing with elementary particles as 'basic building blocks' out of which everything is made, but it also promotes an image of the old materialistic idea of Democritus. In contrast, megascopic quantum phenomena can be seen as reflections of Plato's Forms [140]. As recognized, the world of Plato and his a-spatial and a-temporal Forms are transcendental to our own world of substances, i.e. space and time respectively. It is superordinate to matter.

Since megascopic quantum jumps are not directly measurable, one may wonder whether the claim has any meaning at all. For instance do they proceed at some given time as a microscopic quantum jump or is there any relationship to time at all? An answer might emerge by looking carefully what the attribute "a-temporal" in world of Plato exactly means: A Form does not exist within any time period, rather it provides the formal basis for time. One should perhaps ask whether quantum physics uses a similar time concept as classical/relativistic physics. Here is the view of Lee Smolin on the nature of time [141]: "More and more, I have the feeling that quantum theory and general relativity are both deeply wrong about the nature of time. It is not enough to combine them. There is a deeper problem, perhaps going back to the beginning of physics." In other words it means going back to Plato and Aristotle. In Plato's Forms we will finally find the true relationship between megascopic quantum jumps and time, i.e. these jumps do not proceed in time, rather time is created by them. At the megascopic level this gives rise to the concept of quantum of time, which otherwise is unreachable at the microscopic level. Microscopic quantum theory yields quanta of energy, momentum, angular momentum etc., but never a quantum of time or of space. Different objects have different energies, momenta etc., but the Universe is one, the time is one, and the space is one. In this sense megascopic quantum jumps are the true clock of the Universe.

Focusing now on the relationship time—events. In Newtonian physics time is foundational, and events that can happen at any moment of time are subordinate. Despite the fact that microscopic quantum physics shares the concept of time with classical physics, events do not play the subordinate role there, but are original as well as time and independent on it. On one hand we have time evolution of the wave function, representing only Aristotelian potentiality, and on the other we have reality created during the measurement process by timeless events—microscopic quantum jumps. And just this separation of the events from time is the source of almost all philosophical problems of contemporary quantum physics. Bell wrote a very poignant parody relevant to this topic [136]: "It would seem that the theory is exclusively concerned with "results of measurement" and has nothing to say about anything else. When the "system" in question is the whole world where does one find the "measurer"? Inside, rather than outside, presumably. What exactly qualifies some subsystems to play this role? Was the world wave function waiting to jump for thousands of millions of years until a single-celled living creature appeared? Or did it have to wait a little longer for some more highly qualified measurer—with a Ph.D.? If the theory is to apply to anything but idealized laboratory operations, are we not obliged to admit that more or less "measurement-like" processes are going on more or less all the time more or less everywhere? Is there ever then a moment when there is no jumping and the Schrödinger equation applies?"

Bell was certainly correct in his view regarding "measurement-like" processes above. Our "megascopic quantum jumps" do correspond to Bell's request, and since the origin of time is rooted in megascopic events, we are able to finally reunite time and events on the quantum level. However, unlike events on the classical level those in the quantum case will be foundational and time subordinate. This idea, in fact, goes back to ancient times. Aristotle expressed this opinion in his Physics, Book I, Part 14, where he defined time as "the number of movement in respect of before and after". Thus it cannot exist without a succession, and it does not exist on its own rather it is relative to the motions of things [142]. Moreover, Aristotle's concept of time is more suitable for quantum theory than Newtonian classical mechanics, when he says that time, in order to exist, requires the presence of a soul capable of "numbering" the movement. This statement complies with the von Neumann–Wigner rule.

One might wonder why the Newtonian and Aristotelian concepts of time are so different. Pondering this issue, one must not forget that the Greek language has two expressions for time: χρόνος, chronos, and καιρός, kairos, while the Latin language only one expression *tempus*. Our way of thinking is to a large extent influenced by our spoken language. The Newtonian time concept is basically pragmatic, but it is unable to explain the time arrow problem (see Sect. 3). If we accept the premise that all processes in the Universe consist only of microscopic and megascopic events and therefore that irreversibility is foundational, then the arrow of time naturally develops from Aristotle's definition of time as "the number of movement in respect of before and after". Emergent time reversibility can only appear in subsystems with no SSB (mostly adiabatic systems), where a one-to-one correspondence between the mechanical and the field states holds. These subsystems can then be described by reversible evolution as formulated by the Schrödinger equation. This contradicts Santilli's claim, see Sect. 3, since irreversibility in the Universe has its origin in irreversible microscopic and megascopic events.

Before ending this section we will focus on the second Bohr complementarity asking: why did only Jordan call for it and not Bohr himself? Actually, Bohr did put forward another type of complementarity between observational conditions of animate and inanimate nature [61]: "In this promising development we have to do with a very important and, according to its character, hardly limited extension of the application of purely physical and chemical ideas to biological problems, and since quantum mechanics appears as a rational generalization of classical physics, the whole approach may be termed mechanistic. The question, however, is in what sense such progress has removed the foundation for the application of so-called finalistic arguments in biology. Here we must realize that the description and comprehension of the closed quantum phenomena exhibit no feature indicating that an organization of atoms is able to adapt itself to the surroundings in the way we witness in the maintenance and evolution of living organisms. Furthermore, it must be stressed that an account, exhaustive in the sense of quantum physics, of all the continually exchanged atoms in the organism not only is infeasible but would obviously require observational conditions incompatible with the display of life."

This can undoubtedly be understood as the third complementarity, which, although Bohr mentions here teleological (finalistic) arguments, has nothing to do with the second one. Life is a mystery, and no hitherto available knowledge, as e.g. the discovery of DNA, and no finalistic arguments are sufficient to explain it. However, Bohr becomes here aware of the teleological origin of Darwinian adaptation, in stark contrast to his contemporary quantum scientists who did not see it at all. He, on the other hand, was convinced that teleological adaptation only appeared in living organisms and he never considered the possibility of its occurrence in inanimate matter, such as in all SSB phenomena. Yet by realizing the distinction between causal and teleological phenomena, Bohr was very close to formulating the second kind of complementarity [61]: "In biological research, references to features of wholeness and purposeful reactions of organisms are used together with the increasingly detailed information on structure and regulatory processes that has resulted in such great progress not least in medicine. We have here to do with a practical approach to a field where the means of expression used for the description of its various aspects refer to mutually exclusive conditions of observation. In this connection, it must be realized that the attitudes termed mechanistic and finalistic are not contradictory points of view, but rather exhibit a complementary relationship which is connected with our position as observers of nature."

Bohr states correctly, that attitudes termed mechanistic (causal) and finalistic (teleological) are not contradictory, but fails, in my opinion, when saying that they are in a complementary relationship. Actually there is no complementarity between them because they do co-exist side by side. Summarizing, the first complementarity represents a causal microscopic relationship between mechanical and field descriptions of elementary entities (the particle-wave dualism), the second complementarity epitomizes the teleological megascopic relationship between the mechanical and the field descriptions of the whole Universe, where these two descriptions deal with the centre of gravity in different ways. This conclusion is a direct consequence of the Goldstone theorem, proved by Goldstone, Salam, and Weinberg, incidentally in the same year when Bohr died. Unfortunately, as already pointed out, a year later the Goldstone theorem was misconstrued by Anderson, see Sect. 10, and as a result there appeared an inconsistent Higgs mechanism, which would never reveal a second complementarity.

14 The Second Quantum Floor

The transition from classical to microscopic quantum physics is nicely distinguished if one compares the Rutherford and the Bohr atomic models. In the former an electron needs to accelerate its movement in order to achieve a higher energy level around the nucleus. In the Bohr model there is no need for any movement or acceleration; all electrons occupy certain quantum states and transitions to another level occur by way of quantum jumps.

In addition to the microscopic quantum domain, mentioned above and which we name the first quantum floor, one also has the whole scale of megascopic phenomena, operating on a second quantum floor. Surprisingly we are here in a similar situation compared to a century ago. The microscopic quantum description all of a sudden has become insufficient. Note that we have repeatedly discussed this issue in previous sections, e.g. for transitions between isomers and distorted J-T states in molecules, and in superconductors, and found that microscopic tunnelling, represented by the direct movement of elementary particles, is entirely outside the scope of the given problem. We need an additional concept of megascopic quantum jumps appearing either on the microscopic or the macroscopic level, which is in a similar relation to microscopic tunnelling as the concept of microscopic quantum jumps, in the Bohr model, relates to the classical electronic movement in the Rutherford model.

Concentrating briefly on superconductivity and its megascopic origin, we recapitulate the three main reasons why superconductivity does not have any microscopic explanation, see also Sects. 9 and 11:

(1) Every microscopic theory must result in Born's rule for the probabilities related to the density and the velocity of the superconducting carriers, and these quantities have to be experimentally measurable. But as we know, they are in principle non-measurable.
(2) No microscopic theory allows us to avoid the universal concept of Bloch states for the description of superconducting carriers. It means that the carrier mass must take the effective mass of the electrons into account. This, however, is in direct contradiction with measurements of the London moment, where only bare electronic masses are reported.
(3) According to the second form of van Fraassen's argument, asymmetry cannot arise ex nihilo. Although we know that the original asymmetry around the superconductor can be present in the form of an external magnetic field, no microscopic theory is able to explain the Meissner effect mechanism, when the superconductor is cooled below the critical temperature and the constant magnetic field cannot bring about any acceleration of the superconducting carriers. It means that a microscopic theory is unable to implement the original asymmetry, and this fact contradicts the mentioned second form of the van Fraassen argument.

Nevertheless, superconductors can be described macroscopically on the quantum level. The concept of macroscopically-occupied quantum states was proposed by

London [66]. The macroscopic wave function obtains as

$$\Psi(\mathbf{r}) = |\Psi_0(\mathbf{r})| \exp(i\theta(\mathbf{r})) = \sqrt{n_s} \exp(i\theta(\mathbf{r})) \tag{14.1}$$

with the phase θ and the amplitude Ψ_0 characterizing the density, n_s, of the superconducting carriers. From this equation, using the well-known relation between the kinetic and the canonical momentum, one gets the expression for the velocity of superconducting carriers

$$\mathbf{v}_s = \frac{\hbar}{m_s} \nabla.\theta - \frac{e_s}{m_s c} \mathbf{A} \tag{14.2}$$

and after taking the curl of the velocity

$$\nabla \times \mathbf{v}_s = \frac{-e_s}{m_s c} \mathbf{B} \tag{14.3}$$

and substituting from (9.1) one finally obtains the phenomenological London Eq. (9.5). It is interesting to note that from London's macroscopic wave function (14.1) one can derive Josephson's equations [143]. Hence in this formulation one does not need any microscopic concepts of superconductivity, as was already shown by Feynman [144]. Considering the solution of an easier case of a SIS (Superconductor-Insulator-Superconductor) junction, both "superconductor bodies" start to 'feel' each other at a certain distance, e.g. of the order of nanometres, because of the estimated coherence length consistent with their wave functions. Feynman proposed two coupled time dependent equations

$$i\hbar \frac{\partial \Psi_1}{\partial t} = E_1 \Psi_1 + K \Psi_2; \quad i\hbar \frac{\partial \Psi_2}{\partial t} = E_2 \Psi_2 + K \Psi_1 \tag{14.4}$$

where K is a phenomenological parameter which describes the properties of the insulating barrier. From Eqs. (14.1) and (14.4) one simply obtains the first Josephson equation

$$I(t) = I_c \sin(\theta_2(t) - \theta_1(t)) \tag{14.5}$$

and the second Josephson equation

$$U(t) = \frac{\hbar}{2e} \frac{\partial (\theta_2(\mathbf{r}) - \theta_1(\mathbf{r}))}{\partial t} \tag{14.6}$$

where $U(t)$ and $I(t)$ are the voltage across and the current through the Josephson junction, $\theta_2 - \theta_1$ is the phase difference across the junction, and I_c is a phenomenological constant, representing the "critical current" of the junction.

One should observe that the ground and the excitation states of the superconductors have a microscopic explanation, as well as all solids etc. Previously we have derived the J-T distorted ground state of superconductors, represented by Eq. (11.52), and their excited states, Eqs. (11.54), (11.55). However, superconductivity itself, as caused by the transitions between distorted J-T states, has no microscopic explanation. At this point microcosm does shake hands with megacosm.

Whereas Eqs. (11.52), (11.54), (11.55) are of the Bloch type, for dealing with the transitions between the distorted J-T states, the Fourier mirror of the Bloch functions—the Wannier functions—is more transparent. From the Bloch functions

$$\psi_{\mathbf{k}}(\mathbf{r}) = \exp(i\mathbf{kr})u_{\mathbf{k}}(\mathbf{r}) \tag{14.7}$$

and after the application of the Fourier transformation we get the Wannier functions

$$\chi_{\mathbf{R}}(\mathbf{r}) = \frac{1}{\sqrt{N}}\sum_{\mathbf{k}} \psi_{\mathbf{k}}(\mathbf{r})\exp(-i\mathbf{kR}) \tag{14.8}$$

where N is the number of primitive cells in the crystal and $\mathbf{R}$ is any lattice vector. There is one Wannier function for each Bravais lattice vector. The sum on $\mathbf{k}$ includes all the values of $\mathbf{k}$ in the Brillouin zone. The Wannier functions fulfil the relation

$$\chi_{\mathbf{R}}(\mathbf{r}) = \chi_{\mathbf{R}+\mathbf{R}'}\left(\mathbf{r} + \mathbf{R}'\right) \tag{14.9}$$

Using Wannier functions instead of the Bloch ones, the electronic wave function can be then rewritten as

$$\Psi(\mathbf{r}) = \frac{1}{\sqrt{N!}}\left\|\prod_{I}^{N} \chi_I(\mathbf{r})\right\| = \frac{1}{\sqrt{N!}}\left\|\prod_{i\sigma}^{N} \chi_{i\sigma}(\mathbf{r})\right\| \tag{14.10}$$

In this notation the spin orbitals I (or orbitals i and spins σ) are related to the lattice vectors $\mathbf{R}$.

Figure 1 shows two microscopic solutions for the superconductor ground state: the mechanical one based on the Monkhorst-Cafiero-Adamowitz approach with no broken symmetry and the same lattice constant $\mathbf{a}$ as in the conducting state above the critical temperature; and the field one with J-T broken symmetry and the lattice constant $2\mathbf{a}$. As said in previous sections, one needs a new rule allowing direct transitions from a mechanical state reflecting the whole Universe, into the field states that descends from the fragmentation of the Universe—and vice versa. During these megascopic quantum jumps the electron pairs, valence orbitals occupied by two electrons with opposite spins (no Cooper pairs) have the chance either to be relocated in the right or left directions, or to return into their original position, depending on the external magnetic field that determines the einselection factor μ. In the simplest one-dimensional case there are only three possible discrete values of μ, namely -1, 0 and 1.

In passing we note that only the idea of megascopic quantum jumps is able to overcome the three above mentioned main problems regarding any attempt to explain

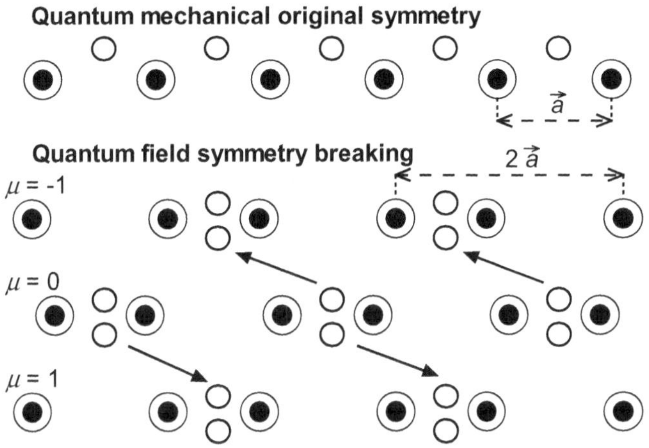

Fig. 1 Megascopic quantum jumps between the quantum mechanical and quantum field states of the Universe, leading to the megascopic tunnelling of superconducting carriers. ⦿ nuclei with electronic core, ○ valence electrons in symmetric quantum mechanical and asymmetric quantum field positions, μ einselection (environment-induced superselection) factor, $\vec{a}$ original lattice constant before the symmetry breaking

superconductivity microscopically. First, immeasurable megascopic quantum jumps imply non-measurability of the density and the velocity of superconducting carriers. Second, the "teleportation" of superconducting carriers, a new kind of non-local quantum phenomenon, guarantees the bare electronic mass in the London moment measurements. Third, only the holistic incorporation of the environment by way of megascopic einselection is capable of responding to the asymmetry in the form of an external magnetic field. Henceforth there is also no contradiction to the second form of van Fraassen's argument, and the Meissner effect can thus be explained.

Returning to the field solutions of Fig. 1, depicting distorted nuclear positions with the adequate electronic redistribution one realizes that the electrons form some kind of "chemical bonds" occupied by two electrons with opposite spins. As it was explained in the previous section, megascopic quantum jumps do not occur in time, but time is created by them. Denoting the quantum time scale as τ, the time $t = n\tau$ will be understood as the interval of n megascopic events, i.e. n transitions from the field state to the complementary mechanical state and back. After two megascopic events, considering only a one-dimensional chain, the valence orbitals occupied by the two electrons will be relocated to new positions, which are in the following relation to the original positions

$$\left(\chi_{i\uparrow}(\mathbf{r})\chi_{i\downarrow}(\mathbf{r})\right)_{2\tau} = \left(\chi_{i\uparrow}(\mathbf{r}-2\mu\mathbf{a})\chi_{i\downarrow}(\mathbf{r}-2\mu\mathbf{a})\right)_0 \exp(i\theta(\mathbf{r})); \quad \mu \in \{-1, 0, 1\} \tag{14.11}$$

Here the phase θ appears at the macroscopic level, reflecting the "teleportation" of superconducting carriers, i.e. the valence orbitals occupied by two electrons. In our megascopic interpretation, unlike the BCS theory which associates this phase with the motion of carriers in Bloch's **k** space, the phase is associated with the macroscopic **l** space, orthogonal to the Bloch **k** space:

$$\theta(\mathbf{r}) = \mathbf{l}\,.\,\mathbf{r}; \quad \mathbf{l} = \frac{m_s \mu \mathbf{a}}{\hbar \tau}; \quad m_s = 2m_e \tag{14.12}$$

This equation gives us a relation between the macroscopic momentum, and the carrier mass and the velocity. The carrier mass corresponds to the double valued electronic mass, as well as the carrier charge to the double valued electronic charge. The superconducting carriers appear without the mentioning of Cooper pairs; they are identical with the smallest relocated entities—double occupied valence orbitals. Using Eq. (14.9) one gets

$$\chi_{i\sigma}(\mathbf{r} - 2\mu\mathbf{a}) = \chi_{j\sigma}(\mathbf{r}); \quad \mathbf{R}_j = \mathbf{R}_i + 2\mu\mathbf{a} \tag{14.13}$$

and after two megascopic events Eq. (14.10) transforms into the new expression:

$$\Psi(\mathbf{r})_{2\tau} = \frac{1}{\sqrt{N!}} \left\| \prod_{j\sigma}^{N} \chi_{j\sigma}(\mathbf{r}) \right\| \exp(i\theta(\mathbf{r})) = \Psi(\mathbf{r})_0 \exp(i\theta(\mathbf{r})) \tag{14.14}$$

Finally generalizing Eq. (14.13) for an arbitrary time $t = 2n\tau$ and to the three-dimensional case, where instead of discrete values of the einselection factor we have the real interval $\mu \in \langle -1, 1 \rangle$, one obtains the relations

$$\chi_{i\sigma}(\mathbf{r} - 2n\mu\mathbf{a}) = \chi_{j\sigma}(\mathbf{r}); \quad \mathbf{R}_j = \mathbf{R}_i + 2n\mu\mathbf{a}; \quad t = 2n\tau; \quad \mu \in \langle -1, 1 \rangle \tag{14.15}$$

Note that Eq. (14.14) can be generalized for arbitrary times, meaning, that in this way we have arrived at London's macroscopic wave function (14.1).

There arises an interesting question, whether the quantum of time from Eq. (14.12) can be measured. Since we do not know the value of the einselection factor μ, an extremely large magnetic field should be necessary in order to achieve the maximal value of μ, i.e. -1 or 1. Unfortunately there is a limit for the magnitude of applied external magnetic fields, because the induced field destroys the superconducting state above the critical value of the current. This critical value is certainly much smaller than the one given by the maximal value of the einselection factor μ.

We should perhaps salute Peirce's prophetic statement, see Sect. 12, regarding classical mechanics and symmetry breakings, namely that the three laws of motion draw no dynamical distinction between right-handed and left-handed screws, and that there are physical phenomena, standing entirely beyond all corpuscular philosophy, yet absolutely inexplicable by mechanical action. Though his claim dates back to the pre-quantum era, it is surprisingly still valid for microscopic quantum physics, based on Bohr's correspondence principle. Therefore contemporary quantum physics has never succeeded to explain the Meissner effect, and if not the Meissner effect, it also concerns superconductivity. Any attempt to find their microscopic solutions inevitably implicates the breakdown of Bohr's correspondence principle, see Sect. 9. Thus the BCS theory and all theories, based on the microscopic dynamics of the superconducting carriers, whatever these carries may be, whether unpaired or Cooper paired electrons, polarons, bipolarons etc., are still inadequate. Only a holistic megascopic quantum theory is able to overthrow the traditional corpuscular philosophy, and to explain the Meissner effect and superconductivity.

Most physicists agree, that superfluidity should have an analogous explanation as compared to superconductivity. However if superconductivity has no microscopic explanation, superfluidity does not, too. Superfluidity shares with superconductivity the macroscopic wave function (14.1). In a similar way, cf. Eq. (14.2), follows from (14.1) in the case of superconductors, that we can write

$$\theta(\mathbf{r}) = \frac{M_s \mathbf{v} \, . \, \mathbf{r}}{\hbar} \tag{14.16}$$

where the magnetic field **A** is not present as in Eq. (14.2) and the superconducting carrier mass m_s is replaced with the superfluid particle mass M_s. This leads to the quantization of the superfluid hydrodynamic circulation κ:

$$\kappa = \oint \mathbf{v} \, . \, d\mathbf{r} = \frac{2\pi n \hbar}{M_s} \tag{14.17}$$

By analogy physicists attempt to see the common feature of both phenomena employing some type of Bose-Einstein condensation, involving atoms or pairs of atoms or pairs of electrons. In this way the interpretations of superfluidity in helium-4 and helium-3 atoms differ: The first one is treated as a direct Bose-Einstein condensation of bosonic particles, whereas the latter one deals with the formation of bosons only by pairing of two fermionic atoms similar to the BCS mechanism of electronic Cooper pairing. This sounds a bit odd—why should the different isotopic number change the whole mechanism of superfluidity?

We may actually impose a stronger requirement for superconductivity and superfluidity: Let they have, not only a common explanation, but let they also be two manifestations of the same phenomenon. Superfluids consist of two components—normal and superfluid. Let us suppose that the latter can form microscopic superconducting grains. Consider a model of a caterpillar truck. The superconducting carriers—"teleported" valence orbitals are temporarily "glued" on the surface of tube or capillary, exactly like the caterpillar on the ground. The nuclei then act like the truck. It is all about the relative relocation of the nuclei with the electronic core and the superconducting carriers: In superconductors the nuclei with the electronic core are macroscopically in rest and the superconducting carriers are relocated. In superfluids the opposite is true.

Let us now discuss other megascopic phenomena. Ferromagnetism can be understood as a macroscopic analogy of microscopic isomerism. Both phenomena have a microscopic quantum explanation. But the tunnelling between different isomers and between ferromagnets with different spin orientation is of megascopic origin and has no microscopic clarification. The latter case is known as the Einstein-de Haas effect [118] or Richardson effect [145]. Richardson was the first to conclude, from the principle of the conservation of angular momentum that the change in the magnetic moment of a free body causes this body to rotate. At that time no electron spin was known, so Richardson derived a prequantum relation between the magnetization $\mathbf{M}_o$ and total orbital angular momentum $\mathbf{J}_o$ of the motion

$$\mathbf{M}_o = \frac{e}{2m}\mathbf{J}_o \tag{14.18}$$

where e and m stand for electron charge and mass. But as soon as the electron spin was discovered in 1925, it was clear that its contribution to the Einstein-de Haas effect in ferromagnets significantly predominates over the contribution of the angular momentum. So instead of Eq. (14.18) one has

$$\mathbf{M}_s = \frac{e}{m}\mathbf{J}_s \tag{14.19}$$

with a gyromagnetic factor equal to 2. For deeper insight into the exact value of the gyromagnetic factor, see e.g. Maruani's article [146]. The similarity between Eqs. (14.18) and (14.19) demonstrates that the spin momentum should be of the same nature as the angular momentum of rotating bodies as conceived in classical mechanics. Quoting Maruani regarding the explanation he says in his work [147]: "The Dirac equation, which was derived by combining the relativistic invariance condition with the quantum probability principle, showed its fecundity by explaining the half-integer spin of fermions and by predicting antiparticles. In previous papers, we conjectured that the spinning motion of the electron was that of a massless charge moving at light velocity, this internal motion being responsible for the electron rest mass involved in external motions and interactions."

Proceeding further to non-equilibrium phenomena with the focus on chemical reactions. Quantum chemists usually and successfully formulate the problem and solve them with the help of microscopic quantum physical equations often without further thinking. They calculate potential energy surfaces (PES) of interacting molecules finding energy landscapes including energy minima and maxima, construing their results to interpret the investigated chemical reactions. This concept, however, does not answer the question, what a chemical reaction really amounts to. The critical point comes when PES's cross. This metaphysical concept is a fundamental consequence of the application of the B-O approximation, exactly as the misunderstood J-T effect. True quantum mechanical solutions, treating electrons and nuclei on the same footing, never lead to any symmetry breaking in J-T systems, as well as to any chemical reactions, since they do not recognize the individuality of molecules and their constituents. Only the true field solution based on the Goldstone theorem implies symmetry broken J-T states and recognizes the individual reacting molecules. If we have a reaction of the type A + B $\leftarrow\ \rightarrow$ C (in both directions), then the left side contains $n_A + n_B - 12$ phonons, 6 rotons and 6 translons, and the right side n_C phonons, 3 rotons and 3 translons. The number of Goldstone bosons during the reaction is preserved, and the chemical reaction happens only if 3 rotons and 3 translons are annihilated and 6 phonons are created, or vice versa. This is a megascopic irreversible process—a quantum jump from the fragmented Universe, containing individual molecules A and B, into the whole Universe, where the molecular individuality is lost, and then again in the fragmented Universe with the molecule C of the same energy as that of A + B in the same configuration at which the reaction proceeds, or vice versa.

We know well the irreversible nature of chemical reactions, and Santilli was a half century ago absolutely right with his "no reduction theorem", see Sect. 2. It means, that contemporary quantum physics with its reversible equations can never justify this irreversibility, and therefore is unable to explain a single chemical reaction. It is able to explain only the physical pre- and postprocesses occurring during reactions, but not the chemical reactions themselves. Santilli still believes in the microscopic explanation of chemical reactions, and has proposed an irreversible mathematical framework for physical equations. We have shown that chemical reactions belong to the group of megascopic phenomena, where irreversibility descends from the megascopic quantum jumps, and thus no irreversible equations are necessary.

If physical entropy is assigned to the multitude of equivalent isomeric molecular ground states there appears another paradox of residual entropy arising from chemistry. One of the first examples of residual entropy was already pointed out by Pauling to describe water-ice [148]. In water, each oxygen atom is bonded to two hydrogen atoms. For a large number of water molecules in this configuration, the hydrogen atoms have a large number of possible configurations that meet the 2-in 2-out rule, since each oxygen atom must have two 'near' (or 'in') hydrogen atoms, and two far (or 'out') hydrogen atoms. This freedom exists down to absolute zero. The existence of these multiple configurations amounts to randomness, or in other words, entropy. Thus systems that can take multiple configurations at or near absolute zero are said to have residual entropy. Of course, residual entropy is in strong contradiction with the third law of thermodynamics that states: the entropy of a perfect crystal of any pure substance approaches zero as the temperature approaches absolute zero. Here one might suggest a megascopic solution to this paradox. Realizing, that possible tunnelling between different isomeric configurations of ice is of megascopic origin, i.e. microscopically inaccessible, then it is convenient to introduce a separate "chemical" entropy for megascopic processes, so that the physical third law of thermodynamics remains still valid.

A second example of the illusive violation of the third law of thermodynamics is the liquid–glass transition with actually a negative residual entropy. There are several differences between liquid crystal and liquid–glass transitions. In the first case the arrangement of atoms and molecules differs from that of a liquid, but in the second case this arrangement is indistinguishable from it. In the first case there is a first-order phase transition involving discontinuities in thermodynamic and dynamic properties such as volume, energy (latent heat), and viscosity. This is a transition between states of thermodynamic equilibrium at a single, well-defined temperature and is time reversible. In the second case there is no first-order phase transition, no latent heat and no well-defined temperature: the slower the cooling, the lower the transition temperature. This is a non-equilibrium and irreversible process forming a quenched disorder state, kinetically locked. Its entropy, density and structure depend on the thermal history. By extrapolating the heat capacity of the supercooled liquid below its glass transition temperature, it is possible to calculate the Kauzmann temperature [149] at which the difference in entropy between the liquid and solid phase becomes zero. Cooling below this temperature leads to the Kauzmann paradox: A supercooled liquid displays a lower entropy than the crystal phase. This paradox was not resolved till now, although there are several proposals how to deal with it. Kauzmann himself postulated that all supercooled liquids must crystallize before the Kauzmann temperature is reached. But the scientific community has no unique opinion.

The most striking enigma of the liquid–glass transition is this: If we use classical physics for its calculation, i.e. classical thermodynamics and thermodynamics of irreversible processes, we end up with the Kauzmann paradox and the negative residual entropy at absolute zero. But on the other hand if we perform the computation of energy landscapes based on the concept of broken ergodicity in the configurational phase space with an entropy loss, a glass does not appear to have any residual entropy at absolute zero. Anderson was inspired by this controversy so utterly that he wrote [150]: "The deepest and most interesting unsolved problem in solid state theory is probably the theory of the nature of glass and the glass transition."

Different results from macroscopic and microscopic calculations imply that the physical microscopic description in this case does not reveal all facts about the macroscopic world. Simply the macroscopic and microscopic entropies are not the same. The irreversible and the non-equilibrium processes behind the glass transition are chemical reactions, giving rise to the creation of anisotropic covalent bonds in polymorphic forms of glass. Analogous to the case of Pauling's ice, we need here a separation of the physical microscopic entropy from the megascopic chemical one, in order not to violate the third law of thermodynamics.

Another paradox arises from the query: do chemical reactions exist on the macroscopic level? A nice example of the type $C \rightarrow A + B$ is brittle fracture. Unlike ductile fracture this effect has no microscopic quantum explanation and is described phenomenologically only by classical physics. Once a crack starts to propagate in a brittle solid its velocity may quickly reach unbelievable several thousands of m/s [151], whereas in ductile fracture there is only a slow propagation of the crack. Further differences are [152]: Ductile material fractures after plastic deformation and its surface obtained at the fracture is dull or fibrous in appearance. A ductile crack will usually not propagate unless an increased stress is applied and generally cease propagating when loading is removed. On the other hand, brittle fractures are characterised as having little or no plastic deformation prior to failure and the fracture surface of a brittle failure is usually reasonably smooth. The cracks that propagate in a brittle material will continue to grow once initiated. Although brittle fracture

Table 1 Megascopic phenomena

Megascopic phenomena	Equilibrium processes		Non-equilibrium processes
	Adiabatic systems	Non-adiabatic systems	
Microscopic level	Isomeric transitions	Jahn-Teller effect	Chemical reactions
Macroscopic level	Einstein-de Haas effect	Superconductivity and superfluidity	Brittle fracture

has no microscopic explanation, there are nevertheless two categories of cracks from the microscopic perspective [152], (i) a crack that passes through the grains within the material undergoing transgranular fracture, (ii) a crack that propagates along the grain boundaries termed an intergranular fracture.

The most interesting fracture feature process is a ductile to brittle transition. At low temperatures some materials that would be ductile at room temperature become brittle. It means that at higher temperatures the fracture is of microscopic origin, but under the critical temperature it is of megascopic origin. Compare for instance the relation as microscopic normal conductivity under the critical temperature becomes megascopic superconductivity. This similarity is certainly not accidental, since materials that usually fracture in a brittle manner are glasses, ceramics, and some polymers and metals. This also explains why good superconductors are usually formed with brittle ceramics on the verge of rupture.

In Table 1 we have summarized the mentioned teleological megascopic phenomena, appearing on both the microscopic and the macroscopic level. This can be considered the fulfilment of Norton's request to overcome the restricted domain of contemporary physics that concentrates solely on causal processes [52]: "In the physical sciences, an important reason for choosing a restricted domain is to fence off processes that are acausal. A second reason to which I gave less attention, is that different domains manifest different sorts of causes… I expect the case of final causes to be prevalent in chemistry and non-physical sciences—that the restriction to different domains will divide different types of causation, as opposed to fencing off acausal processes."

This is also a fulfilment of Jordan's request, see Sect. 13, to introduce the second complementarity in order to explain classicality. Megascopic chemical processes are responsible for the quantum decoherence of the wave function, and the classical appearance of the silver grain on the photographic plate follows accordingly. From the same reason Schrödinger's cat can never be in some state of superposition. Either the irreversible megascopic event, representing the chemical reaction between the poison and the cat happens or it does not, and nothing in between.

To conclude, this might also be an answer to von Weizsäcker's problem, cited in Sect. 6, where he correctly recognized irreversibility as a necessary condition for classicality, but left open the question of sufficiency. We know now, that this condition is not fulfilled in the framework of the microscopic Copenhagen interpretation, but within the conception of megascopic irreversible processes it is fully sufficient.

15 Matter-Mind Dualism

We start with a short recapitulation of the megascopic mirrors of the microscopic quantum concepts and associated rules as they were introduced in Sect. 13.

1. The first complementarity versus the second: while microscopic quantum theory is based on Bohr's complementarity between the mechanical (particle) and field (wave) picture of elementary entities, megascopic quantum theory is founded on two megascopic complementary accounts of the whole Universe, namely the mechanical description representing its total wholeness, and the field formulation representing its total fragmentation. The implementation of a second complementarity was for first time requested by Jordan.
2. Causality versus teleology: microscopic quantum physics inherits causality (or upward causality) from classical physics due to Bohr's correspondence principle. Its megascopic mirror is teleology (downward causation or the final cause), mentioned already for first time by Maxwell, see Sect. 7, as it proceeds from a principle that is extra physical, but not extra natural, and is acting without exerting any force or spending any energy.
3. Projections versus injections: transitions between different states in microscopic quantum physics are expressed as vector projections in Hilbert space. The megascopic mirror of these projections is represented by injections, the latter responsible for the transitions from the fragmented Universe into its wholeness and vice versa. It was proposed for the first time by Bohm.
4. The Born rule versus the einselection rule: The probability of transitions based on microscopic projections is calculated by means of Born's law. Its megascopic mirror is the einselection rule, which takes into account the influence of all environments. The concept of einselection was originally coined by Zurek in his quantum decoherence program with the aim to replace the Born rule. As we have shown in Sect. 7, it cannot be replaced, and in Sect. 13 we also revealed the true meaning of einselection as the megascopic counterpart of the microscopic Born rule. It means that the latter remains to be tied to the microscopic projections, while the einselection principle determines the probabilities of the megascopic injection transitions.

The above mentioned four concepts and rules form, in the microscopic world, the basic axioms of the Copenhagen interpretation of quantum physics. But one axiom is still missing, i.e. the von Neumann–Wigner rule. At this point we have reached a very important but most commanding point: What is its megascopic mirror? Before we try to respond to this challenge, one must first deliberate more deeply on the problem of the matter-mind dualism that is necessary for the understanding of the von Neumann–Wigner rule. The matter-mind dualism had a most significant historical development after the Cartesian split. Heisenberg paid special attention to this problem in the Chapter 5 entitled "Die Entwicklung der philosophischen Ideen seit Descartes im Vergleich zu der neuen Lage in der Quantentheorie" in his book "Physik und Philosophie". Quoting directly the original version in German [153]:

„Der Ausgangspunkt der cartesianischen Philosophie ist völlig verschieden von dem der antiken griechischen Philosophen. Hier wird nicht mit einem Grundprinzip oder einem Grundstoff begonnen, sondern mit dem Versuch, ein grundlegendes, unbestreitbares Wissen zu erwerben. Descartes erkennt, daß unser Wissen über unser eigenes Denken sicherer ist als unser Wissen über die äußere Welt. Aber schon seine Ausgangsposition mit dem Dreieck: Gott, Welt und Ich vereinfacht die Grundlage für das weitere Philosophieren in einer gefährlichen Weise. Die Spaltung zwischen Materie und Geist oder zwischen Körper und Seele, die mit Platons Philosophie begonnen hatte, ist jetzt vollständig. Gott ist sowohl vom Ich als auch von der Welt getrennt. Gott wird tatsächlich so hoch über die Welt und die Menschen erhoben, daß er schließlich in der Philosophie des Descartes nur als ein gemeinsamer Bezugspunkt erscheint, der die Beziehung zwischen dem Ich und der Welt herstellt."

The Cartesian split affected the progress of the whole of classical physics. Heisenberg writes: „Andererseits war diese Spaltung in der Naturwissenschaft für einige Jahrhunderte außerordentlich erfolgreich. Die Newtonsche Mechanik und alle anderen Teile der klassischen Physik, die nach ihrem Vorbild aufgebaut waren, beruhten auf der Annahme, daß man die Welt beschreiben kann, ohne über Gott oder uns selbst zu sprechen. Diese Möglichkeit galt beinahe als eine notwendige Voraussetzung für alle Naturwissenschaft."

Quantum mechanics, however, is entirely different from classical physics: We have first of all to overcome the Cartesian split. Heisenberg comments on its consequences on the ways many physicists think, and that this prevent them to fully understand the Copenhagen interpretation: „Wenn man an die großen Schwierigkeiten denkt, die selbst so bedeutende Naturwissenschaftler wie Einstein bei dem Verständnis und der Anerkennung der Kopenhagener Deutung der Quantentheorie hatten, so kann man die Wurzeln dieser Schwierigkeit bis zur cartesianischen Spaltung verfolgen. Diese Spaltung ist in den drei Jahrhunderten, die auf Descartes gefolgt sind, sehr tief in das menschliche Denken eingedrungen, und es wird noch lange Zeit dauern, bis sie durch eine wirklich neue Auffassung vom Problem der Wirklichkeit verdrängt ist."

For the Copenhagen (and the von Neumann-Wigner) interpretation the conscious mind has a special relevance, but the Cartesian split also imparts consequences for the understanding of the unconscious mind. In Whyte's 1960 book "The Unconscious Before Freud" dealing with the history of ideas about the unconscious mind, he defines the words "conscious" and "unconscious" [154]: "Conscious will be used to mean directly present in awareness (or "immediately known to awareness"). This adjective will be applied only to discrete aspects or brief phases of mental processes. No distinction will be made between "conscious" and "aware". Except for a few doubtful extreme cases (e.g., processes of logical or mathematical deduction) there appear to be no causally self-contained processes of which all aspects directly enter awareness. "Conscious" is a subjective term without, as yet, (i) interpretation in terms of physiological structure, or (ii) explanation of its function."

And further: "Unconscious, in the term "unconscious mental processes" will be used to mean all mental processes except those discrete aspects or brief phases which enter awareness as they occur. Thus "unconscious mental processes" (or the "unconscious", for short) is here used as a comprehensive term including not only the "subconscious" and "preconscious", but all mental factors and processes of which we are not immediately aware, whatever they be: organic or personal tendencies or needs, memories, processes of mimicry, emotions, motives, intentions, policies, beliefs, assumptions, thoughts, or dishonesties."

Whyte describes Descartes' philosophical motivation: "Descartes recognized that the "union of soul with body" is intuitively known, and that their pervasive connections are a matter of everyday experience. However, in his doctrine these connections are not ones of true causation, but an illusion resulting from some kind of parallelism between the two modes of being. It is therefore, in Descartes' view, abortive to study these connections in detail."

The basic comparison of different attitudes towards the unconscious mind from the positions of materialism, idealism and the Cartesian school follows: "During the late seventeenth century three main attitudes dominated European philosophical thought, corresponding to three interpretations of the nature of existence. Materialism treated physical bodies and their motions as the primary reality; idealism took it to be spirit or mind; while Cartesian dualism postulated two independent realms: the mental res cogitans and the material res extensa. For the first two schools there was no difficulty in recognizing unconscious mentality, though under other names… But to the third, Cartesian, school the admission of the existence of unconscious mental processes presented an acute philosophical challenge, for it demanded the discarding of the original conception of the dualism as one of two independent realms, matter in motion and mind necessarily aware. For those who were loyal to Descartes, all that was not conscious in man was material and physiological, and therefore not mental."

Whyte continues with the post-Cartesian chronology of the conceptual development of the unconscious mind up to Freud: "As often happens in the history of thought, an idea may be fashionable and even transform society before it is properly understood… The unconscious mind, in the post-Cartesian sense, was discovered around 1700; it is now transforming Western thought; who would dare to claim that its basic laws are yet understood?… I have given sufficient evidence to show that the general conception of unconscious mental processes was conceivable (in post-Cartesian Europe) around 1700, topical around 1800, and fashionable around 1870–1880. …when Freud was forty-two he was unaware that at least fifty writers (probably many more) had been developing similar assumptions for over two hundred years. Finally in 1925, at the age of sixty-nine, Freud wrote: "The overwhelming majority of philosophers regard as mental only the phenomena of consciousness. For them the world of consciousness coincides with the sphere of what is mental." This curious mistake shows how narrow his reading had been, and how wrong a conception he must then have had of his own originality. The inference is that not only Freud but most of us are largely unaware of what has made us what we are and led us to think as we do, and it is sometimes as well that we should be ignorant."

From the above quote it is apparent that Freud might have been overemphasized at the expense of other researchers. His famous psychoanalysis is claimed to be a method for the treatment for mental-health disorders, but does actually in many cases also become misinterpreted disrupting individuals, families and societies.

Unfortunately Whyte's book ends up with Freud leaving out Jung, who in his later life opposed Freud. His primary disagreement with Freud stemmed from their differing conceptions of the unconscious, but also Freud's anti-religious attitudes mattered. Whereas Freud conceived the unconscious solely as a repository of repressed emotions and desires, Jung called this limited model "personal unconscious", and proposed the existence of a second, far deeper form of the unconscious underlying the personal one. In his work "The Archetypes and the Collective Unconscious" Jung explains the difference between personal and collective unconscious [155]: "A more or less superficial layer of the unconscious is undoubtedly personal. I call it the "personal unconscious". But this personal layer rests upon a deeper layer, which does not derive from personal experience and is not a personal acquisition but is inborn. This deeper layer I call the "collective unconscious". I have chosen the term "collective" because this part of the unconscious is not individual but universal; in contrast to the personal psyche, it has contents and modes of behaviour that are more or less the same everywhere and in all individuals."

The final definition of the collective unconscious acquiesces from the same work [155]: "My thesis then, is as follows: in addition to our immediate consciousness, which is of a thoroughly personal nature and which we believe to be the only empirical psyche (even if we tack on the personal unconscious as an appendix), there exists a second psychic system of a collective, universal, and impersonal nature which is identical in all individuals. This collective unconscious does not develop individually but is inherited. It consists of pre-existent forms, the archetypes, which can only become conscious secondarily and which give definite form to certain psychic contents."

On one hand the collective unconscious dictates how the structure of the psyche autonomously organizes experience, and on the other hand the collective unconscious collects and organizes those personal experiences in a similar way for all members of a particular species. This finding, due to Jung, appears particularly important for us as physicists. In addition to the Schrödinger cat paradox, resolved by the introduction of megascopic quantum jumps, the Copenhagen interpretation suffers from yet another conundrum—the paradox of Wigner's friend, see Sect. 6. The Copenhagen view that the quantum state represents available knowledge to a community of communicating observers was set ad hoc and does not follow from any of the basic axioms of the Copenhagen interpretation. We live today in the time of Internet and mobile phones, and we can be proud of how ingeniously we are all connected. But this is a special type of connection mediated by the material world. If we accept that we are parts of the material world and communicate via this world, then Wigner's conclusion that each conscious being is able to collapse one single objective quantum state, regardless of whether the information is actually physically shared, is correct. Therefore Wigner looked for a way out of the crisis with his "desire for a less solipsistic theory". Only if we accept Jung's concept, we will understand that our conscious minds are all primarily connected in this Jung-like collective unconscious entirely outside the external material world, and this explains and provides the legitimacy of the shared knowledge of a community of communicating observers.

Incidentally sharing the knowledge of a community of communicating observers is not just one single form of sharing that descends from the collective unconscious. The communication and sharing of our discoveries is a similar case. Whyte presents a motto due to Goethe in his book [154]: "No one can take from us the joy of the first becoming aware of something, the so-called discovery. But if we also demand the honour, it can be utterly spoiled for us, for we are usually not the first. What does discovery mean, and who can say that he has discovered this or that? After all it's pure idiocy to brag about priority; for it's simply unconscious conceit, not to admit frankly that one is a plagiarist."

However, many of Jung's works relate to alchemy, though his interests were focused only on its arcane aspects. According to Linden [156], esoteric aspects of alchemy prevailed among psychologists, spiritual and new age communities, hermetic philosophers, and historians of esotericism, while exoteric aspects occurred among historians of the physical sciences, who did examine subjects in terms of protochemistry, medicine, and charlatanism. In this sense Jung's ideas and inspirations are full of occultism, gnosticism and spiritism. On the other hand, if we extract the scientific truth from the Jung's considerations, we immediately recognize that Jung was correctly aware of the collective unconscious as a mind component of alchemy, after the Cartesian split, whereas classical chemistry became a matter component. Wikipedia, in the year 2018, speaks about the origin of chemistry in this way: "Chemistry was preceded by its protoscience, alchemy, which is an intuitive but non-scientific approach to understanding the constituents of matter and their interactions… Chemistry as a body of knowledge distinct from alchemy began to emerge when a clear differentiation was made between them by Robert Boyle… Chemistry is considered to have become an established science with the work of Antoine Lavoisier…"

The comments above raise one of the most important questions of this article: is chemistry, as we know it today, the only true matter branch of alchemy after the Cartesian split? Looking first at the customary definition of chemistry, see e.g. Wikipedia, 2018: "Chemistry is the scientific discipline involved with compounds composed of atoms, i.e. elements, and molecules, i.e. combinations of atoms: their composition, structure, properties, behaviour and the changes they undergo during a reaction with other compounds. Chemistry addresses topics such as how atoms and molecules interact via chemical bonds to form new chemical compounds." By tradition, science dealing with material objects from elementary particles up to atoms is called physics, and from atoms up to molecules it is chemistry, however, from molecules up to the macroscopic level of fluids and solids it is shared with (condensed matter) physics again. At the quantum level the first and third parts essentially form quantum physics, and the second part mostly quantum chemistry. In this way quantum chemistry is a result of an embedding of post-Cartesian classical chemistry ideas into the pre-Cartesian quantum physics.

As Boyle and Lavoisier were fathers of classical chemistry, Löwdin is often regarded as father of quantum chemistry. A recent biography of Per-Olov Löwdin, see Brändas [157], describes his significant achievements for the development of the new sub-discipline of quantum chemistry. Assessing the new field and its specific goals it was stated: "He [Löwdin] referred to a so-called pure theory in a restrictive sense, 'if it derives for instance chemical data from the knowledge of only the physical values of the electronic mass and charge, Planck's constant, the atomic numbers and the form of the Schrödinger Equation, which itself represents the quintessence of a great deal of physical experience'...". Löwdin was of course well aware of the impossibility to derive the Coulomb Hamiltonian from first principles as well as of other limitations and possibilities offered by a structural approach. Even if the concept of quantum chemistry is very expedient and practical, it does not, as already stated, give any answers to the Sutcliffe-Woolley clamped-nuclei paradox, see Sect. 2, and to the Santilli's "no reduction theorem" paradox, see Sect. 3. Authentic pre-Cartesian quantum chemistry differs from the pre-/post-Cartesian mix, i.e. standard quantum chemistry. The former deals with teleological megascopic phenomena, and in addition to chemistry and chemical reactions, it also comprises such phenomena that have no quantum physical explanation, such as the Einstein-de Haas effect and the brittle fracture phenomenon and those with false microscopic interpretation, like superconductivity and superfluidity.

Standard quantum chemistry has encouraged the idea that after centuries of progress physics and chemistry can finally 'sit under one roof' in the sense that these two disciplines of natural sciences refer to the same microscopic quantum laws. In contrast, however, authentic pre-Cartesian quantum chemistry forms a megascopic complement to microscopic quantum physics. Metaphorically, one might aver: what elementary particles and the conscious mind mean for pre-Cartesian quantum physics, is analogous to what the whole Universe and the collective unconscious mean for pre-Cartesian quantum chemistry. In the material world, elementary particles represent the smallest one and the Universe the Greatest One. There is a parallelism in the mind world: Observers' conscious minds represent the smallest one and the collective unconscious the Greatest One.

In microscopic quantum physics the von Neumann–Wigner rule binds together elementary particles with the conscious mind, and the megascopic rule binds together the whole Universe with the collective unconscious. An observer's activity, i.e. a measurement, causes microscopic quantum jumps and hence a discontinuity of the deterministic Schrödinger equation. All events in our world must have a reason. Therefore it is reasonable to postulate a similar megascopic analogy: the existence of megascopic "measurements" as an activity of the collective unconscious must cause megascopic quantum jumps. Due to Jung's celebrated formulation of the concept of the collective unconscious, we might venture to call the megascopic mirror of the von Neumann–Wigner rule as the Jung rule. Having finally found the last requested megascopic mirror, the Jung rule, we can summarize all mirrors in Table 2.

From our previous thoughts about the free will, formulated in Sect. 8, two additional mirrors can be revealed, namely the reflections of microscopic quantum physical laws in the collective unconscious and the reflections of megascopic quantum chemical laws in the observer's conscious mind constituting a form of free will decisions. They will be called the crosswise-law mirrors. The first mirror, mentioned above, is supported by physicists such as Poincaré, Penrose and Compton, i.e. Poincaré's unconscious disorder born on chance, Penrose's largely unconscious

Table 2 Microscopic-megascopic mirrors

First quantum floor: microscopic quantum physics	Second quantum floor: megascopic quantum chemistry
First complementarity	Second complementarity
Causality	Teleology
Projections	Injections
Born rule	Einselection rule
von Neumann–Wigner rule	Jung rule

'putting-up', first phase of the two-stage model, and Compton's relationship between free will and quantum indeterminism and the randomness on the first phase of the two-stage model.

The second mirror is not apparent until we have introduced megascopic quantum chemistry. For instance, Doyle's concept of "adequate determinism" in his Cogito model, attempting to explain free will only on the basis of quantum physics, cannot be right. The physicist Compton and psychiatrist Flanagan were right in claiming that quantum physics is insufficient for the justification of free will. Compton argued that freedom involves an additional determining factor of choice, about which science tells us nothing. Moreover Flanagan stated that free actions need to be caused by me, in an undetermined and non-random manner. The combination of non-determinism and non-randomness follows only from the megascopic einselection processes of quantum chemistry, and not from microscopic physical processes that are indeterministic and random.

Moreover, the crosswise-law mirrors follow directly from the two-stage model formulated by Eccles and Popper, see Sect. 8, which unifies nature's laws of biology (the matter world) and psychology (the mind world). All six mirrors, two microscopic-megascopic, two matter-mind, and two crosswise-law mirrors are pictured in Fig. 2.

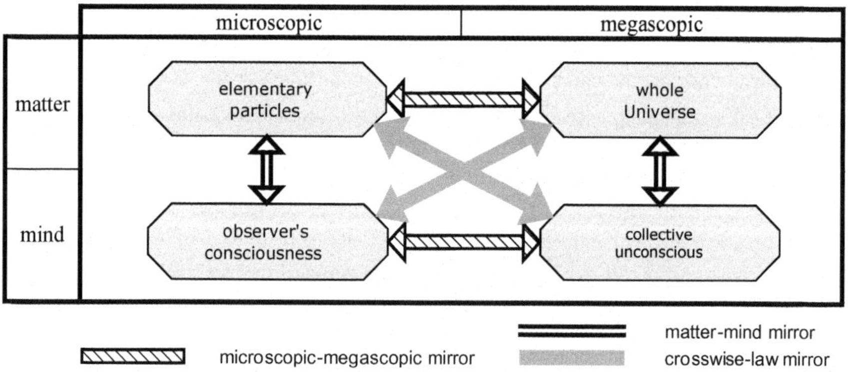

Fig. 2 Matter-mind relationship

The crosswise-law mirrors have tremendous impact on our choice of the correct interpretation of quantum theory. They rule out all "objective" interpretations. The matter-mind dualism can never be removed and replaced by materialistic monism. Megascopic quantum chemistry adds "unconscious" measurements to "conscious" measurements as is known from microscopic quantum physics and this fact disqualifies the MWI interpretation, see Sect. 7, and the arguments of its advocates, such as those of Zeh [158]: "He who considers this conclusion of an indeterminism or splitting of the observer's identity, derived from the Schrödinger equation in the form of dynamically decoupling ('branching') wave packets on a fundamental global configuration space, as unacceptable or 'extravagant' [159] may instead dynamically formalize the superfluous hypothesis of a disappearance of the 'other' components by whatever method he prefers, but he should be aware that he may thereby also create his own problems: Any deviation from the global Schrödinger equation must in principle lead to observable effects, and it should be recalled that none have ever been discovered [160]. The conclusion would of course have to be revised if such effects were some day to be found."

One can also find a nice response to this problem, i.e. of our false impression that all processes in the material world must be measurable, Norton [52]: "It has long been recognized that human action is the prototype of cause and effect. At its simplest, we identify processes as causal if they are sufficiently analogous… Using human action as a prototype, we identify terms in the cause and effect relation whenever we have one that brings about or produces the other; and we identify the process of production as the causal process… We restrict a science to some hospitable domain."

Norton's view is correct. If we confine our knowledge of the material world to causal quantum physics and fully ignore teleological phenomena, then we have restricted science to 'some hospitable domain'. From all known interpretations only the Copenhagen view has the ability to incorporate teleological phenomena via a megascopic mirroring of each of its basic axioms.

16 The Emerald Tablet

It is worth pointing out that important contributions to the concept of the collective unconscious of Jung and those of Darwin to the teleological concept of biological adaptation were not motivated by some true authentic science, but by esoteric occult ideas preserved since ancient times in various clandestine masonic, theosophical and new age societies. Predominantly in these societies, the ancient Emerald Tablet of Hermes Trismegistus has always drawn attention in special ways. Darwin did fulfil his grandfather's wish to apply the content of the Emerald Tablet to biological processes. From Mackey's Encyclopedia of Freemasonry we can cite this information [161]: "Before coming to Derby in 1788, Dr. [Erasmus] Darwin had been made a Mason in the famous Time Immemorial Lodge of Cannongate Kilwinning, No. 2, of Scotland. Sir Francis Darwin, one of the Doctor's sons, was made a Mason in Tyrian Lodge, No. 253, at Derby, in 1807 or 1808. His son Reginald was made a Mason in Tyrian

Lodge in 1804. The name of Charles Darwin does not appear on the rolls of the Lodge but it is very possible that he, like Francis, was a Mason."

Most biologists believing that Darwinian evolution is a scientific theory do not know that it originated in occultism in the disguise of a scientific theory. Applications of occultism, without a firm scientific background, are of course vulnerable to misunderstandings and misinterpretations of all kinds. There was no quantum physics at the time of Darwin, to e.g. vindicate random mutations, and even if Darwin was right with his teleological interpretation of the process of adaptation, the problem of teleology has not been resolved to this day. One might study how the central characters of Darwinian evolution, such as Mayr or Madrell, see Sect. 7, have advanced the meaning of teleology, not in full accord with Darwin. The same applies to protagonists of quantum decoherence who attempt to replace the causal Born rule by a teleological einselection rule. They were all inspired with Universal Darwinism, but they must also realize that animate nature obeys the laws of inanimate nature and not conversely. It means that the Darwinian adaptation process must be first revealed in inanimate nature with the possibility of a consequent application to living organisms.

Philosophers like Popper regarded Darwinism as a tautology. Popper changed his mind after he had noticed similar patterns, see Sect. 8, in the two-stage model, applicable both to biology and the psychology of free will. Maybe such arguments are convincing for philosophers, but not for physicists, since the latter will always look for primary patterns in inanimate nature. We have nevertheless shown that microscopic quantum physics and megascopic quantum chemistry form together the same two-stage model as was applied to biology and psychology by Eccles and Popper.

Note that once we have incorporated teleology into megascopic quantum chemistry, the Darwinian adaptation process becomes identical to megascopic quantum jumps that represent transitions between different symmetry broken states, i.e. between different Platonic Forms, see Sect. 13. This view is synoptic with the prevailing view of Western biologists and philosophers until the 19th century, i.e. that the differences between individuals of a species are uninteresting departures from their Platonic idealism of created kinds. In this way the true Darwinian evolution can be applicable only to questions like why we have so many races of cats and dogs, but the cat will be forever a cat and the dog will be a dog. Darwin himself was aware of the problem of missing transitional forms [162]: "Why, if species have descended from other species by insensibly fine gradations, do we not everywhere see innumerable transitional forms? Why is not all nature in confusion instead of the species being, as we see them, well defined?... But, as by this theory innumerable transitional forms must have existed, why do we not find them embedded in countless numbers in the crust of the earth?... Why then is not every geological formation and every stratum full of such intermediate links? Geology assuredly does not reveal any such finely graduated organic chain; and this, perhaps, is the most obvious and serious objection which can be urged against my theory."

A few decades later Darwin was brainwashed by the illusory "success" of his own theory when he said [163]: "At some future period, not very distant as measured by centuries, the civilized races of man will almost certainly exterminate, and replace, the savage races throughout the world. At the same time the anthropomorphous apes… will no doubt be exterminated. The break between man and his nearest allies will then be wider, for it will intervene between man in a more civilized state, as we may hope, even than the Caucasian, and some ape as low as a baboon, instead of as now between the Negro or Australian and the gorilla."

Today these lines are far from being politically correct particularly since history has demonstrated for us the danger of disgusting "ape theories" and sickening pagan concepts. Returning instead to a closer look at the Emerald Tablet, Darwin's inspiration. It contains 14 verses but only two of them are relevant for an alignment with our current knowledge of natural sciences, i.e. the second and eighth ones. Citing the English translation of the standard 1613 Latin Beato's Aurelium Occultae Philosophorum [164].

Verse 2: *Whatever is below is similar to that which is above. Through this the marvels of the work of one thing are procured and perfected.*
Verse 8: *This ascends from the earth into the sky and again descends from the sky to the earth, and receives the power and efficacy of things above and of things below.*

Verse 2 seems to be intelligible and conceivable for us as physicists. Maruani [147], see also papers quoted therein, compares the microcosm and the macrocosm, and excerpts the same two verses from a similar 1680 Newton's translation and discusses some interesting relations between microphysics and astrophysics, electromagnetism and gravitation, and the "magical" relations between the fundamental physical constants.

Verse 8, on the other hand, has no support in contemporary quantum physics. Yet, based on this verse, Darwin proposed his two-stage model, and even if this proposition was successful, it is still occultism, since there is no scientific base, and hence we do not know the range, the limitations and the critical points of its applicability. But, if comparing verse 8 with Table 2, depicting microscopic-megascopic mirrors of all basic axioms, it is obvious that we may identify a decrypted version of the Emerald Tablet, expressed in terms of a clear and unambiguous language of quantum theory.

One might ask where such a magnificent wisdom of the Emerald Tablet comes from. We do not even know whether Hermes Trismegistus was a real person or only the name of the book dealing with the 'Emerald Tablet'. Most probably this is only a torso of some higher knowledge of antediluvian origin. There is a Jewish testimony of Flavius Josephus who in his Antiquities, Book I, Chapter 2: Concerning the Posterity of Adam, and the Ten Generations from Him to the Deluge [165], wrote about the significant discoveries of the descendants of Adam's son Seth that had survived the flood: "And that their inventions might not be lost before they were sufficiently known, upon Adam's prediction that the world was to be destroyed at one time by the force of fire, and at another time by the violence and quantity of water, they made two pillars, the one of brick, the other of stone: they inscribed their discoveries on them both, that in case the pillar of brick should be destroyed by the flood, the pillar of stone might remain, and exhibit those discoveries to mankind; and also inform them that there was another pillar of brick erected by them."

The philologist and historian van Bladel [166] also mentions an Arabic testimony in which Hermes Trismegistus is identified with Idris, which is the Arabic name of Enoch from the bloodline of Adam's son Seth, mentioned in the Bible for first time in Gen 5:18–24, and then in 1 Ch 1:3, Luk 3:37, Heb 11:5, Jud 1:14–15, and Sir 44:16,49:14. We can assume that Enoch indeed fully understood the content of his teachings, which was part of the true original religion given to the first man by God. After the confusion of tongues in Babel, the Emerald Tablet became a source of occultism in many secret societies, since the key to its understanding has been lost.

It is interesting that this occultism has influenced not only secret societies, but also Folk sayings. One simple example: from all the megascopic phenomena summarized in Table 1, one of them, the brittle fracture, has been familiar with the whole population throughout the history of mankind. And each of us has been subject to attendant negative experiences in everyday life, such as broken dishes, china, porcelain, window glass, etc. However, there is a Folk saying, expressed by the same words in all Central European languages, entirely in opposition to impairments due to brittle fracture. In Czech this saying sounds: *Střepy přinášejí štěstí*, in German: *Scherben bringen Glück*, and the English literal translation is: *Shards bring luck*. Actually, brittle fractures are nightmares for mechanical and building engineers. It caused the loss of Titanic in 1912, the 1962 King Street Bridge collapse in Melbourne, and the 1967 Silver Bridge collapse in West Virginia, all resulting from the brittle fracture effect. There is a riddle: how can shards bring any luck?

No one knows the origin of this Folk saying, except that our pagan ancestors threw shards to drive off evil spirits, or that it was a heathen ritual of shattering the holy altars of clay after completing a sacrifice to the gods. All traces end in ancient Greek and Roman coffin tombs where often find shards donated to the deceased. In verse 8 of the Emerald Tablet, we have used the word "sky" as the English translation of the Latin "coelum". It could also be translated as "heaven", used e.g. in Newton's version. One should here be aware of the danger of capitalizing of the letter "h". Originally brittle fracture was understood as a "consecration of heaven", i.e. a megascopic effect. But after the confusion of tongues in Babel, when the key to the Emerald Tablet was lost, future generations explained it as a "consecration of Heaven", i.e. a supernatural or at least a preternatural phenomenon that brings the blessings of gods. In this way occultism infiltrates Folk sayings.

17 Conclusion

The core of this work originates in an independent proof of Goldstone's theorem [91] for molecules and solids. To better explain the significance of our message, assume at the outset that we have no prior knowledge of this theorem, simply rewriting the total molecular Hamiltonian of interacting nuclei and electrons into quantum field form. Instead of the usual formulation of the original electrons and the associated nuclear vibrational modes/phonons, we will introduce a new set of fermionic and bosonic operators in the most general way, and then optimizing the total transformed

Hamiltonian. We have put forward five sets of well-known equations as limiting cases of the resulting equations of our procedure; a test that our solutions must pass and exactly agree with each other

(1) Pople's equations for ab initio calculations of vibrational frequencies [104]
(2) Fröhlich's expression for the correction of the ground state energy [102, 105]
(3) Fröhlich's effective two-electron interaction [102]
(4) Lee-Low-Pines' polarons and their self-energy [106]
(5) Born-Huang's ansatz [7].

Starting from the field electron-vibrational (electron-phonon) Hamiltonian [101, 103], only the first four sets pass the test successfully. For instance, in small molecules such as H_2 one obtains only a fragment of the whole contribution from the Born-Huang ansatz, and this is a major failure that unfortunately is not well-known in quantum chemistry and solid state physics. In quantum chemistry field methods are not often used, and usually only on the electronic level [99, 100]. However, in solid state physics electron-phonon field mechanisms are frequently employed, and in most cases the contribution of the Born-Huang ansatz is negligible to such an extent that many physicists do not even realize any differences between the Born-Oppenheimer (B-O) and the adiabatic approximations and mistakenly even regard them as synonyms.

Hence the concept of the electron-vibrational field theory is clearly insufficient, and it must be replaced by a more rigorous field theory comprising two major upgrades [107, 109]:

(a) Together with the phonons two new types of quasiparticles appear: rotons and translons descending from rotational and translational degrees of freedom. The contribution from the roton and the translon quanta occurs even if the molecule as whole does not rotate nor move.
(b) In analogy with Lorentz covariance, binding together space and time coordinates, a new covariance is necessary, binding together internal and external degrees of freedom, without the centre-of-mass (COM) separation, which normally applies in both classical and quantum mechanics.

Let us compare these results with the content of Goldstone's theorem [91]: "The spontaneous breaking of a continuous symmetry can be associated with a massless and spinless particle". According to this theorem there are three types of broken symmetries in molecules and crystals, composed of N nuclei ($3N$ degrees of freedom): the Galilean associated with $3N - 6$ phonons (or $3N - 5$ for diatomic molecules), translational associated with 3 translons, and rotational associated with 3 (or 2) rotons. Unfortunately the scientific community does not seem to be aware of this fundamental fact with its origin in the Goldstone theorem, cf. the picture reflected in the (2018) Wikipedia phrase "Goldstone boson" that states the following incongruity: "In solids, the situation is more complicated; the Goldstone bosons are the longitudinal and transverse phonons and they happen to be the Goldstone bosons of spontaneously broken Galilean, translational, and rotational symmetry with no simple one-to-one correspondence between the Goldstone modes and the broken symmetries."

First of all the Goldstone theorem is a complement to the first Noether theorem where a one-to-one correspondence between the conservation laws and associated symmetries always exists. Therefore the unique and unambiguous one-to-one correspondence between the Goldstone modes and spontaneously broken symmetries in solids must necessarily exist as well. It means that vibrational modes/phonons in molecules/solids do not represent the full set of Goldstone bosons. The errors caused by disregarding the rest of the Goldstone modes are usually negligible in adiabatic cases, but in non-adiabatic situations they might lead to fatal consequences, i.e. a deficient view of spontaneously broken systems on the quantum level.

We have demonstrated that a field theory based only on phonons is not sufficient for the description of Born-Oppenheimer (B-O) degenerate states, such as Jahn-Teller (J-T) systems and superconductors, where a general field COM covariant theory incorporating all Goldstone bosons, i.e. phonons, rotons and translons is unavoidable. We have derived the formulae for the ground state energy and the excitation spectra of B-O degenerate quantum chemical and solid state systems together with a complete set of Goldstone bosons.

Within this background the Born-Huang ansatz plays the same role for the field COM covariance as Maxwell's equations for Lorentz covariance. Even though we have here only presented a special proof of the Goldstone theorem for interacting systems of electrons and nuclei, based on the field COM covariance resulting from the Born-Huang ansatz, we have nonetheless arrived at the correct set of Goldstone bosons. However, the general Goldstone-Salam-Weinberg proof [92] of Goldstone's theorem, based on the conserved currents of the first Noether theorem, appears not transparent enough in the case of a correct identification of the Goldstone bosons. The physicists inaccurately identify them only with phonons.

In fact this mistake has led to unfortunate consequences, i.e. an inadequate BCS theory [78] using only the electron-phonon interaction which has never been revisited. Moreover, Anderson in his article [95] added to this misunderstanding claiming that "the solid crystal violates translational and rotational invariance, and possesses phonons". This paper soon became a fateful marker, leading to the 'celebrated' Higgs mechanism, that was incorporated in the Standard model attempting to explain, how particles in the electro-weak interaction gain the mass. Higgs named his Nobel lecture "Evading the Goldstone theorem" [97]. This was alas no 'evading', but rather a misrepresentation of the theorem, since Goldstone bosons cannot after all be suppressed. We have summarized Comay's work [88], where he, after a careful analysis of Higgs' equations, did come to the conclusion that they do violate Bohr's correspondence principle, and therefore cannot be approved as valid equations of quantum physics.

Goldstone bosons, arising as the result of symmetry violations, i.e. rotons and translons, produce singular behaviour in the original symmetrical positions. The system is forced to avoid them by removing the degeneracy, and after apt one-particle transformations, finding itself in new asymmetric positions. This principle unifies the formation of ground states of J-T molecules and superconductors. It should be noted how the quantum field formulation deals with the virtual degeneracies originating from approximative B-O solutions. They are profoundly different from the real degeneracies, where the quantum principle of superposition takes place, after being removed by some external perturbation, such as the Stark or the Zeeman effect. The quantum field simply does not share the centre of mass, defined in standard quantum mechanics, rather it solves the problem of the centre of gravity "in its own way".

We have also offered a reformulated version of the J-T theorem as it follows directly from the Goldstone theorem. Molecular and crystalline entities in a particular geometry of an electronically degenerate ground state are unstable except for the case when all matrix elements of electron-rotational and electron-translational interactions equal zero. This is a striking result, as it shows how a true quantum field, respecting fully the Goldstone theorem, copes with B-O degenerate systems, such as J-T molecules and superconductors. The superposition principle for the removal of the degeneracy is simply bypassed, since the rotons and the translons are actually responsible for the violation of symmetry. For instance in a superconductor, the symmetry breaking produces several geometrically different symmetry broken states, splitting the original half-occupied conducting band of the conductor into two bands, one fully occupied valence band and one empty conducting band, in such a way that the original conductor under the critical temperature becomes a multi-ground-state insulator.

The results discussed above clearly disqualify all theories based on B-O degeneracies, such as the usual solutions of the J-T effect [11, 12] and the BCS theory of superconductivity [78]. Their validity was never justified from simple reasons, i.e. one query of extraordinary importance was never answered, cf. the question of Sutcliffe and Woolley [4]: "The interesting question is how to get from the quantum theory of an Isolated Molecule to a quantum theory of an individual molecule by rational mathematics." We have further concluded that quantum mechanics describing a system of nuclei and electrons can never resolve this task. The quantum theory of an Isolated Molecule is quantum mechanics with a solution avoiding the B-O concept, and this is the method of Monkhorst [5, 6] that "takes a very atomic view of a molecule: instead of fixing the nuclei as in the B-O approximation, the electrons and nuclei are both described quantum-dynamically within a centrosymmetric shell structure." Unfortunately this approach leads neither to any B-O degenerate states nor to any symmetry breaking.

A valid quantum theory of an individual molecule should be a true quantum field theory of fermions represented by renormalized electrons and Goldstone bosons standing for vibrational, rotational and translational modes, i.e. a system where nuclei from the mechanical description are replaced by the Goldstone bosons in the field description. This means, that true quantum mechanics deals solely with isolated systems, while true quantum field theory works with individual systems. Only a quantum field formulation can produce symmetry broken states and, unlike the B-O model, removing degeneracies bypassing the superposition principle.

Returning to the question of teleological principles in science, one might first deliberate over how classical physics copes with the spontaneous symmetry breaking

(SSB). In the classical phenomenological picture of superconductors, the Ginzburg-Landau equation [77] exhibits quadratic and biquadratic functions of the order parameter, forming the well-known Mexican hat. We have also discussed similar Mexican hat-like situations, i.e. the Norton dome [52] and the hydrodynamic problem of Boussinesq [124]. These three cases have one thing in common, viz. the solutions of the corresponding equations are not unique. However, in addition to the standard causal solution also a teleological one is permitted. Maxwell was impressed by the third case of Boussinesq to such an extent that he for the first time spoke about teleological principles that exist outside the known laws of physics and called it [51] "some determining principle which is extra physical (but not extra natural)".

Physicists, contrary to biologists, usually abandon the teleological principle as unscientific, and they do therefore not agree with the problem raised regarding the classical description of the Meissner effect and its relation to the phenomenon of superconductivity. One discerns two camps of physicists. The first camp, following Meissner and Ochsenfeld [64] and London [66], claims that superconductivity unlike ideal conductivity has no classical explanation, and that quantum physics is necessary in order to explain it. The second camp following Cullwick [71] identifies superconductors with perfect conductors. But they cannot both be right. Superconductors are not identical to perfect conductors and quantum physics (BCS theory) does not explain the Meissner effect, without properly explaining superconductivity first. Electromotive forces, described by Faraday's law of induction, are equal to zero in the stationary conditions of the Meissner effect, whereas existing theories do not suggest any other electric forces needed to accelerate the electrons until the steady state supercurrent is achieved. From the view of telicity, one concludes that classical physics provides two solutions, one for perfect conductors and one for superconductors. Undeniably classical physics can formulate, but not explain superconductivity. According to Peirce's philosophical analysis [125] no physical theory based on corpuscular philosophy can solve the problem of the SSB bootstrap. Even if this analysis was written before the era of quantum physics, Peirce's argument is also applicable today. It builds on Democritus' atomic ideas, which is essentially the same corpuscular idea as in classical physics. It means that quantum physics in the Copenhagen interpretation is incapable of solving the problem of the SSB trigger. As we have seen, the teleological problem, which relates to Leibniz' principle of sufficient reason [114], to the second form of van Fraassen's argument [119], and Peirce's corpuscular philosophy argument [125], can be discounted in classical physics. Unfortunately quantum physicists usually ignore this problem. But this attitude is uncalled for, since it is motivated by the conviction, that quantum physics uses the same SSB archetype as the classical one.

In comparing classical and quantum physics, we conclude: classical physics does not distinguish superconductors from ideal conductors and sees them as one system with two possible solutions, one teleological and one causal, while quantum physics does recognize the difference, but, on the other hand does not separate them from multi-ground-state insulators, i.e. is only able to describe the ground states of superconductors, but never superconductivity itself, since the Bohr's correspondence principle does not account for teleological phenomena. Hence the application of the classical SSB, i.e. the Mexican hat, to quantum equations, the J-T effect, superconductivity or the Higgs mechanism becomes a mismatch. Quantum equations, based on a classical SSB, lead to paradoxical situations. With reference to SSB, the scientific community is again divided, i.e. whether SSB here is really present or not. We have, in this work, discussed the arguments both for and against in connection with the J-T effect, superconductivity, the Higgs mechanism and ferromagnetism.

From the addition of the lost Goldstone bosons, rotons and translons, found to be fully responsible for the mechanism of forming quantum states with spontaneously broken symmetries, it follows that the spontaneous symmetry breaking (SSB), only described phenomenologically at the level of classical physics, becomes misinterpreted on the quantum level, where every physical system can be described dually, either as a mechanical system or as a field theoretical system, in accordance with Weinberg's statement (repeated again) [126]: "If it turned out that some physical system could not be described by a quantum field theory, it would be a sensation; if it turned out that the system did not obey the rules of quantum mechanics and relativity, it would be a cataclysm." For molecules and solids treated equivalently, we have presented true quantum field solutions fully respecting the Goldstone theorem.

In a previous paper [109] I introduced the notion of property-object dualism: in quantum mechanics, nuclei and electrons represent objects, and vibrational modes are their common property. In quantum field objects, however, they are represented by electrons and Goldstone bosons, and the "nuclear" positions or "clamped nuclei" are properties of the pertinent field equations. We are now in a similar situation, with the description of the whole many-body system, as one was a century ago regarding the different aspects of a mechanical formulation and a field theory description of elementary entities that finally led to Bohr's complementarity. Obviously one needs a second type of Bohr complementarity on the many-body level as was already requested a long time ago by one of the co-founders of quantum mechanics, Jordan [134], even though, from a different reason, the Copenhagen interpretation based on the first Bohr complementarity, cannot explain classicality as an emergent property.

Linking Sutcliffe's and Woolley's [4] problem of transitions between an isolated and the individual systems with Bohm's [15] question of fragmentation and wholeness, implies that the second complementarity, embracing the whole Universe, is the megascopic mirror of the first Bohr microscopic complementarity. We have just touched upon an ancient knowledge, presenting the Universe as a mosaic where the smallest one resembles the Greatest One. We need to postulate a new axiom that is necessary for the justification of the quantum jumps of the Universe from its state of total wholeness to its total fragmented states and vice versa. This request certainly goes beyond the scope of the Copenhagen interpretation. It neither contradicts nor denies this interpretation, but rather appears to be its natural extension.

While microscopic processes are causal, megascopic phenomena are teleological. Teleology as the final cause can then be comprehended as downward causality, descending from the Greatest One the whole Universe, in contrast to upward causality, ascending from the smallest entities, the elementary particles. The first complementarity represents the causal microscopic relationship between the mechanical and the field descriptions of elementary entities (particle-wave dualism), and the second complementarity represents the teleological megascopic relationship between the mechanical and the field descriptions of the whole Universe, where these two descriptions in particular deal with the centre of gravity in different ways.

Whereas microscopic quantum physics, dealing with elementary particles as 'basic building blocks' out of which everything is made, reflects the ancient materialistic ideas of Democritus, megascopic quantum phenomena can be seen as a reflection of Plato's Forms. In the latter one can finally find the true relationship between megascopic quantum jumps and time: these jumps do not proceed in time, but time is created by them. At the megascopic level this gives rise to the concept of quantum of time, which otherwise is unreachable at the microscopic level. In this sense megascopic quantum jumps are a true clock of the Universe.

Restating that on one hand we have time evolution of the wave function, representing only Aristotelian potentiality, and on the other reality created during the measurement process by timeless events, i.e. microscopic quantum jumps. This separation of events from time, as a classical entity, is the source of almost all philosophical problems of contemporary quantum physics, and since the origin of time is rooted in megascopic events, we are able to finally reunite time and quantum events, but in contrast to processes on a classical level these case events will be foundational and time subordinate. This idea goes back to Aristotle who defined time as, we repeat, "the number of movement in respect of before and after". Since it cannot exist without succession, it does not exist on its own but only relative to the motions of things.

To complete the formulation of megascopic quantum theory one needs to identify megascopic mirrors of all basic axioms of the microscopic quantum theory in its Copenhagen interpretation. There are five fundamental micro-megascopic mirrors:

1. First complementarity versus the second: While microscopic quantum theory is founded on Bohr's complementarity between mechanical (particle) and field (wave) descriptions of elementary entities, megascopic quantum theory is based on two megascopic complementary descriptions of the whole Universe, namely a mechanical one representing its total wholeness, and the field one representing its total fragmentation. The implementation of the second complementarity was for the first time mentioned by Jordan [134].
2. Causality versus teleology: Microscopic quantum physics inherits causality (or upward causality) from classical physics due to the Bohr principle of correspondence. Its megascopic mirror is teleology (or downward causality or final cause), mentioned for the first time by Maxwell et al. [51] as resulting from a principle which, quote, unquote "is extra physical but not extra natural, and is acting without exerting any force or spending any energy".
3. Projections versus injections: Transitions between different states in microscopic quantum physics are expressed as vector projections in Hilbert space. The megascopic mirror of these projections is represented by injections that are responsible for the transitions from the fragmented Universe into its wholeness and vice versa, a statement given for the first time by Sheldrake and Bohm [16].

4. Born rule versus einselection: The probability of transitions based on microscopic projections is calculated from the Born rule. Its megascopic mirror is einselection, taking into account influences from all environments. The concept of einselection was originally coined by Zurek [45] in his quantum decoherence program aimed to replace the Born rule. As we have shown, the Born law cannot be replaced, and we have revealed the true meaning of einselection as the megascopic counterpart of the microscopic Born rule. It means that the latter remains tied with the microscopic projections, while einselection determines the probabilities of the megascopic injection transitions.
5. von Neumann–Wigner rule versus Jung's rule: In microscopic quantum physics the von Neumann–Wigner rule binds together elementary particles with the conscious mind, and its perfect megascopic analogy must exist as well, binding together the whole Universe with the collective unconscious. An observer's activity, i.e. a measurement, causes microscopic quantum jumps and hence a discontinuity of the deterministic Schrödinger equation. Since all events in our world must have their reason, it is reasonable to postulate a similar megascopic analogy: the existence of megascopic "measurements" as the activity of the collective unconscious causing megascopic quantum jumps. Due to Jung's success formulating the concept of the collective unconscious [155], we have suggested to call the megascopic mirror of the von Neumann–Wigner rule as the Jung rule.

We have also applied the above mentioned megascopic rules to superconductivity, and we will give three main reasons why superconductivity (BCS) will not provide a microscopic explanation:

(1) Every microscopic theory must result in the Born rule for estimating the probabilities of the density and the velocity of superconducting carriers, i.e. demonstrate that these quantities have to be experimentally measurable. But as we know, they are in principle non-measurable.
(2) Microscopic theories do not allow us to avoid the universal concept of Bloch states for the description of superconducting carriers. It means that the mass of the carriers must take the effective mass of electrons into account. But this is in direct contradiction with measurements of the London moment, where only the bare electronic mass and not the effective one is reported.
(3) According to the second form of van Fraassen's argument, asymmetry cannot arise ex nihilo. We know that the original asymmetry can be present in the form of an external magnetic field around the superconductor, but no microscopic theory is able to explain the mechanism of the Meissner effect, when a superconductor is cooled below the critical temperature. Moreover, a constant magnetic field cannot cause any acceleration of the superconducting carriers. It means that no microscopic theory is capable to implement the original asymmetry, contradicting the second form of van Fraassen's argument.

Nevertheless, superconductors can be, and have been, described macroscopically on the quantum level. Such a wave function representing macroscopically-occupied quantum states was proposed already by London [66], from which one could derive the Josephson equations [143] without the need for any microscopic concept of superconductivity, as it was also shown by Feynman [144]. Only the ground state and the excitation states of the superconductors have a microscopic explanation, in the same way as all solids must be microscopically described. However, microscopic theory yields merely the symmetry broken states but not transitions between them. We therefore need the concept of megascopic quantum jumps appearing either on the microscopic or the macroscopic level, standing in the same relation to the microscopic tunnelling as the concept of microscopic quantum jumps in the Bohr model stands to the classical electronic movement in the Rutherford model.

During these megascopic quantum jumps the pairs of electrons (valence orbitals forming some kind of "chemical bonds" occupied by two electrons with opposite spins) have a chance either to be relocated in diverse directions, or to return into their original position, depending on the external magnetic field that determines the einselection factor. The superconducting carriers do not recognize Cooper pairing; they are identical with the optimal transformed doubly occupied valence-like orbitals. Unlike BCS theory, which associates this phase with the motion of carriers in Bloch's **k** space, our megascopic interpretation associates the phase with the macroscopic **l** space, orthogonal to Bloch's **k** space.

Hence it is evident that a microscopic theory (BCS), based on the microscopic dynamics of superconducting carriers, whatever these may be, and whether unpaired or Cooper paired electrons, polarons, bipolarons etc.—is, in view of our megascopic interpretation, wrong. To overthrow the corpuscular philosophy in the Peirce's argument, and to explain the Meissner effect and superconductivity a holistic megascopic quantum theory is necessary.

In addition to superconductivity and superfluidity, we have also analysed other megascopic phenomena on both the microscopic and the macroscopic levels, taking into account equilibrium processes of both adiabatic and non-adiabatic systems as well as non-equilibrium processes, see e.g. the complete list in Table 1: isomeric transitions, the Jahn-Teller effect, chemical reactions, the Einstein-de Haas effect, superconductivity and superfluidity, and the brittle fracture.

Note that post-Cartesian classical chemistry deals with material objects from atoms up to molecules, while standard quantum chemistry is a result of an implantation of post-Cartesian classical chemistry ideas into pre-Cartesian quantum physics. On the other hand, one can alternatively build an authentic pre-Cartesian quantum chemistry on quantum megascopic axioms, embracing in addition to chemistry and chemical reactions also all the above mentioned megascopic phenomena.

Furthermore, the concept of megascopic quantum jumps resolves many of the paradoxes of quantum physics: accepting that all processes in the Universe consist of only microscopic and megascopic irreversible events and that the origin of time is rooted in megascopic events, then the problem of the arrow of time is automatically solved. Emergent reversibility can only appear in subsystems with no SSB (most adiabatic systems) where a one-to-one correspondence between the mechanical and the field states holds, with the result that these subsystems can be fully described by the time-reversible Schrödinger equation. This is commensurate with the answer to von Weizsäcker's problem [27], where he correctly recognized irreversibility as a necessary condition for classicality, but he left open the question if this condition is also sufficient. We know now, that this condition is not sufficient in the framework of the microscopic Copenhagen interpretation, but is fully sufficient after the inclusion of megascopic irreversible processes.

This establishes also a fulfilment of Jordan's request [134] to introduce a second complementarity to explain classicality. Megascopic chemical processes are responsible for the decoherence of the wave function, and the classical appearance of the silver grain on the photographic plate, discussed in Jordan's article, ensues. On the same basis Santilli's "no reduction theorem" [13] paradox is explained and Schrödinger's cat can never be in some state of superposition: Either the irreversible megascopic event representing the chemical reaction of the poison with the cat's body happens or does not, and nothing in between.

We also know that our conscious minds are all primarily connected in Jung's collective unconscious entirely outside the external material world, explaining the legitimacy of shared knowledge of a population of communicating observers resolving the paradox of Wigner's friend [38].

Finally we have also discussed the paradox of the residual entropy arising in chemistry, if physical entropy is assigned to the multitude of equivalent isomeric molecular ground states. We have quoted two examples of residual entropy: the first one pointed out by Pauling [148] to describe water-ice, and the second one, the Kauzmann paradox [149], related to the liquid-glass transition with actually negative residual entropy. One needs here to introduce a separation of the physical microscopic entropy from the megascopic chemical one, in order to avoid violating the third law of thermodynamics.

A microscopic, ductile to megascopic, brittle transition is analogous to a transition between microscopic normal conductivity and megascopic superconductivity under critical temperatures. Materials that usually fracture in a brittle manner are glasses, ceramics, and some polymers and metals. This also explains why good superconductors are usually formed with brittle ceramics on the verge of rupture.

The most interesting result of the present work is the aspect that microscopic quantum physics and megascopic quantum chemistry together form a two-stage model, ontologically compatible with the two-stage model formulated by Eccles and Popper [60], which unifies nature's laws of the material world of biology with the mind world of psychology. One can now perform a unification of this two-stage model that embraces the realm of inanimate nature, the realm of animate nature, and the realm of psychology.

Moreover, we have revealed the crosswise-law mirrors that follow directly from the two-stage model formulated by Eccles and Popper, namely the reflections of microscopic quantum physical laws in the collective unconscious and the reflections of megascopic quantum chemical laws in the observer's conscious mind, the latter in a form sustaining his free will decisions. The crosswise-law mirrors have a tremendous impact on our choice of a correct consistent interpretation of quantum theory, disqualifying all "objective" interpretations. Basically, matter-mind dualism cannot be removed and replaced by materialistic monism. Megascopic quantum chemistry adds "unconscious" measurements to the "conscious" ones known from microscopic quantum physics, eliminating also Everett's interpretation [17], since its basic stipulation presupposes a deterministic quantum mechanical wave function of the whole Universe. The Bohm—de Broglie interpretation [14] was already outdated by Bohm himself, when he formulated several additional requirements [15, 16] that a true quantum interpretation must fulfil.

So the final verdict reads: From all the known conceptions of microscopic theories of nature only the Copenhagen interpretation has the ability to incorporate teleological phenomena via the megascopic mirroring of each of its basic axioms, and in that way squaring up with all the universal paradoxes discussed here.

Acknowledgements The author would like to express his greatest appreciation to E. Brändas for his careful reading of the manuscript, offering many linguistic improvements, as well as extensive constructive suggestions for valuable reformulations.

References

1. Einstein A, Podolsky B, Rosen N (1935) Phys Rev 47:777–780
2. Bell JS (1964) Physics 1:195–200
3. Born M, Oppenheimer R (1927) Ann Phys (Leipzig) 84:457
4. Sutcliffe B, Woolley RG (2013) Progr Theoret Chem Phys 27(Part 1):3–40 (Springer)
5. Monkhorst HJ (1999) Int J Quantum Chem 72:281
6. Monkhorst HJ (1987) Phys Rev A 36:1544
7. Born M, Huang K (1954) The dynamical theory of crystal lattices. Oxford University Press, London
8. Kutzelnigg W (1997) Mol Phys 90:909
9. Cafiero M, Adamowicz L (2004) Chem Phys Lett 387:136–141
10. Jahn HA, Teller E (1937) Proc R Soc Lond A 161:220
11. Köppel H, Domcke W, Cederbaum LS (1984) Adv Chem Phys 57:59
12. Bersuker IB, Polinger BZ (1983) Vibronic interactions in molecules and crystals. Nauka, Moscow (in Russian)
13. Santilli RM (1967) Nuovo Cimento 51:570
14. Bohm D (1952) Phys Rev 85:166–179
15. Bohm D (1980) Wholeness and the implicate order. Routledge Classics, London, New York
16. Sheldrake R, Bohm D (1982) Morphogenetic fields and the implicate order. ReVision 5:41
17. Everett H (1957) Rev Mod Phys 29:454
18. Norton JD (2015) The measurement problem. Selected courses in history and philosophy of science. University of Pittsburgh
19. Einstein A (1949) Reply to criticisms in Albert Einstein: philosopher-scientist. The library of living philosophers series. Cambridge University Press
20. Carpenter RHS, Anderson AJ (2006) Ann Fond Louis Broglie 31:1
21. Schreiber Z (1995) The nine lives of Schrödinger's cat. arXiv:9501014 [quant-ph]
22. Schlosshauer M, Camilleri K (2011) AIP Conf Proc 1327:26–35
23. Bohr N (1985) Collected works. In: Kalckar J (ed) Foundations of quantum mechanics I (1926–1932), vol 6. North-Holland, Amsterdam
24. Heisenberg W (1958) Physics and philosophy. Harper & Bros, New York
25. Heelan P (1975) Z Allg Wiss 6:113–138
26. Heisenberg W (1952) Questions of principle in modern physics. Philosophic problems in nuclear science. Faber and Faber, London, pp 41–52

27. von Weizsäcker CF (1971) The Copenhagen interpretation. In: Bastin T (ed) Quantum theory and beyond. Cambridge University, Cambridge, pp 25–31
28. von Neumann J (1955) Mathematical foundations of quantum mechanics (1932). Princeton University Press
29. London F, Bauer E (1983) La théorie de l'observation en mécanique quantique (1939). In: Wheeler JA, Zurek WH (eds and trans) Quantum theory and measurement. Princeton University, Princeton
30. Davies PCW, Brown JR (eds) (1986) The ghost in the atom. Cambridge University Press
31. Wigner EP (1964) Two kinds of reality. Monist 48:248–264
32. Faye J (2014) Copenhagen interpretation of quantum mechanics. In: Stanford encyclopedia of philosophy
33. Howard D (2004) Who invented the 'Copenhagen interpretation?' a study in mythology. Philos Sci 71:669–682
34. Heisenberg W (1955) The development of the interpretation of the quantum theory. In: Pauli W (ed) Niels Bohr and the development of physics. Pergamon, London, pp 12–29
35. Henderson JR (2010) Classes of Copenhagen interpretations: mechanisms of collapse as a typologically determinative. Stud Hist Philos Mod Phys 41:1–8
36. Primas H, Esfeld M (1997) A critical review of Wigner's work on the conceptual foundations of quantum theory. philsci-archive.pitt.edu/1574
37. Wigner EP (1973) Epistemological perspective on quantum theory. In: Hooker CA (ed) Contemporary research in the foundations and philosophy of quantum theory. Reidel, Dordrecht
38. Wigner EP (1962) Remarks on the mind-body question. In: Good IJ (ed) The scientist speculates. Heinemann, London
39. Wigner EP (1977) Physics and its relation to human knowledge. Hellenike Anthropistike Hetaireia, Athens, pp 283–294
40. Einstein A (1955) Einleitende Bemerkung über Grundbegriffe, Louis de Broglie und die Physiker. Hamburg, Claassen Verlag
41. Wigner EP (1969) Are we machines? Proc Am Philos Soc 113:95–101
42. Mach E (1886) Die Analyse der Empfindungen und das Verhältnis des Psychischen zum Physischen. G. Fischer, Prag
43. Goethe JW (2006) Maximen und Reflexionen. Helmut Koopmann (Hrsg.), Deutscher Taschenbuch Verlag und C. H. Beck, München
44. Tegmark M (2007) Found Phys 11(07):116
45. Zurek WH (1982) Phys Rev D 26:1862
46. Kastner RE (2014) Studies in history and philosophy of modern physics. Preprint, philsci-archive.pitt.edu/id/eprint/10757
47. Zurek WH (2009) Quantum darwinism. Nat Phys 5:181–188
48. Lennox JG (1993) Darwin was a teleologist. Biol Philos 8:409–421
49. Mayr EW (1992) The idea of teleology. J Hist Ideas 53:117–135
50. Madrell SHP (1998) Why are there no insects in the open sea? J Exp Biol 201:2461–2464
51. Maxwell JC, Garber E, Brush SG, Everitt CWF (1995) Maxwell on heat and statistical mechanics: on "avoiding all personal enquiries" of molecules. Lehigh University Press, Bethlehem
52. Norton JD (2003) Causation as Folk Science. Philosopher's Imprint 3(4):1–22
53. Doyle B (2011) Free will: the Scandal in philosophy. I-Phi Press, Cambridge, MA
54. James W (1956) The Dilemma of determinism. Republished in The will to believe. Dover
55. Poincaré H (1914) Science and method. In: Mathematical discovery (Chapter 3)
56. Penrose R (1989) The emperor's new mind: concerning computers, minds, and the laws of physics. Oxford University Press, USA
57. Compton A, Johnston M (ed) (1967) The cosmos of Arthur Holly Compton. A. A, Knopf, New York
58. Flanagan O (2003) The problem of the soul. Basic Books, New York
59. Popper K (1978) Natural selection and the emergence of mind. Dialectica 32:339–355

60. Eccles JC, Popper KR (1977) The self and its brain. Springer, Berlin, Heidelberg, London, New York
61. Bohr N (1958) Atomic physics and human knowledge. Wiley, New York
62. Kamerlingh-Onnes H (1911) Comm Phys Lab Univ Leiden, pp 122b & 124c
63. Becker R, Heller G, Sauter F (1933) Z Phys 85:772
64. Meissner W, Ochsenfeld R (1933) Ein neuer Effekt bei Eintritt der Supraleitfähigkeit. Naturwissenschaften 21(44):787–788
65. London F, London H (1935) Proc Roy Soc Lond A 149:71
66. London F (1950) Superfluids, vol I. Wiley, New York
67. Hildebrandt AF (1964) Phys Rev Lett 12:190
68. de Gennes PG (1999) Superconductivity of metals and alloys. Perseus Books, Reading, MA, pp 4–7
69. Essén H, Fiolhais MCN (2012) Meissner effect, diamagnetism, and classical physics—a review. Am J Phys 80:164–169
70. von Laue M (1952) Theory of superconductivity. Academic Press, New York, pp 6–7
71. Cullwick EG (1956) Magnetic energy and electron inertia in a superconducting sphere. Proc IEE Part C Monogr 103:441–446
72. Pfleiderer J (1966) Meissner effect in a classical electron gas. Nature 210:614
73. Karlsson PW (1984) Inductance inequalities for ideal conductors. Arch Elektrotech 67:29–33
74. Badía-Majós A (2006) Understanding stable levitation of superconductors from intermediate electromagnetics. Am J Phys 74:1136–1142
75. Kudinov EK (2006) Thermodynamics of current in superconductors and superfluids. Phys Solid State 48:1447–1454
76. Mahajan SM (2008) Classical perfect diamagnetism, expulsion of current from the plasma interior. Phys Rev Lett 100:075001-1–075001-4
77. Ginzburg VL, Landau LD (1950) Zh Eksp Teor Fiz 20:1064
78. Bardeen J, Cooper LN, Schrieffer JR (1957) Phys Rev 108:1175
79. Kozynchenko A (2005) Apeiron 12(3):330–350
80. Hirsch JE (2016) Europhys Lett 113:37001
81. Hirsch JE (2012) Phys Scr 85:035704
82. Hirsch JE (2016) Europhys Lett 114:57001
83. Frenkel J (1933) Phys Rev 43:907
84. Smith HG (1935) University of Toronto studies. Low Temp Ser 76:23; Smith HG, Wilhelm JO (1935) Rev Mod Phys 7:237
85. Slater JC (1937) Phys Rev 52:214
86. Hirsch JE (2014) Phys Scr 89:015806
87. Hirsch JE (2016) What is the speed of the supercurrent in superconductors? arXiv:1605.09469
88. Comay E (2015) Inherent contradictions in Higgs boson theory. Am J Mod Phys. 5(1–1):18–22 (Special issue: Physics without Higgs and without supersymmetry)
89. Dirac PAM (1978) Mathematical foundations of quantum theory. A. R. Marlow, New York
90. Nambu Y, Jona-Lasinio G (1961) Dynamical model of elementary particles based on an analogy with superconductivity. I. Phys Rev 122:345–358
91. Goldstone J (1961) Nuovo Cimento 19:154–164
92. Goldstone J, Salam A, Weinberg S (1962) Broken symmetries. Phys Rev 127:965–970
93. Schwinger J (1962) Phys Rev 125:397+128:2425
94. Yang CN, Mills R (1954) Conservation of isotopic spin and isotopic gauge invariance. Phys Rev 96(1):191–195
95. Anderson PW (1963) Plasmons, gauge invariance, and mass. Phys Rev 130:439–442
96. Earman J (2002) Laws, symmetry, and symmetry breaking; invariance, conservation principles, and objectivity. Philsci-archive.pitt.edu/id/eprint/878
97. Higgs PW (2014) Nobel lecture: evading the Goldstone theorem. Rev Mod Phys 86:851
98. Goldstone J (1957) Proc R Soc Lond Ser A 239:267
99. Brueckner KA (1955) Phys Rev 97:1353; Phys Rev 100:36 (1955)
100. Paldus J, Čížek J (1975) Advan Quant Chem 9:105

101. Svrček M (1986) Ph.D. thesis. Faculty of Mathematics and Physics. Comenius University, Bratislava
102. Fröhlich H (1952) Proc R Soc Lond A215:291
103. Svrček M. (1992) In: Methods in computational chemistry, vol 4. Molecular vibrations. Plenum Press, New York, pp 150–202 (Chapter 2–6.2) & 204–229 (Chapter 7)
104. Pople JA, Raghavachari K, Schlegel HB, Binkley JS (1979) Int J Quant Chem Symp 13:225
105. Fröhlich H (1950) Phys Rev 79:845
106. Lee TD, Low FE, Pines D (1953) Phys Rev 90:192
107. Svrček M (2012) Prog Theor Chem Phys 22(Part 7):511–552 (Springer)
108. Lenz P, Wegner F (1996) Nucl Phys B 482:693–712
109. Svrček M (2013) Prog Theor Chem Phys 27(Part 1):41–52 (Springer)
110. Handy NC, Lee AM (1996) Chem Phys Lett 252:425
111. Weinberg S (1996) The quantum theory of fields, vol 2: modern applications. Cambridge University Press, Cambridge
112. Bersuker IB (2006) The Jahn-Teller effect. Cambridge University Press, Cambridge
113. Newton I (1686) Philosophiæ Naturalis Principia Mathematica. S. Pepys, Reg. Soc. Præses. Londini, 5 Julii 1686
114. Weyl H (1952) Symmetry. Princeton University Press, Princeton, NJ
115. Leibniz GW (1875) Philosophische Schriften (in particular Leibniz' third letter, §5). Gerhardt, Berlin, pp. 352–440
116. Curie P (1894) Sur la symétrie dans les phénomènes physiques, symétrie d'un champ électrique et d'un champ magnétique. J Phys Theor Appl 3(1):393–415
117. Brading K, Castellani E (2003) In: Brading K, Castellani E (eds) Introduction in symmetries in physics: philosophical reflections. Cambridge University Press
118. Einstein A, de Haas WJ (1915) Experimenteller Nachweis der Ampereschen Molekularströme, Deutsche Physikalische Gesellschaft, Verhandlungen, p 17
119. van Fraassen BC (1989) Laws and symmetry. Oxford University Press, Oxford, New York
120. Stewart I, Golubitsky M (1992) Fearful symmetry. Is God a geometer?. Blackwell Publishers, Cambridge, MA
121. Norton JD (2008) The dome: an unexpectedly simple failure of determinism. Philos Sci 75(5):786–798
122. Malament D (2008) Norton's slippery slope. Philos Sci 75:5
123. van Strien M (2014) J Gen Philos Sci 45:167
124. Boussinesq MJ (1879) Conciliation du véritable déterminisme mécanique avec l'existence de la vie et de la liberté morale. In: Mémoires de la société des sciences, de l'agriculture et des arts de Lille, t. VI, 4e série, pp 1–257
125. Collected works of Charles Sanders Peirce (1964), vol 45–6. Cambridge, MA
126. Weinberg S (1995) The quantum theory of fields, vol 1: foundations. Cambridge University Press, Cambridge
127. Brading K, Castellani E (2013) Symmetry and symmetry breaking. The Stanford Encyclopedia of Philosophy
128. van Dam S (2012) Spontaneous symmetry breaking in the Higgs, philsci-archive.pitt.edu/9295
129. Elitzur S (1975) Impossibility of spontaneously breaking local symmetries. Phys Rev D 12:3978–3982
130. Peierls R, Kaplan TA, Anderson PW (1991) Reflections on Broken symmetry. Phys Today 44(2):13
131. Unzicker A (2013) The Higgs Fake: how particle physicists fooled the nobel committee. Createspace Independent Publishing Platform
132. Primas H, Müller-Herold U (1984) Elementare Quantenchemie (Stuttgart: Teubner), p 147 ff
133. Berry MV (1984) Proc R Soc Lond A 392:45–57
134. Jordan P (1949) Philos Sci 16(4):269–278
135. Vaidman L (2002) Many-worlds interpretation of quantum mechanics. In: Zalta EN (ed) Stanford encyclopedia of philosophy

136. Bell JS (1987) Speakable and unspeakable in quantum mechanics. Cambridge University Press, Cambridge
137. Rowlands P (2015) The foundations of physical law. World Scientific Publishing Co. Pte. Ltd
138. van Vleck JH (1939) J Chem Phys 7:61
139. Low W (1960) Paramagnetic resonance in solids. Academic Press, New York
140. Plato in twelve volumes (1925) (transl: Lamb WRM). Cambridge, MA, Harvard University Press, London, William Heinemann Ltd
141. Smolin L (2006) The trouble with physics. Houghton Mifflin Co., New York
142. Hardie RP, Gaye RK (2014) Aristotle's physics: entire. The University of Adelaide, ebooks.adelaide.edu.au/a/aristotle
143. Josephson BD (1962) Possible new effects in superconductive tunnelling. Phys Lett 1:251–253
144. Feynman R, Leighton R, Sands M (1965) The Feynman lectures on physics, vol III: quantum mechanics. Addison-Wesley, Reading, MA
145. Richardson OW (1908) A mechanical effect accompanying magnetization. Phys Rev Ser I 26(3):248–253
146. Maruani J (2018) Progress in theoretical chemistry and physics, vol 31. Springer
147. Maruani J (2016) The Dirac electron: from quantum chemistry to holistic cosmology. J Chin Chem Soc 63:33–48 and references therein
148. Pauling L (1970) General chemistry. W. H. Freeman & Co., San Francisco
149. Kauzmann W (1948) The nature of the glassy state and the behavior of liquids at low temperatures. Chem Rev 43(2):219
150. Anderson PW (1995) Through the glass lightly. Science 267:1616
151. Field JE (1971) Brittle fracture: its study and application. Contemp Phys 12(1):1–31
152. Campbell FC (2012) Fatigue and fracture: understanding the basics. ASM International, Materials Park, Ohio
153. Heisenberg W (1977) Physik und Philosophie. Ullstein Verlag, Ebner, Ulm
154. Whyte LL (1960) The unconscious before Freud. Basic Books, New York
155. Jung CG (1991) The archetypes and the collective unconscious. Collected works of C. G. Jung, vol 9, Part 1. 2nd edn. Routledge, London
156. Linden SJ (1996) Darke Hierogliphicks: alchemy in english literature from Chaucer to the restoration. University Press of Kentucky
157. Brändas EJ (2017) Per-Olov Löwdin—father of quantum chemistry. Mol Phys 115(17–18):1995–2024
158. Zeh HD (1993) Phys Lett A 172:189
159. Bell JS (1981) In: Isham CJ, Penrose R, Sciama DW (eds) Quantum gravity II. Clarendon Press
160. Weinberg S (1989) Phys Rev Lett 62:485; Polchinski J (1991) Phys Rev Lett 66:397; Squires EJ (1992) Phys Lett A 163:356
161. Mackey AG, McClenachan CT (1884) An encyclopædia of freemasonry and its kindred sciences: comprising the whole range of arts, sciences and literature as connected with the institution. L. H. Everts & Co., Philadelphia
162. Darwin C (1859) On the origin of species by means of natural selection, or the preservation of favoured races in the struggle for life, 1st edn. John Murray, London
163. Darwin C (2013) Descent of man, vol 1, p 201. The Project Gutenberg EBook #2300
164. Beato G (1613) Aurelium Occultae Philosophorum. In: Theatrum Chemicum, vol IV. Zetzner, Strasbourg
165. Josephus F (2009) The antiquities of the Jews. Project Gutenberg, EBook #2848
166. van Bladel K (2009) The Arabic Hermes. From pagan sage to prophet of science. Oxford University Press

Abiogenesis and the Second Law of Thermodynamics

Erkki J. Brändas

Abstract Abiogenesis, life arising from non-living matter, is reconsidered. A historic account is given from Darwin to the present with special consideration of the incipient relationship between quantum theory and life processes. Recent ideas relating to the spontaneous emergence of life are considered and evaluated, including the work on self-organizing systems by Ilya Prigogine and the Brussels-Austin School. The second law is commensurate with chemical evolution, but is incapable of explaining the mechanism behind the selection of specific polymers from an infinitude of molecular specimens. We attest abiogenetic life processes originating from a quantum chemical material world. A Gibbs free energy formulation is derived ab initio conceding non-equilibrium evolutions and open system self-organization. Correlated Dissipative Structures, CDS, combine quantum-thermal correlations at precise temperatures commensurate with their adaptive time scales, constituting CDE, a Correlated Dissipative Ensemble. The theory suggests an objective communication principle reminding of a Call-Centre-Poisson point process. The self-referential structure applies directly to biological systems, suggesting a fundamental quantum-chemistry life-principle incorporating evolution of the whole biosphere without contradicting the objectivity of physical laws.

Keywords Teleonomic evolution · Abiogenesis · Molecular communication · Complex enough system · Correlated dissipative structure · Correlated dissipative ensemble · Life-principle

E. J. Brändas (✉)
Department of Chemistry, Uppsala University, Uppsala, Sweden
e-mail: Erkki.Brandas@kemi.uu.se

L. Mammino et al. (eds.), *Advances in Quantum Systems in Chemistry, Physics, and Biology*, Progress in Theoretical Chemistry and Physics 32,
https://doi.org/10.1007/978-3-030-34941-7_15

1 Introduction

The mystery of the origin of life and the physical basis of evolution has always been a profound theme in the history of science. Yet, from Darwin's "warm little pond"[1] to the famous Miller-Urey experiments [1–3], the scientific deliberations about the plausibility of abiogenesis, i.e. life arising from non-living matter, and habitually contemplated by leading authorities on the subject, has in a deeper sense been peculiarly disguised in the public debate. Granting that the conceivable possibility of a prebiotic reducing atmosphere and its impact on the spontaneous emergence of life supporting molecular complexes, have repeatedly been discussed at length, see e.g. [4] for a recent update, there is nonetheless a befalling lack of critical discussions between leading scientific thinkers regarding the key issue whether a complex enough system would be considered alive or not and furthermore that organic compounds could without mediation lead to live organisms.

Now, Charles Darwin himself was reluctant to discuss this issue, since he insisted that the question of spontaneous generation was not manageable by the existing scientific understanding of the day. Nevertheless, he seems to have accepted the concurrent Victorian archebiosis-view of life, originating from non-living materials, writing in his notebook [1]: *The intimate relation of Life with laws of chemical combination, & the universality of latter render spontaneous generation not improbable.*

Retrospectively the application of the second law of thermodynamics, although not preventing the emergence of modern biology as a genuine discipline, did impart further stumbling blocks since it revealed that spontaneous generation of biological life processes turns out to be highly improbable. The illustrious sage of science, John Desmond Bernal,[2] when asked by Sir William Bragg, whether viruses were alive or not, replied that he would rather not say [5], then quoting a colleague *Life is the mode of action of proteins*, a credo reflecting many research programmes before the appearance of the double helix. The puzzle of life's origin nevertheless stuck with him for the rest of his life, and he devoted considerable time and writings to it [6], settling more or less for the plausibility of the Oparin-Haldane hypothesis [7–9] while advancing alongside his clay-life idea later refined in the Cairns-Smith hypothesis [10].

Since the conclusions from the vital experiments of Pasteur, that *life can only originate from life*, a position also held by the prominent physicist W. Thomson, as mentioned by Oparin [8], paved the way for the theory of Panspermia and later work on cosmic intelligence, see in particular the criticism of Darwinism as expressed in the controversial work *Evolution from Space* [11]. The ensuing polemic that followed and the rumoured bad manners of the senior author is said to have contributed to the exclusion of Sir Fred Hoyle from a deserved share of the 1983 Nobel Prize for his leading role in developing the theory of nucleo-synthesis in the stars.

[1] See e.g. an interesting discussion of Darwin's stance on spontaneous generation in [1].

[2] The story is quoted in Brown's Bernal biography, see Ref. [5].

The work, standpoints and citations mentioned above have provided fodder for diverse scrutinizing investigations regarding the conceivability of abiogenesis [12, 13]. The authors in Ref. [12] re-evaluated the concept of chemical evolution, finding that the thermodynamic costs of work, linked to changes in the configurational entropy associated with polymerization of DNA and proteins, might be sufficient and commensurate with the second law of thermodynamics.[3] Then again after disputing a number of views, e.g. self-organization and autocatalytic activity in non-linear irreversible processes, thermal synthesis involving proteinoids, condensing agents and mineral catalysis, they concluded that the coupling mechanism of selection and the sorting out of a specifically informed macromolecule from a random set was rendered in absentia. Furthermore, they resolved that bonding preferences between various amino acids do not play a significant role in coding the protein questioning the conventional notion of explaining the complexity of life as a result of open system dynamics of mass flow.

Nevertheless, advanced developments in physical chemistry and chemical physics have pushed forward the development and understanding of non-equilibrium phenomena, deliberating on issues such as time irreversibility, entropy production, subdynamics, self-organisation and complexity etc., in order to find a universal natural law that also takes in biology, see e.g. the prize-winning work of Prigogine [14] on complex systems and dissipative structures. The work of the Brussels-Austin School holds an ambitious foundation already displayed in its outline, i.e. to present classical mechanics, CM, and quantum mechanics, QM, within a common super-operator Liouville construction. Its rationale inaugurates a time reversible, fine-grained preparation, which in CM means the study of complete trajectories, with the objective to violate time reversal symmetry, the latter expressed by an equivalent non-unitary transformation that generates irreversible probabilistic behaviour without positing any decisive approximations. The generality of this perspective leads to various realizations [15] and interpretations [16]. Recently a self-referential tenet, that we will return to in more detail, has been proposed and demonstrated to adhere by fiat to authentic teleonomic principles, attained by so-called Complex Enough Systems [17], CES's, that do not contradict the objectivity of physical laws.

To take the examination, implied by the title, into account it is first necessary to identify what characterizes a biological system in relation to ordinary matter and to appraise whether currently established natural laws are sufficient, or not, to guarantee chemical evolution from organic compounds to living organisms. Although similar criticisms have been argued and defended in quite some detail [17], the issues at stage motivate a profound evaluation. The argument will be organized as follows. First some observations are made regarding the views established in Refs. [14, 18], see also Eigen [19], that some systems, obeying non-linear laws driven far from equilibrium with ordering appearing spontaneously, may potentially account for a similar sort of self-organization and complexity that is a focal point in the evolution of living systems. In particular the fundamental question of reversible

[3] According to the reviewer [13], the book comes as a real surprise to those comfortable with the view that the scientific problems of abiogenesis are mostly resolved.

versus irreversible behaviour and the associated process of self-organization will be considered by referring to a sub-dynamics formulation microscopically embedded in QM, while at the same time employing mathematically rigorous dilation techniques [15].

The necessary requirement, i.e. that derived master equations should be commensurate with the generation of a positive preserving contractive semigroups [20] will be reassessed. The section is devoted to modern advances in chemical physics establishing a consistent dissipative evolution incorporating temperature as well as adaptable time scales [21]. A surprising implication of the thermalization process [17, 21] and the instigation of spatiotemporal closure, is the materialization of a Gödel-like self-referential rule that acts as a guide for non-random chemical pathways opening up primary information-rich correlations. The *conditio sine qua non* is a unitary transformation that brings the complex symmetric reduced density matrix to classical canonical form[4] commensurate with a proper stochastic time evolution of the CES. The following sections ensure a positive answer prepared in terms of an authenticated communication hypothesis based on primary teleonomic evolution processes, originally considered and redefined by Mayr[5] [22]. The thermodynamic consequences of molecular communication on the genetic level and beyond will be considered and analysed. The question of teleonomy in biology, reviewed at length in Monod's classic essay[6] *Chance and Necessity* [23], is elucidated, discussed and reinforced as an objective fundamental natural law [24]. This result is of principal importance since it demonstrates that the self-referential property of the sub-dynamical extension induces a teleonomic character to evolution constituting a vital background for bridging the original logical gap in abiogenesis linking also evolution with consciousness. This is not in conflict with the thought-provoking discussions recorded in the recent book *Quantum Aspects of Life* [25]. The panel, although missing input from quantum chemistry and chemical physics, suggested that there should be some sort of quantum-physics-life-principle at work. This, with hindsight, supports, see e.g. footnotes 5–6, an unavoidable teleological component.[7]

The structure of this essay should also reflect the significance of quantum mechanics, QM, as proven in connection with technology, biology, evolution, life, consciousness, the Universe and the possible appearance of quantum black hole entities, QBH.

---

[4]These thermally excited modes are identified as Jordan blocks, converting states to sequences of transitions.

[5]Mayr introduced teleonomic processes, on par, but thus far not accounted for, by known physical laws, as follows: *A teleonomic process or behaviour is one that owes its goal-directedness to the influence of an evolved program.*

[6]Monod struggled with an epistemological contradiction, i.e. objectivity set against the need for teleonomy in biology stating: *In fact the central problem of biology lies within this contradiction.*

[7]The discussion took place at the Canary Islands in 2004, while the transcripts did not appear in print until 2008, see [25] and in particular the contributions by Paul Davies and Howard Wiseman.

2 Sub-dynamics and Quantum Theory

It is clear that a microscopic theory of biological systems by default should be engrained in QM and to originate in the time-dependent Schrödinger equation or its Liouville companion. This demand is by many regarded a tall order, since it is generally felt that the wet and warm environment, such as in a human brain, would be too complex and messy[8] to prohibit decoherence thereby rendering traditional quantum mechanics inapplicable [25]. On the other hand, recent progress in quantum chemistry and chemical physics, see e.g. Refs. [17, 21, 26] for some relevant presentations, do indeed commend a positive and precise response that quantum theory is of crucial importance to unite biophysical and biochemical descriptions of the cell, its constituents, and interactions with the environment. In this evaluation lies a code-forbidden decoherence rule that combines the temperature, the time scales and its underlying teleonomic evolution. In what follows we will essentially characterize a biological structure as complex enough, viz. a CES, that is able to develop a teleonomic apparatus as recognized in [23]. The definition may sound a bit vague so it will be revisited later during this journey.

First some remarks and thoughts regarding the role of irreversible processes in sub-dynamics, see Prigogine [14, 18, 27]. The central idea concerns the problem of going from a time invariant Schrödinger-Liouville equation to a thermodynamically irreversible formulation without invoking any approximations. One reason for the present somewhat stagnant interaction between the various subdisciplines here is most likely semantic-based. Starting off with the Einstein-Bohr discussions on quantum theory, standard QM interpretations have replicated epistemological viewpoints and possible underlying ontological levels. Another divide concerns the way a state of the actual system is encoded. Atmanspacher and Primas [28] offered a detailed account on how epistemic and ontic perspectives are related to scientific knowledge, and that so-called emergent behaviour should be viewed as transitions from ontic to epistemic descriptions.

Although the passage, *an epistemic state refers to the knowledge that can be obtained about an ontic state*, quoted in [28], the actual difference could well recede in a Gödelian self-referential statement. Specifically, ontic significance might become graded fostering ontological commitments such as experienced e.g. in chemistry. As an example, Per-Olov Löwdin,[9] [26], defined the subdiscipline quantum chemistry in terms of a so-called pure theory in a restrictive sense: *if it derives for instance chemical data from the knowledge of only the physical values of the electronic mass and charge, Planck's constant, the atomic numbers and the form of the Schrödinger Equation, which itself represents the quintessence of a great deal of physical experience.* In this

[8]In the foreword to [25] Sir Roger Penrose asks the question: *Is it merely the complexity of biology that gives living systems their special qualities and if so, how does this complexity come about? Or are the special features of strongly quantum-mechanical system in some way essential?* Penrose's answer is well-known, but we will not discuss his "Orch OR" theory, nor in detail Tegmark's decoherence argument.

[9]By many considered the father of quantum chemistry, see [26] for rationales.

domain the corresponding calculations have been labelled ab initio, promoting the important conception of correlation energies as benchmarks for quantum accuracy and scaling. This first principle method[10] should be contrasted with the less time consuming and easier-to-perform semi-empirical approaches that provide important, but less reliable theoretical answers, adding supplementary epistemic contextual data to the table.

These distinctions become important when probing the concepts that go into the sub-dynamics picture. Developing non-equilibrium thermodynamic formulations prompted Prigogine to infer two distinct notions of irreversibility, extrinsic and intrinsic. Basically, the former exemplifies open systems coupled to its surroundings, while the latter does not indicate any environment whatsoever therefore suggesting a more fundamental hypothesis of irreversibility as a viable physical law. This view was central to Ilya Prigogine, since it imparts that irreversibility is in the details of the interactions, rather than being related to the singularity at the origin of our universe.

A detailed examination of this principle has been given by Atmanspacher, Bishop[11] and Amann [16], see also [28, 29], where they appraise the two mentioned kinds of irreversibility based on ontic and epistemic descriptions. Since ontic states usually refer to closed systems which are empirically inaccessible, open systems are clearly different, the latter suggesting the depiction of epistemic states. In other words, Prigogine's objective was to ground onticity, if possible, on a fundamental principle, where in contrast to Monod's *Chance and Necessity* [23] being and becoming should set in motion a new culture about how we understand and experience Nature and its laws [27].

The distinction between extrinsic and intrinsic is not without problem. For instance, a system undergoing radioactive decay, is initially closed with no environment explicitly revealed. Nevertheless, there exists a prepared low entropy initial condition that is embedded in the time span of nucleo-synthesis, which then relates temporal irreversibility with the initial singularity of the universe. A similar problem seems to appear in the case of a box with all the particles collected in one corner. However, here there is no telling whether this initial condition was crafted by outside intervention or by an improbable fluctuation evolved in agreement with time reversible dynamical laws. This uncertainty prompts another fundamental query, viz. does there exist a law, i.e. an unavoidable deficiency rule causing a natural loss of information effectively accompanying sub-dynamics compatible with broken temporal symmetry [30]. We will return to this question further below.

As cited above the critics in Refs. [16, 28, 29] question some of the interpretations and general conclusions of the steps that formed Prigogine's world-view, but still there have not so far been given any serious objections to the derived and applied mathematical formulations, such as those given from classical chaos theory to the

---

[10]The debate whether the Nobel Prize awarded work on density functional theory, DFT, should be considered ab initio will not be discussed here. Yet, DFT is definitely epistemic in nature.

[11]Bishop has analysed the work of the Brussels-Austin Group [29] with particular focus on the definitions and mathematical extensions to rigged Hilbert spaces and its applications to concrete mathematical systems.

rigged Hilbert space formulation of resonances. Nevertheless, some reviewers of the work by the Brussels-Austin Group [31, 32] assert that the local origin of irreversibility has yet to be unambiguously proven, which might seem to be reasonable in view of the philosophical evaluations discussed earlier. However, Bricmont[12] [32], see also [33–35], goes one step further stating that Boltzmann's late-nineteen-century formulation is fully satisfactory[13] and that Prigogine does not pay sufficient attention to the fundamental attributes of the classical explanation of irreversibility, i.e. to the initial conditions and to the distinction between microscopic and macroscopic variables. He also states [32]: *Although the quantum picture may be more complicated, I do not believe that it renders obsolete the basic ideas explained here.* It is evident that the presently offered viewpoints will confront these statements. The quantum picture is not only fundamental but it will by necessity add more options and possibilities as will be demonstrated below.

The general conclusion so far is that the Brussels-Austin School should be essentially correct in that their mathematical formulation does not incorporate any approximations that exposes their sub-dynamical evolution with any unveiled epistemic attributes. However, their work[14] is not entirely able to resolve questions such as: why does nature choose one privileged direction of time, how does time symmetric physics lead to time-asymmetric chemistry, conservative reversible dynamics result in dissipative irreversible evolution and unitary time evolution convert to a contractive semigroup? These topics were already discussed at length during the Uppsala Nobel Jubilee Meeting, Frontiers of Science, in 1991 [36], and in particular at the associated Satellite Nobel Symposium, Resonances and Microscopic Irreversibility [30]. Although full agreement between the various groups could not be attained some tentative lines of thought were documented [30]: (i) *there exist intrinsically irreversible systems with the chosen time direction compatible with the cosmological arrow of time*, (ii) *there exists some kind of informity rule, i.e. a certain unavoidable natural loss of information, which is compatible with broken temporal symmetry.*

One should be reminded that the key problem related to abiogenesis, i.e. the biological machinery behind evolution or rather to find an understanding of the coupling mechanism behind the selection of a functional macro-molecular structure from a big molecular library, has yet to be resolved. An answer will be given in the following sections, coming as an additional bonus of the quantum theoretical mathematical formulation. In more detail we will derive, for each CES, the boundary condition that links the micro- and the macroscopic levels finding the relation between the temperature, the time scales and the dimension of the active degrees of freedom.

---

[12]Bricmont refers to Lanford's original derivation of Boltzmann's equation [33].

[13]Lanford's theorem in the case of N hard spheres was extended in [34]. The review [35] nevertheless concludes: *Boltzmann's 1872 effort, to conciliate microscopic time-reversible dynamics with increase of entropy and trend to equilibrium, still remains as a source of challenging mathematical problems.*

[14]It is hard to evaluate general statements made here, since the fundamental difference between CM and QM at the end of the day turns out to be significant.

The derivation has a logical framework[15] commensurate with a stochastic Poisson point process engendering communication protocols for teleonomic evolution. The theoretical analysis is not complicated, yet perhaps a bit extraordinary, if one simply respects non-Hermitian extensions of quantum mechanical methods and their interpretation [37], see also [21, 38, 39].

3 Biology and Quantum Theory

In his formulation of the goals of quantum chemistry [26], Löwdin called specific attention to quantum biology and the theory of life processes on the quantum molecular level [40]. The conception concerned the importance of mobile electrons, the role of protons and the hydrogen bonds integrating transport phenomena related to transfer of energy, momentum, charge etc. in living systems. Theoretical issues addressed the treatment of statistical ensembles, density matrices, their evolution, the phase problem including irreversible processes in so-called random phase systems. However, proton tunnelling exchanges, contingent upon the potential barrier in DNA suggesting spontaneous stability-error mechanisms in the genetic base sequence, could not be accurately determined by the quantum chemical computational techniques at the time. Recent advances are promising as quantum chemistry today is used in practically all branches of chemistry. Yet the universal origin of life forms in biology is not settled by any means, requiring as already said a quantum-chemistry-life-principle.[16]

Another fundamental shortcoming involves the quest for a realistic first principles formulation of the all-embracing sensitivity to temperature variations in physical systems and particularly in biological organisations. The theory of density matrices, established by Husimi in his doctoral thesis [41], included already in 1940, the reference to general Liouville formulations as Bloch's equation[17] formally linking time and temperature. In the following section we will generalize Bloch thermalization to incorporate spatio-temporal evolution within a background of quantum-thermal fluctuations and further to investigate the ensuing stochastic time evolution. This portrait elucidates the unitary decomposition of strong correlation effects in quantum systems as derived by the Japanese chemist Fukashi Sasaki[18] [42]. Its importance stems from the conceivable emergence of macroscopic quantum phenomena such as those of superconductivity and superfluidity. The Sasaki formula gave an independent proof and verification of Yang's fundamental theory of Off-Diagonal Long-Range

[15] The non-Hermitian extension permits higher order singularities affecting the time evolution of the CES.

[16] See e.g. Wiseman in [25].

[17] Cf. the Koopman representation employed in the Prigogine sub-dynamics. For more details see [29].

[18] The original Technical Note, dated May 1, 1962, from the Uppsala Quantum Chemistry Group is available on internet and authenticated as ASTIA DOCUMENT No: AD No. 296970. The story behind the lost manuscript and the progress of ODLRO has been told at numerous occasions, see [26] for more details.

Order [43], ODLRO, identified by a sudden appearance of a large eigenvalue in the fermionic second-order reduced density matrix, providing the mathematical conditions for Coleman's concept of so-called extreme states [44]. We leave these issues briefly, but will return further below.

Viewing from the unfolding Plenary Debate: *Quantum Effects in Biology: Trivial or Not?* [25], the general understanding transpired as collaborations between experimental and theoretical physics,[19] cosmology, electrical- electronic- and biomedical engineering, with inputs from structural biology, medicine and psychology. Since the panellists were partitioned for the purpose of a lively debate, wittily with those in favour of quantum effects in biology to be the no-team, it foreshadowed an entertaining and outspoken encounter, however not always reflecting the true opinions of the contributors, yet with some of the main actors sincerely contesting the pros and cons. It was agreed that the crux of the matter was the word *trivial* mostly interpreted here to mean *quantum mechanics is the default position*, see Davies [25], while *nontrivial* was indicated as *something that would convince biologists that they need to learn quantum mechanics*, Wiseman [25]. Except for pioneering experimental and theoretical advances related to quantum entanglement, see e.g. Zeilinger [45],[20] many technological advances in material science, chemistry and genetics, being the source of dramatic progress in human understanding and the manufacturing of medicines and pharmaceuticals, were actually missing in the debate, which therefore ran the risk of being trivialized by the panellists. Hence, one of the conclusions emerged that, if life nontrivially should emerge from the quantum mechanical material world complex, intermediate chemistry would not be necessary.[21] This view will be challenged in this work.

To reconnect with the fundamental work of Eigen, Prigogine and others, let us first take a short look at some recent advances in non-equilibrium statistical mechanics. An interesting concept, known as Crooks' fluctuation theorem [46] relating free energy differences and irreversible work along an ensemble joining the states, also known as the Jarzynski equality [47], has recently been applied to self-replicating nuclear acids and bacterial cell division by England [48]. The derivation shows macroscopic irreversibility from microscopic reversibility relating the populations of microscopic trajectories in the forward direction with the time reversed ones. The ensuing master equation becomes

$$\dot{P}_m(t) = W_{mm-1}(P_{m-1}(t) - P_m(t)) - W_{mm+1}(P_m(t) - P_{m+1}(t)) \tag{3.1}$$

where $P_m(t)$ is the probability of a population of m at time t, W_{mn}, the transition probability per unit time between populations m and n, given by $W_{mm-1} = gm$ and $W_{mm+1} = \delta m$, where g and δ are birth and decay rate parameters with $g > 0; \delta > 0$.

---

[19]Other physical areas represented in the book were also laser physics, quantum optics, quantum information and subatomic physics.

[20]Ref. [45] was originally published in 2005 in different form in German by C. Bertelsmann, Germany, as *Einstein Spuk: Teleportation und weitere Mysterien der Quantenphysik*.

[21]Cf. the statements by Paul Davies in [25].

In passing we note that (3.1), providing the general constraints on self-replication, has a form that could be derived from Pauli's master equation, if one applies the condition of microscopic reversibility,[22] $W_{mn} = W_{nm}$,

$$\dot{P}_m(t) = \sum_n W_{mn} P_n(t) - \sum_n W_{mn} P_m(t) \tag{3.2}$$

Equation (3.1) leads to a deterministic exponential growth path that sets a positive lower bound on the total entropy provided $g > \delta$. This entails thermodynamic constraints, relevant for growth and division of whole single cell organisms, including exponential build-up through self-replication, doubling times and heat bounds, staging a deeper understanding of the information content of inherent molecular structures. The bounds reveal notable enthalpy differences between RNA and DNA as regards the participation in self-catalysed replication reactions, with RNA operating close to the limit of thermodynamic efficiency, while for DNA in this case a higher per-base cost being paid in entropy production matching the RNA growth rate all-things-equal.

By implication Crooks' theorem proves that the dynamics of a system that satisfies classical microscopic reversibility during a non-equilibrium transformation develops macroscopic irreversibility. Comparing the work dissipated in the forward transformation with those trajectories running backwards it is clear how irreversibility sneaks in. It demonstrates the fundamental difference between the treatment of microscopic- and macroscopic variables, yet, while providing important thermodynamic extensions of the second law, will not add anything new to the fundamental questions posed here.[23] It is important, however, to observe that $W_{mn} = W_{nm}$ does not generally hold in QM. Hence a quantum mechanical proof of (3.2) will be given below, see [40], derived without using the equality above by and large referred to as the law microscopic reversibility.

Starting with a many-particle system subject to the Schrödinger equation, i.e. initially being in a low entropy state, we will study its approach towards equilibrium. The time evolution of an initial wavefunction $\Psi(0)$ at time zero to time t is given by the unitary operator U given by $\Psi(t) = U(t)\Psi(0)$, which in the (time independent) basis[24] $\{\phi_k\}_1^r$ writes $U_{kl} = \langle\phi_k|U(t)|\phi_l\rangle$. One obtains the general relations between the quantum mechanical entities as

$$\Psi(t) = U\Psi(0) = \sum_k \phi_k c_k(t); \; c_k(t) = \sum_l U_{kl} c_l(0) \tag{3.3}$$

with the interpretation that

---

[22]Note that the notion of microscopic reversibility means different things in CM and QM.

[23]Cf. Bricmont's comments in the previous section and associated footnotes.

[24]Although the complete state space dimension might be infinite, the relevant degrees of freedom using partitioning technique turns out to be finite. Employing the thermodynamic limit is another source of irreversibility.

$$P_k(t) = |c_k(t)|^2 \tag{3.4}$$

gives the probability for the quantum system to be found in the state k characterized by ϕ_k at time t. Hence one obtains directly

$$P_k(t) = \sum_l |U_{kl}|^2 P_l(0) + \sum_{l \neq m} U_{kl}^* U_{km} c_l^*(0) c_m(0) \tag{3.5}$$

and inferring from the properties of the unitary matrix U, it follows that $\sum_k P_k(t) = 1$ at all times. It is customary to investigate incoherent phenomena by averaging of the phases of the states involved. Alternatively, one may refer to so-called random phase systems reflecting ensembles of equilibrium systems subject to

$$\overline{c_l^*(0) c_m(0)} = 0; \;\; l \neq m \tag{3.6}$$

with the bar indicating a phase average of the assembly. Applying (3.6) to (3.5) one gets with the definition

$$S_{n \to k}(t) = |U_{kn}(t)|^2$$

and noting that the second term of (3.5) average to zero

$$P_k(t) = \sum_{l \neq k} S_{l \to k}(t) P_l(0) + |U_{kk}|^2 P_k(0)$$

Since U is unitary one finds

$$|U_{kk}|^2 = 1 - \sum_{l \neq k} S_{l \to k}(t) = 1 - \sum_{k \neq l} S_{k \to l}(t) \tag{3.7}$$

and

$$P_k(t) - P_k(0) = \sum_{l \neq k} S_{l \to k}(t) P_l(0) - P_k(0) \sum_{k \neq l} S_{k \to l}(t) \tag{3.8}$$

or using (3.7), despite that in general $S_{k \to l}(t) \neq S_{l \to k}(t); \; k \neq l$

$$P_k(t) - P_k(0) = \sum_{l \neq k} S_{l \to k}(t) P_l(0) - P_k(0) \sum_{k \neq l} S_{l \to k}(t) \tag{3.9}$$

Dividing (3.8) by t, taking the limit $t \to 0$, one obtains the gain-loss equation known as the Pauli master equation, PME.[25] The limit is taken with the understanding that we are dealing with an ensemble of systems approaching equilibrium, i.e. with a higher entropy than the initial system, i.e. $\Delta S = k_B \ln r$, where ΔS is the change in entropy, k_B is the Boltzmann constant and r the dimension of the state space.[26] Defining the transition rates

$$W_{mn} = \lim_{t\to 0} \frac{S_{n\to m}(t)}{t}$$

one obtains from (3.8) the PME

$$\dot{P}_m(t) = \sum_n W_{mn} P_n(t) - \sum_n W_{nm} P_m(t) \tag{3.10}$$

and similarly, from (3.7) and (3.9) one gets Eq. (3.2), which as we have repeatedly stated, holds without the use of the condition $W_{mn} = W_{nm}$.

The somewhat confusing terminology arises from fundamental classical-quantum differences. Microscopic reversibility characterizes classical trajectories and their time reversed paths, while time invariant quantum mechanical reversibility entails $\sum_n W_{mn} = \sum_n W_{nm}$. A more detailed analysis will be given in succeeding sections, based on rigorous sub-dynamical master equations [49–51] and their extensions.

Rounding off, the equilibrium condition, $\dot{P}_m(t) = 0$, as well as the detailed balance condition

$$W_{mn} P_n(t) = W_{nm} P_m(t)$$

amounts, if $P_n(t) = P_m(t) = P$, to $W_{mn} = W_{nm}$, which again relates back to the issue of microscopic reversibility. One might additionally consider the present question with the concerns related to the proof of Boltzmann's H-theorem [52]. In what follows we will return to the key issue [30], i.e. *how does time symmetric physics lead to time-asymmetric chemistry, conservative reversible dynamics lead to dissipative irreversible evolution and unitary time evolution becomes a contractive semigroup?*

[25] The limit is usually interpreted as the course-grained limit with the time being smaller than a process-defined relaxation time, $\tau_{\rm rel}$, but still large enough to randomize the phases. The Prigogine sub-dynamics does not contain this assumption.

[26] We have naively assumed a finite number of reference states, which in a more refined treatment must be coupled to a continuum, which technically can be formulated by Löwdin's partitioning technique [49] or the Nakajima–Zwanzig equation [50, 51].

4 Technology and Quantum Theory

It is worthwhile to remember that signal processing and quantum mechanics, despite different habits and routines, share a similar mathematical apparatus [53, 54]. In particular image science concerns various systems, methods, procedures, devices and techniques to analyse the physical effects that influence the structure of images. Important advances in the interdisciplinary domain of quantum mechanical signal- and spectral analysis have been developed and exploited in medical practice by Belkić [54] utilizing spectral estimators as Padé approximants and the Lanczos algorithm. Other interdisciplinary applications concern filtering and postprocessing, such as convolution with prolates for optimizing the information contained in the autocorrelation function and its Fourier transform [55].

Even if the mathematical methods are in many respects identical, there is nevertheless a fundamental distinction to be made. First of all, to practise research, we are searching for knowledge by accumulating facts by systematic observation, deliberate experiments and rational theory. In every branch of technology and science, measurement tools link a physical operational process with an observer. The distinction between the apparatus and the environment, including the observer is clear-cut. In quantum mechanics there is the notorious measurement problem due to the fundamental absence of a well-defined fixed observer-system demarcation. While engineering disciplines, using both CM and QM operations, concern procedures that treat, handle and mediate devices and sensors, quantum science models the universe in that each elemental systems are subject to QM. The former constructs tools that in principle reflect processing data at an arbitrary precision and accuracy, while the interpretation of QM sets a limit by the Heisenberg uncertainty relations. Granted that the Fourier transformation between conjugate variables, i.e. between the time correlation function $C(t)$ and the corresponding Fourier transform in the frequency domain $Q(\omega)$

$$C(t) = \frac{1}{2\pi} \int_{-\infty}^{+\infty} e^{-i\omega t} Q(\omega) d\omega \tag{4.1}$$

and

$$Q(\omega) = \int_{-\infty}^{+\infty} e^{i\omega t} C(t) dt \tag{4.2}$$

fulfil uncertainty-like relations, the ordinary CM environment of a measurement system, is not constrained just because the mathematical formulation acquires a limit. We will not discuss conventional convergence properties of Eqs. (4.1)–(4.2) and its

Table 1 Quantum technology

Application area	Mechanism/technique
Quantum information	Nonclassical
Condensed matter	Broken symmetry
Ultracold matter	Dissipation-dispersion
Interferometry	Coherence-decoherence
Correlated dissipative systems	Quantum-thermal correlations
Stochastic dynamical systems	Stochastic resonances

relation to classical harmonic analysis, see also Wiener[27] and references therein [56], but we will return to this issue in connection with similar and corresponding connections between concepts such as propagators and resolvents in QM [57].

While being clear about the different objectives between CM and QM interpretations, let us not forget the dramatic advances in the state-of-the-art technology that depends profoundly on fundamental principles. This progress, based on non-classical traits, often denoted *quantum technology* has been listed by recent topics and techniques, in Table 1, starting from pure quantum systems going towards more "mixed-dirty" fields of applications, where each topic further may contain several relevant subfields [58]. We have not included the highly fragmented subfields usually growing under the name of Artificial Intelligence, AI, and with the aim to simulate Natural Intelligence, the latter saved for Sects. 5–7.

There have been recent movements towards a new production of knowledge [59], where the traditional disciplines, Mode 1, is prognosticated to generate a new branch, Mode 2, the latter being embedded in a broader, transdisciplinary milieu. In brief, Mode 2 is heterogeneous, not hierarchical and carried out in a context of applications incorporating a wider set of practitioners compared to Mode1. Although it is often declared: *there is no transdisciplinary work without disciplines* it is also true that *there are no disciplines in nature only phenomena.*[28] Although it is an onerous task to describe where evolution might take these Modes, there seem to be no problem to engage universities and other teaching and research organizations with a so-called public outreach [60]. For instance the Uppsala Graduate Programme on Advanced Instrumentation and Measurements [61, 62], attracted specific industrial interest by selling a many-dimensional education map characterized by theme/application, such as space, accelerators, energy, bio, medicine, environment and information technology, with technique/skills such as sensors, computational methods, simulations and modelling, signal processing and complex systems, exhibiting in detail the scientific

[27] Wiener, the father of *Cybernetics*, introduced feed-back and learning as mechanisms for information and communication in evolving organizations. See more on *Communication Simpliciter* in later sections.

[28] Stated by the Swedish Chemistry Nobel Laureate The Svedberg, see also the AIM magazine in Phase below.

Fig. 1 Conceptual design of a Universal Device, UD, and its environment, where the outgoing flux has a lower entropy than the input flux

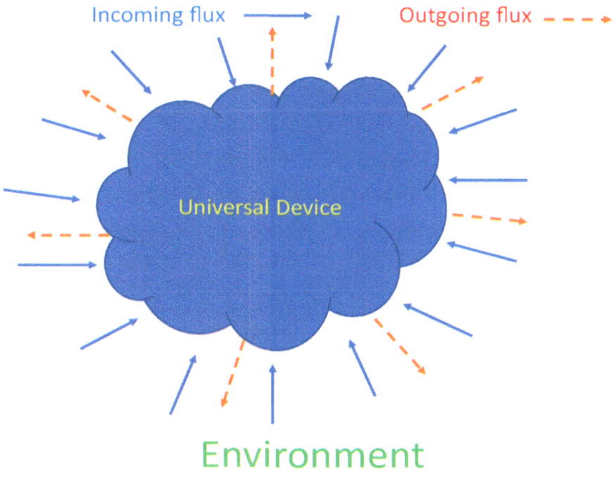

muscles delivered in each Ph.D. program presented by the research groups of the seven participating departments.[29]

In order to explore the distinctions made here, we will take a closer look at the notion of a device or an entity that performs a special function. It is clear what it means in connection with a measurement system, where the latter provides the relation between the physical process and the spectator, who by definition needs a certain information from the environment. We assume that the device is in equilibrium with its environment before being interfered by an ingoing perturbation. After a certain time, τ_{rel}, the set-up responds, delivering a specific result generated by the system. With the intention to extend the concept to apply to the generation of life processes in biology, we will introduce a general definition as follows: *A Universal Device, UD, is a physical system where the ingoing entropy flux is higher than the outgoing one*, see Fig. 1. It is obvious that this definition fits engineering operations, since signals generally contain noise, but it should also apply to biology, e.g. a human organ, but, as will be seen, with a fundamentally distinct quality. In terms of entropy changes the condition for *UD* writes [18].

$$dS = dS_{\text{UD}} + dS_{\text{Q}} \tag{4.3}$$

with

$$dS_{\text{UD}} < 0 \tag{4.4}$$

In Eqs. (4.3), (4.4) dS is the total entropy change, dS_{UD} the entropy flux due to exchanges of energy-matter between the Universal Device and the environment and dS_{Q} the entropy production due to irreversible processes inside the device. The

[29]The graduate program, producing 35 Ph.D's, was sponsored by the Swedish Foundation for Strategic research. The AIM Magazine, InPhase: Bridging the Gap between Academia and Industry is available from the author on request.

second law says $dS \geq 0$ and $dS_Q \geq 0$ implying $dS_{UD} \geq -dS_Q$. The *UD* condition thus becomes $dS < dS_Q$. In particular the steady state condition, $dS = 0$, implies $dS_{UD} = -dS_Q$, which is negative as long as any entropy production subsists. We will preferentially refer to QM processes, since pure CM studies of conceivable *UD*'s may yield different answers, see e.g. the analysis of the Cyclotron Maser Concept, CMC, in [57]. The QM view will be shown to be crucial, while making a profound derivation of the characteristics of our Universal Device. The relation with Schrödinger's classic treatment "What is Life?" and the practice of the associated concept of negentropy is deferred to the Conclusion.

Consider an *UD*, by its definition, an open system, depicted in Fig. 1, and defined by n degrees of freedom[30] signifying interactions between the environment and the device during τ_{rel}. Utilizing conventional collision theory, one obtains the outgoing flux, i.e. the ensuing reactant particles/waves. The experiment is controlled at the temperature T with a time scale τ_{rel}, in general much larger than the thermal time scale $\tau_{corr} = \hbar/k_B T$, where $\hbar$ is the reduced Planck's constant. The operation of the device is defined by a process that, on the average, detects one degree of freedom,[31] in the differential solid-angle element $d\Omega$ during τ_{corr} at T K. Straightforwardly one gets, for the incoming flux, N_{inc}, being the number of degrees of freedom per unit area and time, and $N_s d\Omega$, the outgoing flux scattered into the solid angle $d\Omega$ per unit time, the standard relations between the differential- and the total cross sections σ_Ω and σ_{tot} respectively, the following formulas

$$N_{inc} = \frac{n}{\sigma_{tot}\tau_{rel}}; \; N_s d\Omega = \frac{d\Omega}{\tau_{corr}}; \; \sigma_{tot} = \int \sigma_\Omega d\Omega = \int \frac{N_s}{N_{inc}} d\Omega \tag{4.5}$$

which yields the simple relationship below[32]

$$\frac{n}{4\pi} = \frac{k_B T}{\hbar}\tau_{rel} = \frac{\tau_{rel}}{\tau_{corr}} \tag{4.6}$$

For purposes to be evident later we rewrite (4.6), for a different derivation see [21],

$$\tau_{rel} = (l-1)\tau_l = \tau_2 = \frac{n\tau_{corr}}{4\pi}; \; l = 2, 3 \ldots n \tag{4.7}$$

---

[30] The nature of these degrees, paired fermionic or bosonic etc., will be discussed later.

[31] This might appear as an unnecessary limitation. However, assigning instead g degrees of freedom, the result becomes simply that n will be replaced by n/g in Eq. (4.6). Without restricting the formulation one might refer to $n' = n/g$ as the relevant degrees of freedom defining $UD(n')$, provided $n'/4\pi \gg 1$.

[32] Note that σ_{tot} cancels out in the final relation and that the factor 4π occurs since the outgoing flux of $UD(n)$ is defined per solid angle. In biological systems the number n gets important since it conveys crucial information regarding the performance of each CES. In such cases the quality index n will be explicitly denoted separating it from other technological or artificial intelligence devices.

introducing a series of time scales τ_l intrinsic to the dynamics of the device. This relation may also be expressed in terms of the widths ϵ_l via the standard relation between life times and energy widths $\epsilon_l = \hbar/2\tau_l$

$$\beta\epsilon_l = \frac{2\pi(l-1)}{n} \tag{4.8}$$

where $\beta = 1/k_B T$, T the absolute temperature, k_B the Boltzmann constant. Condition (4.8) will provide a surprising inner quality to the regulations of biological CES's. Finally, we note that formulas (4.6)–(4.8) relate the temperature T, with the inherent time scales of the UD and the number of correlated degrees freedom, or particles/waves, n. We will label the particular device as $UD(n)$.

From the Gibbs free energy, where μ is the chemical potential follows

$$\begin{aligned} G &= n\mu = nk_\mathrm{B}T \\ S &= \left(\frac{\partial G}{\partial T}\right)_n = nk_\mathrm{B} \end{aligned} \tag{4.9}$$

and from Eq. (4.6)

$$dS = \frac{Sdn}{n} = \frac{Sd\tau_\mathrm{rel}}{\tau_\mathrm{rel}} + \frac{SdT}{T} = dS_\mathrm{UD} + dS_\mathrm{Q} \tag{4.10}$$

and for the steady state $dS = 0$ one finds, using (4.7)

$$dS_\mathrm{UD} = \frac{Sd\tau_l}{\tau_l} = -\frac{SdT}{T} < 0 \tag{4.11}$$

General examples are e.g. the laser, i.e. the incoming flux of photons pumping electrons in a lower energy level of atoms or molecules in the device to a higher one, inverting the population, which then may be stimulated to radiate with the modes phase correlated, or in the case of the hypothetical Maxwell Demon, separating fast and slow molecules at the expense of raising its own entropy. Another primordial case is our planet Earth receiving energy in terms of sunlight, making photosynthetic activities transforming carbon dioxide and water to chemical free energy in form of carbohydrates and molecular oxygen, producing high entropy waste and dispersing heat. As there are countless examples of devices, such as sensors, detectors and computers in science and engineering, we will carry on the narrative in a different direction by posing the question: can it be applied to life systems in biology and if so under what conditions?

5 Evolution and Quantum Theory

A living system is characterized by (a) its dissipative coupling to the environment, (b) its metabolic processes, (c) the genetic function or higher order molecular function and (d) homeostasis for appropriate spatio-temporal regulation. Without going into detail how a biological system, such as a cell obtains energy through the various biochemical pathways, we will begin to model the biological process via the concept of the Universal Device $UD(n)$, but with some added properties commensurate with (a-d) above. Obviously, the device agrees with (a) and (b) yielding an unambiguous output. However, point (c) imparts a self-referential adaptation that is a fundamentally trait of all life forms, missing in the present laws of physics. Moreover, what is not known here is how energy exchanges, related to metabolism, transduce the information that emerges for instance in the genetic code and its associated spatio-temporal regulation mechanism. Whereas the appliances in engineering eventually wears out after good and loyal service, and while these dimension-determining limitations are usually not displayed in the mandatory parameters that controls the behaviour of the device, an $UD(n)$ adapts its generic approach thermally and spatio-temporally as will be seen in more detail in the next section.

According to the transcripts of the Gran Canaria debate [25], the relevancy of primary quantum aspects must be proved to be essential for the understanding of all life processes and its quantum chemical origin. Since in vivo systems are dissipative, pioneering QM is not adequate to describe an $UD(n)$. In fact, the resonance picture of so-called unstable states in the continuum, such as e.g. Gamow waves associated with quantum tunnelling or the resonance formation of quasi-bound states appearing in total- or differential cross sections, are, strictly speaking, outside the domain of Hilbert Space quantum mechanics. Conceding that scattering theory approaches belong to traditional quantum mechanics, yet the cross-section peaks, for details see Ref. [63] and Fig. 2, associated with the continuous part of the energy spectrum, reflect the existence of complex poles in the Greens function, located on a second sheet of the complex energy plane.[33] The location of the resonance structure embedded in the continuum commands non-trivial extensions to non-Hermitian quantum mechanics [21, 37–39]. The latter is a consequence of the Balslev-Combes theorem for dilation analytic families of Hamiltonians [37] leading to the popular complex scaling technique in quantum chemical applications [38, 39]. The idea is very simple. It starts with the concept of scaling, a unitary operation in quantum mechanics and a fundamental method in proving quantum virial relations.[34] Making the dilation parameter complex yields surprising consequences: Hermitian operators, after complex scaling, become non-Hermitian and importantly non-normal, i.e. not commuting with their own adjoints; time evolution is no longer a unitary operation,[35] and the poles of the resolvent, see Eq. (4.11) below, may leave the real energy axis.

[33]This is usually referred to as analytic continuation onto the second unphysical Riemann sheet.

[34]For instance the relation between kinetic and potential energies, see relevant discussions in Ref. [26].

[35]Cf. the concept of star-unitary transformations introduced by the Brussels-Austin School.

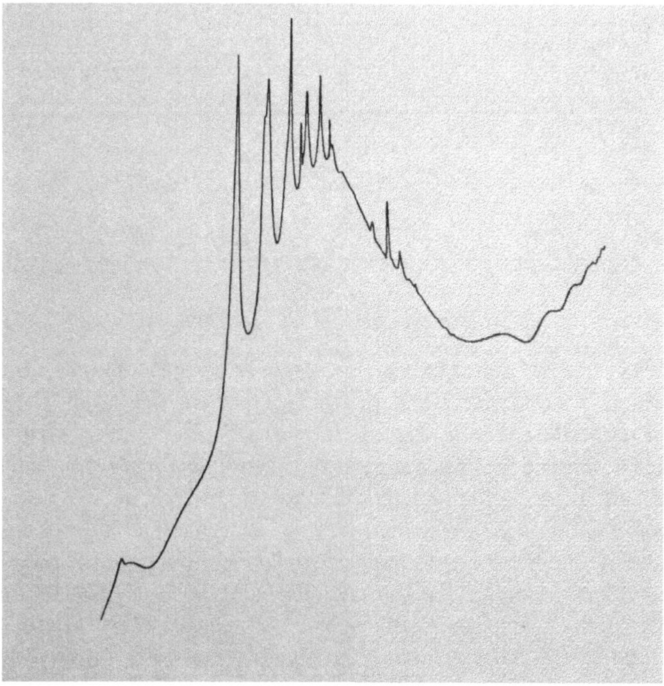

Fig. 2 Display of a characteristic logarithmic charge transfer cross section plot in a logarithmic centre-of-mass energy range. The theoretical query is to analyse the composition of the resonance structure, in this case in terms of orbiting and vibrational quantum numbers. For details see Ref. [63]

However, the most significant result is a somewhat ignored property, namely that the poles of the analytically continued resolvent may no longer be simple.[36] This follows from the trivial insight that imposing analyticity with respect to the scaling parameter demands complex symmetric matrix representations[37] of the Hamiltonian rather than Hermitian ones. In addition, the domains and ranges of the operators have to be checked carefully while properly extended. Still, these highly unwanted complications will surprisingly enough turn out to be a blessing in disguise. The left part of Fig. 3 displays a representative situation, where the Hamiltonian, exhibits a point spectrum and a completely continuous spectrum. After complex scaling, leaving the bound states, denoted with a + sign, intact, exposed complex resonances will often be located along a typical trajectory on the second Riemann sheet. The right part of Fig. 3 shows the situation for a related Liouville operator [15]. In addition to the

[36] This imparts the occurrence of the irreducible Jordan normal form of the operator representation, where the order of the largest block defines the Segrè characteristics corresponding to the actual degeneracy.

[37] Any matrix can be brought to complex symmetric form.

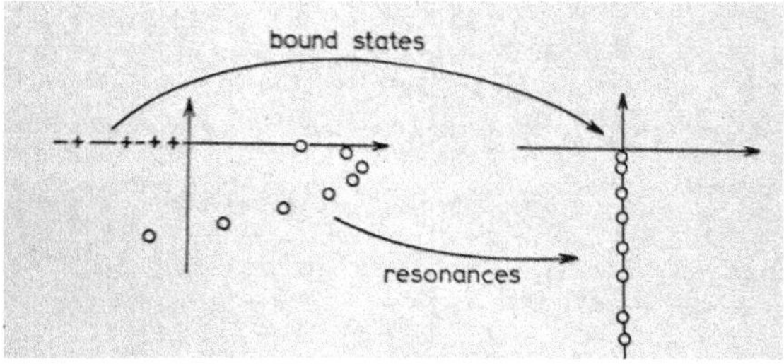

Fig. 3 Typical behaviour of bound states and resonances, in the complex energy plane, the latter embedded in the continuum, related to a complex-scaled Hamiltonian, left, and the associated Liouville operator, right

observation that resonances appear in the lower part of the complex energy plane,[38] singularities might, as pointed out above, also be of various multiplicities. The original derivations of sub-dynamics from a QM perspective, see [15, 64] demonstrated the generic properties inherent in the dynamical conditions and showed the rigour and usefulness of partitioning techniques [49–51] in extending sub-dynamical master equations beyond coarse-graining approximations.

A proven feature of the popular complex scaling method is the combination of mathematical rigour in defining the concept and the numerical precision of a physical resonance, with its position and width, the latter inversely proportional to its lifetime. Nevertheless, there has appeared amongst some mathematicians a certain heretical view, see e.g. Ref. [65], that *a resonance is a much more basic phenomenon than the subtleties of analyticity*, further posing: *Define and study resonances using less analyticity (and less rotational symmetry)*. This general view, although certainly practical in many physical situations, where sufficient data is missing, will be contested below.

To support our claim, let us first return to the transformations (4.1) and (4.2). Using the spectral expansions of the retarded-advanced propagator[39] $G_P^{\pm}(t)$ and the resolvent $G_R(z)$

$$G_P^{\pm}(t) = \mp \mathrm{i}\Theta(\pm t)\mathrm{e}^{-iHt};\; G_R(z) = (z - H)^{-1} \tag{5.1}$$

where $\Theta(y)$ is the Heaviside step function, which equals one if the argument is positive otherwise zero, z is the energy, or frequency comprising the choice $\hbar =$ 1. From the spectral expansion of the Hamiltonian it is easy to establish that the propagator and the resolvent are related via a generalized Fourier transformation[40] [66, 67]. In Fig. 4, the contour for the integration path C_+ corresponding to $t > 0$ is indicated for the general relation

[38]Note that a resonance may come arbitrarily close to the real axis and still be classified as being located on the second sheet.

[39]The partition of the propagator into its retarded-advanced parts will be crucial as it imposes a separation of positive-negative times corresponding to the lower-upper parts respectively of the complex energy plane.

[40]The Swedish mathematician Torsten Carleman extended the Fourier transformation by splitting the integral into two parts with the variable z being complex. This generalization is well described in Lützens's Studies in the History of Mathematics and Physical Sciences No. 7 [67].

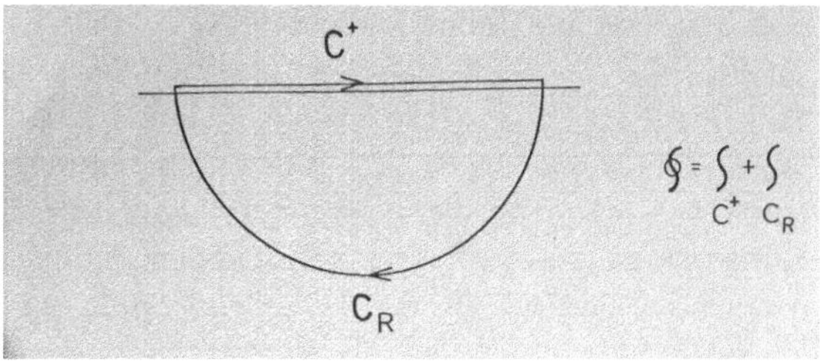

Fig. 4 Integration path for Eq. (5.2) relating the retarded time propagator with the resolvent. The spectrum on the real axis is located within the curve, which is closed in the lower complex plane. The contribution from C_R becomes zero when $R \to \infty$

$$G_P^{\mp}(t) = \frac{1}{2\pi} \int_{C^{\pm}} G_R(z) e^{-izt} dz \tag{5.2}$$

An analogous path would be C_-, in the lower halfplane, where C_R closes in the upper complex plane going with $t < 0$. Since the integral does not depend on C_+ or C_- one can let the paths arbitrarily approach the real axis. If the Hamiltonian is self-adjoint the spectrum lies on the real axis and combining the two contours via a principle value analysis one obtains agreement with the transformation (4.1) in the previous section.

Although the rule above is uncomplicated, i.e. positive time corresponds to closing the contour in the lower part of the complex plane, see Fig. 4, and negative time constrains closing in the upper half, there are some additional technical details that one must bring up. There are essentially three things, which are interrelated, engendering the conditions that are crucial for the conversion of reversible to irreversible dynamics. First, there is the domain-range issue of the unbounded scaling operation when the argument is complex. Second, the scaled propagator must be positive preserving. Third the dissipative evolution must be contractive

$$\left\| e^{-iH(\eta)t} \right\|^2 \underset{t \to \infty}{\to} 0 \tag{5.3}$$

where $\eta = e^{-i\vartheta}$; $\vartheta > 0$ is the complex scaling parameter.

Summarizing: The characteristic features of a life system, see points (a) to (d) above, reveals that we must extend the dynamics to non-Hermitian QM. As a consequence

(i) time evolution induces a separation of positive and negative times
(ii) the conjugate energy spectrum relocates to corresponding complex halfplanes
(iii) complex orthogonal transformations replace unitary descriptions
(iv) complex symmetric matrices replace Hermitian representations
(v) range and domain characteristics must be exercised
(vi) observables become non-normal operators
(vii) Jordan blocks[41] with Segré characteristics $m > 1$ do emerge [68].

[41] For a simple introduction see Ref. [68] and footnote 36.

Without going into technical details, it is clear that rigorous analytic continuation [37] implicates an inevitable loss of information due to the mandatory modifications (i–vii). Take e.g. point (v). The complex scaling operation is unbounded and its domain must be restricted to a smaller class of functions, or to a dense subspace of Hilbert space [69], before the dilated non-normal operator again is extended back to the full Hilbert space,[42] see [70]. This is the essential reason behind the previously mentioned deficiency rule that governs the conversion of stationary QM to the dynamics of evolution.

For the interested reader one should mention, as previously brought up, the issues related to the positivity condition in the theory of irreversibility [71]. It was soon realized that the advantage of rigorously applying complex scaling techniques to atomic and molecular systems revealed a possible drawback in that the quantum chemical Hamiltonian, containing the attractive Coulomb potential, does not generate a contractive semigroup or in essence would violate the positivity condition (5.3). The problem was discussed and solved at the time [20] although the topic unfortunately appeared a bit confused for a while. The proof involves the knowing that the scaling operator can be identified as a Lyapunov converter on the Nelson class [72], i.e. mapping an isometric- to a contractive semigroup.[43] The theorem is of principal importance, since the long-range Coulomb potential is inevitable for the emergence of so-called coherent-dissipative structures in condensed amorphous phases [21, 73].

6 Life and Quantum Theory

In the discussions of the physical origin of the time asymmetry in the Universe,[44] a general solution to the problem is often suggested as the appearance of a low entropy initial condition of a system governed by time symmetric mathematical equations [74]. A different view is proposed and discussed in Ref. [75] and references therein. We will not discuss these matters in any detail here, however, as an illustration of our intentions one should say that a piece of ice in a glass of water, an example of Reichenbach's notion of branch systems, would not qualify as an *UD*, since it would not satisfy $dS_{UD} < 0$.

With the cosmological aspect in mind, it is inevitable that the issue of the appropriate boundary condition turns up. Traditional QM is based on Schrödinger's quantization as an eigenvalue problem, which, combined with the unique behaviour of the

[42]This is a technical problem, yet of crucial importance. Suffice it to say that Nelson's class of analytic vectors, dense in Hilbert space, is a key ingredient in the proof of the Balslev-Combes theorem [37], see also Ref. [70] for more details on its use.

[43]The proof involves the notion of quasi-isometry. The problem is due to the fact that the numerical range of a complex scaled attractive Coulomb potential cannot be proven to be restricted to the lower complex halfplane, a necessary and a sufficient condition for the fulfillment of the Hille-Yosida theorem [20].

[44]The discussions occur in an interdisciplinary poll [74] that concerns the reconciliation of profoundly conflicting facts regarding time and the fundamental processes going on in the world.

wavefunction at singularities in the potential and the general requirement of square integrability, avoid any arbitrariness or subsequent additional criteria in the formulation. This understanding prevails also in the extension to resonances and unstable states in the continuous spectra, such as Gamow waves and the Stark effect [21, 76]. Generalizing the question to include the Universe and the Universal Device it becomes a daunting problem as it must include at least a realistic idea of what characterizes a Quantum Black Hole, QBH. Some suggestions will be offered in the last section, but here we will be content with a picture commensurate with Eqs. (4.5), (4.6) and Fig. 1.

Obviously, it is reasonable to assume that the device is brought into some kind of equilibrium with its environment, i.e. subsisting in a steady state configuration as it carries out its functionality. In principle the prehistory, i.e. the making of the device, might be part of the incident flux. However, since the steady state exhibits stochastic traits[45] there appears no restriction or loss of generality to formulate the initial condition as carried out previously for $UD(n)$.

The present formulation holds in principle for both the Schrödinger-, the Liouville- and the Bloch equation[46] with Ψ, ρ, ϱ being the wavefunction, the evolving- and the thermalized density matrices.[47]

$$H\Psi = i\hbar\frac{\partial\Psi}{\partial t} \tag{6.1}$$

$$\mathcal{L}\rho = i\hbar\frac{\partial\rho}{\partial t};\ \mathcal{L}\rho = [H\rho - \rho H] \tag{6.2}$$

$$\mathcal{L}_{\mathrm{B}}\varrho = -\frac{\partial\varrho}{\partial\beta};\ \mathcal{L}_{\mathrm{B}}\varrho = \frac{1}{2}(H\varrho + \varrho H) \tag{6.3}$$

appropriately continued to the complex plane [15]. The extended spectral modifications corresponding to (6.1) and (6.2) are exemplified in Fig. 3. The Bloch Eq. (6.3) permits a thermalization procedure [21, 57, 70, 73], invoking the formal solutions

$$e^{-\frac{i}{\hbar}Ht}\Psi;\ e^{-\frac{i}{\hbar}\mathcal{L}t}\rho;\ e^{-\beta\mathcal{L}_{\mathrm{B}}}\varrho \tag{6.4}$$

The system, characterized by a suitable sub-dynamics formulation, provides the necessary machinery for the mechanism of the Universal Device, originally in equilibrium with the environment, and performing under steady state conditions with an associated entropy production see Fig. 5. A microscopic description of an *UD* starts with the second order, reduced density matrix for *N* fermions, characterized by the

[45]The evolution of $UD(n)$ turns out to be Poissonian the loss of memory is here implicit.

[46]See the definitions introduced by Husimi [41].

[47]In the reduced representation one could employ the reduced Hamiltonian [44], and in a thermalized picture one might also include the chemical potential.

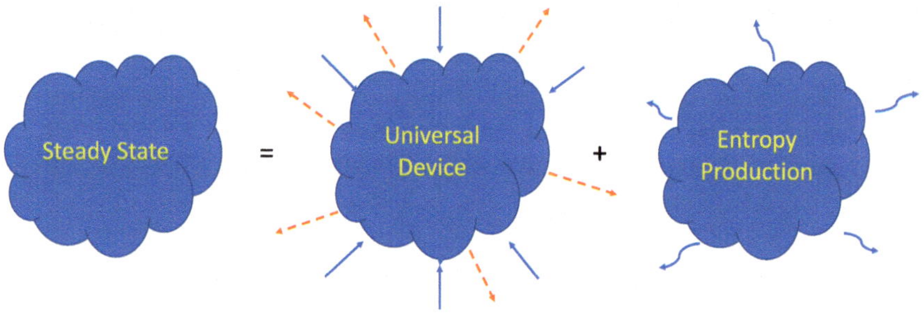

Fig. 5 Display of the Universal Device as an open system exchanging entropy across the boundaries of the system and the entropy produced within the system. For the case of a steady state, with $dS = 0$, one gets $dS_{\mathrm{UD}} = -dS_{\mathrm{Q}}$

space-spin coordinates[48] x_k normalized to the number of pairings (for details see [17, 21])

$$\Gamma^{(2)}\left(x_1, x_2 | x_1^{'}, x_2^{'}\right) = \binom{N}{2} \int \Psi(x_1, x_2, x_3, \ldots x_N)\Psi^*\left(x_1^{'}, x_2^{'}, x_3, \ldots x_N\right)dx_3, \ldots dx_N \tag{6.5}$$

with

$$E = \mathrm{Tr}\left\{H_2\Gamma^{(2)}\right\} \tag{6.6}$$

for a suitable reduced Hamiltonian—so far all embedded in a standard setting of quantum mechanics. For simplicity the nuclear coordinates are not displayed here, but we will return to this question below.[49] Without restricting the formulation, we choose $E = 0$.

To adapt the formulation to concur with Fig. 5, we will make an abridged account of the formulation. The theory has been presented in many versions and will not be detailed here. Since our starting point relates to Yang's Off-Diagonal Long-Range Order [43], or ODLRO, see alternative derivations [42, 44], we will start with the degenerate form of $\Gamma^{(2)}$, which, if the temperature is sufficiently cold, might develop superconductivity. It is notable that the partial trace over the N-2 fermion degrees of freedom in Eq. (6.5) can in the present case be derived from a statistical analysis [77], at the same time offering a sub-dynamics representation identifying the partial trace as a projection operator.

Imagining a general system of N electrons, in a nuclear skeleton making up a specific molecular structure. If the temperature is low enough we might consider

[48]We will denote the spatial coordinates with a vector notation, i.e. $\vec{x}_k$.

[49]The electronic- and the nuclear systems are entangled, which, as we will see, provides a way to go beyond the Born-Oppenheimer approximation [17, 21, 70].

the system to be subject to Schrödinger's equation with a wavefunction that can be represented as an anti-symmetrized product [44] of $N/2$ geminals g_1, i.e. $\Psi = g_1 \wedge \cdots \wedge g_1 = g_1^{N/2}$ with[50]

$$g_1 = \frac{1}{\sqrt{n}} \sum_{i=1}^{n} | \phi_i, \phi_{i+n}\rangle \tag{6.7}$$

In Eq. (6.7) the system of N fermions are described in a preferred basis $\{\phi_i\}$ of $2n$ orthonormal spin orbitals, where $| \phi_i, \phi_{i+n}\rangle$ is a normalized Slater determinant of ϕ_i and ϕ_{i+n} with the pairing of n spatial orbitals with two opposing spin functions. The typical "Box and Tail" form reads [44, 73, 77].

$$\Gamma^{(2)} = \lambda_L | g_1\rangle\langle g_1 | + \lambda_S \sum_{i,k=1}^{n} | \phi_i, \phi_{i+n}\rangle(\delta_{ik} - \frac{1}{n})\langle\phi_k, \phi_{k+n}| + \lambda_T \gamma_T$$

$$\gamma_T = \sum_{\substack{i < j \\ i+n \neq j}} |\phi_i, \phi_j\rangle\langle\phi_i, \phi_j| \tag{6.8}$$

with the "Tail" given by the unpaired orbitals above.[51] The n-dimensional "Box" can be fully diagonalized via

$$g_k = \sum_{l=1}^{n} | \phi_l, \phi_{l+n}\rangle B_{lk}; \;\; k = 1 \ldots n$$

$$B_{lk} = \frac{1}{\sqrt{n}} e^{\frac{i\pi}{n}(2l-1)(k-1)} \tag{6.9}$$

Hence, omitting the "Tail" the diagonal form becomes[52]

$$\Gamma^{(2)} = \lambda_L | g_1\rangle\langle g_1 | + \lambda_S \sum_{k=2}^{n} | g_k\rangle\langle g_k | = \Gamma_L^{(2)} + \Gamma_S^{(2)} \tag{6.10}$$

The emergence of a large eigenvalue was independently recognized by Yang [43], Sasaki [42] and Coleman [44, 78].[53] The eigenvalue[54] $\lambda_L \to \frac{N}{2}$; with $\lambda_S \to$

---

[50]The normalization factor is not explicitly shown in the antisymmetric wedge product symbolized by $\wedge$.

[51]In fact $\lambda_T = \lambda_S$, displaying that all eigenvalues are degenerate except λ_L.

[52]The transformation $\boldsymbol{B}$ becomes a key quantity, see also footnote 55 and the next section.

[53]The story is detailed in Ref. [78].

[54]As can also be deduced from statistical arguments [77], one finds $\lambda_S = \frac{N(N-2)}{4n(n-1)}$; $\lambda_L = \frac{N}{2} - \frac{N(N-2)}{4n}$.

0, as $n \to \infty$, exhibits a dominating eigenvector of $\Gamma^{(2)}$ yielding the condensate wavefunction, i.e. reflects the population of the same pair state analogous to Bose-Einstein condensation [79, 80]. But Coleman [81] concedes: *However, Yang's paper went much farther and showed that it is when $\Gamma^{(2)}$ has a "large" eigenvalue that a fermion system manifests long-range order such as presented in superconductivity.*

We will assign a universal statistical extreme configuration (6.8) as a general initial condition for the thermal evolution of the molecular assembly corresponding to an *UD*. To operate Bloch thermalization, (6.3) and (6.4) must conform to complex symmetry. Analyticity requirements require a complex conjugate in the bra position, see e.g. [82] for details,

$$\tilde{\varrho}_{kl} = |\phi_i, \phi_{i+n}\rangle\langle\phi_k^*, \phi_{k+n}^*| \tag{6.11}$$

Hence one obtains a simple expression for $e^{-\beta\mathcal{L}_B}\tilde{\varrho}$, with each determinantal wavefunction assigned a complex energy, $-i\epsilon_i$ on the second Riemann sheet (note that we have chosen a zero-energy reference).

$$e^{-\beta\mathcal{L}_B}\tilde{\varrho}_{kl} = e^{-\beta H}\tilde{\varrho}_{kl}e^{-\beta H} = e^{\frac{i\beta}{2}(\epsilon_i+\epsilon_k)}\tilde{\varrho}_{kl} \tag{6.12}$$

Employing the *UD* boundary conditions (4.8) one gets

$$e^{-\beta\mathcal{L}_B}\tilde{\varrho}_{kl} = e^{\frac{i\beta}{2}(\epsilon_i+\epsilon_k)}\tilde{\varrho}_{kl} = e^{\frac{i\pi}{n}(i+k-2)}\tilde{\varrho}_{kl} \tag{6.13}$$

The thermalization of $\Gamma^{(2)}$ with the *UD* boundary condition offers a surprise (the large part similarly)

$$e^{-\beta\mathcal{L}_B}\tilde{\Gamma}_S^{(2)} = \lambda_S \sum_{i,k=1}^{n} |\phi_i, \phi_{i+n}\rangle e^{\frac{i\pi}{n}(i+k-2)}(\delta_{ik} - \frac{1}{n})\langle\phi_k^*, \phi_{k+n}^*| \tag{6.14}$$

since the matrix $\boldsymbol{Q}$ with $Q_{ik} = \left(\delta_{ik} - \frac{1}{n}\right)e^{\frac{i\pi}{n}(i+k-2)}$ is nothing but a Jordan block with Segré characteristic n disguised as a complex symmetric matrix representation [83, 84], or

$$\boldsymbol{B}\boldsymbol{Q}\boldsymbol{B}^{-1} = \boldsymbol{J} \tag{6.15}$$

where $\boldsymbol{B}$ is given in Eq. (6.9) and $J_{ik} = \delta_{i+1,k}$; $i, k = 1 \ldots n$. One notes that $\boldsymbol{B}^{-1} = \boldsymbol{B}^\dagger$ transforms the symmetric matrix $\boldsymbol{Q}$ to classical canonical Jordan form, exhibiting the nilpotent property[55]

$$\boldsymbol{J}^n = 0; \ \boldsymbol{J}^{n-1} \neq 0 \tag{6.16}$$

[55] It is interesting to note that $\boldsymbol{B}$ and $\boldsymbol{B}^\dagger$ occur both in connection with (6.9) and (6.15). This coincidence is not obvious.

Equation (6.14) shows that thermalization of a molecular system under the requirements of complex symmetry and the conditions (4.8) transcends from a state representable matrix to a transition density matrix. This condition further imparts that γ_T vanishes in the process.[56] Hence thermalization erases $2n(n-1)$ dimensions and $\frac{1}{2}N(N-2)$ fermion pairings linked to the "Tail" displaying the following matrix structure: state $\rightarrow$ transition

$$\begin{pmatrix} \lambda_L\ 0 & \cdots & 0\ 0 \\ 0\ \lambda_S & & 0\ 0 \\ \vdots & \ddots & \vdots \\ 0\ 0 & \cdots & \lambda_S\ 0 \\ 0\ 0 & & 0\ \lambda_S \end{pmatrix} \rightarrow \begin{pmatrix} 0\ \lambda_S & \cdots & 0\ \lambda_L \\ 0\ 0 & & 0\ 0 \\ \vdots & \ddots & \vdots \\ 0\ 0 & \cdots & 0\ \lambda_S \\ 0\ 0 & & 0\ 0 \end{pmatrix}. \tag{6.17}$$

Finally, some words on the treatment of the nuclear coordinates and the interrelated Born-Oppenheimer approximation. As mentioned earlier the full density matrix should theoretically contain both the electronic and the nuclear degrees of freedom. One set could be traced away in order to study the remaining sub-dynamics and vice versa by a suitable master equation. Since the two sub-systems are entangled through trace projections they are subjected to the mirror theorem[57] [68, 85, 86]. One concludes from this theorem that the mappings between the electron- and the nuclear degrees of freedom, and back, exhibits the same classical canonical form as the one mapping the nuclear- and the electron degrees and back. This reveals the possibility of a dual interpretation of the actual matrix representations of either the quantum-thermal correlated electronic motion or the associated thermal nuclear motion, in principle bypassing the adiabatic formalism.

Hence the classical canonical form Eq. (6.14) can be interpreted as an electronically correlated steady state system, mirrored by the corresponding thermally excited nuclear dynamics parametrized as an $UD(n)$. Note that there are N fermionic degrees of freedom and n nuclear centres. In brief we have proven that an $UD(n)$, as defined by Eqs. (4.5)–(4.7) and Fig. 1, commensurate with Yang's ODLRO, reads

$$e^{-\beta\mathcal{L}_{\mathrm{B}}}\tilde{\Gamma}^{(2)} = e^{-\beta\mathcal{L}_{\mathrm{B}}}\tilde{\Gamma}_L^{(2)} + e^{-\beta\mathcal{L}_{\mathrm{B}}}\tilde{\Gamma}_S^{(2)} + e^{-\beta\mathcal{L}_{\mathrm{B}}}\lambda_T\gamma_T \tag{6.18}$$

and utilizing the transformation $\boldsymbol{B}^{-1}$, see Eqs. (6.13)–(6.16), and the disappearance of the "Tail", assumes the matrix form

[56] Cf. a random phase system.

[57] The mirror theorem was revived by Carlson and Keller [85] and generalized by Löwdin [68, 78]. The mapping between the two-particle electronic coordinate, characterized by the internal and the center-of-mass of the pair, and the apposite nuclear degree of freedom is studied in [86].

$$\begin{pmatrix} 0\ \lambda_S & \dots & 0\ \lambda_L \\ 0\ 0 & & 0\ 0 \\ \vdots & \ddots & \vdots \\ 0\ 0 & \dots & 0\ \lambda_S \\ 0\ 0 & & 0\ 0 \end{pmatrix} = \lambda_L \boldsymbol{J}^{\mathrm{n}-1} + \lambda_S \boldsymbol{J} \tag{6.19}$$

The matrix $\boldsymbol{B}$, see Eq. (6.9), can simply be represented as

$$\mathbf{B} = \frac{1}{\sqrt{n}} \begin{pmatrix} 1 & \omega & \omega^2 & \cdot & \omega^{n-1} \\ 1 & \omega^3 & \omega^6 & \cdot & \omega^{3(n-1)} \\ \cdot & \cdot & \cdot & \cdot & \cdot \\ \cdot & \cdot & \cdot & \cdot & \cdot \\ 1 & \omega^{2n-1} & \omega^{2(2n-1)} & \cdot & \omega^{(n-1)(2n-1)} \end{pmatrix} \tag{6.20}$$

with $\omega = e^{i\pi/n}$. The transition density matrix, Eq. (6.19), signifies a steady state with $dS = 0$. Since erasing information, see the discussion in the previous section and Eq. (6.18), according to Eq. (4.11) generates entropy, one obtains an over-all agreement with Fig. 5 and the conditions for an $UD(n)$. The steady state carries information in terms of the matrix $\boldsymbol{B}$, in analogy with the knowledge given by the eigenvectors of the Schrödinger equation, its symmetries and other properties obtained in quantum chemical applications for instance accurately predicting the properties of a hypothetical polyatomic molecule.

The matrix in (6.20) has a special property that we will return to in the next section. The matrix $\boldsymbol{B}$, see Eqs. (6.9) and (6.10), transforms the second order reduced density matrix to diagonal form, which under specific circumstances such as a sufficiently low temperature leads to a dominating high eigenvalue that signifies Yang's ODLRO and superconductivity. At the same time, $\boldsymbol{B}^{-1} = \boldsymbol{B}^{\dagger}$ transforms the correlated dissipative structure, CDS, at temperature T, Eq. (6.18), to classical canonical form in terms of a Jordan block of Segré characteristic n. Defining the dual preferred basis of n nuclear oscillators, $|\boldsymbol{h} = |h_1, h_2, h_3 \dots h_n\rangle$, mirroring the $N/2$ electron pairs, the canonical form (6.19) is represented by the transformation[58]

$$|\boldsymbol{f} = |\boldsymbol{h}\rangle \boldsymbol{B}^{\dagger} \tag{6.21}$$

In terms of $|\boldsymbol{f}\rangle$ the irreducible part of the CDS, see (6.18), (6.19), obtains as

$$|\boldsymbol{h}\rangle \boldsymbol{Q} \langle \boldsymbol{h}| = |\boldsymbol{f}\rangle \boldsymbol{J} \langle \boldsymbol{f}| = \sum_{k=1}^{n-1} |f_k\rangle\langle f_{k+1}| \tag{6.22}$$

[58]The present mapping appears incompatible since in general $N < 2n$. This is automatically taken care of by the mirror theorem, since the extra eigenvalues in the larger space dimension are filled out by zeros.

The dimension n, see Eq. (4.6), of the irreducible representation of the thermalized density matrix, suggests a unique identification of the molecular system, to be used in the build-up of a particular cell or fragment of an organ.[59] The properties of $\boldsymbol{B}$ and $\boldsymbol{B}^{-1} = \boldsymbol{B}^{\dagger}$ demonstrate fundamental symmetries including a factorization quality of the former uniquely dependent on the number n of active nuclear centers with $|\boldsymbol{h}\rangle = |\boldsymbol{f}\rangle\boldsymbol{B}$ while correlated by the associated dual paired fermionic degrees of freedom. The prime numbers of the factorization remind of the Gödel numbering of mathematical logic indicating a teleonomic trait to evolution. The intimate relation with the stochastic Poisson distribution derives accordingly, see the next section.

7 Consciousness and Quantum Theory

The fundamental aspects of "all there is" contains in addition to the mystery of life, i.e. the theme of this article, the conundrum of consciousness and the contradictions related to the running of the universe. It is often construed that life, consciousness and cosmos are knowingly facing different kind of challenges. We will devote the final sections to these subjects in order to demonstrate that they actually are strongly linked together.

The "premise of the mind is the brain" is a statement indicating that mental states are the same as the physical ones of the brain.[60] Hence every physical brain state determines every subsequent physical state, all within the frame of the second law. This is essentially the Rosenberg dictum of physicalism implying "the physical facts fix all the facts" [87, 88]. However, as we have seen mental- and physical states are not encoded in this standard way rather they occupy a self-assembling cumulative spatio-temporal CDS structure that is based on sequences of transitions. To pick up where we left off in the last section we will add to the previous definition of life: the possible evolution of higher meta-levels, above the genetic code, for linking diverse levels of organization, e.g. between genotypic and phenotypic processes, and possibly to link our knowledge towards an understanding of consciousness and Natural Intelligence [89].

The tangible level of explanation, i.e. suggesting the neural basis of consciousness, detailing an attentional mechanism, was presented by Crick and Koch [90] in their monumental paper of 1990. In this article the authors formulate their thesis: *It seems probable, however, that any one moment some active neuronal processes correlate with consciousness*. This hypothesis is framed today under the name of Neural Correlates of Consciousness, NCC, see Ref. [91] for a recent update, and the

[59]Note that we have left the original coherent wavefunction in (6.5) for the statistical derivation [77] that provides the necessary link with an original ensemble representative density matrix technically derivable through an appropriate sub-dynamics.

[60]The stringent philosopher, Rosenberg [87], views biological issues within the same epistemological framework as the philosophy of physics, primarily based on the second law. It is of course beyond this summary to render a fair and ample account of the detailed ingredients that goes into such a stance.

working rationale: *The minimum neural mechanisms jointly sufficient for any one specific conscious experience*. In particular Crick and Koch focus on the binding problem and the way a responding neuron can be considered to be binding a set of points in relation to working memory and appropriate responses of objects in the visual field as an emerging process during evolution. A critical detail in the proposed theory is the prevailing neural oscillations of about 40 Hz (35–75 Hz) that occurs in cortical areas, serving as an underlying mechanism for neural synchronization and phase locking.

Since our leitmotif is to provide a bottom-up argument for a possible understanding of general life processes, including microscopic self-organization of fundamental quantum-thermal correlations all the way from DNA-RNA and the genetic code to conscious communication, it is necessary to explain the appearance of the above-mentioned neural activity, its meaning and how it emerges, perhaps also being able to do some linkage with established top-down knowledge. The oscillations appear as a stochastic Brownian-like background for conscious communication. Consider a human cerebrum, right or left, of about 600 g each, containing a high percentage (about 85%) of water imparting $x_A \approx 27$ moles of brain entities and $N_B = x_A N_A$ Brownian particles, where N_A is Avogadro's number. Close to equilibrium one obtains the flux from Gibbs distribution

$$\frac{\Delta N_B}{\bar{N}_B} = \frac{1}{\sqrt{\bar{N}_B}} \tag{7.1}$$

where $\bar{N}_B$ is the average number of particles and $\Delta N_B^2 = \overline{(N_B - \bar{N}_B)^2}$ is the variance.[61] Application of Eq. (4.6) and (4.10) gives the flux of Brownian-like motion, noting that in Eq. (4.6) the factor 4π occurs because n on the left-hand side concerns total fluxes while the right-hand side refers to flux per solid angle,

$$\frac{\Delta n}{n} = \frac{1}{\sqrt{\bar{N}_B}} \tag{7.2}$$

identifying the outgoing flux as in Eq. (7.2). Hence one obtains from[62]

$$\frac{\sqrt{x_A N_A}}{4\pi} = \frac{k_B T}{h} t_{\text{rel}} \tag{7.3}$$

a Crick-Koch frequency of about 20 Hz corresponding to $t_{\text{rel}} = 50$ ms. This agreement is of course fortuitous beyond expectation, but one might nevertheless anticipate that there should lie an ounce of truth in these orders of magnitudes.

---

[61] For the Poisson distribution the variance equals the mean.

[62] Equations (4.6) and (4.10) do not guarantee that the *UD* condition holds. In this particular case the Brownian-like motion contains no information.

With such a Brownian-type motion as background for a higher order dynamics, it is a straightforward task to implement the correlated dissipative structures,[63] CDS's, as represented by (6.21), (6.22), or (6.19). For instance, choosing the form (6.22)

$$\rho = \frac{1}{\sqrt{n-1}} \sum_{k=1}^{n-1} | f_k\rangle\langle f_{k+1} |; \ \mathrm{Tr}\{\rho^{\dagger}\rho\} = 1 \tag{7.4}$$

one can build a Liouvillian, subject to Eq. (6.2) properly continued to the second Riemann sheet in the complex energy plane. The various CDS entities may correspond to molecular aggregates, cells etc. characterized by a Quality Value n and a given relaxation time $\tau = \tau_{\mathrm{rel}}$ subject to condition (4.6), with $\omega_0 = 1/\tau_{\mathrm{corr}}$

$$\int \omega_0 \tau d\Omega = n \tag{7.5}$$

We will not repeat the full deduction of the super propagators and resolvents as there is a close analogy with the derivations in Sect. 6. Suffice it to say that an irreducible operator appears with the nilpotent properties $J^m = 0$ and $J^{m-1} \neq 0$ yielding the results

$$\mathcal{P} = (\omega_0\tau - i)I + iJ \tag{7.6}$$

$$\mathcal{G}(t) = \mathrm{e}^{-i\mathcal{P}\frac{t}{\tau}}; \ \mathcal{G}_R(\omega) = (\omega\tau J - \mathcal{P})^{-1} \tag{7.7}$$

$$\mathrm{e}^{-i\mathcal{P}\frac{t}{\tau}} = e^{-i\omega_0 t} e^{-\frac{t}{\tau}} \sum_{k=0}^{m-1} \left(\frac{t}{\tau}\right)^k \frac{1}{k!} J^k \tag{7.8}$$

$$(\omega\tau I - \mathcal{P})^{-1} = \sum_{k=1}^{m} [(\omega - \omega_0)\tau + i]^{-k} (-iJ)^{(k-1)} \tag{7.9}$$

The expansions are finite, limited by the dimension m of the irreducible part, generally different from n, with the occurrence of higher order poles in Eq. (7.9) providing a build-up of a polynomial in front of the decay factor in (7.8). Hence the usual microscopic law of evolution $dN(t) = -1/\tau N(t)dt$ modifies according to the highest power $m-1$ of J, see e.g. [17, 21, 24, 89, 92] for more details

$$dN(t) = t^{m-2}\left(m - 1 - \frac{t}{\tau}\right) N(t)dt; \ dN(t) > 0; \ t < (m-1)\tau \tag{7.10}$$

One notes that the irreducible part of (7.6) results in sequences of transitions between m CDS's, each of dimension n, which according to (7.10) generates a new

[63] Note that different situations lead to various Jordan forms and their Segré characteristics as well as differing magnitudes of λ_L, λ_S. For simplicity we have made the choice (6.22).

"communication" time scale $\tau_{com} = (m-1)\tau_{rel}$ or $m\tau_{rel}$ adding the decay time. We will denote the super propagator a Correlated Dissipative Ensemble, CDE, i.e. a statistical ensemble associated with a perturbation out of equilibrium representing assembled stochastic features adapted for Complex Enough Systems, CES's, with evolved size- and time scales at precisely given temperatures.

The formulation suggests "molecular communication", i.e. the transmission of crucial molecular traits providing cells with their functional duties to be extended from the microscopic level to generating significant cellular information. Obviously, there might be focal relations between the primal factorization properties of n and m. It is also interesting to realize the importance of the transformation $\boldsymbol{B}^\dagger$, bringing the CDS, or the CDE, to classical canonical form, and $\boldsymbol{B}$ documenting the ODLRO information via transforming the steady state flow back to its localized binding sets of points.

To display the symmetries of the matrix $\boldsymbol{B}$ consider $n = 12$, showing the result below as an abridged diagram, in which only the dimensions of the vector cycles are listed. To simplify the diagram one may remove the first column of ones. The subsequent graph comprises 11 columns that looks like

$$\begin{array}{ccccccccccc}
 & & & & & 2 & & & & & \\
 & & & 3 & & 2 & & 3 & & & \\
 & & 4 & & & & & & 4 & & \\
 & 6 & & 3 & & 2 & & 3 & & 6 & \\
12 & & 4 & & 12 & & 12 & & 4 & & 12 \\
 & 6 & & 3 & & 2 & & 3 & & 6 & \\
 & & 4 & & & & & & 4 & & \\
 & & & 3 & & 2 & & 3 & & & \\
 & & & & & 2 & & & & &
\end{array} \tag{7.11}$$

The highly specific characteristics above suggests a teleonomic process, i.e. a natural law governed by an evolved program [22], and an established communication protocol. We have in previous work proved the fundamental basis and relevance of the Poisson distribution in this context, proposing a communication concept,[64] *Communication Simpliciter*, that shares some resemblance with the dynamics of a Call or Contact Centre. Such analogies have been carried out in some detail particularly in relation to Trehub's celebrated Retinoid Model of the cognitive brain [93, 94], in terms of autapses, specific neurons with chemical synapses connected onto themselves. This model, although originally entirely classical, have played an important role in our Gödelian[65] self-referential network [92, 95]. Teleonomic principles have also been analysed and compared with the views of Jacques Monod as expressed in his classic essay "Chance and Necessity" [23, 24]. Some surprising applications and explanations concern the Necker Cube and Spinning Dancer illusions and the paradox of upside-down vision [92].

---

[64] Communication taken in a restricted sense, i.e. without semantics employing only syntax and pragmatics.

[65] Attaching the name "Gödel" to the network has been explained in several articles [89, 95], see e.g. Chap. 4 in [95].

8 The Universe and Quantum Theory

Dirac derived a consistent relativistic quantum theory for electrons, showing in particular the entanglement between matter and anti-matter; for a recent history and update, see [96]. However, Einstein's theory of general relativity and quantum mechanics still remain incompatible. Hence it would seem to be an impossible undertaking to bring together the topics expressed in the header above. This task becomes even more difficult in view of the existing cosmological contradictions, such as the accelerating expansion of the universe, the formation of black holes via a collapsing star, the information paradox, the nature of the black hole singularity etc., see e.g. Mersini-Houghton [97, 98] for a critical appraisal. Solutions to these quandaries should most likely be found in appropriate modifications of classical black hole dynamics to incorporate fundamental quantum mechanical effects.

Our present incipient aim has been to promote an evolutionary procedure, representing and interpreting abiogenesis, and encompassing the whole biosphere from first principles. A first step in this direction was the development of ZEUS, a Zero Energy Universe Scenario, conveying a plausible picture [95] commensurate with Darwin's theory of evolution, from the microscopic ranks to the cosmological domain, see also [70]. A natural starting point for a quantum mechanical treatment à la ZEUS would be to extend the relativistic energy-momentum formulation to a non-positive definite metric, while keeping the fundamental commutation relations with their conjugate variables intact. This simple directive is a reminder of Kant's dictum more than two hundred years ago: *space and time are the two essential forms of human sensibility*. In other words, the aphorism states that evolved life forms perceive and experience phenomena in our environment, set in this perspective. Hence physics, the knowledge of nature, connect material degrees of freedom such as energy-momentum, with their conjugate immaterial partners space-time. In QM they are correlated through the usual commutation relations recognized by every physics student today.

In terms of the energy given by the mass relation $E = mc^2$, m the mass, c the velocity of light, τ the time, and $\vec{v}$ the velocity, the classical Klein-Gordon equation for a non-zero rest-mass particle reads, with $E_0 = m_0c^2$, m_0 the rest-mass and $\vec{p} = m\vec{v}$,

$$-\frac{E_0^2}{c^2} = \vec{p}^2 - \frac{E_{\text{op}}^2}{c^2} \tag{8.1}$$

The matching relation for the conjugate variables introducing the familiar eigentime expression τ_0 obtains

$$-c^2\tau_0^2 = \vec{x}^2 - c^2\tau^2 \tag{8.2}$$

In order to be consistent with QM, we extend the definition of the usual conjugate pairs energy-time and momentum-space to their operator equivalents; the time operator is defined when the energy and time intervals are $(-\infty, +\infty)$,

$$E_{\mathrm{op}} = i\hbar\frac{\partial}{\partial t}; \quad \vec{p}_{\mathrm{op}} = -i\hbar\vec{\nabla}_{\vec{x}} \tag{8.3}$$

and

$$\tau = t_{\mathrm{op}} = -i\hbar\frac{\partial}{\partial E}; \quad \vec{x}_{\mathrm{op}} = i\hbar\vec{\nabla}_{\vec{p}} \tag{8.4}$$

Using Dirac's trick of "taking the square root of the equation" one can rewrite the observables in matrix form with (8.1), (8.2) corresponding to the associated secular equations. One crucially important fact is that relativity theory necessitates a presentation based on non-positive definite metrics [68]. As shown in [99], the latter is commensurate with complex symmetric matrix representations and corresponding dissipative dynamics. For this reason, one admits a complex symmetric ansatz, with the negative sign, $-i$, in the off-diagonal elements, chosen by convention,

$$\begin{pmatrix} E_{\mathrm{op}} & -i\vec{p}_{\mathrm{op}}c \\ -i\vec{p}_{\mathrm{op}}c & -E_{\mathrm{op}} \end{pmatrix} \tag{8.5}$$

The secular equation corresponding to (8.5) is precisely (8.1) with the two eigenvalues $\pm\lambda$, given by $\lambda^2 = m_0^2c^2$, with $m_0 \neq 0$. Similarly, one obtains from (8.4) that Eq. (8.2) turns into the secular equation of the matrix

$$\begin{pmatrix} c\tau & -i\vec{x} \\ -i\vec{x} & -c\tau \end{pmatrix} \tag{8.6}$$

with the eigenvalues $\pm c\tau_0$ displaying the two-time directions, not forgetting, the left- and right-handed coordinate systems linked by space inversion symmetry.

From (8.1), (8.2) one further gets the relativistic space-time-scales of special relativity, valid both in CM and QM

$$m = \frac{m_0}{\sqrt{1-\beta^2}} \tag{8.7}$$

with $\beta = v/c = p/mc$, here with the momentum $\vec{p}$ directed along the x-axis, and

$$\tau = \frac{\tau_0}{\sqrt{1-\beta^2}}; \quad x = \frac{x_0}{\sqrt{1-\beta^2}} \tag{8.8}$$

It is interesting to consider the limit $m_0 \to 0$ in (8.5), since something unusual occurs when inserting the relation $E = pc$ resulting in the degenerate matrix

$$\begin{pmatrix} p/c & -ip/c \\ -ip/c & -p/c \end{pmatrix} = p/c\begin{pmatrix} 1 & -i \\ -i & -1 \end{pmatrix} \tag{8.9}$$

Since two column vectors in the matrix are linear dependent, it is similar to the classical canonical form

$$2p/c\begin{pmatrix} 0 & 1 \\ 0 & 0 \end{pmatrix} \tag{8.10}$$

Since the photon complies with $c = |\vec{x}|/\tau$, the conjugate matrix, (8.6), writes

$$\begin{pmatrix} c\tau & -i|\vec{x}| \\ -i|\vec{x}| & -c\tau \end{pmatrix} = c\tau\begin{pmatrix} 1 & -i \\ -i & -1 \end{pmatrix} \rightarrow 2c\tau\begin{pmatrix} 0 & 1 \\ 0 & 0 \end{pmatrix} \tag{8.11}$$

Hence the complex symmetric matrices (8.10) and (8.11) are all Jordan blocks with Segré characteristic 2. The interpretation is evident as the linear dependency constrains space-time with one less spatial dimension, i.e. along the x-axis. Consequently, the longitudinal degree of freedom is absent for a zero rest-mass particle with speed c, such as the photon! This indicates that zero- and non-zero rest-mass particles behave fundamentally different, which will become of crucial importance in the general case.

It is not difficult to extend our conjugate operator formulation, to the general case. With μ the gravitational radius, G the gravitational constant, $v = p/m$, $r = |\vec{x}|$, M a spherically symmetric (non-rotating) mass, independent of m, one finds

$$\begin{pmatrix} m(1-\kappa(r)) & -iv \\ -iv & -m(1-\kappa(r)) \end{pmatrix} \tag{8.12}$$

with[66]

$$m\kappa(r) = \frac{m\mu}{r}; \; \mu = G\cdot\frac{M}{c^2} \tag{8.13}$$

Since the area velocity multiplied by m is a constant of motion, one obtains for local circular motion the boundary condition

$$v = \kappa(r)c \tag{8.14}$$

which incorporated in Eq. (8.12) yields the similarity, provided $\kappa(r) \neq 1/2$,

$$m\begin{pmatrix} (1-\kappa(r)) & -i\kappa(r) \\ -i\kappa(r) & -(1-\kappa(r)) \end{pmatrix} \rightarrow m\begin{pmatrix} \sqrt{1-2\kappa(r)} & 0 \\ 0 & -\sqrt{1-2\kappa(r)} \end{pmatrix} \tag{8.15}$$

[66]The matrix (8.12) has a direct link with Gödel's self-referential paradox, see footnote 65.

and straightforwardly the so-called Schwarzschild[67] line element, leaving out the term $r^2 d\Omega^2$, and $m_0 \neq 0$,

$$-c^2 ds^2 = -c^2 d\tau^2 (1 - 2\kappa(r)) + dr^2 (1 - 2\kappa(r))^{-1} \tag{8.16}$$

However, at $\kappa(r) = 1/2$, i.e. at the Schwarzschild radius $r = 2\mu$, one encounters an old 'friend', i.e. a Jordan block of Segré characteristic 2 (independent of $m \neq 0$)

$$\frac{1}{2} m \begin{pmatrix} 1 & -i \\ -i & -1 \end{pmatrix} \rightarrow \begin{pmatrix} 0 & m \\ 0 & 0 \end{pmatrix} \tag{8.17}$$

The singular behaviour at the Schwarzschild radius suggests a limit, where material particles, $m_0 \neq 0$, loose its material properties and spatiotemporal characteristics, defining black-hole-like objects with at most rotational degrees of freedom.[68] For photons at $r \neq 2\mu$, $ds = 0$. With $A = \sqrt{1 - 2\kappa(r)}$, see [100] for details,[69] one obtains

$$cAd\tau \begin{pmatrix} 1 & -i \\ -i & -1 \end{pmatrix} \rightarrow 2cAd\tau \begin{pmatrix} 0 & 1 \\ 0 & 0 \end{pmatrix} \tag{8.18}$$

The energy formula for photons, consistent with Einstein's law of gravitational bending, shows that the metric is conditionally singular for both mass and massless particles at $\kappa(r) = 1/2$, and with $A = 0$ for photons.

In brief we have proved that the a priori selection of conjoining quantum related conjugate partners, such as energy-time, lead to simple operator relations, in direct agreement with the theory of relativity. The formulation incorporates the appearance of higher order mathematical singularities at precise physical conditions in agreement with such constraints as the limit velocity of light, the associated loss of the longitudinal spatial dimension of electromagnetic waves and the emergence of a Quantum Black Hole, QBH. Energetic particles lose their rest-mass at the event horizon, $r = 2\mu$, see (8.17).

9 Black Holes and Quantum Theory

In the discussion above we have assumed a spherical non-rotational QBH which in principle could rest in its quantum mechanical ground state. However, the QBH will not rest in isolation since it exchanges matter and entropy with its surroundings. It

---

[67] As it is usually projected today, for more details, see [70, 95, 99]. Note that (8.16) also may be written as a secular equation of a complex symmetric matrix, see [99, 100].

[68] In retrospect the singularity shares the same self-referential conundrum as we associate with Gödel's incompleteness theorem [89, 95].

[69] The formula (28) in [99] has been corrected in [100].

is important to remember that pioneering quantum mechanics, although it started as a revelation of quantization in the microscopic province, also tows a continuum. Obviously, our aim here has been focused on incorporating the latter in which there are embedded resonances, similar to bound states, permitting a new freedom of interpretation such as principal time scales, necessary temperature dependences, thermal- and quantum correlations, microscopic self-organization, Einstein's laws of relativity, teleonomic communication and Poisson statistics.

To honour the theme in this article the obvious suggestion would be to formulate an all-inclusive, universal thermodynamics from the notion of the Universal Device, $UD(n)$. Since a QBH only exhibits rotational degrees of freedom, it is reasonable to assume a naïve rotor model and that a simple spectral analysis might be informative.[70] For instance, the symmetric rotor approximates as

$$E(J, M_J, K) = BJ(J+1) + (A-B)K^2$$

with M_J and K specifying the components of the angular momentum along the laboratory- and the principal molecular axis and with the rotational constants

$$A = \frac{h}{8\pi^2 I_{||}}; \; A = \frac{h}{8\pi^2 I_{\perp}}$$

where $I_{\perp}$, $I_{||}$ are the moments of inertia, orthogonal and parallel, respectively to its figure axis. For various values of A and B one may describe oblate and prolate objects and the $(2J+1)^2$ degenerate spherical rotator when $A = B$.

Note that the structure displayed by a QBH acquiesces to the $UD(n)$ analysis[71] in Sects. 4, and 6. The rotational spectrum should mimic Eqs. (6.6) and (6.10), with a large eigenvalue of J. Since the structure is almost spherical one expects $M_J = 0$ and $K = 0$. Hence the quantum state is characterized by the quantum numbers $2J+1$, $M_J = K = 0$. After Bloch thermalization one obtains the familiar transitions between the spectrally identified states, i.e. the first term cf. (6.10) and (6.19), corresponds to a transition between $K = J$ and $K = -J$, while the second term display all the other transitions. Since the eigenvalue corresponding to the thermalized density matrix, (6.6) is zero, $K = M_J = 0$, the QBH shows an unusual structure. It is organised as a rotational system, exhibiting an enormous rotational energy, large J, but with the rotational z-component in any direction of space equal to zero. The interpretation of this is quite surprising in view of the conjugate relationship that holds between angular momentum and orientation. For instance, an object subject to immensely high quantum numbers J, but with a zero z-component attached to any arbitrary direction in space imparts a fixed orientation with the consequence that an observer in our material Universe experiences to be enfolded by the rotational QBH,

[70]There is no relative motion here, then again Sects. 6 and 7 applies. A discussion, including the Kerr metric and a possible quantum analogue of Penrose's cosmic censorship hypothesis, can be found in [70].

[71]The mirror relation between N, the number of fermions, and n, the number of nuclear centers naturally disappears in the QBH, but a bosonic analysis nevertheless yields a number, see [77].

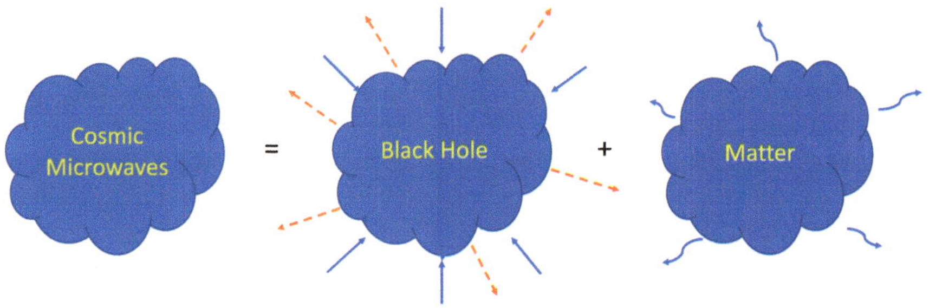

Fig. 6 Display of the quantum black hole as a universal device

i.e. a Quantum White Hole, QWH, producing a cosmological or particle horizon more or less equivalent to the event horizon. A possible scenario is illustrated in Fig. 6.

One of the stumbling blocks in formulating a consistent cosmological perspective of the evolving Universe refers to the attractive characteristics of gravitational interactions. This enigma is excellently articulated by Penrose [101]. Not only does he discuss pre-Big Bang proposals, but his main contention concerns the conundrum related to the second law of thermodynamics and the amount of disorder introduced by Big Bang and what happens with the passage of time. His solution is a new testimonial, Conformal Cyclic Cosmology, CCC, employing conformal and angle-preserving diagrams by bringing in a smooth conformal geometry in relation to the second law[72] and the entropy concept. It is not always easy to follow the Penrose argumentation, although they, despite sometimes being called heretical, such as the Penrose-Lucas argument, the cosmic censorship idea and the Orch OR theory, are original, ingenious, imaginative and truly inspirational for his readers.

To demonstrate an alternative to the cosmological paradoxes above, we return to the Zero Energy Universe Scenario, ZEUS [9]. As shown in Fig. 6, we select the incoming degrees of freedom from the Cosmic Microwave Background, CMB, that sweep across cosmos as a relic from the past. Note also that material particles according to (8.17) lose their material properties at $r = 2\mu$ imparting that the QBH singularity is protected from massive particles. The steady state character of the non-material degrees of freedom passing the event horizon impacting the QBH by definition prompts $dS = 0$. The QBH receiving the incoming flux through the particle horizon becomes an $UD(2J + 1)$, with $dS_{UD} = -dS_Q$, where dS_Q acquires from significant entropy productions, based on the ejected baryonic and leptonic matter[73] that is occupying the visible Universe. In this formulation dark energy associates with the QBH or rather the QWH, while dark matter most likely consists of material or non-material degrees of freedom not in the visible part of the Universe. Another

[72]Penrose discusses space-like futures with massless particles calling our Universe an aeon in an endless sequence of aeons.

[73]In this formulation concepts such as Hawking radiation and the Bekenstein-Hawking entropy have not been considered here. Whether and how they might become applicable also in a quantum theoretical description remain a topic for the future.

conundrum is the matter- antimatter asymmetry. A simple explanation could be the symmetry characteristics of Eqs (8.6) and the equivalent form including gravity, remembering that there appear the two-time directions each with a proper link to spatial handedness. Matter spatially evolving forward in time, is by symmetry, parity-linked to evolving antimatter backwards in time, entangled via a QBH. Combining two-time directions and two parity related space dimensions, imparts about 75% hidden energy entangled with the preferred space-time through the QBH.

10 Conclusion

One might justly ask oneself whether the present undertaking and analysis should be developed without referring to Schrödinger's classic [102]. In the foreword of the Canto imprint collection, Roger Penrose, like many of his legendary historic predecessors in science, professed his penchant for the book's impact on his life and future thinking. Although being quickly outdated, shortly after its first edition appeared in 1944, its focus expressed in the query: *how can the events in space and time which take place within the spatial boundary of a living organism be accounted for by physics and chemistry*? His prophetic anticipation just preceded the birth and proliferation of a whole new profession of bio-molecular science. It begs no explanation to insert its legitimacy in the present narratorial, even if it has led to many side tracks and crucial misinterpretations of the appropriacy of the realm, where quantum mechanics reigns uncontested. The possible explanation of e.g. the genetic code, or at the time of the publication of the classic, the storing and the communication of information in complex molecular aggregates, within the demarcation of the second law of thermodynamics, did not offer any justifiable inroads to the provenance and evolution of learning processes.

Most thermodynamical treatments of biological systems do not explicitly consider temperature and associated time scales. One interesting exception, in addition to the work of England [48], mentioned in Sect. 3, is a recent study of Brownian motion and the temperament of living cells [103]. Although not specifically oriented towards quantum interpretations, its theoretical framework is the Nakajima-Zwanzig version of the Liouville equation, providing the notion of cell temperament as a biophysical parameter driving the motion of living biological entities, in close analogy with the standard concept of temperature.

Although the second law of thermodynamics sustains that life could arise from non-living matter, it is clear, as elucidated here, that it does not explain chemical evolution or the appearance of abiogenesis. It does not yield a strategy or algorithm in selecting a particular protein from an infinitude of specimen. Natural Selection, NS, requires a fundamental physical principle that provides Nature with an indigenous composition for Life. Our present formulation supports abiogenesis as an innate process that is ingrained with the gradual transformation of non-living to living entities.

We have proceeded to establish a teleonomic approach to describe strongly correlated biological systems with germane time scales at precise temperatures. The thesis extends fundamental correlations, via entropy production and the performance of a Universal Device, with the objective to set up communication between Complex Enough Systems, CES's. $UD(n)$, subjected to *Communication Simpliciter*, operates under a Poissonian Call-Centre stochastic background, where the syntax and possibly pragmatics,[74] provide authentic physical channels for exchange of information. Examples of biological $UD(n)$'s are cells, organs, brains, human beings, cities, societies and black holes.

An essential ingredient of the quantum-chemistry-life principle is a physical rule that maps the self-referential Gödel-like formulation onto a higher order singularity through a logical negation and its associated bias table, which converted to a complex symmetric matrix, exhibits irreducible block diagonal matrices with Segré characteristic larger than one.[75] Extensions to high dimensional CES's suggest the appearance of such Jordan blocks of arbitrary high orders leading to the transformations $\boldsymbol{B}^{\dagger}$ and $\boldsymbol{B}$. It has also been demonstrated that the dimension of these transformations provides a quality index n for the $UD(n)$, such as a cell defining its position in the hierarchy of the organism. Furthermore $\boldsymbol{B}$ conjectures a coding protocol for *Communication Simpliciter* [17, 92, 95, 100], including an elemental rendition of our brain functions as expressed by the Necker cube illusion and the paradox of upside-down vision [92].

As Darwinism supports the evolution of awareness, perception, cognition, feelings and emotions, i.e. consciousness, a brief mentioning of David Chalmers[76] distinction between the so-called Hard Problems and the Easy ones might have been on its place [104]. Solving the latter, see above, will no doubt contribute to understanding the former, nevertheless this philosophical question is best left elsewhere.

Our second, likewise significant, aim has been, not just to address QM as a default option, but rather to demonstrate its fundamental importance in current areas such as technology, biology, evolution, life, consciousness, and the Universe. Our final conclusion is that the second law of thermodynamics, commensurate with the evolution of life bearing processes, cannot explain abiogenesis within the existing paradigms of physics. This necessitates a quantum-chemistry-life-principle with a built-in self-referential physical law as derived and demonstrated here.

Acknowledgements I am grateful to the Chair of QSCP XXIII, Prof. Liliana Mammino, and the Co-Chair Jean Maruani for generously allowing me to present this work in the present proceedings from the meeting. My indebtedness for constructive suggestions and recommendations goes to Prof. Alia Tadjer and Prof. Jean Maruani. This work has over time been supported by the Swedish Natural Science Research Council, the Swedish Foundation for Strategic Research, The European Commission and the Nobel Foundation.

[74]Note that communication is here taken in a restricted sense in concert with the concept of semiology which excludes semantics, one of three components of semiotics, see also [24, 92].

[75]Regarding self-references and Gödel see footnotes 65, 66, 68 and in particular chapter 4 in [95].

[76]Note that Chalmers' concept of the Hard Problem presupposes a dualistic attitude. Various recent suggestions display 'the long and winding road' from monism to dualism.

References

1. Peretó J, Bada LB, Lazcano A (2009) Charles Darwin and the origin of life. Orig Life Evol Biosph 39(5):395
2. Miller SL (1953) A production of amino acids under possible primitive earth conditions. Science 117(3046):528
3. Miller SL, Urey HS (1959) Organic compound synthesis on the primitive earth. Science 130(3370):245
4. Trainer MG (2013) Atmospheric prebiotic chemistry and organic hazes. Curr Org Chem 17:1710
5. Brown A (2005) J. D. Bernal: the sage of science. Oxford University Press, Oxford
6. Bernal JD (1967) The physical basis of life. Wiedenfeld and Nicholson, London
7. Oparin AI (1924) Proiskhozhdenie zhizny. Izd. Moskowhii Rabochii, Moscow
8. Oparin AI (1938) The origin of life. Macmillan, New York
9. Haldane JBS (1929) The origin of life. The Rationalist Annual, Watts CA, pp 3–10
10. Cairns-Smith AG (1982) Genetic takeover: and the mineral origins of life. Cambridge University Press, Cambridge
11. Hoyle F, Wickramasinge NC (1981) Evolution from space. J M Dent, London
12. Thaxton CB, Bradley WL, Olson RL (1984) The mystery of life's origin: reassessing current theories. Philosophical Library Inc, New York
13. Jekel JF (1985) The mystery of life's origin: reassessing current theories. Review. Yale J Biol Med 58(4):407
14. Prigogine I (1996) The end of certainty: time, chaos, and the new laws of nature. The Free Press, New York
15. Obcemea CH, Brändas EJ (1983) Analysis of Prigogine's theory of subdynamics. Ann Phys 151:383
16. Atmanspacher H, Bishop RC, Amann A (2001) Extrinsic irreversibility in probabilistic dynamical laws. In: Khrennikov A (ed) Proceedings of the conference foundations of probability and physics, QP-PQ: Quantum probability and white noise analysis, vol XIII, p 50
17. Brändas EJ (2017) The origin and evolution of complex enough systems. In: Tadjer A, Pavlov R, Maruani J, Brändas EJ, Delgado-Barrio G (eds) Quantum systems in physics, chemistry, and biology, vol 30. Springer, Berlin, p 413
18. Nicolis G, Prigogine I (1977) Self-Organization in nonequilibrium systems. From dissipative structures to order through fluctuations. Wiley, New York
19. Eigen M (1971) Self-organization of matter and the evolution of biological macromolecules. Die Naturwissenshaften 64:465
20. Kumicak J, Brändas EJ (1993) Complex scaling and Lyapunov converters. Int J Quant Chem 46:391
21. Brändas EJ (2012) Examining the limits of physical theory: analytical principles and logical implications. Adv Quant Chem 63:33
22. Mayr E (2004) What makes biology unique?. Cambridge University Press, New York
23. Monod J (1971) Chance and necessity: an essay on the natural philosophy on modern biology. Alfred A. Knopf Inc., New York
24. Brändas EJ (2018) Molecular theory of the genetic code. Mol Phys 116(19–20):2622
25. Abbott D, Davies PCW, Pati KP (2008) Quantum aspects of life. Imperial College Press, London
26. Brändas EJ (2017) Löwdin—father of quantum chemistry. Mol Phys 115(17–18):1
27. Prigogine I (1980) From being to becoming time, and complexity in physical sciences. W. H. Freeman and Company, San Francisco
28. Atmanspacher H, Primas H (2003) Epistemic and ontic quantum realities. In: Castell L, Ischebeck O (eds) Time, quantum and information. Springer, Berlin, Heidelberg, p 301
29. Bishop RC (2004) Non-equilibrium statistical mechanics, Brussels-Austin style. Stud History Philos Mod Phys 35:1

30. Brändas EJ (1993) Resonances and microscopic irreversibility: an introduction. Int J Quant Chem 46:339
31. Earley JE Sr (2006) Some philosophical influences on ilya Prigogine's statistical mechanics. Found Chem 8(3):271
32. Bricmont J (1995) Science of chaos or chaos in science. Physicalia Magazine 17:159
33. Lanford OE (1975) Time evolution of large classical systems. Dynamical systems, theory and applications. Lecture notes in physics, vol 38, p 1
34. Gallagher I, Saint-Raymond L, Texier B (2013) From Newton to Boltzmann: hard spheres and short-range potentials. In: Zürich lectures in advanced mathematics. European Mathematical Society, Zürich
35. Simonella S, Spohn H (2015) Bull Am Math Soc 52(3):533
36. Brändas EJ, Chatzidimitriou-Dreismann CA (1995) Fundamentals, logical structure and unification of natural sciences. Int J Quant Chem 53:95
37. Balslev E, Combes JM (1971) Spectral properties of many-body Schrödinger operators with dilatation-analytic interactions. Commun Math Phys 22:280
38. Nicolaides CA, Brändas EJ (eds) (2010) Unstable states in the continuous spectra, part I: analysis, concepts, methods, and results. Adv Quant Chem 60; ibid. (2012) Unstable states in the continuous spectra, part II: interpretation, theory and applications. 63
39. Moiseyev N (2011) Non-Hermitian quantum mechanics. Cambridge University Press, New York
40. Löwdin P-O (1964) Some aspects of quantum biology. In: Biopolymers symposia, vol 1, p 293
41. Husimi K (1940) Some formal properties of the density matrix. Proc Phys Math Soc Jpn 22:246
42. Sasaki F (1965) Eigenvalues of fermion density matrices. Phys Rev 138B:1338
43. Yang CN (1962) Concept of off-diagonal long-range order and the quantum phases of liquid helium and of superconductors. Rev Mod Phys 34:694
44. Coleman AJ (1963) Structure of fermion density matrices. Rev Mod Phys 35:668
45. Zeilinger A (2010) Dance of the photons from einstein to quantum teleportation. Farrar Strauss and Giroux, New York
46. Crooks GE (1999) Entropy production fluctuation theorem and the nonequilibrium work relation for free energy differences. Phys Rev E 60:2721
47. Jarzynski C (2007) Equilibrium free-energy differences from nonequilibrium measurements: a master equation approach. Phys Rev E 56:5018
48. England JL (2013) Statistical physics of self-replication. J Chem Phys 138:121923
49. Löwdin P-O (1951) A note on the quantum-mechanical perturbation theory. J Chem Phys 19:1396
50. Nakajima S (1958) On quantum theory of transport phenomena—steady diffusion. Prog Theor Phys 20:948
51. Zwanzig R (1960) Ensemble method in the theory of irreversibility. J Chem Phys 33:1338
52. Huang K (1963) Statistical mechanics. Wiley, New York
53. Barrett HH, Myers KJ (2004) Foundations of image science. Wiley, Hoboken New Jersey
54. Belkić D (2005) Quantum-mechanical signal processing and spectral analysis. Institute of Physics Series in Atomic- and Molecular Physics, Institute of Physics Publishing, Bristol
55. Levitina T, Brändas EJ (2009) Filter diagonalization: filtering and postprocessing with prolates. Comp Phys Comm 180:1448
56. Wiener N (1948) Cybernetics or control and communication in the animal and the machine. Wiley, New York
57. Brändas EJ (1997) Resonances and dilatation analyticity in Liouville space. Adv Chem Phys 99:211
58. Brändas EJ (2003) New perspectives in theoretical chemical physics. Adv Quant Chem 42:383
59. Gibbons M, Limoges C, Nowotny H, Schwartzman S, Scott P, Trow M (1994) The new production of knowledge. The dynamics of science and research in contemporary societies. Sage Publications Ltd., London

60. Feder T (2000) Physics graduate programs train students for industrial careers. Phys Today 53(8):39
61. Boström R, Brändas E, Kullander S (1998) Promoting university—industry links by a new type of Ph.D. education: The AIM Graduate School. In: Conference proceedings, Vienna, 4–6 November 1998 ESA, vol 432. Academic and Industrial Cooperation in Space Research, p 181
62. Brändas EJ (2003) Advanced instrumentation and measurements: a research education programme at Uppsala University. Eur Phys Educ Netw 7:147
63. Rittby M, Elander E, Brändas E, Barany A (1984) Resonance structure in charge transfer cross sections: an application to the $N^{3+} + H \rightarrow N^{2+} + H^{+}$ reaction. J Phys B: At Mol Phys 17:L677
64. Brändas EJ (1984) Time evolution of chemical systems far from equilibrium. In: Lecture Notes in Physics, vol 211, p 307
65. Enss V (1984) Summary of the conference and some open problems. In: Lecture Notes in Physics, vol 211, p 351
66. Carleman T (1944) L'Intégrale de Fourier et questions que s'y rattachent. Mittag-Leffler Institute of Publications, Uppsala
67. Lützen J (1982) The prehistory of the theory of distributions. Springer, New York
68. Löwdin P-O (1998) Linear algebra for quantum theory. Wiley, New York
69. Nelson E (1959) Analytic vectors. Ann Math 70:572
70. Brändas EJ (2012) Time asymmetry and the evolution of physical laws. In Hoggan PE, Brändas EJ, Maruani J, Piechuch P, Delgado-Barrio G (eds) Advances in the theory of quantum systems in chemistry and physics, vol 22, Springer, Berlin, p 3
71. Brändas EJ, Antoniou IE (1993) On the positivity condition in the non-unitary transformation theory of irreversibility. Int J Quant Chem 46:419
72. Kumicàk J, Brändas E (1993) Some aspects of approach to equilibrium in classical and quantum systems. Int J Quant Chem 32:669
73. Brändas EJ, Chatzidimitriou-Dreismann CA (1989) Creation of long range order in amorphous condensed systems. Lecture Notes in physics, vol 325. Springer, Berlin, p 485
74. Halliwell JJ (1994) Quantum cosmology and time asymmetry. In: Halliwell JJ, Pérez-Mercader J, Zurek WH (eds) Physical origins of time asymmetry. Cambridge University Press, Cambridge, p 369
75. Brändas EJ (2013) Arrows of time and fundamental symmetries in chemical physics. Int J Quant Chem 113:173
76. Hehenberger H, McIntosh HV, Brändas E (1974) Weyl's theory applied to the Stark effect in the hydrogen atom. Phys Rev A 10(5):1494
77. Brändas EJ, Chatzidimitriou-Dreismann CA (1990) On the connection between certain properties of the second-order reduced density matrix and the occurrence of coherent-dissipative structures in disordered condensed matter. Int J Quant Chem 40:649
78. Löwdin P-O (1987) Some aspects of the development of the theory of reduced density matrices and the representability problem. In: Erdahl R, Smith VH Jr (eds) Density matrices and density functionals. Reidel, Dordrecht, p 21
79. Dunne LJ, Brändas EJ, Cox H (2017) High-temperature superconductivity in strongly correlated electronic systems. Adv Quant Chem 74:183
80. Leggett AJ (2006) Quantum liquids, bose condensation and cooper pairing in condensed matter systems. Oxford University Press, Oxford
81. Coleman AJ (1987) Reduced density matrices: 1929–1989. In: Erdahl R, Smith VH Jr (eds) Density matrices and density functionals. Reidel, Dordrecht, p 5
82. Brändas E, Froelich P (1977) Continuum orbitals, complex scaling and the extended virial theorem. Phys Rev A 16:2207
83. Reid CE, Brändas EJ (1989) On a theorem for complex symmetric matrices and its relevance in the study of decay phenomena. In: Lecture notes in chemistry, vol 325. Springer, Berlin, p 475

84. Brändas EJ (2009) A theorem for complex symmetric matrices revisited. Int J Quant Chem 109:28960
85. Carlson BC, Keller JM (1961) Eigenvalues of density matrices. Phys Rev 121:659
86. Brändas EJ, Hessmo B (1998) Indirect Measurements and the mirror theorem. In: Lecture notes in physics, vol 504, p 359
87. Rosenberg A (1985) The structure of biological sciences. Cambridge University Press, Cambridge
88. Rosenberg A (2011) The atheists guide to reality: enjoying life without illusions. W W Norton & Co., New York
89. Brändas EJ (2011) Gödelian structures and self-organization in biological systems. Int J Quant Chem 111:1321
90. Crick F, Koch C (1990) Towards a neurobiological theory of consciousness. Semin Neurosci 2:263
91. Koch C, Massimini M, Boly M, Tononi G (2016) Neural correlates of consciousness: progress and problems. Nat Rev Neurosci 17:307
92. Brändas EJ (2019) Molecular foundation of evolution. Adv Quant Chem 78:193
93. Trehub A (1991) The cognitive brain. MIT Press, London
94. Trehub A (2007) Space, self, and the theatre of consciousness. Conscious Cogn 16:310
95. Brändas EJ (2015) A zero energy universe scenario: from unstable chemical states to biological evolution and cosmological order. In: Nascimento MAC, Maruani J, Brändas EJ, Delgado-Barrio G (eds) Frontiers in quantum methods and applications in chemistry and physics, vol 29. Springer, Dordrecht, p 247
96. Maruani J (2016) The Dirac equation: from quantum chemistry to holistic cosmology. J Chin Chem Soc 63:33
97. Mersini-Houghton L (2014) Backreaction of Hawking radiation on a gravitationally collapsing star I: black holes? Phys Lett B 738:61
98. Mersini-Houghton L (2015) Back reaction of the Hawking radiation flux in Unruh's vacuum on a gravitationally collapsing star II. arXiv:1409.1837v2 [hep-th]
99. Brändas EJ (2016) A comment on background independence in quantum theory. J Chin Chem Soc 63:11
100. Brändas EJ (2018) A simple communication hypothesis: the process of evolution reconsidered. In: Wang YA, Thachuk M, Krems R, Maruani J (eds) Concepts, methods and applications of quantum systems in chemistry and physics, vol 31. Springer, Cham, p 381
101. Penrose R (2010) Cycles of time an extraordinary new view of the new universe. The Bodley Head, London
102. Schrödinger E (1992) What is life? with mind and matter & autobiographical sketches. Cambridge University Press, Cambridge
103. Tsekov R, Lensen MC (2013) Brownian motion and temperament of living cells. Chin Phys Lett 30:070501
104. Chalmers DJ (1996) The conscious mind in search of a fundamental theory. Oxford University Press, Oxford

Can Quantum Theory Concepts Shed Light on Biological Evolution Processes?

Jean Maruani

Abstract The Lamarckian theory of biological evolution assumed that changes induced by the environment on what we now call a 'phenotype' could be transferred to its 'genotype', while the Darwinian theory was based on natural and sexual selection of those 'genotypes' that showed up in fit 'phenotypes'. Darwinian mechanisms have proven rather efficient in explaining microevolution processes, but not sufficient for understanding macroevolution, speciation and radiation, which involve highly coordinated, structural and functional biological changes. The fact that the nucleic acids bearing the genetic material are amenable to quantum theory, as foreseen by Schrödinger, raises the question as to whether the physico-chemical internal dynamics of these biomolecules over extended time scales could not play a crucial role in these coordinated changes. As wave mechanics took its roots in de Broglie's analysis of the Maupertuis and Fermat extremum integral principles, one may go further and wonder if a similar formulation involving also an extremum integral principle may not be derived for biological evolution as well, which would make its traditional finalist flavour compatible with the modern determinist views. The present paper is a recollection of these ideas aimed at a possible formulation of a quantum theory of biological evolution.

Keywords Quantum theory · Biological evolution · Lamarck · Darwin · Complexity · Fitness · Entropy · Natural selection · Sexual selection · Microevolution · Macroevolution · Phenotype · Genotype · Nucleic acids · Proteins · Descartes · Fermat · Lagrange · Hamilton · Maupertuis · Huygens · de Broglie · Schrödinger · Feynman · Derivative · Integral · Determinist · Finalist · Wave interference · Extremum principle · Intelligent design

---

J. Maruani (✉)
Laboratoire de Chimie Physique - Matière et Rayonnement, CNRS & UPMC, 4 Place Jussieu, 75005 Paris, France
e-mail: jean.maruani@upmc.fr

L. Mammino et al. (eds.), *Advances in Quantum Systems in Chemistry, Physics, and Biology*, Progress in Theoretical Chemistry and Physics 32,
https://doi.org/10.1007/978-3-030-34941-7_16

1 Introduction

For a long time, in the Western world, it was generally believed that, in accordance with what was taught in the Holy scriptures, but contrary to what was thought by Ancient philosophers, the world had a beginning and would also have an end, while the living species had remained unchanged since they were created 'after their kind' [1]. These beliefs prevailed from the early Christian times to the Renaissance.

The idea that the Universe may actually be eternal—and infinite—reappeared at that time. But it was with later progress in the natural sciences (mainly geology and paleontology) that one started to realize that the Universe was much older than it had earlier been surmised. About the end of the 17th century, Whiston suggested that the six days of the Creation were in fact extended eras. A century later, Cuvier distinguished several eras which were populated by now extinct, fossile species—the present, living forms being survivors of geological devastations, supposedly ending with the Flood.

About the same epoch, ancient beliefs according to which the living forms had undergone thorough changes through the ages aroused new interest, although interpretations avoided challenging the Holy scriptures. But when Lamarck [2] then Darwin [3–5] developed their consistent—though competing—theories, the idea of *transformism* became definitely established. In §2 we present a short survey of the developments that occurred in the 19th century and led to the modern theories of biological evolution [6–13].

Another challenge to the views of the Middle Ages was Copernicus' proposal that the Earth was just one of many planets rotating around the Sun, then Galileo's discovery that the Solar system was more complex than it was thought. Still later, the Sun itself appeared as being but one of a hundred billion stars in our Galaxy. However, contrary to what had happened in the biological sciences, modern cosmologies lead to a revival of the biblical idea that the Universe has a beginning—and a limit [14, 15].

Until Lamarck, biology tended to use a *finalist*, holistic, teleological language, the living organisms being viewed *as if* they were *aiming at* some goal. After Darwin, this language started to be proscribed and *determinist*, reductionist, mechanistic formulations became the rule in biology. Until the recent prospects on demographic, climatic, and ecological changes, human history had come to be viewed as a series of chaotic events ruled by *chance and necessity*, not by some eschatologic design. As a result, the scientific interest in biological evolution was focussed on the past, not on the future, and speculations on *intelligent design* were discarded. Research on an encompassing, predictive theory of biological evolution was limited. On the opposite, modern cosmologies built on the relativistic and quantum theories came to speculate not only on the past, but also on the future of the Universe, data gathered from its remote limits being related to its early history [14, 15].

The problem raised in this paper is whether the traditional, finalist, and the modern, determinist views of biological evolution are compatible, and whether a mathematical formulation can be derived to reconcile the two viewpoints. In §3.1 we recall how, in classical mechanics, the *Lagrange* (or *Hamilton*) *differential* equations, derived from the Newton laws of motion for material bodies, which have a *determinist* flavour, are physically equivalent to the *Hamilton* (or *Maupertuis*) *integral* principle, which has a *finalist* flavour. In §3.2 we also recall how, in classical optics, the *Snell-Descartes* laws of reflection and refraction for light rays, which have a *determinist* flavour, are equivalent to the *Fermat* principle, which has a *finalist* flavour. In §3.3 we recall how the analogy between the two equivalences can be understood through the *de Broglie* assumption of a wave character of matter—as that of *Huygens-Fresnel* for light—and resulting wave interference effects in *Feynman*'s path integral formulation.

In §4.1 we recall the conditions for life to emerge on any planet, the main steps of the biological evolution on ours, and the universal rules of self-organization processes leading to increase in complexity. In §4.2 we search for hints and criteria for a biological extremum principle, using concepts from the Lamarck and Darwin former theories and from the modern synthetic theory of biological evolution [2–13]. In §4.3 we recall *Schrödinger*'s views on the impact of the quantum properties of genome molecules on the specificity of living beings [16] and make proposals to set a link between the determinist and finalist views. In the Conclusion, we summarize our overview and propose a prospective quantum theory of biological evolution.

2 A Short Survey of Theories of Biological Evolution

> 'When the god Anou had created the Heaven, that the Heaven had created the Earth,
> that the Earth had created the ditches, that the ditches had created the quagmires,
> that the quagmires had created the worms, then Life appeared on the Earth.'
> Sumerian cosmogony tablet, *Louvre Museum* (ca. 3000 BC)

> 'The production of living beings was completed from the origin only in its principle,
> for God did not create at once all Nature. He endowed the earth and the waters,
> when he took them out of nothingness, with the potential to bring up one day,
> at a fixed epoch, all beings that were to spread life and motion over the Earth.'
> Saint Augustine, *Commentary on Genesis* (ca. 450 AD)

> 'The present contains nothing more than the past,
> and what is found in the effect was already in the cause.'
> Henri Bergson, *L'Evolution Créatrice* (Alcan, Paris, 1907), chap. 1

The concept of *evolutionism*, as opposed to *fixism*, assumes that different species appeared at different epochs; while *transformism*, as opposed to *creationism*, explains that modern species emerged by 'transmutation' of archaic forms. Evolutionism won acceptance before transformism due to the progress of natural sciences around the 17th century. The transition between ancient beliefs and modern views involved scientists such as Whiston, Cuvier, Sidgwick, Saint Hilaire, who believed that new living forms may appear as 'special creations' (*macromutations* we would say).

The more specific doctrine of *transformism* has ancient roots. In the 5th century BC, Empedocles assumed that all structures were built from the combination of four 'elements' (*states of matter* we would say): *earth*, *water*, *air*, and *fire*, under the action of two opposed 'trends': *love* and *strife* (*attraction* and *repulsion* we would say). At the beginning, elements were merged under the sole action of love. Then strife arose and initiated heterogeneous structures, both inorganic and organic. *Afterwards there came sexes and species.* The different parts of the bodies were first built separately and then gathered at random: *only survived those beings that were best fit to their environment.*

A century later Aristotle, to whom Darwin gives credit in his *Origin of species*, stated that 'Nature evolves in a continuous manner from inanimate to animal forms', that 'in plants one finds a progressive ascension to animal life', and that 'animal life involves a gradual differentiation in vitality and motricity'. The same beliefs led Aristotle to teach that insects can arise from mud or manure, and later Pline claimed that they can also arise from snow or fire. The belief in 'spontaneous generation' went through the Middle Ages with no visible conflict with biblical creationnism. It was five years after Darwin's *Origin of species* that Pasteur magistrally refuted this millenary belief.

A more relevant precursor of scientific transformism is Erasmus Darwin, the very grandfather of Charles Darwin, who foresaw Lamarck's theory and evoked sexual selection in his *Zoonomia*. But it was Jean-Baptiste Monet de Lamarck who proposed the first consistent system of explanation of biological evolution by means of *transmissible adaptative variations leading to increased complexity* [2]. But Lamarck also believed in permanent, spontaneous generation and *orthogenetic progression on the scale of life*. Although Lamarck greatly contributed to spread transformism, his theory is now viewed as obsolete for being too closely linked to the *transmission of acquired characters*. However, it should be noted that Darwin himself, while putting the stress on *natural* (and *sexual*) *selection of individual variations* [3], attributed their genesis *to the environment* and also believed in the transmission of acquired characters. It was in the light of later discoveries [6–13] that the major importance of his specific contributions appeared, and it was neodarwinists as Jacques Monod [17] who expurged the theory of biological evolution of every trace of Lamarckism.

While Lamarck had focused his interest on the evolution through successive *eras* (*vertical* evolution), Darwin first tried to understand how different species emerge in different *areas* (*horizontal* evolution). The divergence between him and Lamarck was most clearly expressed by Wallace, the codiscoverer of natural selection. In his book *Darwinism* (1889), he presented a theory free of the vitalist elements that Darwin had maintained. But Wallace differed from Darwin—and from Haeckel, a codiscoverer of sexual selection and a supporter of Lamarck's theory—on a major point: he refused to admit that Man is a simple product of natural and sexual selection; the brain and hand of Man, he wrote, 'were in advance on the needs of their possessor'.

Other findings improved our knowledge of the living world. (1) Permanence of hereditary characters and statistical laws of their hybridation by Austrian monk Gregor Mendel (experiments on peas) in 1865. (2) Distinction of *soma* and *germen* 'plasmas' by German biologist August Weismann in 1892. (3) Spontaneity and heredity of mutations by Dutch botanist Hugo de Vries in 1903. (4) Chromosome theory of heredity by American zoologist T. H. Morgan (experiments on drosophiles) in 1910.

(5) Molecular structure of the genetic material (*DNA*) par Crick, Watson and Wilkins in 1953. The *synthetic theory of biological evolution* [6], founded in the 1940s to 70s by paleontologist Simpson, biogeographer Mayer, genetician Dobzhansky, and others [e.g. 7–10] endeavors to integrate the findings from various fields, including embryology, ethology, population genetics, and molecular biology. But it maintains natural selection as the keystone of the structural and functional order of the living world.

For sociobiologists as Dawkins [11], natural selection acts even at the gene level: social behaviors such as altruism are governed by 'selfish genes' that presented some selective advantage during evolution, each gene trying to expand even at the expense of its own carriers. But Dawkins also investigated *cultural evolution* and introduced the concept of a *meme* (unit of cultural behaviour), a cultural analog to the *gene* in *biological evolution*. In *Mind and Matter* [16], Schrödinger had earlier noticed that, if not to the *first* (biological) degree, to the *second* (cultural) degree, evolution appears as Lamarckian:

> 'It is the physical change in the parents that modifies … their behaviour; and this change is … transmitted to the progeny, along with the physical change carried by the genome.'

For Schrödinger, new behaviours induced by genome mutations, especially in human societies, can change environments in such a way as to make natural selection less (or more) effective (*dynamic retroaction*).

Scientific criticisms of Darwin's theory started very early (including from his half cousin Sir Francis Galton), and were renewed in recent times. They bore essentially on: (1) the gradualism of Darwinian evolution (for Darwin followed geologists Lyell and Owen in holding Aristotle's view that 'Nature does not make jumps'); (2) the key roles attributed to natural and sexual selection. In the 1970s, paleontologists Eldredge and Gould [12] proposed a scheme of ***stepwise progression*** with alternate periods of large mutations and others of relative stability, *evoking the light and dark interference fringes in wave theories*. In the 1980s, genetician Kimura [13] proposed a ***neutralist scheme*** in which, *at the molecular level*, most evolutive changes result from random shift of selectively equivalent mutated genes. Recent schemes involve an evolution programmed by ***architect genes*** similar to those monitoring the embryo development, reminding of 'intelligent design'.

The strongest criticisms to Darwin's theory were that, in spite of findings by Oparin in the 1920s and Miller in the 1950s, natural selection cannot explain the *emergence* and the *specificity* of life and also that, if numerous races and varieties have been produced by man, no new species has ever appeared. In spite of its improvements, it still does not properly account for such phenomena as *speciation* and *radiation* (the formation and divergence of species) or for the remarkable *permanence* of specific species over geological durations.

Regarding the human species, paleontology does not seem very relevant to reveal much of its past. As for the future, *if evolution is monitored by external effects rather than by internal processes* (*functionalism*), it seems unlikely that man will continue to evolve under natural selection. This is because he is now able to create *new tools expanding his body functions* and *making him an actor competing with nature*, for better or for worse. Darwin himself was actually inspired by the artificial selection practised by farmers and breeders and by the struggle for life in his society. The evolution of man will now likely proceed in the fields of culture, science, and technology [18], possibly using mechanisms similar to those of biological evolution. The concept of *transhumanism*, in its different forms, results from this novel paradigm [19].

3 Differential and Integral Formulations in Physical Sciences

The laws of motion of material bodies took a long time to be established [14, 15]. As events on Earth seemed to occur at random, one first looked for predictive rules in the regular motions of celestial objects. In the 6th century BC, *Pythagoras* deviced a system where the Sun, the Moon, and the then known planets were revolving on circles around a spherical Earth, forming a *cosmic lyre* ruled by *harmonic numbers*. Two centuries later, *Aristotle* distinguished between *sublunar* and *supralunar* worlds, motions being angular and chaotic in the lower one but circular and regular in the upper one. When observations were refined and circular motions could no longer cope with results, auxiliary circles (*epicycles*) were added to the main circles, culminating in the complex geocentric system of *Ptolemy* in the 2nd century AD.

In the 16th century AD, *Copernicus* came to assess *Aristarque*'s heliocentric model with updated numerical tables. A century later, *Kepler* managed to explain the distribution and motion of planets by using a model inspired by the five platonician polyhedra and searching for a musical fifth in the ratio L^3/T^2. He was led to three laws that summarize celestial mechanics, putting an end to the dogma of circular and uniform motions and linking area and time. Later on, *Galileo* derived laws for terrestrial dynamics, introducing the concepts of inertia and acceleration and the principle of motion relativity. *Descartes* played a role in the emergence of this new paradigm by founding analytical geometry, which implies homogeneity and isotropy of space; he was also first to propose replacing the concept of empty vacuum by that of a space endowed with geometric and kinematic properties in the vicinity of massive objects. But it was in the 17th century that *Newton* developed the theory of universal gravity, thus unifying terrestrial and celestial mechanics.

Further refinements came with *Einstein*'s relativistic mechanics, which led to the intrication of space and time and to the equivalence of mass and energy, then to the intrication of mass and gravity and to novel cosmology theories. But the increasing number of phenomena unexplained by these theories led to corrective adjunctions, reminding of Ptolemy's epicycles. Even in this most advanced field of science, a new paradigm appears necessary.

While the human-scale world is well understood in terms of *Newton*'s unified mechanics and *Maxwell*'s electromagnetism, the micro- and macroscopic worlds are governed by theories that both emerged from the analysis of light properties: quantum theory stems from the exchange of radiation occurring in the form of energy packets, expressed as a relation between energy and frequency, while relativity theory rests on the velocity of light remaining invariant in all inertial frames.

At the lower end of the material world, the atomic theory was conjectured as early as the 6th century BC. It was shown to be consistent with increasing experimental and theoretical evidence starting from the 17th century AD, and definitely proven during the 20th century. Today, the 'standard model' assumes that all particles are made from four families of fermions, interacting through four basic forces mediated by bosons. But here also, the increasing number of unexplained phenomena led to epicyclic complexities, which makes a new paradigm necessary.

We believe that present-day theories of biological evolution are in a state of elaboration comparable to that of the physical sciences somewhere between Ptolemy and Newton, and we conjecture that further progress may come from the search of analogies with the more advanced quantum theories.

3.1 From Lagrange's Equations to Hamilton's Principle

'Nature … always acts through the simplest paths.'

Pierre Louis de Maupertuis (*Essai de Cosmologie*, 1750)

Newton's main, second law states that, for a material point of mass m and position $\boldsymbol{r}$ submitted to a resultant force $\boldsymbol{F}$, motion is governed by a second-order, differential equation: $\boldsymbol{F} = m(d^2\boldsymbol{r}/dt^2)$. The first law simply explicits the limiting case $\boldsymbol{F} = 0$. From the concepts of force and momentum emerged those of potential and kinetic energies: $V(\boldsymbol{q})$ and $T(\boldsymbol{p})$, $\boldsymbol{q}$ designating the set of $3N$ Cartesian coordinates (N being the number of material points in the system) and $\boldsymbol{p}$ that of momentum coordinates.

A century after Newton, *Lagrange* generalized his formulation by introducing the function: $L = T(\boldsymbol{v}) - V(\boldsymbol{q}, \boldsymbol{v}, t)$, of n generalized coordinates q_i and respective generalized velocities v_i and of time t. The Lagrangian coordinates may be Cartesian, curvilinear, or independent combinations of them if one wished to take account of c constraints or symmetries between the N particles, i.e.: $n = 3N - c$. The potential V may involve conservative (e.g. gravitational) or non-conservative (e.g. electromagnetic) forces, and may depend explicitly on time t. Newton's laws can then be derived from the generalized, n-component, second-order differential *Euler-Lagrange equations*:

$$d[\partial L/\partial \boldsymbol{v}]/dt = \partial L/\partial \boldsymbol{q}. \tag{1}$$

Still later, *Hamilton* reformulated the Lagrangian mechanics into a more convenient form by using conjugate momenta: $p_i = \partial L/\partial v_i$, and introducing a total energy function: $H = T(\boldsymbol{p}) + V(\boldsymbol{q}, \boldsymbol{p}, t)$, which is the *Legendre* transform of the Lagrangian L. The time evolution of the system is then given by the $2N$ coupled, first-order differential *Hamilton equations*:

$$dp/dt = -\partial H/\partial q; \quad dq/dt = \partial H/\partial p. \tag{2}$$

The Hamiltonian formulation of Newtonian mechanics has allowed various mathematical improvements and physical extensions. In addition to being easier to use in *complex systems* involving constraints and symmetries, it could be extended to *statistical mechanics* (through *Liouville*'s equation), *quantum mechanics* (in *Heisenberg*'s formulation), and *relativistic mechanics* (replacing *T* and *V* by their relativistic expressions). For instance, if $\boldsymbol{r}$ is a quantum observable and $\boldsymbol{\rho}$ is a density distribution one can write, with a properly defined Hamiltonian *H*:

$$[H, \boldsymbol{r}] = -i\hbar\partial \boldsymbol{r}/\partial t; \tag{3}$$

$$[H, \boldsymbol{\rho}] = \partial \rho/\partial t. \tag{4}$$

This ***differential*** formulation is ***determinist*** in that sense that the positions and momenta describing a system can be known at each instant *t* if their values at time t_0 are given. Here, the *past* determines the *future* once the internal structure of the system—its 'genetic code' so to say—and its interactions with the environment—its 'natural selection' conditions—are properly defined through a Hamiltonian $H(\boldsymbol{q}, \boldsymbol{p}, t)$.

In the 17th century, *Leibnitz* conjectured that there might be many universes, governed by different laws, and that we lived in 'the best possible world' (in the physical sense, of course). This idea was underlying *Fermat*'s principle, formulated a century earlier for optics, and was to lead to what we now call the *Anthropic principle* [20, 21]. A similar principle was formulated in mechanics by *Maupertuis*, who used as a stationary function the integral of *vis viva* over the duration of motion, and by *Euler*, who proposed the integral of the momentum over the distance travelled, both having the dimension of an action: $A \sim M\,L^2\,T^{-1}$. According to the virial theorem: $2\,\underline{T} = -\,\underline{V}$, this *principle of least action* implies that a system at rest will adopt a configuration that minimizes its potential energy.

Maupertuis was also interested in practical matters (he measured the ellipticity of the Earth) and philosophical issues: it was through his feeling that 'Nature is thrifty in all its actions' that he came to the 'least action principle', which he tried to apply also to *biology* and even to *psychology*! There were other extremum principles proposed in mechanics, such as *Gauss*' 'least constraint principle' or *Herz*' 'least curvature principle'. But it was *Hamilton* who gave the exact form to the idea.

Given a physical system described by *n* generalized, time-dependent coordinates $\boldsymbol{q} = \{q_1, \ldots, q_n\}$ and associated velocities $\boldsymbol{v} = \{v_1, \ldots, v_n\}$, and defining the action functional *S*(*a*, *b*) between two states $\boldsymbol{q_a}$ and $\boldsymbol{q_b}$, reached at times t_a and t_b, as the integral of the Lagrangian function between these two states, the evolution of the system is such that the action be stationary:

$$S[\boldsymbol{q}] \equiv \int L(\boldsymbol{q}, \boldsymbol{v}, t)dt(t_a \to t_b); \quad \delta S/\delta \boldsymbol{q} = 0. \tag{5}$$

This ***integral*** formulation looks ***teleological*** in that sense that the positions and velocities describing the system can be known at each instant t by minimizing the action integral $S[\boldsymbol{q}]$ between the initial and final states. Here the *future* determines the *past* since the system will, throughout its motion, 'choose' the steps that will minimize the action integral.

This *Hamilton principle* is invariant under coordinate transformations and yields the same results as the *Hamilton equations*. But it can readily be extended to deformable systems and to relativistic and quantum theories, where it finds an interpretation in Einstein's curved gravitational field and Feynman's path-integral formulation.

3.2 From Snell-Descartes' Laws to Fermat's Principle

'The principle … that Nature always acts through the shortest and simplest paths is merely a moral, not a physical principle, which is not and cannot be the cause of any effect in Nature.'

Claude Clerselier, *reply to Fermat's letter* (1662)

The early observations of celestial objects also aroused interest in the propagation of light. In the 4th century BC, *Euclid* investigated *reflection* and, in the 2nd century AD, *Ptolemy* studied *refraction*. In the 11th century, *Alhacen* gave an early expression of the 'principle of least path' and applied it to both reflection and refraction. Some time earlier, the law of refraction had been accurately described and used to devise lenses by *Ibn Sahl*. In the 17th century, this law was reformulated by *Harriot* then by *Snell*. *Descartes* gave a different derivation of the law and used it to solve optical problems. But it was *Fermat* who made the correct assumptions and used the 'principle of least time' to derive both laws of reflection and refraction.

These laws, which apply to isotropic or specular media, can be written as follows. If θ_1 is the angle of incidence of a light ray through medium 1 with refractive index n_1 and θ_2 is its angle of refraction (or reflection) through medium 2 (or 1) with refractive index n_2 (or n_1), then the two angles are related by:

$$\sin\theta_2/\sin\theta_1 = n_1/n_2 \,(\text{for refraction}); \tag{6}$$

$$\theta_2 = \theta_1 \,(\text{for reflection}). \tag{7}$$

Again these laws are ***determinist*** in that sense that the angle of refraction (or reflection) can be known if the angle of incidence and the properties of the media are given. Here the internal structure of the system is defined by a set of refractive indices that, according to *Fermat's principle*, determine the velocities of light in the media: $n_1 v_1 = n_2 v_2 = c$.

This principle, which was clearly formulated much after the earlier studies on reflection and refraction, but was used to provide the first exact proof of their laws, can be expressed as follows: if $T(n)$ is the observer's time needed by light to cover an optical path $S(n)$ between points A and B, then T—or equivalently S—is an extremum:

$$T \equiv \int dt(t_A \to t_B) = c^{-l} \int nds(r_A \to r_B) \equiv c^{-1}S; \quad \delta S = 0. \tag{8}$$

The Fermat principle can be given a form similar to that of the Hamilton principle, Eq. (5), by introducing an optical Lagrangian involving a refraction index field $n(\boldsymbol{r})$. Its integral form again looks ***teleological*** in that sense that light will, throughout its motion, 'choose' the steps that will minimize the resulting optical path.

3.3 The Interpretation of Fermat's and Hamilton's Principles in Terms of Constructive Wave Interferences

For matter points ruled by the Newton laws, Hamilton's (differential) equations and (integral) principle have been shown to be equivalent. And for light rays ruled by the (determinist) Snell-Descartes laws, Fermat's (finalist) principle gives identical results. But light is known to behave sometimes as *rays* (in geometrical optics, governed by Fermat's principle), sometimes as *waves* (in the interference and diffraction patterns, studied by Huygens, Fresnel, and Kirchhoff), more precisely *electromagnetic* waves (governed by the Lorentz-invariant, Maxwell equations), and sometimes as *particles* (in light emission and absorption phenomena, which led *Planck* to the quantum hypothesis, *Einstein* to the explanation of the photoelectric effect, and *Compton* to that of X-ray inelastic scattering) [15].

Newton's model of *light particles* is compatible with rays, which can be seen as their trajectories: hence the similarity between the Hamilton and Fermat principles. But rays have also been shown to be derivable from the interference of *light waves*, using the Huygens-Fresnel principle which states that *every point on a wave front is a source of dephased and damped spherical wavelets*. The Fermat principle can then be derived from the superposition of these wavelets, *those with stationary paths contributing more to constructive interferences*.

It was from the analogy between the Fermat and Maupertuis principles, and the relations between frequency, energy, and mass that *de Broglie* conjectured that waves are also associated with particles external motions [22], which led Schrödinger to his wave mechanics. And it was by consistently associating the relativistic and quantum requirements that Dirac explained the electron spin motion [23] and laid the ground for quantum electrodynamics [24]. One has:

$$\lambda_C = h/m_0c, \tag{9a}$$

$$\lambda_2 - \lambda_1 = \lambda_C \cos\theta, \tag{9b}$$

$$\lambda_B = h/mv, \tag{9c}$$

$$V = c^2/v. \tag{9d}$$

Here λ_C and λ_B are the *Compton* (internal) and *de Broglie* (external) wavelengths, respectively, m_0 is the rest mass and m, the moving mass of the particle, v is its velocity and c, the speed of light. λ_C appears in Eq. (9a) for inelastic scattering. It can be seen that the *spin* (light) velocity c [23] is the geometric average of de Broglie's 'group' v and 'phase' V velocities [22]. Hamilton's principle can be then derived from the interference of the waves associated with the particles, using Feynman's path integral formulation [25, 26]. As the Huygens-Fresnel principle is ultimately related to the homogeneity of space (and conservation of linear momentum), the determinist and finalist formulations of the laws of optics and mechanics are conditioned by this invariance.

We shall return when needed to specific details of the above theories in our search for a quantum theory of biological evolution. Two guidelines already appear:

(1) *The whole Universe having a history, part of which is that of Life on the Earth, there should be some common laws linking the Universe history to the history of Life.*
(2) *If space-time homogeneity and isotropy also underly biological processes, and if biological evolution also proceeds by waves, then an extremum principle may be uncovered that will set a correspondence between the determinist and finalist formulations.*

4 Searching for Stationary Variables in Biological Evolution

> 'The laws of movement and of rest deduced from this principle being exactly the same as those observed in Nature, we can admire the application of it to all phenomena. The movement of animals, the growth of vegetals … are only its consequences; and the spectacle of the Universe becomes so much the grander, …, the worthier of its Author, when one knows that a small number of laws, most wisely established, suffice for all movements.'
>
> Pierre Louis de Maupertuis (*Essai de Cosmologie*, 1750)

The first step in a scientific process is to recognize regularities in phenomena and to search for relationships between them. That is how astronomy emerged in the Middle East long ago, and how the modern laws of physics and chemistry were established. In order to find a quantity (if any—and if only one) whose stationary integral would monitor biological evolution on our planet, we shall first briefly recall its main steps as we know them nowadays, then elaborate on some principles of former theories.

4.1 Main Steps of Biological Evolution

Nowadays it is recognized that not only life on the Earth but also the whole Universe has a beginning and a history [14, 15]. A current view is that life emerged as a unique cosmic accident. Nobel Laureate Jacques Monod expressed this view in a form reminiscent of Blaise Pascal's frightened wonder of man's loneliness [17]:

> '… Man at last knows he is alone in the indifferent immensity of the Universe, from which he has emerged by chance. As well as his destiny, his duty is written nowhere ….'

But it appears more and more likely that life can arise, evolve and expand in a large number of planets as soon as it gets a chance, provided that a number of conditions be fulfilled. In the 19th century, Louis Pasteur expressed an even more radical view:

> 'I have been looking for spontaneous generation for twenty years without discovering it …. What allows you to make it the origin of life? You place matter before life and you decide that matter has existed for all eternity. How do you know if the incessant progress of science will not compel scientists [who will live in a hundred, a thousand, ten thousand years] … to consider that *life* has existed during eternity, *not* matter? ….'
>
> René Dubos, *Louis Pasteur: Free Lance of Science* (Da Capo Press, Boston, 1950)

Some of the conditions for the appearance of life on an exoplanet are conjonctural, and probably similar to those that exist on the Earth. But others appear essential for being related to the very structure of our Universe: especially the numerical values of the nuclear, electromagnetic, and gravitational force constants. The fact that these fall precisely in the narrow ranges that make it possible for heavy elements to exist, white stars to emerge, and planets to have stable orbits led to introduce the so-called *Anthropic principle* [20, 21]. This principle has various formulations, but it may be seen as a kind of 'natural selection' (in space/time) of our universe among an infinity of possible universes, the 'fitness criterium' being that of being able to lead to Mankind! It may also be seen as a kind of Fermat or Hamilton principle, with the shortest optical path or lowest action integral being replaced by 'man-fit force-constant values'. The fact that the nuclear, electromagnetic, and gravitational force constants are very close to numbers that are endowed with particular properties [27, 28] pleads for this idea.

The electric force constant, in particular: $a \approx 137.036$, has been shown to relate the weights of the 4 DNA nucleotides, as well as those of the 20 encoded proteinogenic aminoacids, through the relations [28]:

$$GC \approx 613.58,\ AT \approx 612.60 \rightarrow GC \approx AT + 1; \tag{10a}$$

$$(3/2)(GC + AT)(= 1839.27) \approx m_H/m_e(= 1837.15); \tag{10b}$$

$$GC/AT \approx (a/137)^6 (\sim 24\,\text{ppm}); \tag{10c}$$

$$\underline{M}_{Cd} \approx 3^3 a\,(<0.5\%);\ \underline{M}_{Am} \approx a\,(<0.9\%), \tag{10d}$$

where A, T, G and C designate the molecular masses (in dimensionless ratios to m_H) of the adenine, thymine, guanine and cytosine nucleotide DNA monomers; $\underline{M}_{Cd}$ is the average mass of the 20 bicodons, a bicodon being a set of three couples of nucleotides (in pairs AT and GC) coding one of the 20 regular aminoacids; and $\underline{M}_{Am}$ is the average mass of the 20 corresponding aminoacids. A number of other amazing correlations between cosmological and biological data have been reported by Sanchez [28].

Now, it can be recalled that in the 1920s, Oparin then Haldane suggested that our early atmosphere contained essentially (in addition to hydrogen) CH_4, NH_3, and H_2O molecules which, under energy inputs from sun radiation (and also lightnings, radioactivity, volcano heat), gave rise to a number of organic compounds. In the 1950s, Miller submitted this gas mixture to electric sparks and obtained aminoacids and various other organic compounds. In the 1960s, Calvin et al. submitted the same mix to cyclotron electrons and obtained aminoacids as well as adenine and important sugars. All proteinogenic aminoacids were obtained by Fox by heating the mix up to 1000 °C, and all nucleotide bases as well as ribose and desoxyribose sugars entering DNA and RNA by other workers [29]. The sequence of reactions giving rise to both amino-acids and nucleotide bases was shown to be as follows:

$$\begin{aligned}&(\mathrm{A})\, H_2O + CH_4 \rightarrow HCHO;\ NH_3 + CH_4 \rightarrow HCN;\\&(\mathrm{B})\, 5HCHO \rightarrow \text{Ribose};\ 5HCN \rightarrow \text{Adenine};\ HCHO + H_2O + HCN \rightarrow \text{Glycine};\ \ldots.\end{aligned} \tag{11}$$

The point is that, *once the presence of hydrogen and a few light atoms is granted, the formation of basic constituents of life polymers appears unavoidable*.

Now we know that the 4% visible matter that constitute our universe involve ~74% hydrogen and ~24% helium, plus traces of carbon, nitrogen, oxygen, and heavier elements from which all living beings are made (Fig. 1). As nothing prevents the formation of white stars like our Sun, illuminating blue planets like our Earth, there can be plenty of these. But it took over 9 billion years after the *Big Bang* for them to be formed. It then took over 0.5 billion years for the first prebiotic molecules to appear on the Earth. A few of them were selected to serve as bricks for making up all living beings: 22 aminoacids associated to form the 'phenotype' structural and functional (protein) molecules, while 5 nucleotide bases formed the 'genotype' preservation and transmission (DNA and RNA) molecules. It is not yet fully understood how these two sets of macromolecules cooperated in metabolizing aggregates and became endowed with a replication mechanism. 2.5 billion years later, the first pluricellular organisms appeared in primitive waters.

During the following billion years, the living structures progressed first to primitive animal forms (540 million years ago), then to vertebrates (fishes, amphibians, reptiles), then to mammals and birds, and closer to us to primates (which replaced dinosaurians 66 million years ago). Later on emerged the related families of anthropoids (big apes) and hominidae (prehumans), to which belongs our *homo* kind (2 million years ago). The *homo sapiens* species appeared ~200,000 years ago. It has

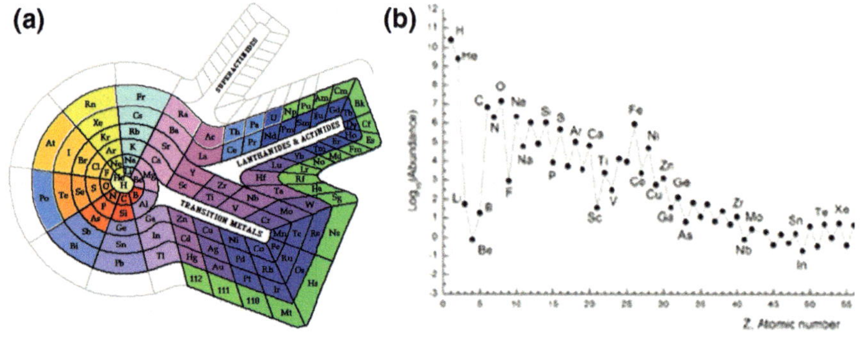

Fig. 1 Elements making up usual matter. Left **a**: Periodic table arranged by Theodor Benfey* showing *progress in complexity* in the building up of atoms. Right **b**: Relative abundance of elements up to xenon in our galaxy, showing *natural selection* according to weight/structure. The main elements involved in living organisms are H, C, N, O, P, S in the structure of proteins and nucleic acids; also F, Mg, Si, Ca, Fe in the building of tissues and bones; Cl, Na, K in physiological fluids; and Cr, Mn, Cu, Zn, Br, Mo, Se, I as oligoelements involved in hormone and enzyme activity. Most other elements before Carbon or beyond Iodine are ineffective or toxic. * In http://chemlab.pc.maricopa.edu/periodic/spiraltable.html

not undergone significant changes, at the biological level, since thousands of generations. But the explosive evolution of its ideas, techniques and societies, made possible by the tremendous capacity of its organs of representation, prehension and communication, changed our world nearly as much as the biological evolution that had preceded it.

This evolving self-organization process [30] shows up as a fractal progress in complexity. Among the so-called elementary particles, some of the most stable (electrons and nucleons) build up a few scores of atoms, which constitute a *kind of alphabet* making up usual matter (Fig. 1). Some of the molecules (words) built up from a few light atoms constitute a *higher-order alphabet* making up the molecules of life (sentences). These assemble into cells (paragraphs), tissues (chapters), organisms (books), species (bookseries), and ecosystems (libraries) ....

This process follows four main rules [31]. **1**. *At each level of complexity*, the forces that drove the lower-level structures become less relevant and *new forces take over*. **2**. *The higher the level of complexity, the weaker the driving force*: nucleons are held together by the strong nuclear force (virtual gluon exchange); atoms and molecules by the weaker, electromagnetic force (virtual photon exchange); and living organisms communicate by exchanging olfactive, auditive and visual signals, which act less by their energy than by their information content (*languages*). **3**. *At each level of complexity, a new variety of structures emerge*, while only a few of those that emerged at the lower level are retained: nucleons and electrons to build atoms, a few light atoms to built organic molecules, 22 aminoacids and 5 nucleotide bases to build proteins and nucleic acids, specific types of eucaryote cells to build pluricellular organisms, a few organisms to assemble in societies ... (*natural selection*). **4**. Subsidiary to this rule, *at critical levels of complexity* there is a *dissymmetry in the entities retained*:

matter over antimatter for elementary particles, levogyre aminoacids in proteins and dextrogyre sugars in nucleic acids. The highest level of complexity presently achieved involves international organizations using electronic communication, which extend the *lithosphere* (matter) and *biosphere* (life) to a *noosphere* (spirit) [18].

The *third rule* given above starts applying at the lowest level: particles may form complex systems without going through molecular structures and living organisms. Hydrogen is the main constituent of complex stars like the Sun, while nucleons make up neutron stars. Some stars drive a complex system of planets and satellites, comets and asteroids, and in turn they belong to galaxies like the Milky Way. At the cosmic scale, the *first two rules* make the very weak gravitational force control the self-organization process, although nuclear forces are still effective in the core of stars and electromagnetic radiation in their vicinity. The rise of complexity in the living systems is in fact conditioned by that in cosmic structures: all elements heavier than hydrogen are synthesized in the core of stars through nuclear fusion, then dispersed in galaxies through supernova explosion, and later gathered in colder bodies by the gravitational force. It seems as if the whole process aims at producing life in the Universe: small changes in the original conditions, motion laws, or universal constants would prevent the process from being completed [20, 21].

The *electrostatic force*, in its quantum version involving *antisymmetry of the wave function* with respect to permutation of coordinates for particles obeying the Fermi-Dirac statistics, plays the main role in *chemical bonding* and in the structure and properties of molecules, *including biomolecules*. However, according to our *second rule*, magnetostatic and weak nuclear forces can also affect specific biological phenomena. On line with this idea, investigations have since long been made on a possible role of the *weak nuclear force* in *biomolecular homochirality* [32]. Regarding the *magnetostatic force*, it is indeed involved in an extra sense due to *magnetosensitive proteins* [33], which allows various species (from bacteria to migrating birds) to detect the magnetic fields resulting from motions in the Earth nucleus or storms in the Sun corona.

Four features emerge from this short survey: (**1**) the role of *coordinated events* (e.g. the joint evolution of proteins and nucleic acids); (**2**) the *increase of complexity* (from prebiotic molecules to procariotic bacteria and then pluricellular organisms); (**3**) the role of *unexpected events* (e.g. the meteorite that provoked the extinction of dinosaurians, *which boosted the expansion of mammals*); (**4**) the fact that *life survives and even progresses through all events*, however catastrophic (e.g. the 5 mass extinctions that occurred between 444 and 66 million years ago), *due to the variety of its forms*. For an individual, this means *homeostasis*; for the phylum of life, this entails *evolution*.

This can be related to Schrödinger's conjecture [16] that, as Man is endowed with a brain and limbs allowing him to replace missing organs or failing functions by evolving artefacts (medicals, protheses, instruments, vehicles), 'we have come to a standstill *as a developing species*, and have little prospect of further biological advance'. This is presently leading to the transhumanist novel paradigm [19].

4.2 *Hints and Criteria for a Biological Extremum Principle from Concepts of Former Evolution Theories*

It was recalled in §2 that Lamarck was the first to propose a consistent theory of biological evolution [2]. During medical studies in Paris, he was interested in meteorology and alchemy, then turned to botanics and zoology. His theory can be summarized in four main principles. (**1**) Life naturally tends to *increase the volume and complexity* of organisms and of their parts. (**2**) New needs generate new functions, which in turn create new organs. (**3**) The development or atrophy of organs depends on their use or nonuse. (**4**) Acquired characters are hereditary.

Point **1** simply describes the development of ordered structures in the living world and could also, in Gould's scheme [12], be applied to whole species. It is compatible with Carnot's principle, because living systems are submitted to energy flows as sunlight. The teleological flavor of Point **2** is discarded in modern evolution theory. Point **3** simply expresses the adaptative character of living structures and has no explaining value. As to Point **4**, which was then current belief among biologists (including Darwin and Haeckel), it lead to hard controversies between neodarwinists and neolamarckists and came to be identified with the whole of Lamarck's theory.

Newton was inspired by astrology and alchemy when he unified the antique lower and upper worlds and postulated the action at a distance, which was then unexplained. Similarly, Lamarck was inspired by meteorology and alchemy when he elaborated his theory. Newton's gravitational force was later understood as resulting from space curvature in Einstein's relativity theory, while Lamarck's first law found a justification in Prigogine's thermodynamics of dissipative systems [30]. One could thus consider using concepts introduced in Lamarck's theory in the search for a biological equivalent to the 'action integral' or optical path'.

Lamarck's first three laws would apply to single individuals in a given species during their growing period as well as to long-lived ecosystems, while the fourth law can only apply to specific species, or to a whole phylum if interspecific mutation is recognized. The second and third laws can be seen as modes of realization of the first law under the influence of an internal structure $\boldsymbol{q}$ and dynamics $\boldsymbol{v}$ of a system and its interactions with the environment, defining a kind of 'biological Lagrangian operator' G. One may then consider choosing as a 'biological Hamilton principle' that of *maximizing* a 'complexity' C to be defined, possibly using the *entropy* concept, either in its *statistical* (Clausius) or *algorithmic* (Kolmogorov) form, both in *extensive* (size) and *intensive* (variety) terms, over a given period of time:

$$C[\boldsymbol{q}] \equiv \int G(\boldsymbol{q}, \boldsymbol{v}, t)dt(t_a \to t_b); \quad \delta C/\delta \boldsymbol{q} = 0. \tag{12}$$

Note that in Hamilton's principle, Eq. (5), the stationary property, the action functional S, has dimensions: $S \sim ML^2T^{-1}$, while in Fermat's principle, Eq. (8), the optical path S has dimension: $S \sim L$. In Eq. (12), the complexity functional C may be defined as dimensionless, as in the statistical interpretation of negentropy. Here it may be recalled that in a 1973 Note [22], de Broglie came to the conclusion that 'the principle of least action would only be an aspect of the second principle of thermodynamics'.

For a biopolymer, complexity can be defined as a function of both *size* (total number of monomers) and *variety* (numbers of different monomers and combinations). At the higher level of a cell, complexity can be defined as a function of the numbers of genes in the nucleus and of organelles in the cytosol (eukaryotes being more complex than prokaryotes). For an organism, one could again use both *size* (the total number of cells) and *variety* (the numbers of different cells and connections): thus an elephant is 'more complex' (in terms of *size*) than a mouse; and a human is 'more complex' (in terms of *variety*) than a chimpanzee.

For an ecological area, complexity may again involve *size* (total number of living organisms) and *variety* (numbers of different species and interplays, whether competitive or cooperative). In Eq. (12), the time-dependent, extended coordinates $\boldsymbol{q}$ and associated velocities $\boldsymbol{v}$ would express the *structure* and *dynamics* with respect to both *size* and *variety* at each level of organization. The overall complexity functional $C(a, b)$ between two states $\boldsymbol{q}_a$ and $\boldsymbol{q}_b$, reached at times t_a and t_b, is the integral of the function G between these two states.

In §2 it was recalled that Darwin's theory came to prevail over Lamarck's theory. Darwin had studied medicine at Edimbourg and theology at Cambridge, and was influenced by ideas from Malthus and Spencer, which had developed at an epoch of free capitalism, as well as by the artificial selection commonly practiced by farmers and breeders. But it was during his five-year journey on the *Beagle* around the world that he became fully aware of the Nature wilderness and developed his theory, which he published twenty years later [3]. In other works, he expressed novel views on geological, biological, and even psychological evolution, in which he uncovered homologic and symbiotic links between the mineral, vegetal, animal, and human realms (e.g. the fertilizing role of earthworms or the symbiosis between flowers and insects).

Regarding the philosophy of science, Darwin introduced existential and populational approaches which replaced the platonician essences and aristotelician categories that underlined Linné's and Cuvier's classifications. By reintroducing the concept of chance in biology, which had been extirped from physics par Descartes' and Newton's rationalism and determinism, Darwin was a precursor of the epistemologic revolution provoked by the advent of quantum mechanics.

The expressions 'struggle for life' and 'survival of the fittest' are titles of chapters of the first volume. In the second volume he conjectured that 'man is a codescendant, with other species, of some extinct inferior form'. His theory, as it took form over the years, can also be summarized in four main points.

(**1**) *Individual variability*: organisms of a same species show individual differences. Darwin attributed these small variations partly to *direct effects* of the environment 'on the whole organisation or on certain parts only' of the adult or the embryo, and partly to *indirect effects* of the environment 'on the reproductive system'.

(**2**) *Differential heredity*: an important number of the small variations are transmissible. Darwin included among the transmissible variations those resulting from 'direct effects' (through a mechanism which he called *pangenesis*).
(**3**) *Fight for survival and for procreation*. The first idea, independently discovered by Alfred Wallace [4], was inspired by the famous *Essay on the Principle of Population* (1796) by Thomas Malthus, who conjectured that for a given species on a given territory more individuals are born than can survive until getting able to procreate, limited subsistence setting a limit to population increase. The second idea was also proposed by Ernst Haeckel [5] (the founder of *Ecology*) in order to explain the differentiation of sexes and races: there is 'competition between individuals of a same sex, usually males, to insure possession of the other sex', through seduction or fighting.
(**4**) *Natural selection and sexual selection*, the second often prevailing over the first: since all individuals of a given species cannot survive and procreate, only do so those whose transmissible variations allow better adaptation to their environment: weather conditions, food supply, resistance to predation, reproduction. To insure procreation all means are good, even those that make their user less apt to find subsistence and/or more vulnerable to predation (as the woods of male deers or the tail of peacocks).

Darwin also believed in the role of *neutral variations* in the explanation of the *polymorphism of species*, and in 'laws of increase and reciprocal action of the parts' to account for 'coordinated changes'. The neutral hypothesis was later developed by Kimura [13]. The combination of these factors suggests that randomly occurring neutral mutations may remain 'silent' *until their number reaches a threeshold*, above which a consistent macromutation becomes possible.

Gould [12] proposed to complete Darwin's scheme by two more points. (**5**) Single genes do not fully control the capacity of interaction with the environment: organisms are not 'the product of additive effects of individually optimized genes' but, through development, an *emerging product of nonlinear interactions* between genes. (**6**) In addition to the selection of individual organisms within species, there is an *overall selection of whole species*, which can be viewed as 'collective organisms'. This helped him understand recorded punctuated equilibria.

Point **1** simply expresses the variability of living structures under effects of the environment. Point **2** does not differ much from Point **4** in Lamarck's theory. Points **3** and **4** propose mechanisms by which living organisms adapt to their environment. It can be noticed that the concept of *survival of the fittest* is an *extremum principle*. One may then consider choosing as a 'biological Hamilton principle' the maximization of a 'fitness' F to be defined, both in internal (*homeostasis*) and external (*adaptability*) terms, over duration $t_a \rightarrow t_b$:

$$F[\boldsymbol{q}] \equiv \int V(\boldsymbol{q}, \boldsymbol{v}, t)dt(t_a \rightarrow t_b); \quad \delta F/\delta \boldsymbol{q} = 0. \tag{13}$$

The 'biological Lagrangian operator' V used to define the 'fitness functional' F is different from that used to define the complexity functional C in Eq. (12). The same remarks apply as to the successive levels of organization, from biopolymers to ecosystems. Now we have to relate the requirements expressed in Eqs. (12) and (13).

On one hand, it is known from numerous examples [3–11] that, when submitted to natural selection under changing environment, a species has better chance to survive when it shows more varieties, more overall complexity. This means that 'complexity' is, as least for species, and also partly for individuals, a general condition for 'fitness'. But the increase of complexity is also, as we saw in §4.1, a general trend in Nature. The natural trend then seems to be not 'complexity for fitness' (there are many other conditions for fitness), but rather 'fitness-selected complexity' (from an interplay between 'chance and necessity'). Assuming there is 'complexification' from t_a to t then 'fitness selection' from t to t_b, one could propose the set of conditions:

$$C[\boldsymbol{q},t] \equiv \int G(\boldsymbol{q},\boldsymbol{v},t)dt(t_a \to t);\ \delta C/\delta \boldsymbol{q} = 0 \to \boldsymbol{q}_o; \tag{14a}$$

$$F[\boldsymbol{q}_o,t] \equiv \int V(\boldsymbol{q}_o,\boldsymbol{v}_o,t)dt(t \to t_b);\ \delta F/\delta \boldsymbol{q}_o = 0 \to \boldsymbol{q}_r. \tag{14b}$$

First, the 'biological Lagrangian' G is integrated from a time origin t_a until some step time t (relatively close to t_a according to geological standards), and the result $C[\boldsymbol{q},t]$ is optimized over $\boldsymbol{q}$ (maximal effect of 'chance'); then, $V[\boldsymbol{q}_o,\boldsymbol{v}_o,t]$ is integrated up to the observation time t_b, and the result is optimized over $\boldsymbol{q}_o$ (maximal effect of 'necessity'). The duration $t_b - t_a$ depends on the model used for the system dynamics. These equations could apply to the varieties of a given species in a time-dependent environment, or to an ecosystem as a whole, or to the whole phylum of life.

This stepwise procedure is comparable to that used in computing molecular structure in the Born-Oppenheimer approximation, starting with an electronic wavefunction in a clamped-nuclei model, then proceeding with the nuclear wavefunction in the resulting electronic potential, and iterating if appropriate.

A more consistent procedure would be to introduce an 'evolution functional' E, the *convolution* between 'biological Lagrangian operators' of complexity G and fitness V, as in defining the Voigt profile from Gaussian and Lorentzian shapes:

$$E[\boldsymbol{q}] \equiv (G * V)[\boldsymbol{q}] \equiv \iint G(\boldsymbol{q}-\boldsymbol{q}_a,\boldsymbol{v}-\boldsymbol{v}_a,t)V(\boldsymbol{q}_a,\boldsymbol{v}_a,t)d\boldsymbol{q}_a dt;\quad \delta E/\delta \boldsymbol{q} = 0. \tag{15}$$

Here *selection through fitness plays as and when complexity progresses along time*. Deconvolution can be performed using Fourier transforms, which is especially useful when G and V are themselves convolutions of specific complexity and fitness factors:

$$G(\boldsymbol{q},\boldsymbol{v},t) - \Gamma_i G_i,\ V(\boldsymbol{q},\boldsymbol{v},t) = \Gamma_i V_i;\ F[\Gamma_i G_i * \Gamma_i V_i] = \prod_i F[G_i]\cdot F[V_i], \tag{16}$$

where Γ_i stands for multiple convolution product and F stands for Fourier transform.

Work is in progress to specify the best variables, functionals, and optimization procedures in concrete cases.

4.3 Looking for an Extremum Principle for Biological Evolution from the Quantum Properties of Living Systems

Now we shall look for an extremum principle for biological evolution similar to those of Fermat and Hamilton for trajectories of light rays and matter particles.

Figure 2 is intended to illustrate the analogies and differences between the propagation of light through an inhomogeneous medium, which obeys Fermat's principle *for individual wavelengths*, and the progression of life through variable environments during geological eras. *Trees of life* are mentioned in many traditions [e.g. 1], but they were first used by Darwin [2] and Haeckel [4] as *branching diagrams* to show evolutionary links between biological species. In the sketch shown in Fig. 2b, based on traditional morphologic criteria, living species appear as emerging all from a same root, as in the light dispersion pattern of Fig. 2a. However, white light is a *superposition of wavelengths* which are dispersed in variable directions, while the root genome undergoes structural changes while interacting with the environment, as in the light inelastic scattering expressed in Eq. (9b). Figure 2c displays a more sophisticated branching diagram, based on modern genetic criteria (but not showing horizontal gene transfer between unrelated organisms).

The determinist way to proceed would be to identify the original genome, if any, then to monitor its fractal evolution through the various, extinct or living species, if they are known. However, in searching for an extremum principle it is, as a first step, simpler to focus on its evolution to that of a single species, e.g. *homo sapiens* (which would clarify the detailed working of the Anthropic principle).

In *What is Life* [16], Schrödinger showed that a living organism is a kind of *aperiodic crystal* endowed with *quantum properties* showing up at the macroscopic level. Ever since then, starting with the pioneering work of a few quantum scientists [34], including Löwdin's team [35], there have been advances on both the trivial (physico-chemical) and non-trivial (specific biological) quantum aspects of biological systems and processes [e.g. 36–38]. An overview of these issues can be found in Ref. [39].

Various attempts have also been made to rationalize biological evolution processes through modern thermodynamics [40] or quantum mechanics and statistics [41], or by using information recording, processing, and transfer [42]. Extremum biological principles involving genidentity, entropy or fitness have been searched by extending usual thermodynamic concepts [e.g. 43–45]. Although not antinomic, our approach is somehow different. It rests on the idea that genomes are endowed with 'quantum waves' corresponding to 'higher-order structures' which may interfere either *directly*, due to their internal dynamics, or *indirectly*, through the organisms undergoing natural and sexual selection, species symbiosis, and ecosystem dynamics. This would give rise to interference fringes and resulting extremal paths, in a way similar

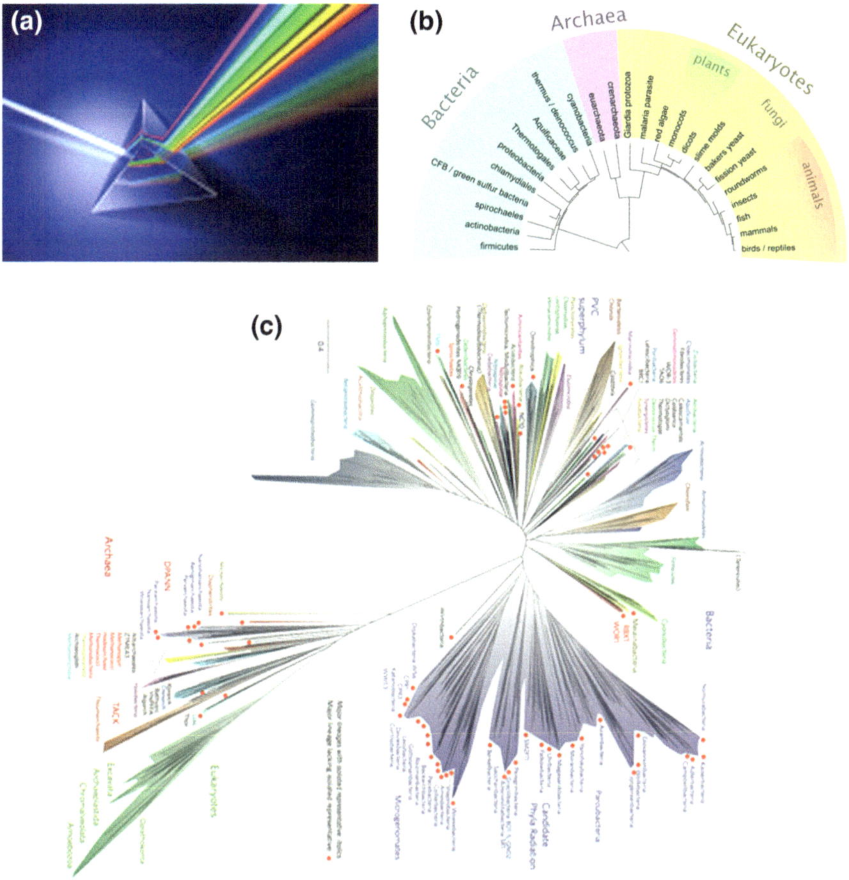

Fig. 2 Comparison of light dispersion and life radiation. Upper left **a**: Dispersion of white light to colourful pattern #. Upper right **b**: Simplified rooted tree of species radiation*. Low picture **c**: A metagenomic representation of the tree of life**. # In https://nancywait.com/2015/01/03/from-prison-to-prism/. * In https://simple.wikipedia.org/wiki/Tree_of_life_(biology). ** L A Hug, B J Baker, et al., Nature Microbiology 1 (2016) 16048

to Huygens-Fresnel's principle for light waves or Feynman's formalism for matter waves.

In a former review paper [31] we presented musical scores defined either from particle or nuclei mass sequences, or atomic or molecular spectral lines, or biopolymer distribution patterns, making use for these latter of de Broglie's formula (Eq. 9a) at body temperature and of Sternheimer's scale invariance protocol [46]. It is these *low-frequency quasi-musical waves* which we shall use by analogy with the light and matter waves. It is to be noted that, as these 'waves' are just transcriptions of the very stable *nucleic-acid or protein patterns* in terms of the 5 nucleotides or 22 aminoacids, the problem of decoherence [47] does not appear on the scale of organisms' lifetimes.

For instance, let us consider the 4 DNA nucleotide bases: A (135.13), T (126.11), G (151.13), C (111.10), the molecular masses given in parentheses not including that of the double-helix backbone desoxyribophosphate. At body temperature (T ~ 310 K), their thermal energy is (half) $mv^2 = 3kT$ ($k = 1.38065 \times 10^{-23}$ J.K^{-1}). The respective de Broglie wavelengths (Eq. 9a, with $h = 6.62607 \times 10^{-34}$ J.s) can be written in the form: $\lambda = h/\sqrt{3mkT} = (58.475/\sqrt{m}) \times 10^{-25}$ (in meters), yielding (omitting the common factor 10^{-25} m): A (5.03), T (5.21), G (4.76), C (5.55).

The ratios of these wavelengths to their smallest one (or, equivalently, the ratios of the largest frequency to the other respective ones) are: $\rho_i = 1.058$ (A), 1.095 (T), 1.0 (G), 1.166 (C), and their relative increments are: $\delta_i \equiv 10^3\,(\rho_i - 1) = 58 \sim 1\underline{l}_b$ (A), $95 \sim 2\underline{l}_b$ (T), $00 \sim 0\underline{l}_b$ (G), $166 \sim 3\underline{l}_b$ (C), where $\underline{l}_b \sim 53$ stands for the average increment (these relations hold within 4–12%). Symbolically, one can write: A (1), T (2), G (0), C (3). The fact that A + T ~ G + C ~ 3 can be related to Eq. (10a). A DNA double nucleotide sequence such as that sketched in Fig. 3 may then be seen as a *musical stave* displaying a double 'pitch wave' built on such numbers.

It should be stressed that the 'pitch waves' built on such *nucleotide sequences* in a nucleic acid are not of the same kind as the *quantum waves* related to the thermal motions of the different bases, each corresponding to a 'note' in the *musical score* of the polymer [31]: they are a higher-level (and lower-frequency) kind of wave. Each of the (related) strings displayed in Fig. 3 could be used to define a Kolmogorov

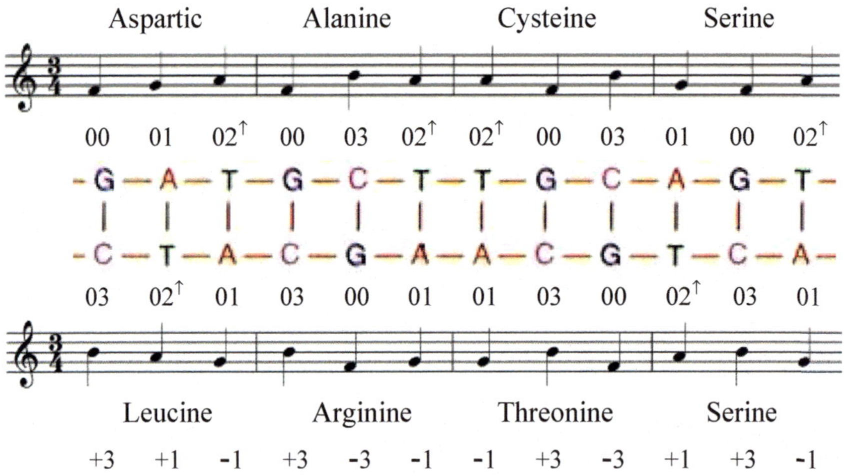

Fig. 3 A DNA sequence involving 12 nucleotide base pairs #. The numbers above or below the nucleotide symbols are defined in the text. The listed amino-acids are coded by the closest 3-base codons. The musical scores display melodies built on the base sequences, assigning the note *F* to the number 00 and a full interval to the unit increment. The sum of two paired base numbers is always equal to 3, while differences follow a kind of spin-3 quantization scheme. The 'sharp sign' ↑ is to recall that, in RNA, T ($2\underline{l}_b$) is replaced by U (112.09 → 1.161 → $3\underline{l}_b$). # From www.biologydiscussion.com/dna/dna-structure-function-packaging-and-properties

complexity [48] that may help specify the 'biological Lagrangian operator' G in Eq. (16).

In Feynman's formalism [26], the Hamilton principle can be derived from constructive/destructive interferences of quantum waves associated with particle trajectories, as the interference of Huygens wavelets associated with light rays were shown to yield the Fermat principle. We shall proceed by analogy in our search for a biological extremum principle, using the *pitch waves* built on the sequences of nucleotides A, T (U), G, and C, carrying the genetic information in the DNA (RNA) molecules, with which the thousands of *genes* embedded in *chromosomes* are built. A *genome* would then be translated into a *complex superposition of pitch waves*, from which Fourier analysis could extract fundamentals and harmonics.

Feynman's proof [26] relies on Dirac's remark (earlier surmised by de Broglie) on the relation of *quantum phase* to *classical action*. One of Feynman's postulates is that 'the (component) paths contribute equally in magnitude, but the phase of their contribution is the classical action in units of $\hbar$'. For a small (virtual) classical path $[x_i, x_{i+1}]$, described in a time interval $t_i \rightarrow t_{i+1} = t_i + \varepsilon$, the probability amplitude can be written as follows:

$$\begin{aligned} \phi_\varepsilon\left[x_i, x_{i+1}\right] &\propto exp.(i/\hbar)\, S_\varepsilon\left[x_i, x_{i+1}\right], \\ S_\varepsilon\left[x_i, x_{i+1}\right] &\equiv min. \int L(\mathbf{q}, \mathbf{v}, t) dt (t_i \rightarrow t_{i+1}). \end{aligned} \tag{17}$$

Feynman's formulation of wave mechanics then yields the path integral formula:

$$\psi\left(x'', t''\right) = \lim(\varepsilon \rightarrow 0) \mathcal{N} \int \psi\left(x', t'\right) \prod_i \phi_\varepsilon\left[x_i, x_{i+1}\right] dx' \ldots dx_i dx_{i+1} \ldots dx''. \tag{18}$$

If one considers a surface $t = t'$ of a progressive wave $\psi\ (x', t')$, the wave $\psi\ (x''\ t'')$ at point x'' on surface $t = t''$ will be a sum of contributions: $exp.\ (i/\hbar)\ S\ [x_i, x_{i+1}]$, on all possible paths from all possible sources $\psi\ (x', t')$. This expression for *matter waves* is similar to the Huygens-Fresnel wavelet superposition principle for *light waves*, recalled in §3.3. It is shown to yield Hamilton's *least action integral principle*, the equivalent of Fermat's *shortest optical path principle*. Here again, wavelets with stationary paths contribute more to constructive interference.

Now the question is: is it legitimate to make the *pitch waves* representing the *DNA polymers* interfere in a manner similar to the *light waves* leading to Fermat's principle at the limit $\lambda \rightarrow 0$ or to the *matter waves* yielding Hamilton's principle when $\hbar \rightarrow 0$? There is a priori no reason why such *pattern waves* could not be used. Let us recall that the waves involved in quantum mechanics are *probability waves* and that in his earlier work [22] de Broglie introduced *group waves* and *phase waves*. The real question is: what would be an equivalent to the *optical path* or the *action integral* in a formalism involving *pitch waves*? As a starting assumption (a question to be investigated in further work) we shall use the *evolution functional E* introduced in Eq. (15), whose virtual paths would interfere similarly as in Eq. (18) to yield the real evolution path. The combination of complexification and selection would then naturally lead to punctuated evolution, and macromutations yielding *speciation* and *radiation* would be understood as *quantum biodynamics* tunneling effects ('shape resonances').

Consider a species S_a endowed with generic genome G_a (e g, Fig. 4a), which evolves along the ages (both through Darwinian progressive selection and unexpected catastrophic events) to two different species S_b (G_b) (e g, Fig. 4b) and G_c (S_c) (e g, *homo sapiens*). Genomes G_a, G_b and G_c can be expressed as complex systems of waves, and there are multiple paths (sequences of mutations) that can lead from 'state' G_a to state G_b or G_c. This situation is more complex than that shown in Fig. 2a, where each light wave diffracts independently of other light waves. In a way, the *Archaeothyris* extinct species potentially includes all further mammals, as the white light in Fig. 2a includes all visible colors (which will diffract differently due to their different wavelengths), or alternatively as x-rays will have their wavelength increase (and energy decrease) as a function of the angle, cf. Equation (9b), in inelastic scattering by electrons.

Here we assume that the *evolution path* followed by S_a (G_a) to S_b (G_b), for instance, is such that the *evolution functional* $E\,[\boldsymbol{q}]$, defined as a convolution of complexity and fitness functionals in Eq. (15), is minimal with respect to the *extended coordinate* $\boldsymbol{q}$ defining the (possibly time-dependent) *structure of the genome interacting* (through its phenotype) *with the environment*. As in the quantum mechanics of matter waves, Eq. (18), the genome 'finds' the *optimal path* by 'feeling' the other paths, due to the *wave character* of its molecular structure.

The probability amplitude $<G_b\, t_b | G_a\, t_a>$ for a genome to be in *state* G_b at time t_b after being in *state* G_a at time t_a is a sum of contributions, one for each path leading from $G_a\, t_a$ to $G_b\, t_b$. The contribution of a given path would be written: $\mathcal{N}$ *exp* $(2\pi i\, E/\mu)$, where $\mathcal{N}$ is a normalization factor independent of the path, E is the *evolution functional* computed along the considered path, and μ is an *evolution quantum* which

Fig. 4 Left **a**: Life restoration of *Archaeothyris*, a synapsid ancestor of mammals that lived 306 My ago #. Right **b**: Leopard on a tree in Kruger National Park, South Africa*. # From https://en.wikipedia.org/wiki/Archaeothyris. * Picture taken by Pr Jan Dillen on the road from Phalabowra airport to Mopani camp

may be defined as an elementary genome mutation (possibly induced by electron or proton tunneling), whether it be positive, negative, or neutral in terms of fitness and whether it is followed or not by overall gene rearrangement. As μ is tiny regarding the aperiodic macromolecular genome complexity, the only paths that contribute to $<G_b\, t_b | G_a\, t_a>$ without cancelling through *pattern pitch wave interference* correspond to some stationary phase, meaning here stationary E, cf. Equation (15).

However, contrary to a light ray or a particle trajectory, which is unique once the frequency or the momentum and the characteristics of the medium are defined (Fig. 2a), the formation of a species (*speciation*), e.g. S_a (G_a), Fig. 4a, and its evolution (either directly through its internal dynamics or indirectly from effects of the medium), may lead to a large number of other species (*radiation*), Fig. 2c. This is comparable to a molecule appearing in several isomer or tautomer forms, each corresponding to a potential well separated from the others by potential barriers. With given numbers of the A, T (U), G and C monomers, a very large number of DNA (RNA) polymers can be built. And the number gets larger as the proportions and positions of the monomers are modified through complexification/selection processes.

Work is in progress to specify the concrete meaning and mechanisms of pitch wave interferences in evolving genome structures.

5 Discussion and Conclusion

It is now common belief that *evolution is a universal phenomenon*, from the birth of the Universe to the appearance of Life. With Man, *cultural evolution is taking over biological evolution* and may even modify its course. It is also established that transformism is the only assumption that can account for the *rise in complexity of living structures* over geological durations, even though neither spontaneous generation nor mutation of a species to another has ever been observed. Of all mechanisms proposed to explain evolution, Darwin's *natural and sexual selection of random variations* prevailed over Lamarck's *transmissible adaptative changes* or other proposed mechanisms. However, natural selection alone cannot explain such *coordinated* events as the building of the eye or the *teleological* aspect of many biological processes.

In fact, natural selection is only one of the rules that govern a universal, *evolutive self-organization process* which goes beyong the living world [30]. At the *microscopic level*, this process obeys *quantum mechanics* while at the *macroscopic level*, it follows *statistical mechanics* [16]. Except in the *living world*, where the control of phylum phenotypes by molecular genotypes makes it a necessary to introduce a new, fundamental *quantum biodynamics* approach [39].

Following Mendel's laws of heredity and Morgan's theory of genes, Schrödinger's argumentation on the quantum nature of living systems (which inspired the discovery of DNA), together with Darwin's theory of biological evolution, reminds of Epicure's conjecture that *matter is made of atoms driven by chance and necessity*. These latter play a role similar in the Darwinian theory of biological evolution and in the Copenhagen interpretation of quantum mechanics. In Gould's extension of Darwin's theory to whole species [12], the *ecosystem* acts as an 'observer' reducing the *species* 'wave packet' (population variety) generated by 'subquantum' (genotype) fluctuations. This is somehow similar to an open quantum system [49] and then possibly relevant to the Lindblad, extended Liouville equation [50]. In addition, the fact that the nucleic acids bearing the genetic material are amenable to quantum theory raises the question as to whether the internal dynamics of these biomolecules over extended durations does not also play a role in the *highly coordinated, structural and functional changes* of speciation or radiation.

In the biological sciences, only determinist approaches to *microevolution* have indeed been rationalized. There is no real understanding of the *macroevolution* in which we are embedded—not just external observers as in the physical sciences. But if biological phenomena are indeed *a manifestation of quantum laws at the macroscopic level*, then *constructive and destructive interferences* may operate also among living systems. The Feynman path integral formulation could then be used to clarify the link between *determinist processes and finalist manifestations*, and possibly uncover *selected trajectories in phylum evolution*.

In this paper, we tried to prepare the ground and lay the foundations for a *quantum theory of biological evolution*. This approach is different from the standard studies of the structure and dynamics of biomolecular systems by using quantum chemical methods, or from the investigations of specific biological processes such as enzymatic activity, photosynthesis, or magnetoreception, which involve quantum mechanisms as electron or proton tunnelling or long-range coherence. It is also different from the application of concepts borrowed from thermodynamics or cybernetics in order to rationalize teleological aspects of biological processes. It is a provisional attempt to elaborate an understanding of phylum trajectories that would reconcile *the finalist and determinist views* through a quantum path integral approach.

Fermat's shortest optical path and Hamilton's minimal action integral do not imply *intelligent design*, except perhaps in the very laws of wave mechanics! We temptatively proposed as a quantity to be optimized *for a biological trajectory in a space-time ecosystem* a convolution of 'complexity' and 'fitness', which have to be defined in a way allowing *pitch waves* expressing nucleic-acid structures to produce *interference fringes*. Investigations are in project on the relevance of various definitions of these concepts for applying effectively to available paleontological data.

As recalled earlier, unexpected catastrophes also play an important role in biological evolution. In addition, as pointed out by Henri Poincaré, small uncertainties in initial conditions may lead to unpredictable effects. This led to the theory of deterministic chaos, which involves *nonlinear equations* [51]. However, while modern celestial mechanics, an ancient fortress of determinism, shows numerous examples of distinct, regular and chaotic motions, little was done in biology, except in *population dynamics* [52]. Gould's assumption that organisms are selected by the environment not only by additive (linear), but also by combined (nonlinear) effects of gene mutations [12] may make it necessary to incorporate chaotic effects in a full understanding of biological teleology, at least at departures of punctuated equilibria.

Acknowledgements I wish to thank the colleagues who helped me clarify these ideas, especially Erkki Brändas, Roland Lefebvre, Christian Marchal, Denis Chadebec, and referees for suggesting improvements. Thanks are due to my wife, Marja Rantanen, for stimulating my speculations with inspiring piano playing.

References

1. Genesis 1:11–25
2. Monet de Lamarck J-B, Recherches sur l'organisation des corps vivants (Paris, 1802); Philosophie zoologique (Paris, 1809); Système analytique des connaissances positives de l'homme (Paris, 1820); Histoire naturelle des animaux sans vertèbres (Paris, 1822)
3. Darwin C (a) Beagle Diary (1831–1836); (b) On the origin of species by means of natural selection (1859), J. Murray, London; The descent of man and selection in relation to sex (1871), J. Murray, London
4. Darwin C, Wallace A (1958) Evolution by natural selection (a compilation by Sir Gavin de Beer, Cambridge U P)
5. Haeckel E: Generelle Morphologie der Organismen (1866), Berlin; Entwickelungsge-schichte des Menschen (1874), Berlin; Die systematische Phylogenie (1894), Berlin
6. E.g. (a) Huxley J (1942) Evolution: a modern synthesis, Allen & Unwin. (b) Kutschera U, Niklas KJ (2004) The modern theory of biological evolution: an expanded synthesis. The Science of Nature 91:255–276; and references therein
7. Dobzhansky T: Genetics and the origins of species (1937), Columbia U P; Genetics of the evolutionary process (1970), Columbia U P
8. Mayer E: Systematics and the origins of species (1942), Harvard U P; Animal species and evolution (1963), Harvard U P; What makes biology unique (2004), Cambridge U P
9. Simpson GG (1944) Tempo and mode in evolution, Columbia U P
10. Ayala FJ (1997) Genetics and the origin of species. Freeman, San Francisco
11. Dawkins R (1976) The selfish gene, Oxford U P, followed by The Extended Phenotype (1999), Oxford U P
12. (a) Eldredge N, Gould SJ (1972) Punctuated equilibria: an alternative to phyletic gradualism. In: Schopf TJM (ed) Models in paleobiology. Freeman & Co, pp 82–115. (b) Eldredge N (1991) Fossils: the evolution and extinction of species. Aurum Press, London. (c) Gould SJ (2002) The structure of evolutionary theory. Belknap Press, Harvard; and references therein
13. Kimura M: The neutral theory of molecular evolution (1983), Cambridge U P; Population genetics, molecular evolution, and the neutral theory: selected papers (1994), Chicago U P
14. Charon JE: Vingt-cinq Siècles de Cosmologie (1970) Hachette, Paris; L'Homme et l'Univers (1974), Albin Michel, Paris. These are personal lucid overviews for French readers
15. de la Cotardière P (edr) (2004) Histoire des Sciences, de l'Antiquité à nos Jours. Tallandier, Paris. This is an experts' encompassing overview for French readers
16. Schrödinger E (1944) What is life? (1st edn); Combined publication with Mind and matter (1967), Cambridge U P, 180 pp
17. Monod J, Chance and necessity. Essay on the Natural philosophy of modern biology (first published in French by Seuil, Paris, 1970)
18. de Chardin PT, The human phenomenon (first publshed in French by Seuil, Paris, 1955)
19. More M, Vita-More N (eds) (2013) The Transhumanist reader. Wiley/Blackwell, and references therein
20. Carr BJ, Rees MJ (1979) The anthropic principle and the structure of the physical world. Nature 278:605–612
21. Barrow JD, Tipler FT (1986) The anthropic cosmological principle, Oxford U P, and references therein

22. de Broglie L (1924) Recherches sur la Théorie des Quanta. Thesis, Sorbonne, Paris. Ann Phys 10 III (1925) 22–128. A 1992 facsimile edition by Fondation Louis de Broglie contains a number of other documents, especially the Note in C R A S (Paris) 277 (1973) 71–73
23. Paul AM Dirac (1958) The principles of quantum mechanics (Clarendon Press, Oxford, 1st edn 1930, 4th edn), ch. 11–12; and references therein
24. Feynman RP (1998) Quantum electrodynamics. Addison-Wesley, Reading, MA
25. Feynman RP, Hibbs AR (1965) Quantum mechanics and path integrals. Mc Graw Hill, New York
26. Cohen-Tannoudji C, Haroche S (1966) Lecture notes on quantum mechanics I, ENS, Paris. See also: (a) Feynman RP (1948) Space-time approach to non-relativistic quantum mechanics. Rev Mod Phys 20:367–387; (b) Dirac PAM (1945) On the analogy between classical and quantum mechanics. Rev Mod Phys 17:195–200; and also Ref. 22
27. Maruani J (2018) The Dirac electron and elementary interactions: the gyromagnetic factor, fine-structure constant, and gravitational invariant. In: Wang YA et al (eds) Prog Theor Chem & Phys B 31:361–380. See also: The Dirac electron: From quantum chemistry to holistic cosmology. In: Hsu C-P et al (eds) J Chin Chem Soc 63(2016):33–48; The Dirac electron as a massless charge spinning at light speed: implications on some basic physical concepts. In: Hotokka M et al (eds) Prog Theor Chem & Phys B 27(2013): 53–74
28. Sanchez FM (2017) A coherent resonant cosmology approach and its implications in microphysics and biophysics. In: Tadjer A et al (eds) Prog Theor Chem & Phys B, vol 30, pp 375–407; references therein, and private communication
29. (a) West MW, Ponamperuma C (1970) Space life sciences 2, 225ff, and references therein. (b) Eigen M (1971) Self-organization of matter and the evolution of biomolecules. Naturwissenschaften 64, 465ff
30. Nicolis G, Prigogine I (1977) Self-organization in non-equilibrium systems: from dissipative structures to order through fluctuations, Wiley, NY; Exploring complexity (1989), Freeman, NY, and private communication
31. Maruani J, Lefebvre R, Rantanen M (2003) Science and music: from the music of the depths to the music of the spheres. In: Maruani J et al (eds) Prog Theor Chem & Phys B, vol 12, pp 479–514, and references therein
32. E.g. (a) Garay AS (1978) Superweak interactions and the biological time direction. Origins of Life 9:1–5. (b) Gilat G (1991) Chiral interaction in protein structures. Mol Eng 1:161–178, and private communication. (c) Quack M (2012) Molecular parity violation and chirality: the asymmetry of life and the symmetry violations in physics. In: Nishikawa K et al (eds) Prog Theor Chem & Phys B 26:47–76, and references therein
33. E.g. (a) Winklhofer M (2010) Magnetoreception. J Roy Soc Interface 7-S2:131–134. (b) Hore PJ, Mouritsen H (2016) The radical-pair mechanism of magnetoreception. Ann Rev Biophys 45:299–344; and references therein
34. Joachim L, Freira O, Charbel E-H (2015) Quantum explorers: Bohr, Jordan, and Delbruck venturing into biology. Phys Perspect 17:236–250
35. Löwdin PO (1965) Quantum genetics and the aperiodic solid. Some aspects of biological problems in view of the quantum theory of the DNA molecule. Adv Quant Chem 2:213–260, and further work. See also: Brändas EJ (2018) Molecular theory of the genetic code. Mol Phys 116:2622–2632
36. Davies PCW (2004) Does quantum mechanics play a non-trivial role in life? Biosystems 78:69–79
37. Bordonaro M, Ogryzko VV (2013) Quantum biology at the cellular level. Biosystems 112:11–30
38. Brookes JC (2017) Quantum effects in biology: golden rule in enzymes, olfaction, photosynthesis, and magnetodetection. In: Proc Roy Soc A MPES, vol 473, 2016.0822
39. Abbott D, Davies PCW, Pati AK (eds) (2008) Quantum aspects of life. Imperial College Press, London
40. E.g. (a) Wicken JS (1987) Evolution, thermodynamics, and information: extending the Darwinian program, Oxford U P. (b) Michaelian K (2016) Thermodynamic dissipation theory and the origin and evolution of life, CreateSpace, Amazon

41. E.g. Brändas EJ (2015) A zero-energy universe scenario: from unstable chemical states to biological evolution and cosmological order. In: Nascimento MAC et al (eds) Prog Theor Chem Phys B, vol 29, Springer, pp 247–284; The origin and evolution of complex enough systems in biology. In: Tadjer A et al (eds) ibid B 30, Springer, pp 409–437 (2017); Abiogenesis and the second law of thermodynamics. In: Mammino L et al (eds) ibid B 32, Springer, pp 393–436, 2020; and references therein
42. E.g. (a) Kovac L (2000) Fundamental principles of cognitive biology. Evol Cogn 6:51–69. (b) Brändas EJ (2011) Gödelian structures and self-organization in biological systems. Int J Quant Chem 111:1321–1332
43. Deppert W (1998) Teleology and goal functions—which are the concepts of optimality and efficiency in evolutionary biology. In: Müller F, Leupelt M (eds) Eco targets, goal functions, and orientors. Springer, Heidelberg, pp 342–354
44. Sieniutycz S (2004) Extremum properties of entropy to determine dynamics of growth and evolution in complex systems: Physica A 340:356–363; Entropy-based modeling and simulation of evolution in biological systems. In Moreno-Diaz R, Pichler F, Quesada-Arencibia A (eds) Lecture Notes in Computer Science 4739. Springer (2007), pp 34–41
45. Edwards AWF (2007) Maximisation principles in evolutionary biology. In: Matthen M, Stephens C (eds) Philosophy of biology. North-Holland, pp 335–347
46. Sternheimer J (a) C R A S (Paris) 297(1983): 829–834; (b) Rev Biomath (Paris) 94(1986):1–47; (c) Procédé de régulation épigénétique de la biosynthèse des protéines par résonance d'échelle (Patent # 9206765, INPI, Paris, 1995), and private communication. A review of the history, principles and applications of this approach can be found in: (d) Fukagawa Y, Protein Music (Y F edr., Tokyo, 1999). For the music of nucleic acids, one can see: (e) Hayashi K, Munakata N (1984) Nature 310:96; (f) Ohno S, Jabara M (1986) Chem Scripta 26 B:43
47. E.g. (a) Tegmark M (2000) The importance of quantum decoherence in brain processes. Phys Rev E 61:4194–4209. (b) Engel GS, Calhoun TR, Read EL, et al (2007) Evidence for wavelike energy transfer through quantum coherence in photosynthetic systems. Nature 446:782–786. (c) Panitchayangkoun G, Hayes D, Fransted KA et al (2010) Long-lived quantum system coherence in photosynthetic complexes at physiological temperature. Proc Nat Acad Sci 107:12766–12770
48. Kolmogorov AN (1965) Three approaches to the quantitative theory of information. Problems Inform Transm 1:1–7; Logical basis for information theory and probability theory. IEEE Trans Inform Theor 14(1968):662–664. See also: Solomonoff R (1964) A formal theory of inductive inference. Inform Control 7:1–22 (I) and 224–254 (II)
49. (a) Kossakowski A (1972) On quantum statistical mechanics of non-Hamiltonian systems. Rep Math Phys 3:247–274. (b) Ingarden RS, Kossakowski A, Ohya M (1997) Information dynamics and open systems: classical and quantum approach. Springer, Heidelberg
50. Lindblad G (1976) On the generators of quantum dynamical semigroups. Comm Math Phys 48:119–130; Non-equilibrium entropy and irreversibility (Reidel, Dordrecht, 1983)
51. E.g. (a) Stewart HB, Thompson JM (1986) Nonlinear dynamics and chaos. Wiley. (b) Gleick J (1987) Chaos theory. Viking Press, New York
52. E.g. (a) May RM (1976) Simple mathematical models with very complicated dynamics. Nature 261:459–467; with Oster GF (1976) Bifurcations and dynamic complexity in simple ecological models. Am Nat 110:573–599. (b) Gold HJ (1977) Mathematical modeling of biological systems, Wiley. See also: Heller EJ (1990) Mode mixing and chaos induced by potential surface crossings. J Chem Phys 92:1718–1727

Index

L. Mammino et al. (eds.), *Advances in Quantum Systems in Chemistry, Physics, and Biology*, Progress in Theoretical Chemistry and Physics 32,
https://doi.org/10.1007/978-3-030-34941-7

Printed by Printforce, the Netherlands